This is an informative and well-illustrated guide to planetary observations for amateurs. After a brief description of the Solar System and a chapter on the celestial sphere, readers are shown how to choose, test and use a telescope with various accessories and how to make observations and record results. For each planet and the asteroids, details are given of observational techniques, together with suggestions for how to make contributions of scientific value. From a general description and detailed observational history of each planet, observers can anticipate what they should see and assess their own observations. The chapter on planetary photography includes the revolutionary use of videography, charge coupled devices and video-assisted drawing. There are also chapters on making maps and planispheres and on photoelectric photometry.

The planet observer's handbook

The planet observer's handbook

Fred W. Price

Professor of Biology/Biochemistry
State University College, Buffalo, New York

Published by the Press Syndicate of the University of Cambridge
The Pitt Building, Trumpington Street, Cambridge CB2 1RP
40 West 20th Street, New York, NY 10011-4211, USA
10 Stamford Road, Oakleigh, Melbourne 3166, Australia

First published 1994

Printed in Great Britain at the University Press, Cambridge

A catalogue record for this book is available from the British Library

Library of Congress cataloguing in publication data

Price, Fred W. (Fred William), 1932–
The planet observer's handbook / Fred W. Price.
p. cm.
ISBN 0-521-44257-5
1. Planets – Observers' manuals. 2. Planets – Amateurs' manuals.
3. Astronomy – Amateurs' manuals. I. Title.
QB601.P67 1994
523.2–dc20~93–15286~CIP

ISBN 0 521 44257 5 hardback

To the memory of my beloved Mother and Father
Corona A. and William G. Price,
who first encouraged me in astronomy.

CONTENTS

Foreword		xiii
Preface		xv
Acknowledgements		xvii
Abbreviations used in this book		xix
Introduction. Why observe the planets?		1
1	**The Solar System**	**4**
	General	4
	A scale model of the Solar System	9
	Bode's Law	10
	Kepler's Laws of Planetary Motion	11
	Elements of planetary orbits. Perturbations	12
	Planetary conjunctions, oppositions, phases and transits	12
	The sidereal and synodic orbital periods of the planets	15
	The brightness of the planets	17
	Further reading	19
2	**The celestial sphere**	**20**
	General	20
	Positions on the celestial sphere	21
	The ecliptic and the Zodiac	21
	Celestial latitude and longitude	25
	The precession of the equinoxes. Nutation	26
	Sidereal time (star time)	27
	The apparent motions of the planets on the celestial sphere	27
	Further reading	30
3	**Telescopes and accessories**	**31**
	Types of telescopes	31
	The choice of telescope	40
	Protecting the telescope from dust and atmospheric pollution	65
	Cleaning the mirror of a Newtonian reflector	66
	Housing and care of your telescope	66
	Further reading	68

4	**The atmosphere and seeing**	**69**
	General	69
	Assessing atmospheric seeing conditions	70
	The effect of telescope aperture	70
	Local effects on seeing	71
	Further Reading	72
5	**Mercury**	**73**
	General	**73**
	History of observation	75
	Visibility of Mercury	82
	The axial rotation of Mercury	87
	Observing Mercury	88
	Transits of Mercury	93
	Further reading	98
6	**Venus**	**99**
	General	99
	History of observation	102
	Space probe exploration of Venus	117
	Observing Venus	120
	Transits of Venus	129
	Further reading	129
7	**Mars**	**131**
	General	131
	Orbital characteristics	132
	Predicting oppositions	134
	The retrograde motion of Mars	135
	Martian seasons	137
	Surface features	139
	Atmospheric phenomena	142
	History of observation	144
	Observing Mars	163
	Features for observation	171
	Longitude determination of Martian features	177
	Further reading	178
8	**The minor planets (asteroids)**	**181**
	General	181
	Discovery and history of observation of the minor planets	183
	Visibility of the minor planets	189
	Observing the minor planets	191
	Further reading	200
9	**Jupiter**	**202**
	General	202
	History of observation	206
	Variations in the cloud belts	216

	Surface markings of the satellites	221
	Spacecraft observation of Jupiter	222
	Visibility of Jupiter	225
	Observing Jupiter	226
	Determination of the longitudes of Jovian features by central meridian transit timings	228
	Classification and description of Jovian disc features	232
	Determination of latitudes of Jovian features	235
	Disc drawings, strip and sectional sketches	239
	Determination of rotational periods of Jovian features from longitudinal drift	241
	Observations of the Great Red Spot	243
	Colour changes and intensity estimates of Jovian features	246
	General observing notes	247
	Further reading	259
10	**Saturn**	**260**
	General	260
	History of observation	264
	Spacecraft exploration of Saturn	292
	The satellites of Saturn	296
	Visibility of Saturn	298
	Observing Saturn	298
	Forthcoming oppositions of Saturn	311
	Further reading	312
11	**Uranus**	**314**
	General	314
	The discovery of Uranus	320
	Prediscovery sightings of Uranus	321
	History of observation	321
	Spacecraft exploration of Uranus	330
	Visibility of Uranus	332
	Observing Uranus	332
	Further reading	335
12	**Neptune**	**337**
	General	337
	The discovery of Neptune	337
	Prediscovery sightings of Neptune	343
	History of observation	344
	Spacecraft exploration of Neptune	348
	Visibility of Neptune	352
	Observing Neptune	352
	Further reading	353
13	**Pluto**	**355**
	General	355
	The search for a trans-Neptunian planet	355

The discovery of Pluto 357
History of observation 358
Visibility of Pluto 361
Observing Pluto 362
Further reading 363

14 Constructing maps and planispheres 364
General 364
The horizontal orthographic projection 364
Cylindrical projections 366
The polar projection 369
Further reading 369

15 Planetary photography and videography 370
General 370
The planetary photographer's camera 370
Choice of film 374
Characteristics of some films 375
Black and white film processing 375
Photography of individual planets 375
Exposure times 377
Video and CCD photography (videography) of the planets 379
Using a CCD camera 382
Suppliers of CCD cameras 386
Video-assisted drawing (VAD) of the planets 386
Further reading 388

16 Photoelectric photometry of the minor planets, planets and their satellites 389
General 389
The photoelectric photometer and its components 390
Telescopes for photoelectric photometry 392
Photoelectric photometric procedure 393
Photoelectric photometry of the minor planets 393
Colorimetric photoelectric photometry 396
Photoelectric photometry of the planets and their satellites 396
Further reading 398

Name index 399
Subject index 403

FOREWORD

Many amateur astronomers have a fascination for the objects of the Solar System. This is not surprising; some of these objects are bright enough to observe from urban skies and can look interesting even in small telescopes; also, unlike many vaster and further-distant objects, the bodies that we share our sun with *move* and *change in appearance*.

A person becomes an amateur 'planetarian' because the planets and their cousins the asteroids, meteors and comets interest that person. By definition, amateurs receive no pay, neither do they need formal certificates to pursue their interest. These self-motivated amateur students of the Solar System belong to at least three categories (often overlapping) and I think that *The planet observer's handbook* offers something for each:

The armchair amateur Many amateur astronomers have no telescopes. They educate themselves about the history of this field, perhaps through the most current developments. This book pays due attention to the work that has gone on before from Galileo's 30-power telescope to *Voyager II*, unlike many amateur and most professional works in this field.

The intelligent observer This amateur probably has a telescope, but he or she wishes to know what to look at and to understand what is seen. This observer may well be interested in challenges – finding the rare, the elusive or the transitory object or event in the heavens. Such targets may include eclipses, occultations, faint asteroids, comets or satellites, or perhaps simply something like Jupiter's Great Red Spot. This book can tell you what to turn your telescope (or binoculars, or even your eyes) toward.

The amateur planetary scientist This person wants to contribute to the human knowledge of our Solar System. This is demanding work, but well within the reach of the well-prepared and dedicated amateur. This book describes a variety of scientifically-useful observing projects and how to go about them.

This last form of amateur astronomer is probably the rarest of the three types. The novice in amateur Solar System research needs to understand that many very bright people have already been carefully studying the Solar System for centuries, recently with very 'sophisticated' and expensive equipment. They certainly have not discovered everything there is to be discovered, but it is true that casual observation is no longer sufficient for the scientific study of the Solar System, if indeed it ever was. What the scientifically-inclined amateur most needs is perseverance and organisation; in particular the ability to follow a demanding observing schedule.

What the amateur has above all is telescope time. What *type* of telescope is less important than the dedication to use it effectively. For example, even the unaided eye is a valuable tool in meteor observing. Binoculars can be used to advantage for the brighter asteroids and comets. The next step up, a small telescope in the 4- to 8-inch range, allows its owner to take part in the great majority of the planetary programmes of the Association of Lunar and Planetary Observers or the British Astronomical Association.

Speaking of organizations, the advanced amateur (armchair or observing) needs to become part of the amateur-astronomer information network. First, our knowledge of the Solar system is not static; there are always new discoveries to be aware of and upcoming events to plan for. Even the best books go out of date. Thus, in order to keep abreast, a person needs to subscribe to at least one of the major astronomy magazines, and to belong to an observing-oriented organisation as well. You can obtain first-hand knowledge about instrumentation by attending star parties and the meetings of your local astronomy club. Finally, knowledge must flow both ways; the other purpose of magazines and organisations is as recipients of your observations. However well they are made, observations are useless unless communicated to others.

If you want to observe, you need some equipment, but don't get carried away. This book describes the basic equipment you will need, and this may be all that you will ever need. If, after serious observing, you find that you have exhausted the capabilities of your telescope and its accessories, that is the time to move up, not before. Upgrading may mean a larger telescope, or one of better optical or mechanical design. Or it may mean joining the electronics revolution that is sweeping amateur (and professional) astronomy. Many amateurs own and use photoelectric photometers now. Others are experimenting with telescopic video recording. Still others are beginning to experiment with digital CCD cameras. What is exciting about these newer techniques is that they allow the amateur to make *quantitative* observations that are of professional quality, even with moderate-sized telescopes.

Amateur study of the Solar System is experiencing profound changes. Some once feared that space-probe observations of the planets would make earth-based studies obsolete. However, spacecraft really are just another form of scientific instrument, and every new instrument ends up raising more questions than it answers. So there is a continued need for ongoing amateur scrutiny of the Solar System; to discover, for example, lunar transient phenomena, storms in planetary atmospheres, and new comets, to give a few examples. In addition, new instruments and techniques allow the dedicated amateur *routinely* to make professional-level observations.

The planet observer's handbook is also of use to those who are not interested in spending their leisure time in directed research. There is plenty of room for the casual or even armchair observer in amateur astronomy. The idea is to do what one wants to do. Whatever form your interest in our Solar System takes, I think this book will help you to pursue it.

John E. Westfall,
Executive Director,
Association of Lunar and Planetary Observers

PREFACE

Many books have been written about practical amateur astronomy which cover the entire field of observational astronomy – the planets, sun and moon, stars, galaxies, comets and so on. The space allocated to planetary observation in such books is thus necessarily limited which is a pity because planetary observation has a special appeal and fascination unique to itself. The keen planetary observer deserves something more than a few brief notes about what to look for on each planet. I believe that there is a need for a book-length work on planetary observation that goes much further than this. In compiling this book, which is an attempt to meet this need, I have striven to remedy some of what I perceive to be deficiencies in the sections devoted to planetary observation in the usual books on general amateur astronomy.

First, I believe that an intelligent planetary observer should have a good background knowledge of the relevant observational aspects of each planet (as opposed to the planet's internal structure and composition, mass, presence of a magnetic field, for example). This can, of course, be obtained from the comprehensive treatises dealing with these matters but I consider that it is better to have the relevant information together with the practical observational aspects under one cover in the observer's handbook, for convenient study and reference.

Second, as well as detailed descriptions of observing equipment, accessories and methods of observation, the detailed chronological history of observation of each planet is given so that the significance of results of the observer's own telescopic research can be better appreciated and evaluated.

Third, plentiful illustrative material in the form of planetary drawing by observers of the past and present, using telescopes large and small, is provided. This enables the planetary observer not only to gain a good idea of what he can expect to see when starting out but also to assess the significance and value of his observations on becoming experienced in planetary work.

The first three chapters briefly review the basics of the Solar System, the celestial sphere and celestial coordinates, observing equipment and accessories. In the latter chapter is described the construction of an easily and cheaply made telescopic accessory – the apodising or antidiffraction screen – which significantly improves the clarity and contrast of planetary images and reduces the effects of atmospheric turbulence without reducing the working aperture of the telescope.

The effect of local atmospheric conditions on planetary observation and the assessment and recording of atmospheric 'seeing' and transparency at the time of an observation are described in chapter 4.

Chapters 5 to 13 inclusive treat each individual planet. The telescopic characteristics of each, visibility, best times for observation and recording of observations are described. In addition the text gives a detailed history of observation of each of the planets, the phenomena recorded by the great observers of the past and the history of discovery where applicable as in the case of the planets Uranus, Neptune and Pluto. The planetary observer of today will thus be enabled to put his own observations into better perspective and will thereby be helped to decide what is worth pursuing and what will be merely repetitive. In addition, each planet chapter is illustrated with many drawings made by various observers with telescopes of various apertures. The practical planetary observer will thus gain a good idea of what he may expect to see with his own telescope and also to compare what he sees with observations of other observers of the past and present. The book concludes with a chapter each on planetary photometry, photography and videography.

Although knowledge of the planets of our Solar System has been greatly increased by close-up photography by space probes and in some cases by craft actually landing on planetary surfaces as in the case of Mars and our own Moon, this does not mean that Earth-based telescopic observations of the planets is obsolete or futile. It is certainly true that no Earthly telescope, however large, can hope to compete with the highly detailed views of planetary surfaces and atmospheres that are possible with these space probes. However, planetary research is not entirely a matter of resolving finer and finer detail; the long-term study of the major atmosphere phenomena of the planets Jupiter and Mars and the seasonal and secular changes undergone by the dusky markings on the surface of Mars, for example, are obviously not feasible by orbiting spacecraft. In fact, it would be wasteful to do so. It is best to think of Earth-based telescopic observation of the planets and close-up studies from space probes and orbiting craft not as being in competition but as complementary. The advent of the electron microscope with its great resolving power and the enormous magnifications of microscopic objects thus made possible did not make the light microscope obsolete. The two instruments are complementary and each is adapted to different kinds of microscopic investigation.

If, in writing this book, I succeed in convincing even a few of my readers to take up planetary observation as a scientifically worthwhile pursuit and a fascinating hobby, then it will have been well worth all the time and effort that I have put into it.

Buffalo, NY, USA

Fred W. Price, Ph.D.

ACKNOWLEDGEMENTS

I am deeply indebted to Dr John E. Westfall, Executive Director of the Association of Lunar and Planetary Observers (ALPO), for bravely undertaking the task of reading the drafts of chapters 5–11 inclusive, offering many valuable criticisms and suggestions and correcting factual errors that inevitably creep into the most carefully researched writing. Any further errors of fact elsewhere in the book are entirely my responsibility. Dr Westfall also kindly agreed to write the Foreword.

I am equally indebted to Mrs Maureen Storey, subeditor, for performing the onerous task of reading the entire manuscript in preparation for the printer, for her meticulous weeding out of errors and omissions and for her numerous valuable suggestions.

Ernst E. Both, Director of the Buffalo Museum of Science, Buffalo, NY, USA, and former Curator of Astronomy, supplied his beautiful observational drawing of the planet Saturn made using the Museum's Kellogg Observatory's 8-inch refractor for the illustration in fig. 10.19.

My friend Eugene Witkowski, a clever astrophotographer and a long time member of the Buffalo Astronomical Association, used his 14.25-inch Newtonian reflector to secure the photographs of Mars and Jupiter which appear in figs. 9.13, and 15.5–15.7. His generosity in letting me use these pictures is appreciated.

Special thanks are due to Richard M. Baum, the noted planetary observer, for providing me with several of his splendid observational drawings of planets for fig. 5.18 (Mercury), figs. 6.4, 6.9, 6.22, 6.23, 6.25 and 6.26 (Venus), figs 7.7, 7.24 and 7.26 (Mars), fig. 9.28 (Jupiter) and fig. 10.24 (Saturn).

The Spectra Source Instrument Corporation (Agoura Hills, California) supplied the two photographs of charge-coupled devices (CCDs) for figs. 15.8 and 15.9.

I am grateful to the many publishers, organisations and individuals for permission to reproduce published photographs, drawings and diagrams, whose names are separately acknowledged in the captions attached to the relevant illustrations in the text.

My friend Joseph Provato's support and encouragement helped me considerably during the writing of this book which sometimes seemed a never-ending task – but still a labour of love!

Finally, I wish to thank Dr Simon Mitton, science director of the Cambridge University Press, England, for his encouragement and patience in what proved to be a bigger and more time-consuming project than we at first anticipated; also the Press for the splendid job they did of producing this book.

ABBREVIATIONS USED IN THIS BOOK

BAA	British Astronomical Association
ALPO	Association of Lunar and Planetary Observers
IAU	International Astronomical Union
JBAA	Journal of the British Astronomical Association
JALPO	Journal of the Association of Lunar and Planetary Observers
CCD	Charge-coupled device
RA	Right ascension
UT	Universal time

INTRODUCTION

Why observe the planets?

Astronomy has always been a popular hobby with all kinds of people, especially since the arrival of the 'Space Age'. Telescope sales are brisk and books on astronomy abound. Astronomy is a hobby that can be enjoyed even if the only optical instrument that you have is a pair of binoculars or a very small telescope. Books have even been written on naked eye astronomy.

There is something for everyone in observational astronomy. Some like to study either the Sun or Moon, which are especially suitable for those owning only modest telescopic equipment. Others prefer 'deep sky' observing and love to probe the depths of space with the largest telescopes that they can afford and enjoy the satisfaction of locating and identifying bright and faint star clusters, galaxies and nebulae that abound in our universe. Still others like to hunt for comets or keep track of the brightness changes in variable stars, or plot the paths of meteors ('shooting stars'). Many who are also keen photographers couple their cameras to their telescopes and delight in taking portraits of their favourite celestial objects.

To those who enjoy deep sky observing and the mind-boggling immensities of outer space, planetary observing must seem a little tame. The planets of the Solar System must seem like mere pebbles in their back yards when compared to the immensities of the universe beyond the Solar System. So – why observe the planets? The simple answer to that question is that planetary observation has a fascination of its own just as stellar astronomy has its own peculiar appeal. There is no accounting for taste; we are all different and we must accept this. The telescopic observer who gets a thrill from glimpsing a faint stellar object for the first time after repeated unsuccessful attempts is not likely to get as excited over variable diffuse markings on the surface of a comparatively close planet – yet this is precisely what thrills a planetary observer, who is not likely to feel a strong emotional or other response to a tiny wisp of faintly luminous nebulous fluff just glimpsed at the eye end of a telescope even if it does have a 30-inch mirror or object glass or on being told that what is seen is so many millions of light years distant.

The special fascination of the planets of the Solar System is their nearness to us; they are our closest celestial neighbours. Also, they are worlds more or less like our own and so there is a feeling of intimacy and kinship. The distances and sizes of these bodies are much easier to grasp and visualise than the unthinkable immensities of the universe beyond the Solar System.

Another reason why planetary observation has a special appeal is that you are never quite sure about what you are going to see, for the planets exhibit a continually changing telescopic spectacle – the cloud belts of Jupiter, the rings of Saturn, the surface markings and atmospheric phenomena of Mars and the phases of Mercury and Venus to name the most prominent examples.

Observation of the planets is rewarding because there is always something new to see and record that may be added to the continually growing body of planetary knowledge. There is even a chance that a dedicated amateur may make a discovery. Stellar observation admittedly gives us a much greater variety and number of different objects to explore such as star clusters, double stars, galaxies, nebulae and so on, but apart from variable stars, novae and Solar System comets, the stellar heavens are virtually changeless, year after year, century after century. Deep sky observing can be frustrating too, because of the increasing problem of light pollution in and near urban areas so that enthusiasts are continually yearning for bigger and bigger telescopes to collect as much light, as possible from the remote faint objects that interest them so much. To avoid light pollution deep sky observers often put up with the inconvenience of driving with their telescopes into remote country areas at night in search of darker skies. Because of this and in spite of the enormous cost of purchase or the time and effort required to build them, reflecting telescopes of up to 30 inches of aperture are not uncommonly found in the observatories of amateur astronomical clubs and societies and occasionally found in the hands of private individuals. In contrast, excellent views of the planets are obtained with even quite small telescopes and light pollution poses no serious problems.

Observing versus sightseeing

Because of the essentially unchanging nature of the stars and galaxies, there is in one sense nothing new to see except the different appearances of this or that Messier object, say, in telescopes of different apertures under varying conditions of seeing and light pollution. To say that a given galaxy was barely visible with a particular telescope and that spiral arms were just glimpsed with a much larger telescope is interesting. However, this is not really observing. It is sight-seeing. Telescopic deep sky study thus tends to become a sport (except for supernova searches) perhaps even involving friendly rivalry, rather than being a scientific pursuit. Telescope owners are really only comparing the performances of their telescopes with those of others, rather than making genuine observations.

What exactly do I mean by 'observation'? When something is observed, we do much more than merely look at it. We scrutinise it carefully, noting every detail. We ask questions about it. Finally, the observation is not complete until we have recorded it either by taking notes, drawing it, or both and perhaps photographing it as well. It goes without saying that observation requires practice, patience and perseverance. It is hard work. Some of the best astronomical seeing conditions occur on clear intensely cold winter nights so that observation can be physically trying; observation under these conditions thus demands self-discipline. This book has been especially written for those prepared to go out and endure the rigours of winter evenings in devoting themselves to planetary observing.

No one actually seems to look through the big professional observatory telescopes these days; everything is automated, electronic and computerised and the photographic plate and charge-coupled device (CCD) replace the human retina. For the forseeable future these 'giant eyes' will be trained almost exclusively on stellar objects. No amateur with even a fairly big back-yard telescope can hope to compete with the observatory giants in the field of astrophysics and astrophotography or in attempts to contribute to stellar research. But those same big observatory telescopes which are hardly ever pointed to the planets can be overtaken by the modest telescopes of back-yard amateur planetary observers who have considerable potential to contribute to planetary knowledge in addition to enjoying their fascinating hobby.

For those amateur observers working with modest telescopes who wish to contribute something to our knowledge of the heavenly bodies, as opposed to merely enjoying themselves, planetary observation therefore offers the greatest opportunities. In addition to the sheer love of doing it, which is the best of all reasons for following any pursuit, the opportunity to contribute to astronomical knowledge afforded by planetary observation is my best answer to the question "Why observe the planets?"

1

The Solar System

General

The Solar System consists of a central hot, massive and very large body, the sun, which is a star, with numerous smaller bodies circling around it in orbits varying from nearly circular to very eccentric ellipses. The principal members of this family of bodies orbiting the Sun are the nine planets of which our Earth is one. All of them move in approximately circular orbits (actually ellipses which differ only slightly from true circles) around the Sun and all in the same direction which is the same as the direction in which the sun rotates on its axis (fig. 1.1). The orbits of the planets lie roughly in the same plane as the sun's equator.

The planets rotate on their own axes in the same direction as the sun's rotation, only the axis of the planet Uranus having an unusual tilt (see chapter 11).

It is convenient to express the distances of the planets from the sun in terms of astronomical units (AU). An astronomical unit is equal to the length of the semi-major axis of the Earth's elliptical orbit around the Sun which is 92.9 million miles (149.5 million km). The names of the planets, their distances from the sun and their astronomical symbols are shown in table 1.1.

Table 1.1. *The planets: names, symbols and distances from the sun (AU).*

Planet	Symbol	Mean distance from sun (AU)
Mercury	☿	0.4
Venus	♀	0.7
Earth	⊕	1.0
Mars	♂	1.5
Jupiter	♃	5.2
Saturn	♄	9.5
Uranus	♅	19.2
Neptune	♆	30.1
Pluto	♇	39.5

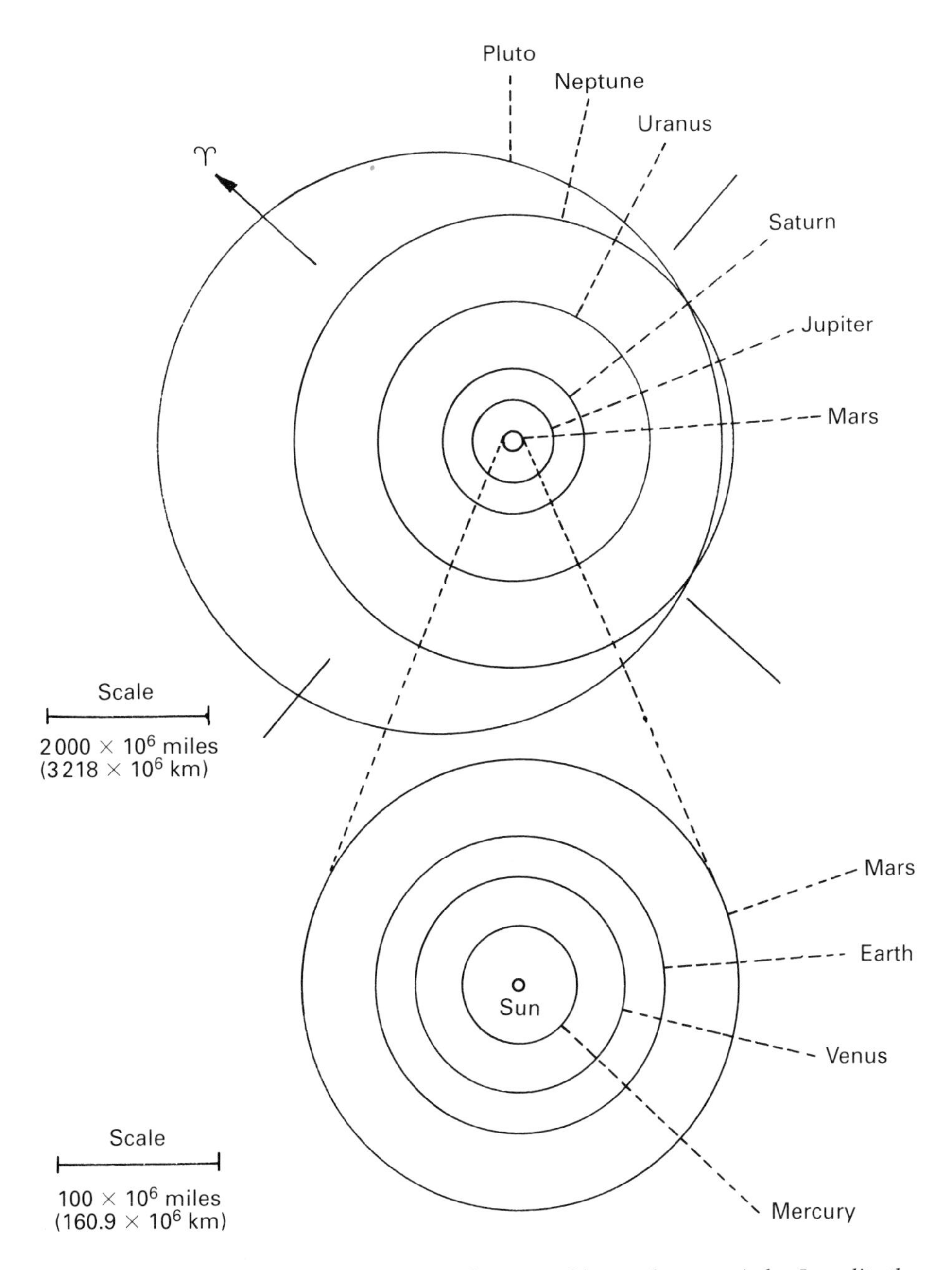

Fig. 1.1 The Solar System. For simplicity the planetary orbits are shown as circles. In reality they are slightly elliptical.

The largest planet is Jupiter with an equatorial diameter of 88 700 miles (142 718 km) and the smallest is Mercury, diameter 3010 miles (4878 km). Detailed data pertaining to each planet such as diameter, distance from the sun, orbital speed and so forth are given individually in the chapters devoted to each planet. The comparative sizes of the planets are shown in fig. 1.2.

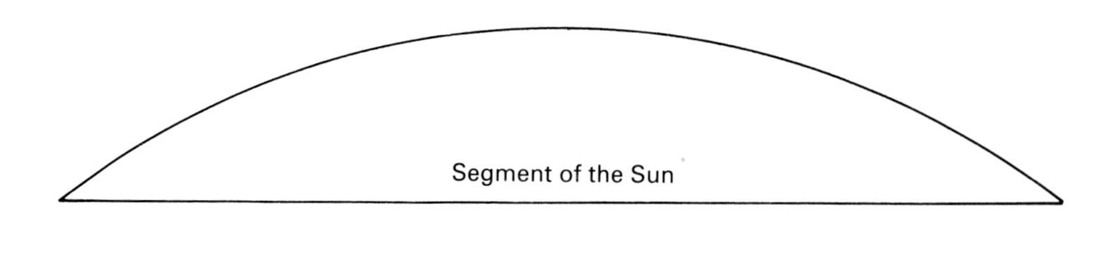

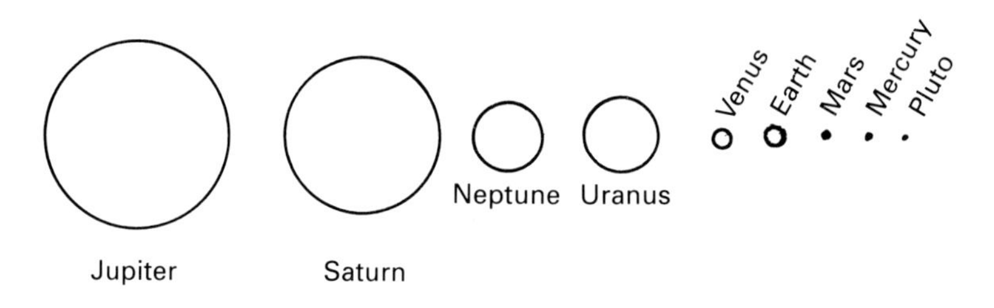

Fig. 1.2 Comparative sizes of the sun and planets.

Between the orbits of Mars and Jupiter is a large gap within which is a swarm of small bodies varying in size from a few hundred miles in diameter down to large boulders. These are the minor planets (planetoids) also known as the asteroids, because of their star-like appearance in the telescope. Most of them lie in a main belt with more or less circular orbits at mean distances from the sun of 2–4 AU but several of them have highly eccentric orbits which carry them out far beyond the confines of the main asteroidal belt (fig. 1.3).

The planets fall naturally into two groups of four each:

(1) *The terrestrial planets*: Mercury, Venus, Earth and Mars. They are all relatively small, have solid surfaces and are all of about the same order of size. Their chemical compositions are all more or less similar. Because they lie within the asteroid belt and are the planets nearest to the sun they are also called the inner planets.

(2) *The 'gas giants'*: Jupiter, Saturn, Uranus and Neptune. These are all much larger than the terrestrial planets and do not have solid surfaces. What is seen of them in the telescope is the top of a cloud-laden atmosphere. Because they lie outside the asteroid belt and are the furthest planets from the sun they are (together with Pluto) also called the outer planets. Pluto doesn't seem to fit easily into either the gas giant or terrestrial planet category. It is more like an asteroidal body.

All of the planets except Mercury and Venus are attended by one or more satellites, most of these revolving around their primaries in the same direction. In addition to satellites the four large outer planets are surrounded by concentric

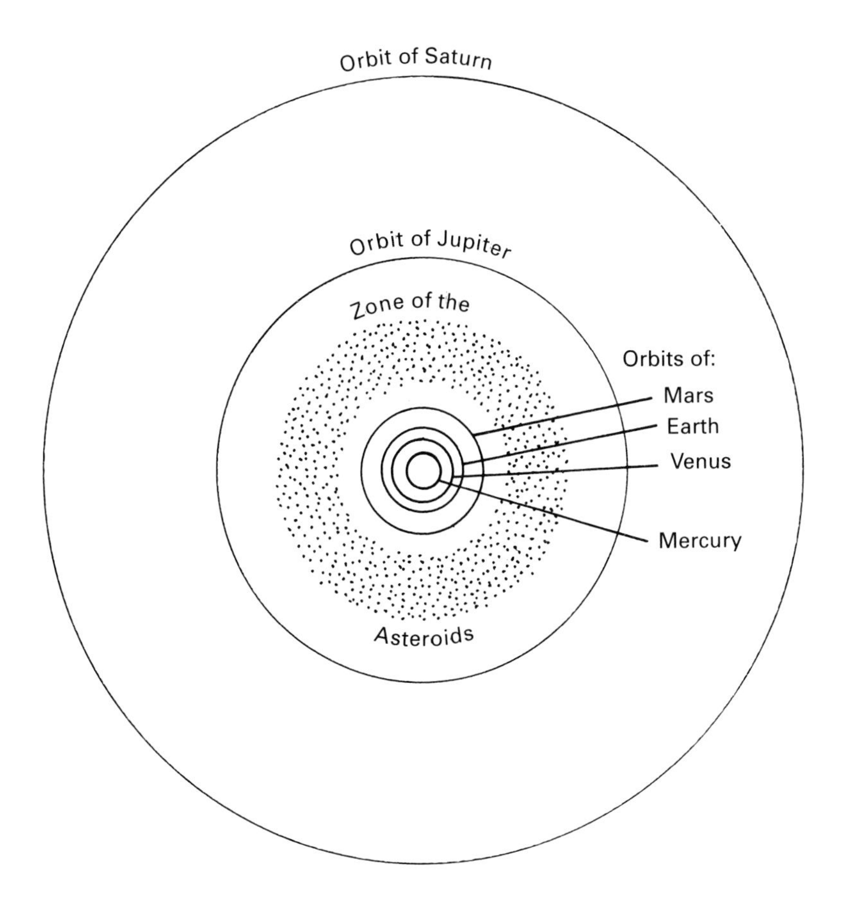

Fig. 1.3 The asteroids.

rings consisting of swarms of countless small bodies ranging in size from kilometre-sized boulders to microscopic particles revolving around their primaries in the plane of the equator. The planet Saturn has the brightest and most prominent system of rings which have been known since telescopic observation of the planets began. Jupiter, Uranus and Neptune have much less well-developed faint rings that were finally detected and then photographed long after the 'Space Age' began.

The plane of the Earth's orbit is called the ecliptic and the orbits of the other planets are all nearly coplanar with it, the majority being tilted to the ecliptic plane by a few degrees at most. The planes of the planets' orbits will therefore intersect the plane of the ecliptic at two points. These are called nodes (fig. 1.4). The node where the planet passes from south to north of the ecliptic plane is called the ascending node and that where it passes from north to south of the Ecliptic plane is the descending node. The line joining them is called the line of nodes.

It would appear that the Solar System family of sun, planets and their satellites is not a chance assemblage of material bodies but that they all had a common origin. This would seem to be very probable in view of the extreme isolation of the Solar System; the nearest star is about 6000 times as far away as the outer-most

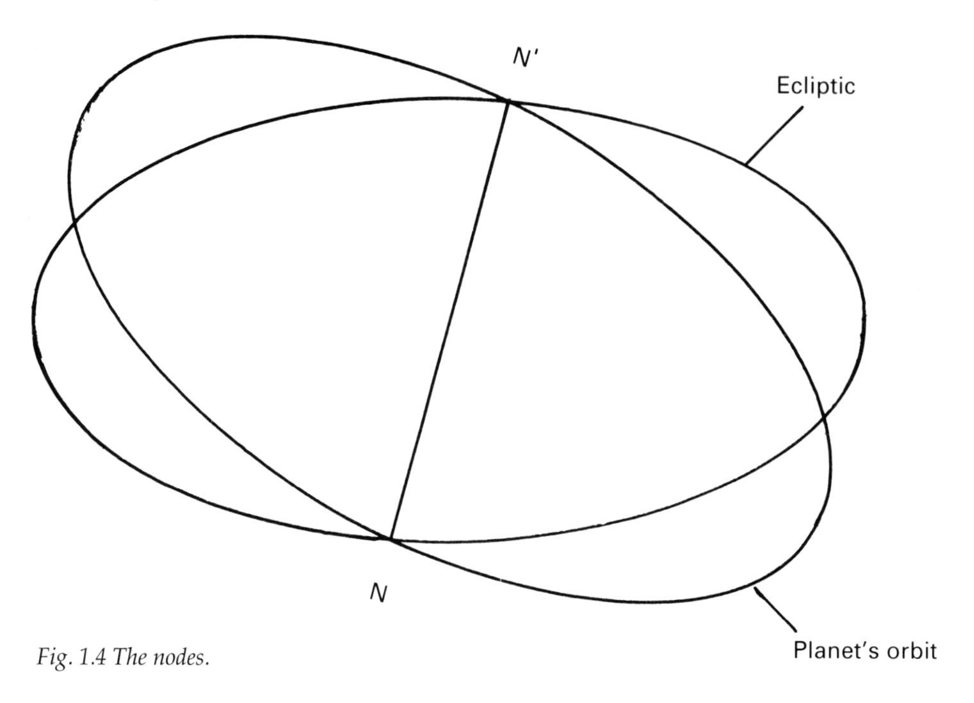

Fig. 1.4 The nodes.

planet at its mean distance from the sun. Many ideas have been put forward to account for the origins of the Solar System. One of these is that it was formed by condensation from a nebula and another is that it arose from the condensation of myriads of small bodies into larger bodies – the process of accretion. A third theory pictures a close encounter of the sun with another star, with or without collision, in which a long filament of material from the sun was pulled out by the gravitational pull of the intruding star. This subsequently broke up and condensed into the planets. That this filament may have been spindle- or cigar-shaped is suggested by the size variation of the planets and their distances from the sun; those nearest and farthest (Mercury, Pluto) are smallest whereas those at middle distances are the largest. None of these and other theories that have been put forward to explain the origin of the Solar System is completely satisfactory.

In addition to the nine principal planets and the asteroids there are numberless other objects within the domain of the sun. Some of these are huge boulders – meteorites – that have fallen to Earth on rare occasions, and some are tiny particles the size of sand grains. Upon entering the Earth's atmosphere at high speed they burn up as a result of the frictional energy generated by flying through the Earth's gaseous envelope and we see them as 'shooting stars'. Then there are the comets, large bodies consisting of rock and ice and circling the sun, most of them in highly eccentric elliptical orbits. On approaching the sun the comet gives off gases and dust which stream away from it in a direction away from the sun as though blown by a wind originating in the sun and resembling in form a feathery tail. The comet looks like a 'hairy star,' from whence the name comet is derived (*comes* = hair). Perhaps the best known is Halley's comet. Its orbit is shown in fig. 1.5. Some comets have parabolic or hyperbolic orbits which are open curves. They approach the sun from the depths of space, recede from it and are never seen again.

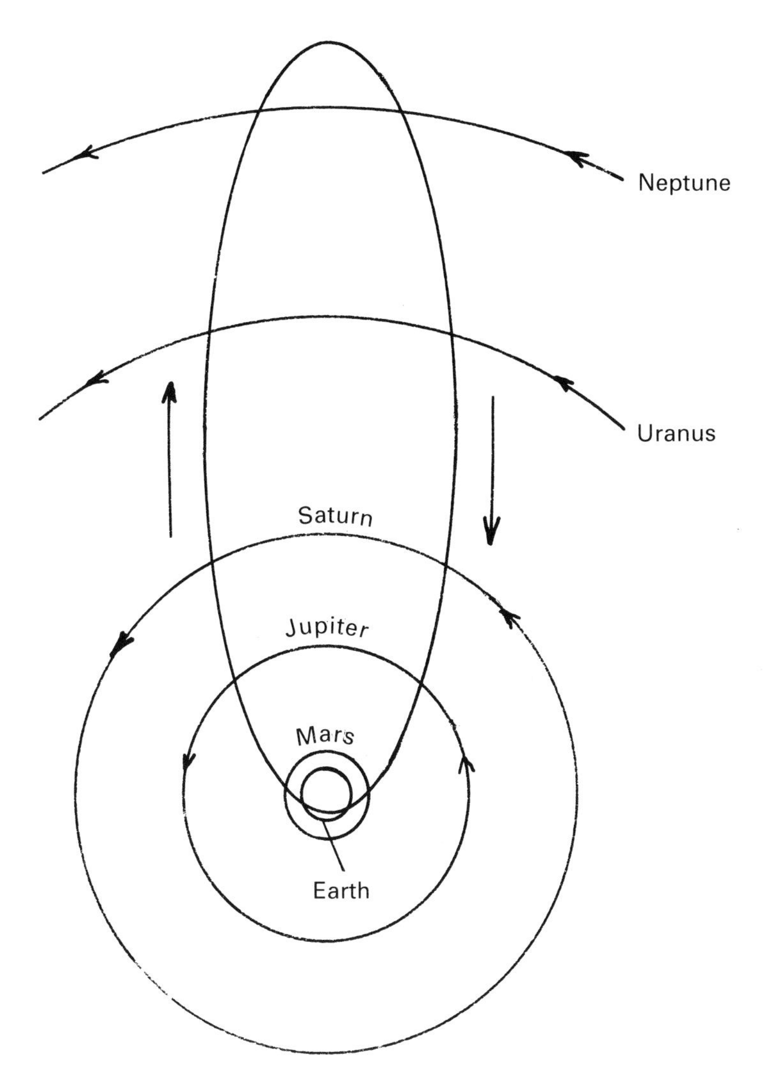

Fig. 1.5 The orbit of Halley's Comet.

A scale model of the Solar System

It is not possible to represent in a single diagram the relative sizes and distances from each other of the sun and planets because the distances are enormously greater than their sizes. The relative sizes of the sun and planets are shown in fig. 1.2 from which it will be appreciated that the sun is many times larger and more massive than all of the planets put together.

To gain a mental image of the size–distance proportions of the Solar System, first visualise a globe 2 feet (0.61 m) in diameter to represent the sun. On this scale, Mercury will be at an average distance of 83 feet (25.3 m) from it and in size will be represented by small shot or a good-sized pin head. Venus, represented by a small pea, will be at a distance of 156 feet (47.6 m) and the Earth will be another small pea at about 215 feet (65.6 m). Mars, a little larger than the shot representing Mercury, will be about 109 yards (99.7 m) from the ball representing the sun and Jupiter represented by a fairly large orange at a distance of about 373 yards (341.3 m). A large plum represents Saturn at a distance of about 0.38 mile (0.61 km). Two cherries at distances of 0.80 mile (1.29 km) and 1.22 mile (1.96 km) represent respectively Uranus and Neptune. Pluto will be a small pea at an average distance from the sun of slightly more than 1.6 miles (2.6 km). On the same scale, the nearest star would be at a distance of 10 000 miles (16 090 km).

Bode's Law

The names of J. B. Titius of Wittenberg (1729–96) and J. E. Bode (1747–1826) are both associated with the formulation of an empirical law that seems to govern the distances of the planets from the sun. Bode and Titius both found that the mean distances of the planets from the sun are not distributed at random but follow a pattern. They derived a law such that the distance of a planet from the sun is given by the formula:

$$\text{distance} = 0.4 + 2^n \times 0.075$$

S. W. Orlow stated that if the constant 0.075 is incorporated in this formula as shown, then the exponent n is the planet's location in order from the sun corresponding to the distance given by the formula, i.e., Mercury = 1, Venus = 2 etc. If n is given whole number values from 1 to 10, then the values for the expression will give the distances of the planets from the sun in terms of astronomical units quite accurately except for Mercury, Neptune and Pluto (see table 1.2).

Another frequently quoted and simpler way to derive Bode's Law is, starting with zero, to write down the following numbers each number being double the previous one:

0 3 6 12 24 48 96

Table 1.2. *Distances of the planets from the sun.*

Planet	Distance from sun (AU)	
	Bode's Law	Actual mean solar distance (AU)
Mercury	$0.4 + 2^1 \times 0.075 = 0.55$	0.4
Venus	$0.4 + 2^2 \times 0.075 = 0.70$	0.7
Earth	$0.4 + 2^3 \times 0.075 = 1.00$	1.0
Mars	$0.4 + 2^4 \times 0.075 = 1.60$	1.5
Asteroids	$0.4 + 2^5 \times 0.075 = 2.80$	2.8
Jupiter	$0.4 + 2^6 \times 0.075 = 5.20$	5.2
Saturn	$0.4 + 2^7 \times 0.075 = 10.00$	9.5
Uranus	$0.4 + 2^8 \times 0.075 = 19.60$	19.2
Neptune	$0.4 + 2^9 \times 0.075 = 38.8$	30.1
Pluto	$0.4 + 2^{10} \times 0.075 = 77.2$	39.5

To each add 4 to give:

4 7 10 16 28 52 100

Dividing each by 10 gives a series of figures representing again the planets' distances from the sun in astronomical units:

0.4 0.7 1.0 1.6 2.8 5.2 10.0

Apart from the discrepancy in the distances of Mercury in the formulae, the two are virtually identical.

The Bode–Titius Law has never been satisfactorily explained. Since the solar distances of the planets in the law follow a simple mathematical progression, the whole thing might be a coincidence, yet it is hard to believe this. However, the law, at least in its simpler form, is so easy to remember that it provides a good way of memorising the planets' distances from the sun.

Astronomers have often wondered at the large gap between the orbits of Mars and Jupiter because Bode's Law appeared to indicate that there should be a sizeable planet there at an average distance from the sun of 2.8 AU. After much search was made, a planetary body was, in fact, found in 1801 at the predicted distance but it was unusually small for a planet. Subsequently a whole host of small bodies, 'planetoids' or 'asteroids' as they were called, was found, most of them journeying around the sun between the orbits of Mars and Jupiter, others having highly elliptical orbits so that some of them travel far beyond the orbit of Mars or approach the sun more closely than Mercury. The four largest and most massive of these planetoids fit the Bode–Titius Law of distance quite closely. On the whole, however, the orbits of the planetoids tend to vary in size and eccentricity to a considerable extent, their mean solar distances ranging from 1.85 to 5.0 AU. It is thought that the planetoids may be fragments of what was once a single large planet that broke up for some unknown reason or that never formed a planet.

Kepler's Laws of Planetary Motion

Using observational data on planetary motions accumulated by Tycho Brahe (1546–1601), Johannes Kepler (1571–1630) attempted to explain planetary motion accurately which the Ptolemaic and Copernican theories (explained in chapter 2) had failed to do with complete success. After many years of unbelievable toil and many wrong guesses, Kepler formulated three laws of planetary motion:

(1) Every planet moves in an elliptical orbit in which the sun occupies one of the foci.

(2) The radius vector of each planet (the line joining the centres of the sun and the planet) sweeps out equal areas in equal times (fig. 1.6).

(3) The cubes of the distances of any two planets from the sun are to each other as the squares of their orbital revolution periods.

These are therefore known as Kepler's Laws of Planetary Motion. Since planetary orbits are elliptical, the distance of a planet from the sun varies. When closest to the sun it is said to be at *perihelion* and when farthest it is at *aphelion*. From Kepler's Second Law it follows that a planet's orbital speed is variable being fastest at perihelion and slowest at aphelion.

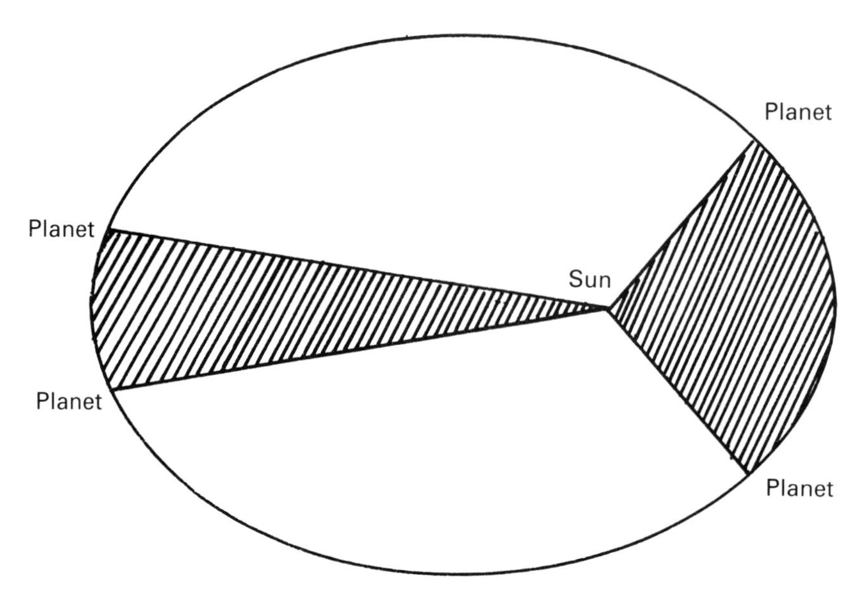

Fig. 1.6 Kepler's Second Law of Planetary Motion. The radius vector (the line joining the centres of the Sun and planet) sweeps out equal areas in equal times.

Elements of planetary orbits. Perturbations

A planet's orbit is completely defined only when certain fundamental quantities called elements have been measured. These orbital elements are:

(1) the mean distance of the planet from the sun;
(2) the orbital eccentricity;
(3) the inclination of the planet's orbit to the ecliptic plane;
(4) the position of the ascending node;
(5) the angle between the ascending node and the planet's perihelion point;
(6) the period of orbital revolution around the sun.

When these have been accurately determined and the precise orbital position of the planet is known at some particular time, then the position of the planet can be computed at any past or future date insofar as that depends only on the sun's gravitational influence. This will be slightly modified by the gravitational attractions of the other planets giving rise to slight disturbances in the planetary orbits. These are called perturbations. They cause slight discrepancies between observed and predicted planetary motions and positions, both short and long term. It was the investigation of perturbations in the motions of the planets Uranus and Neptune that led to the discoveries of Neptune and Pluto respectively.

Planetary conjunctions, oppositions phases and transits

Two of the planets, Mercury and Venus, have orbits inside the Earth's and so they are called inferior planets. The other planets outside the Earth's orbit are called superior or exterior planets.

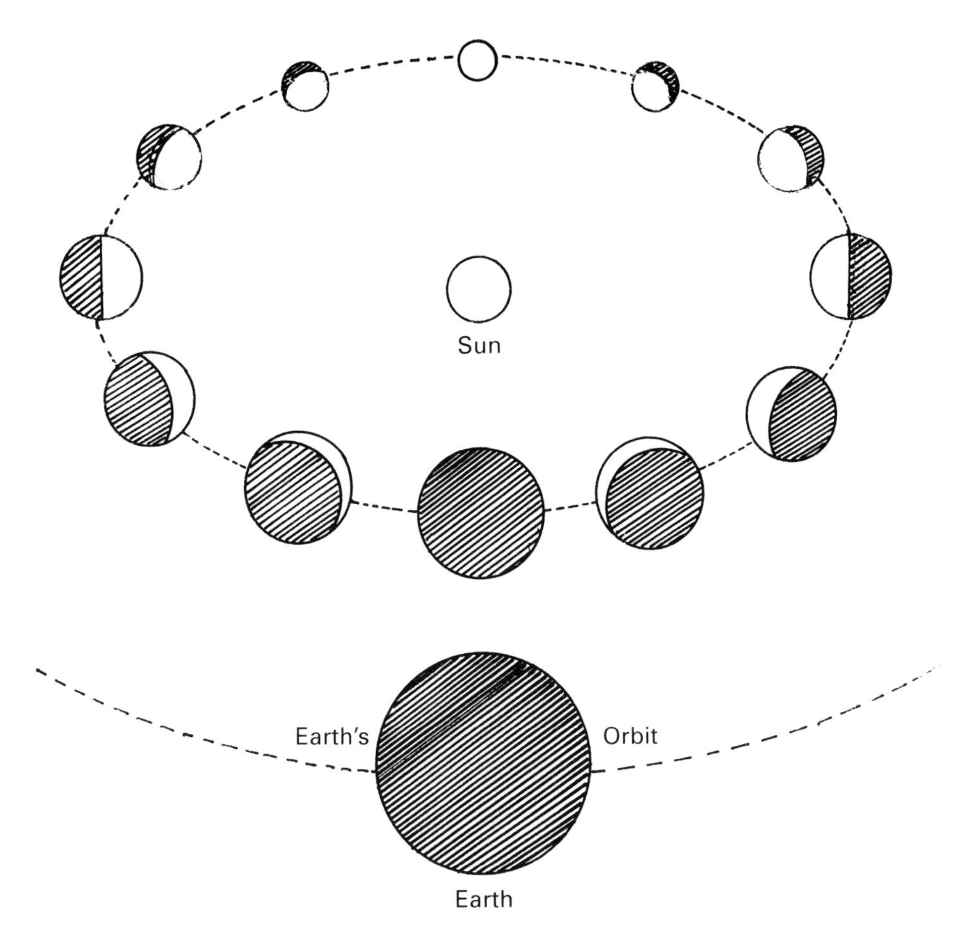

Fig. 1.7 Phases of an inferior planet.

As seen from Earth, Mercury and Venus appear to swing pendulum-like from one side of the sun to the other. When viewed in the telescope they exhibit phases like those of the moon (fig. 1.7) and show considerable variations in apparent angular diameter during an orbital revolution. These phenomena are explained in fig. 1.8. When the planet is between the Earth and the sun it is said to be at inferior conjunction and the dark side of the planet is turned Earthward. It is at its nearest to the Earth and has its maximum angular size when seen in the telescope and appears as a very thin crescent of light. As the planet moves anticlockwise in the diagram its angular size as seen from Earth decreases, its apparent angular distance from the sun in the sky increases and the crescent phase increases in thickness. At greatest western elongation, the planet appears at its greatest apparent angular distance from the sun and telescopically will be at the 'half moon' phase. It will then appear to draw close to the sun again as it continues in its orbit and will progressively decrease in apparent angular size and the phase will be gibbous. At superior conjunction the planet will be on the opposite side of the sun, exhibiting the 'full' phase and least angular size but it will be unobservable because it will be lost in the glare of the sun. After superior conjunction, the planet moves towards its greatest eastern elongation and then to inferior con-

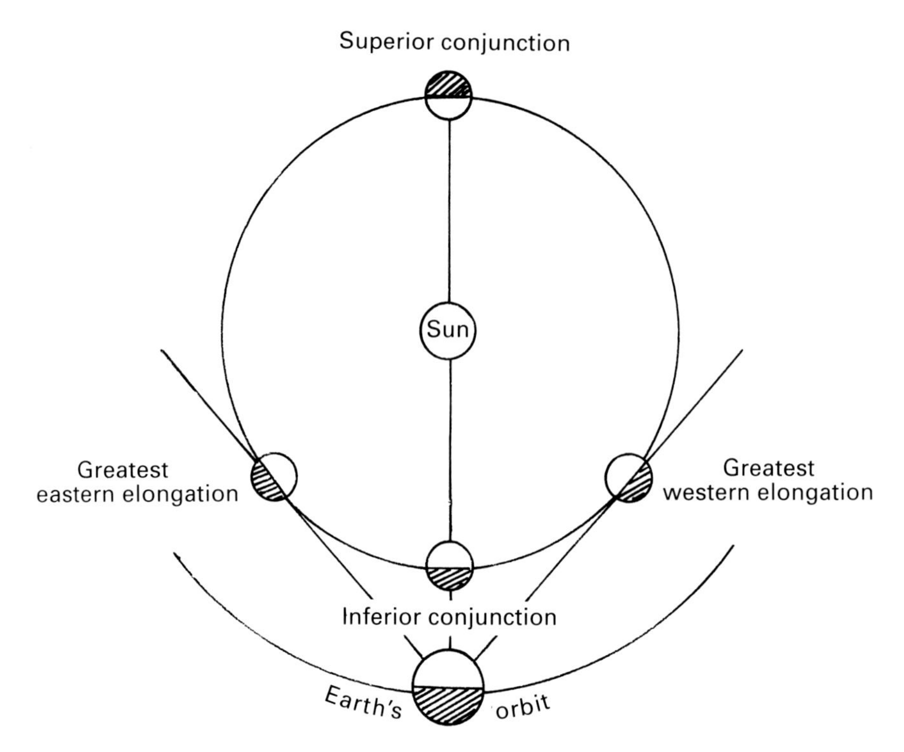

Fig. 1.8 Orbits of the Earth and an inferior planet.

junction, going through the reverse cycle of apparent angular size changes and phases as was seen in going from inferior to superior conjunction.

As the planet moves towards, through and away from its greatest western elongation it rises before the sun and appears as a 'morning star' in the dawn sky. Similarly, as it moves towards, through and away from greatest eastern elongation it will set after the sun and be seen in the sky after sunset as an 'evening star'.

If the orbital planes of the inferior planets were exactly in the same plane as the Earth's they would be seen passing across the sun's disc at every inferior conjunction. This is called a transit. However, the orbits are inclined at such an angle to the Earth's that the inferior planets will not cross the sun's disc at inferior conjunctions unless they also happen to be at an orbital node at the same time. The Earth, planet and sun will then be in a straight line. This only happens very rarely and so solar transits of the inferior planets are very rare. So rare are transits of Venus that a person may never see one during even a long lifetime. For the same reason an inferior planet will rarely pass behind the sun at superior conjunction.

In the case of a superior planet (fig. 1.9) it will be closest to the Earth at opposition when the Earth is exactly between it and the sun. The planet will show the 'full' phase, will have its largest apparent angular size and will cross the meridian at midnight. It is the best orbital position for observing the planet. Oppositions of Mars are especially eagerly awaited by planetary observers as will be explained in chapter 7. At solar conjunction a superior planet will be at its greatest distance from the Earth and will be unobservable as it will be lost in the sun's glare. The quadratures are the orbital positions where the lines joining the centres of the planet and Earth and sun form a right angle. Here the planet will

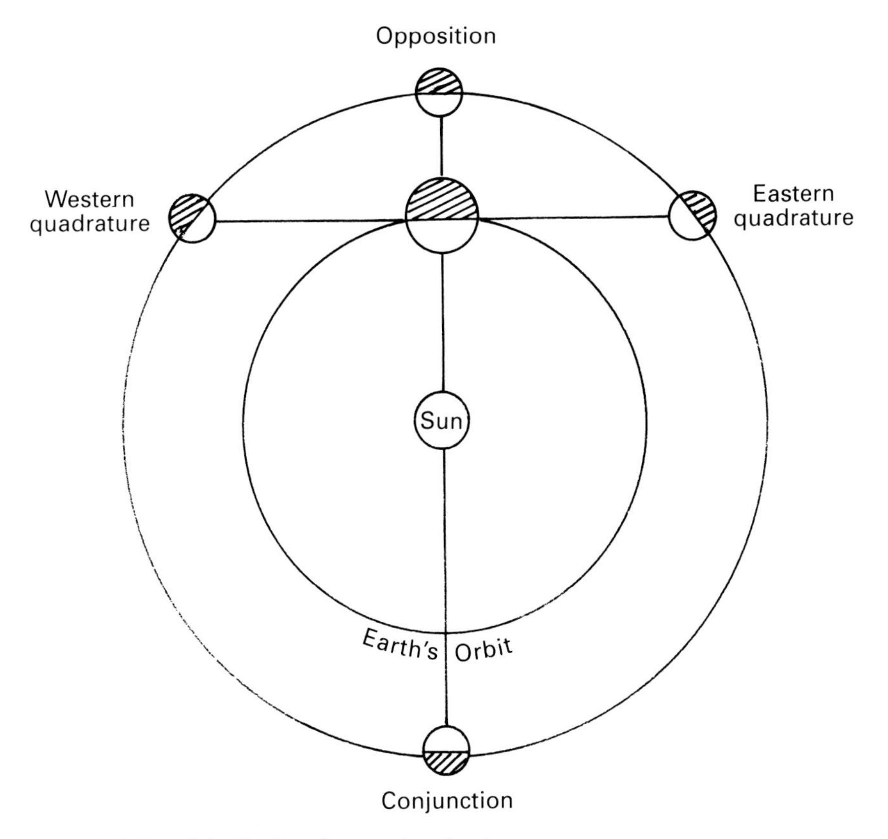

Fig. 1.9 Orbits of the Earth and a superior planet.

show its maximum phase effect which will be gibbous, resembling our moon two or three days after full. As should be evident from fig. 1.9, a superior planet can never exhibit a crescent, 'half moon' or 'new' phase, neither can it transit the sun as seen from the Earth.

The most favourable times for observing the inferior planets are when they are well away from the sun in the sky and appear as morning or evening 'stars' when near to western and eastern elongation respectively. The most favourable times for observing the superior planets are when they are at and near to opposition. They are then at their brightest and exhibit the largest disc angular size. These times when planets are prominent in the sky and well placed for observation are called 'apparitions'.

The sidereal and synodic orbital periods of the planets

The sidereal period of a planet is the time it takes to make one complete orbital revolution. It is so called because as seen from the sun, it is the same time that it would take to move from a specified position against the background of the fixed stars and to return to it again.

The synodic period is the time that a superior planet takes to go from one opposition to the next or, in the case of an inferior planet, from one conjunction to the next, as seen from the Earth.

The synodic periods are variable owing to the variable speeds of the planets in different parts of their orbits. The synodic periods of the inferior planets are larger than the sidereal periods because by the time the planet returns to the position in its orbit where it was at the previous conjunction, the Earth will have moved on in its own orbit and the inferior planet has to 'catch up' with the Earth before the next conjunction can occur (fig. 1.10).

In the case of a superior planet the synodic period is longer than the sidereal period for a similar reason but the Earth has to do the 'catching up' this time. When the Earth returns to the place in its orbit where the last opposition occurred, the outer planet will have moved further on in its orbit and so the Earth will take time to move to the place where the next opposition will occur (fig. 1.11).

The first superior planet, Mars, covers about one half of its orbit during the time that the Earth makes one complete orbital revolution. Therefore, oppositions of Mars occur at approximately two-yearly intervals. As their distances from the sun increase, the synodic periods of the superior planets get progressively shorter than the sidereal periods. This is due to their relatively slow orbital motion; during the time that the Earth makes one complete orbital revolution, the superior planets beyond Mars cover much smaller arcs of their orbits than

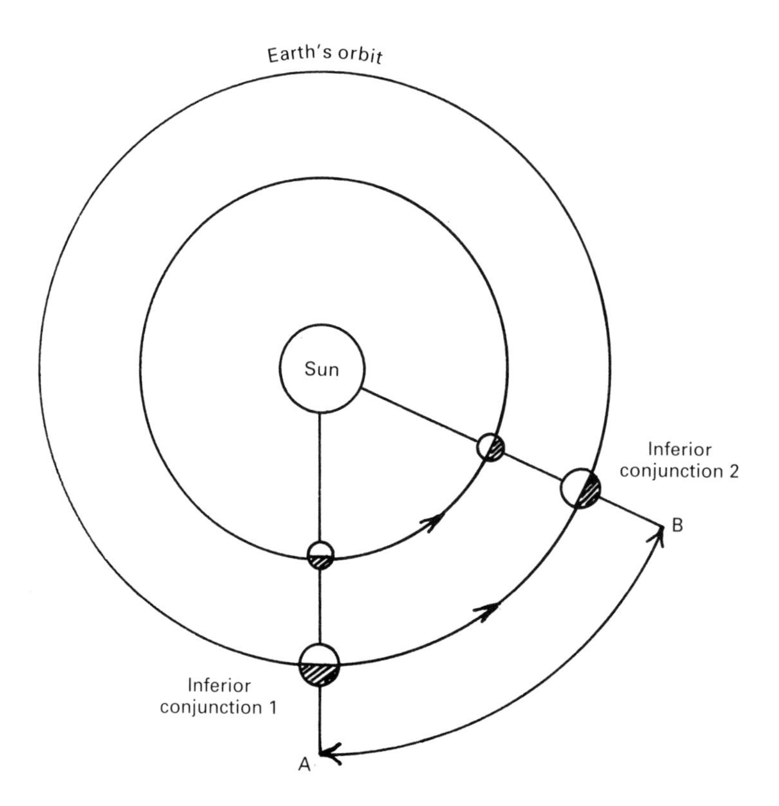

Fig. 1.10 Synodic orbital revolution period of an inferior planet. A–B is the distance moved by Earth from inferior conjunction 1 during one orbital revolution of the inferior planet.

does Mars in this same time and the arcs get progressively smaller the further the planet is from the sun. The Earth therefore has relatively little 'catching up' to do between one opposition and the next. In the case of Neptune this takes only about two days so that Neptune's synodic period (367.49 Earth days) is only about two days longer than the Earth year and oppositions of Neptune therefore occur about two days later each year. Similarly, for Jupiter, Saturn and Uranus, which also come to opposition every year, the difference in time between their synodic and sidereal periods and the length of the Earth's year gives the amount of time later each year that opposition will occur.

The brightness of the planets

The brightness ranges of the planets compared to some well-known stars are shown in fig. 1.12.

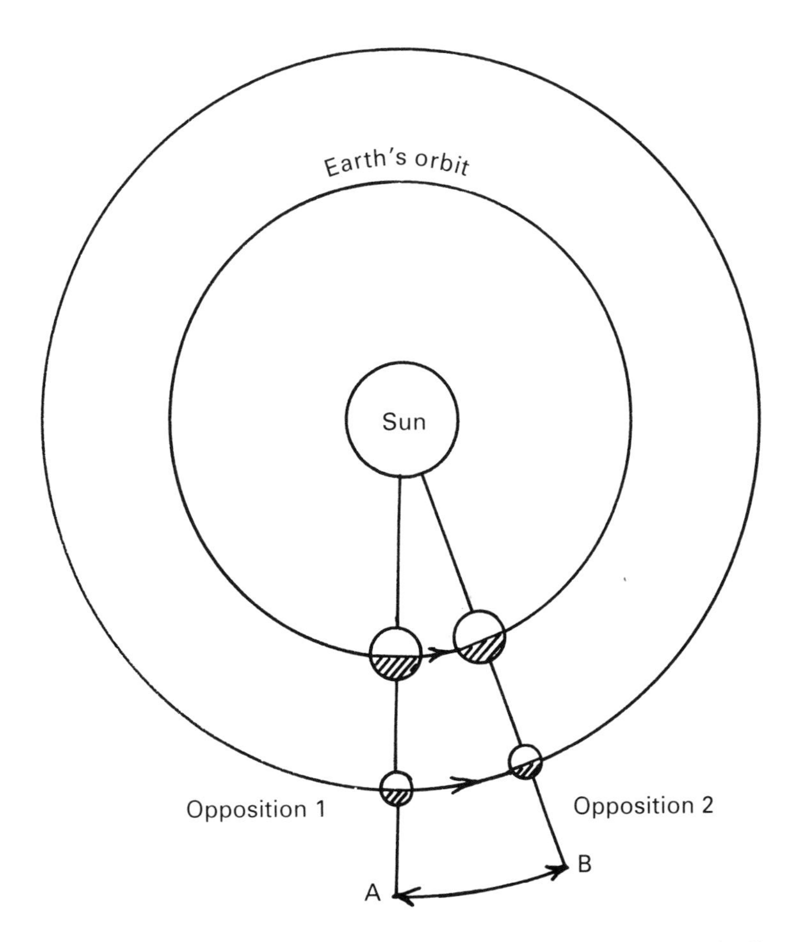

Fig. 1.11 Synodic orbital revolution period of a superior planet. A–B is the distance moved by the superior planet during one orbital revolution of Earth. (In the case of Mars two orbital revolutions of the Earth occur before the next opposition alignment is approached.)

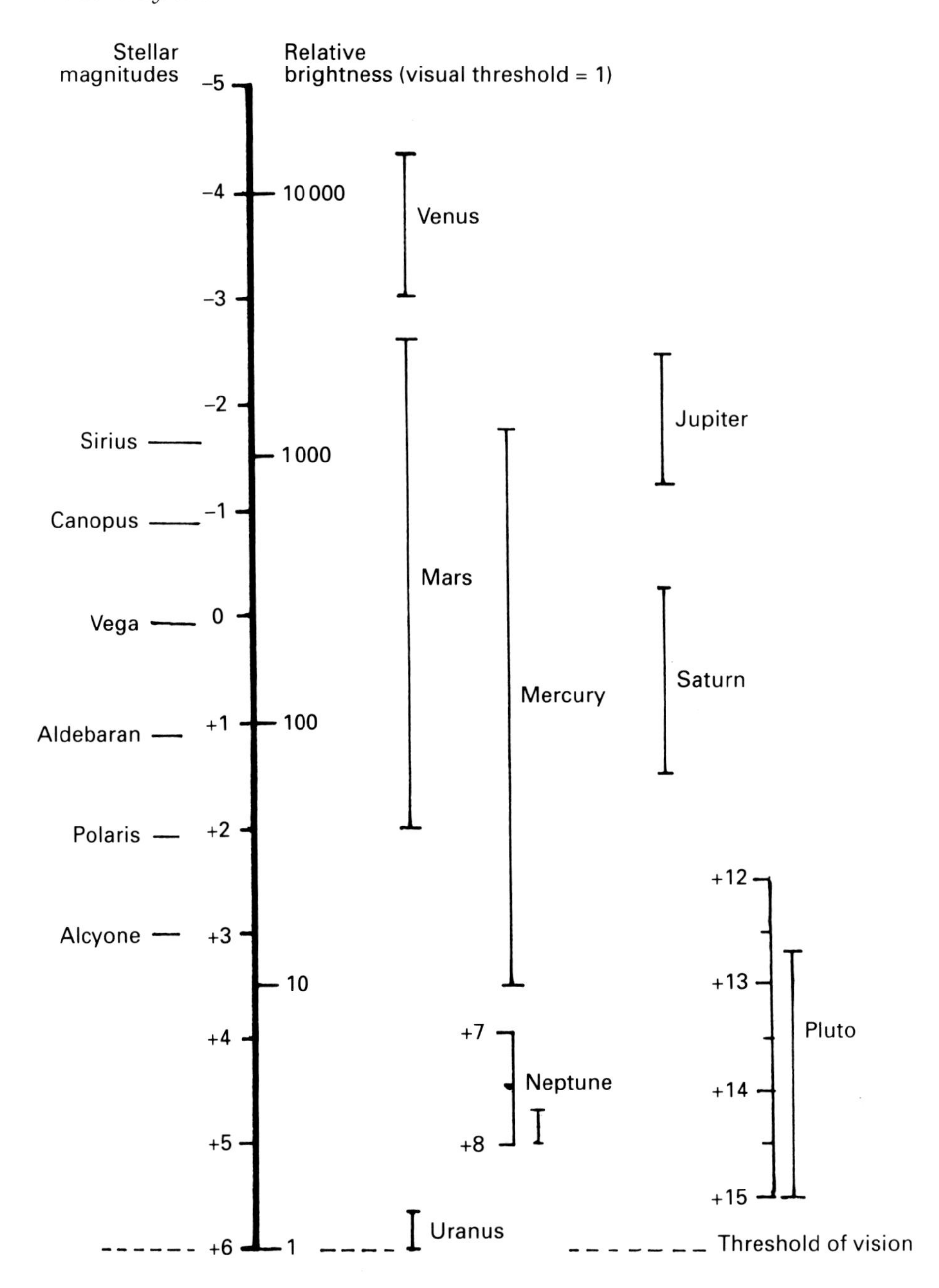

Fig. 1.12 Brightness ranges of the planets with some stars for comparison.

Further reading

Books

The New Guide to the Planets. Moore, P. A., W. W. Norton & Co. Inc., New York (1971).

The New Solar System. Beatty, K. J., O'Leary, B. and Chaikin, A. (eds), Cambridge University Press, Cambridge, England and Sky Publishing Corporation, Cambridge, Mass., USA (1981).

The Planets (Readings from *Scientific American*) W. H. Freeman & Co., San Francisco (1983).

The Atlas of the Solar System. Moore, P. A., Hunt, G., Nicolson, I. and Cattermole, P., Crescent Books, New York (1990).

The Astronomer's Manual. Rükl, A. (ed. S. Dunlop), pp. 15–93, Crescent Books, New York (1988).

Papers and articles

Scientific American: Solar System Issue. **233(1)**, (1975).

2

The celestial sphere

General

Having now looked at the Solar System from the outside, what does it look like to us standing on the Earth's surface? To us, the sky appears like a great inverted hemispherical bowl over our heads – or so it says in most books but to myself and many people it appears subjectively to be shallower or flatter than a true hemisphere would be. At night the stars, planets, moon and other heavenly bodies appear to be fixed to the inner surface of this bowl. We ourselves seem to stand at the centre of a horizontal circular more or less plane surface extending all the way to the horizon. Four points spaced at right angles on the circumference of the horizon mark the well-known north, south, east and west cardinal points. The point directly overhead on the dome of the sky is called the zenith. The other half of the bowl is out of sight beneath us and is continuous with the hemisphere above. The point on the bowl directly beneath us and therefore exactly opposite the zenith is called the nadir. This great globe of the heavens is called the *celestial sphere*.

Watching the clear night sky as the hours pass reveals that the stars retain the same relative positions to one another, forming patterns well known to us as the constellations but collectively they appear to drift slowly in an east to west direction across the sky if we face south. If we face north, the stars seem to revolve in a counter-clockwise direction in circular paths around one particular star that hardly moves at all. In fact the whole celestial sphere appears to be revolving around this star and it takes nearly 24 hours to make one complete revolution.

Of course, there isn't really a celestial sphere at all and the stars and other heavenly bodies are all at various immense distances from us. Their appearance of all being at the same distances as if projected onto the domed ceiling of a planetarium is an illusion.

In reality it is the Earth's west to east rotation on its axis that creates the illusion of the heavens revolving in an east to west direction. The Earth's axis happens to point almost exactly in the same direction of the star called Polaris, the Pole Star, and so this star seems to remain almost motionless while the other stars rotate around it. The stars relatively close to the pole star as seen in the sky will describe complete circles around the pole star and never set below the horizon. These are therefore called *circumpolar stars*. Whether or not a star is circumpolar depends on your location on the Earth's surface. As a moment's reflection will show, the further north that you travel, the more stars will be circumpolar because the pole star will be higher in the sky. In the northern hemisphere of the Earth the further south you are the fewer circumpolar stars there will be. Part of the daily path in the sky of the non-circumpolar stars will be spent below the horizon and so they will rise and set.

Positions on the celestial sphere

Although fictitious, the concept of the celestial sphere is a useful one for describing the positions and movements of the planets and other celestial bodies.

The planet observer should be familiar with the terms used to describe positions of planets and other celestial objects in the sky and the instruments used to determine them. This is important when using an ephemeris to locate the position of a planet at any time during the year. It is especially important in locating the daytime positions of planets like Venus and Jupiter which can be advantageously observed in broad daylight – but you must then know where to look for them, their night-time brilliance being swamped by the light of day! In this connection the use of setting circles supplied with most astronomical telescopes will be described in chapter 3.

There are various points and circles on the celestial sphere that enable us to describe positions on it. First, if we imagine the ends of the Earth's rotational axis – the north and south poles – to be prolonged indefinitely outwards from the Earth, they intersect the celestial sphere at the north and south celestial poles respectively (fig. 2.1). If, say, you are located in the northern hemisphere, the angular height of the Pole Star, which is very close to the north celestial pole, above your horizon is therefore equal to your latitude (fig. 2.2). Unfortunately, there is no prominent star close enough to the south celestial pole to serve as a pole star for southern hemisphere dwellers.

The great vertical circle on the celestial sphere that passes through the north and south celestial poles, the zenith and the nadir is called the *meridian*. It divides the celestial sphere into eastern and western hemispheres.

The celestial equator is the great circle around the celestial sphere whose plane is perpendicular to the line joining the celestial north and south poles. In effect, it is a projection of the Earth's equator onto the celestial sphere. It intersects the horizon at the east and west points and divides the celestial sphere into northern and southern hemispheres.

The ecliptic and the Zodiac

The plane of the Earth's orbit around the sun is called the *ecliptic* and the Earth's axis is tilted at an angle of 23.5° to it. The ecliptic projected onto the celestial sphere therefore appears as another great circle that is tilted at an angle of 23.5° to the celestial equator. The sun therefore appears always to lie on the ecliptic as seen from Earth and is occasionally eclipsed by the Earth's moon passing in front of it.

The moon is also eclipsed if the Earth happens to be exactly between the sun and moon. The moon will then be immersed in the Earth's shadow. At the same time the moon will also be on the ecliptic, as a little thought will show. Hence the name of the ecliptic – the great circle in the sky where *eclipses* of the sun and moon are seen.

In summer in the Earth's northern hemisphere, the north pole is tilted towards the sun and so the ecliptic appears to be on the north side of the celestial equator during the day and south of it at night. Likewise, the north pole is tilted away from the sun during the northern winter and the ecliptic appears south of the celestial equator during the day and north of it at night.

The diametrically opposite points where the ecliptic and celestial equator intersect are called the *equinoxes*. One, the March or vernal equinox is the point on the celestial sphere where the sun appears to cross from the south to the north side of the celestial equator on or about March 21st, marking the beginning of

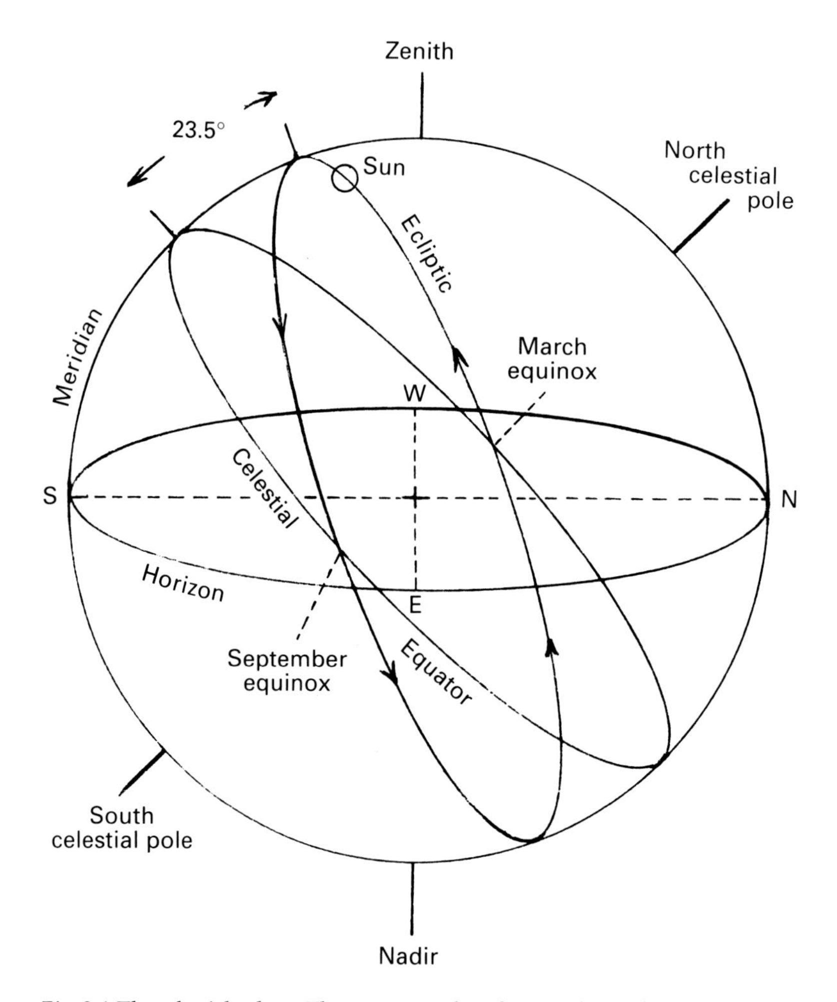

Fig. 2.1 The celestial sphere. The arrows on the ecliptic indicate the apparent eastward drift of the sun during the year. The celestial sphere is shown as it would appear to an observer at lat. 43°N shortly after midday at the summer solstice (June 21st) in the northern hemisphere.

Spring in the northern hemisphere. Similarly, the September or autumnal equinox is the point where the sun appears to cross from the north to the south side of the celestial equator on or about September 21st when Autumn commences in the northern hemisphere.

If we could mark the position of the sun against the background of the fixed stars in broad daylight every day for a year, we would find that the sun appears to drift slowly eastward at the rate of slightly less than 1° a day, i.e., a distance in the sky slightly less than twice its apparent angular diameter which is about 0.5°. (This has nothing to do with the sun's daily *east–west* rising and setting which is due to the Earth's rotation on its axis.) The slow eastward drift against the fixed star background is due to the Earth's orbital motion around the sun, a complete circuit of which takes 365.25 days, the terrestrial 'year'. There are 360° in a circle so the Earth's changing position in space at the rate of slightly less than 1° a day gives rise to the sun's apparent slow west–east drift against the fixed star background.

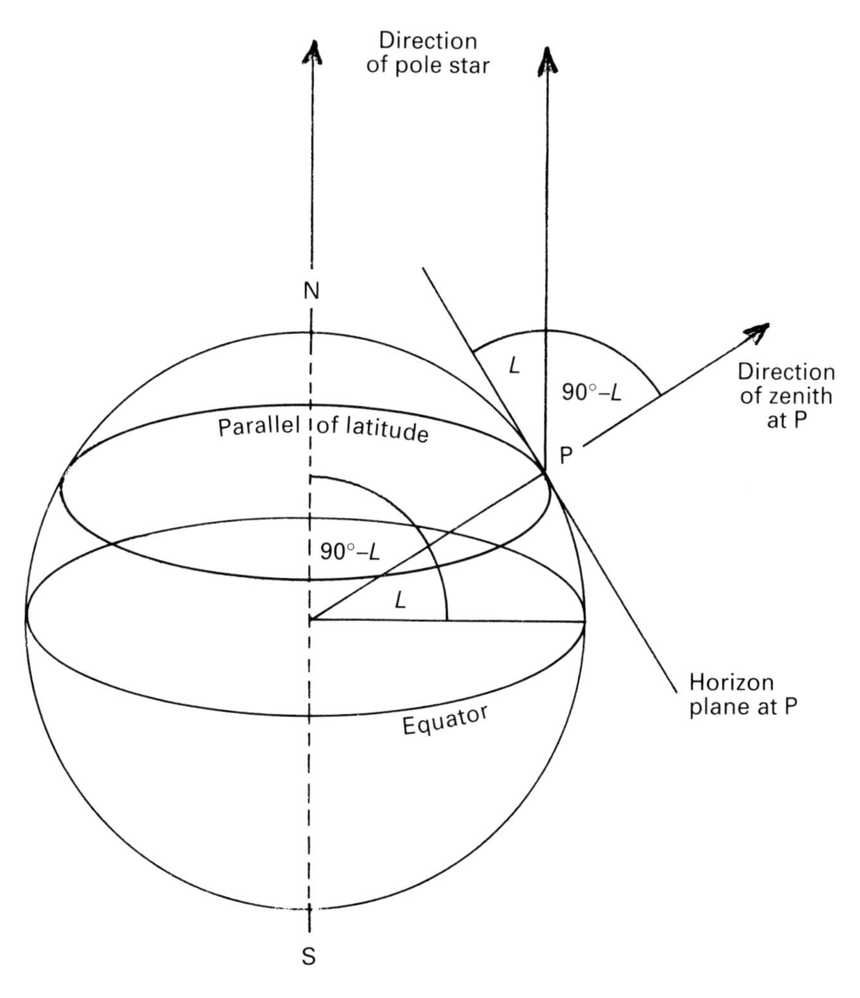

Fig 2.2 Latitude on the Earth's surface. L *is the north latitude of the point P and all other points on the same parallel of latitude.*

During the course of a year in its circuit of the ecliptic, the sun appears to traverse twelve different constellations, i.e., about one a month. Most of these represent and are named for mythical animals. These constellations are: Aries (ram), Taurus (bull), Gemini (twins), Cancer (crab), Leo (lion), Virgo (virgin), Libra (scales), Scorpio (scorpion), Sagittarius (archer), Capricornus (goat), Aquarius (water bearer), Pisces (fishes). The sun's apparent path in the sky through these constellations is therefore sometimes called the Zodiac (Gk. *zoon* = animal).

We can indicate the position of a star or planet at any instant on the celestial

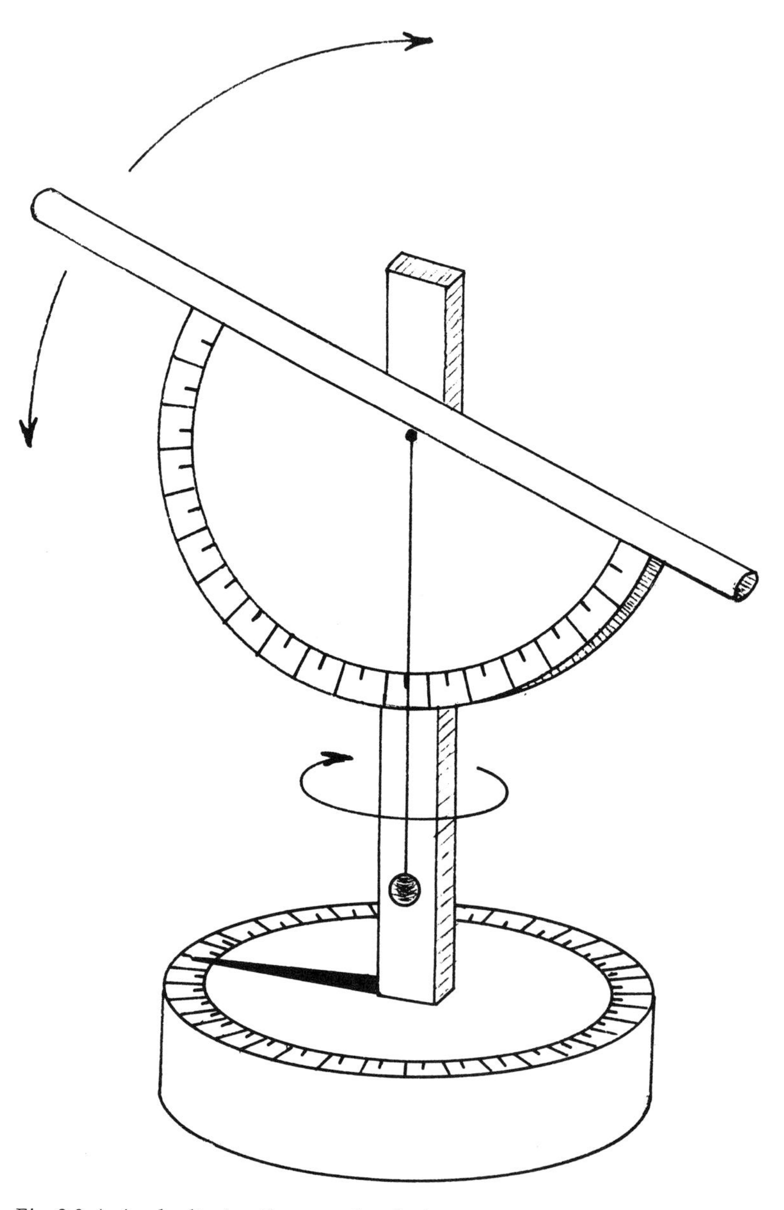

Fig. 2.3 A simple altazimuth measuring device.

sphere by making two measurements, one in altitude which is the angular height of the object above the horizon whose complement, measured from the zenith, is called the zenith distance. The other measurement is the azimuth, the angular distance east or west of some selected point on the horizon, usually the south point. These measurements can be made with a simple altazimuth instrument, as it is called, constructed from two semicircular protractors, a straight piece of cardboard or metal tube, a vertical support that can be swivelled around east–west, a horizontal axis on the top of this to move the tube up and down in a vertical plane and a plumb line (fig. 2.3).

The smaller circles of equal altitude on the celestial sphere and parallel with the horizon are called *almucantars* (fig. 2.4). Vertical circles are those circles other than the meridian that pass through the zenith and nadir and that intersect the horizon at positions other than the north and south points.

The true horizon is the great horizontal circle perpendicular to the zenith–nadir line and 90° from each. It may also be defined as the great circle along which a horizontal level plane meets the celestial sphere. The apparent or visible horizon is the one that we actually see and is irregular on land due to mountains, etc., and at sea it is a small circle located below the true horizon.

Celestial latitude and longitude

The positions of the planets, stars and other celestial bodies on the celestial sphere at any time in a given year are predicted in almanacs and ephemerides. It is especially useful to know the positions of the three faint planets Uranus, Neptune and Pluto so that we know exactly where to point the telescope when

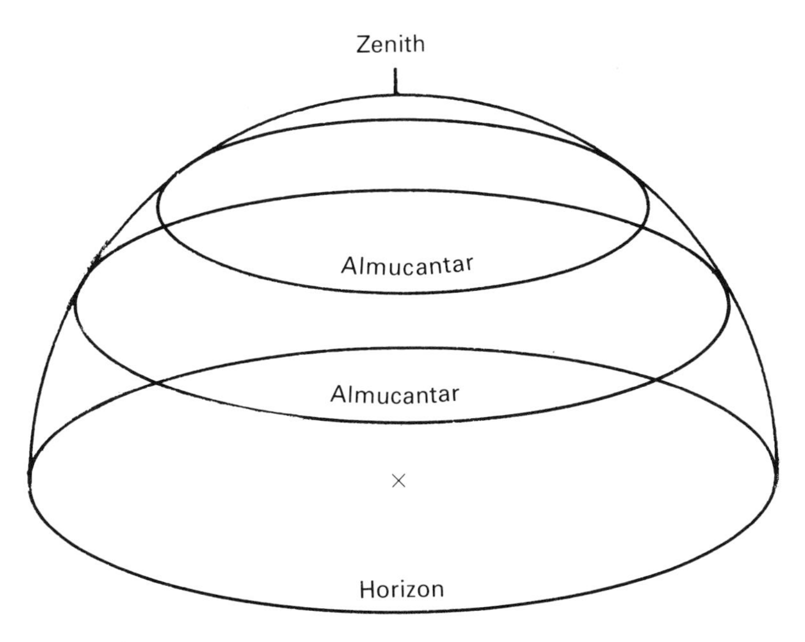

Fig. 2.4 Almucantars.

we wish to observe them. The other five planets are so bright that there is no difficulty in locating them when they are above our local horizon. Mercury is difficult to locate at times owing to its closeness to the sun's glare. The other bright planets can often be spotted in broad daylight but without the benefit of the dark sky background of night we need to know exactly where to point the telescope by looking up the planets' exact sky positions in an ephemeris.

The positions of the planets throughout a given year *could* be expressed as their altitudes and azimuths at various times but the altitude and azimuth of a planet at a given time both depend on the observer's latitude on Earth. Similarly, tables of altitude and azimuth for each planet *could* be computed for each latitude on the Earth, say at 10° or even 5° intervals. However, unless your own location was exactly at one of these latitudes a correction would have to be calculated. Clearly, such a system would be too cumbersome and prodigal of space in the published ephemerides for it to be practical. The system universally adopted in astronomy is that of celestial latitude and longitude. Latitude on the celestial sphere is measured in degrees north or south of the celestial equator just as latitude on the Earth is measured north or south of the Equator. Celestial latitude is called *declination* and is abbreviated to Dec. or symbolised by the Greek letter beta (β).

On the Earth, longitude is measured east or west from an internationally agreed-upon arbitrary zero meridian of longitude. Meridians of longitude are the great circles on the Earth passing through the poles that intersect the Equator at right angles. The standard meridian of zero longitude on the Earth is the Greenwich Meridian. All longitudes on Earth are measured eastward or westward from it. Similarly, on the celestial sphere, meridians of longitude are great circles passing through the celestial poles and intersecting the equator at right angles. The standard meridian of longitude passes through the point in the sky where the centre of the sun's disc crosses the celestial equator at the vernal equinox from the south to the north of the ecliptic. It is called the *first point of Aries* and is symbolised by the sign for the zodiacal constellation Aries. It is the zero point of celestial longitude or *right ascension* as it is usually called, from which the right ascensions of celestial bodies are measured in an eastward direction. It is abbreviated to RA or represented by the Greek letter alpha (α).

RA is customarily measured in units of time (hours) rather than degrees. The advantage of this is that the RA of a star or planet is the local sidereal time (see in this chapter under sidereal time) at which it will cross, or transit, the observer's meridian. Since the celestial sphere appears to rotate at an angular rate of 15° in an hour, an hour of RA is therefore equal to 15° of celestial longitude. The circles of RA are therefore called hour circles. RA is thus measured from the first point of Aries through 24 hours (360°).

The precession of the equinoxes. Nutation

Precession is the slow 'wobble' of the Earth on its axis that causes the poles to describe circles in space much as does the axis of a spinning top when it begins to slow down and wobble. Thus, the positions of the celestial poles shift slowly around a circle, a complete circuit of which takes 26 000 years. Presently, the north celestial pole is close to the star Polaris.

Simultaneously with precession there occurs *nutation* which is analogous to the nodding of a spinning top so that its head describes a wavy circle. Nutation thus causes the Earth's axis to approach and recede from the ecliptic pole.

The combined result of precession and nutation is the slow westerly movement of the equinoxes along the celestial equator at the mean rate of 50.26 seconds of arc per year. This is called *precession of the equinoxes.* The phenomenon was discovered long ago by the ancient Greek astronomer Hipparchus although he did not know the explanation of it. The rate of precession is non-uniform because of nutation. Precession causes the RA of stars to increase by more than 50 seconds of arc every year; the effects of precession on both RA and declination of celestial bodies must therefore be allowed for in accurate work. Published tables of RA are therefore strictly correct only for a given year or 'epoch' but for many practical purposes can be considered accurate enough for many years later.

Sidereal time (star time)

The time taken for the Earth to make one complete axial rotation relative to the fixed stars is the *sidereal day* and is also the time that the celestial sphere takes to make one complete revolution. This time is equal to 23 hours and 56 minutes of clock time and is thus 4 minutes less than the mean solar day of 24 hours. (The solar day is the time interval between two successive transits of the sun across the observer's meridian. Its duration varies depending on the time of year because of the Earth's variable orbital speed and the elliptical figure of the Earth's orbit.) That is why the stars appear to rise 4 minutes later every night by ordinary clock time. Astronomical clocks keep sidereal time as opposed to ordinary clocks that keep mean solar time.

The sidereal day is divided into 24 sidereal hours, each somewhat shorter than the ordinary clock hour. Sidereal noon occurs when the vernal equinox is on the meridian. This occurs at different times of the day or night at different times of the year.

The apparent motions of the planets on the celestial sphere

The planets as seen from the sun would appear as an orderly procession, those nearer to it repeatedly catching up with and overtaking the more distant planets. The view of these motions from the Earth, however, is much more complex since we are one of the several planets circling the sun. By about 2000 BC the ancients had divided the fixed stars on the celestial sphere into the patterns known as constellations. Among these they noticed five bright star-like objects that continually moved among the stars. They therefore called these bodies *'Planetes Asteres'* which means wandering stars. Hence their modern name of planets.

Protracted observation of the movements of the planets revealed that their wanderings are restricted to the narrow belt in the sky known as the Zodiac, within which they and the sun and moon appear to move continually eastward. This belt, of course, is centred on the ecliptic. The planets were observed to make complete circuits of the ecliptic, the longest taking about 30 years (Saturn) and then returning to practically the same starting point to begin the journey all over again. The motions of the planets within the Zodiac were complex and puzzling. Mercury and Venus were seen to swing first to one side and then the other side of the sun within certain limits, continually approaching and receding from it giving rise to the 'morning star' and 'evening star' apparitions. On the other hand, Mars, Jupiter and Saturn were seen to recede from the sun until they were opposite to it so that they were in the southern part of the sky at midnight. They then approached the sun again and were lost in its glare, only to repeat the whole cycle once more. At times, while moving eastward among the stars, they would appear to slow down, stop, move westwards, stop and then start moving eastwards again, thus describing a looped path in the sky. When Mars, Jupiter or Saturn were seen opposite the sun they would appear to be between the two stationary points. This *retrograde* and looped motion of the outer planets is explained fully in chapter 7 which is about the planet Mars.

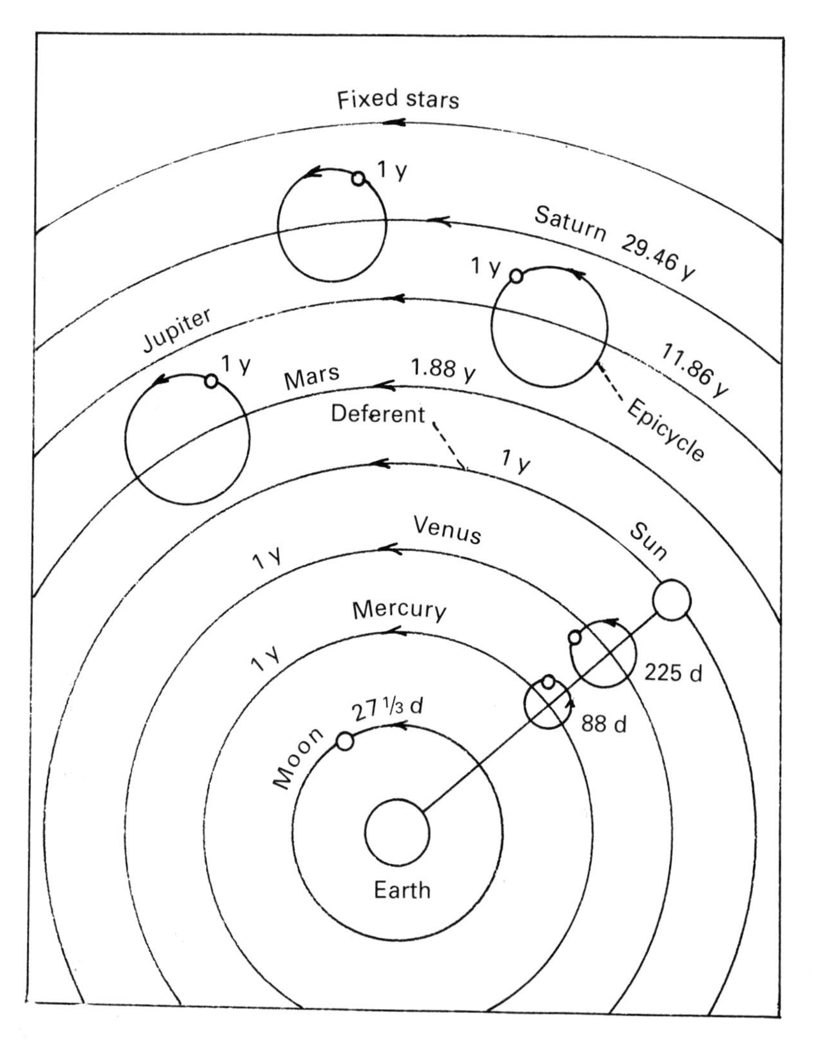

Fig. 2.5 The Ptolemaic system in its simplest form. In order to explain actual observations many more circles were needed. Note that the Sun, Mercury and Venus were always on a straight line.

The Ptolemaic explanation of planetary motions

To account for these puzzling and erratic planetary motions, the ancient thinker and philosopher Claudius Ptolemy (AD 100–170) considered that the Earth was at the centre of the universe. The motions of the sun, moon and planets were in a system of circles. Beyond the realm of the planets and at a great distance was the sphere of the fixed stars. Ptolemy developed a complex system of 'epicycles' and 'deferents' to explain the daily diurnal motion of the sun, moon and planets as well as the other complex movements of the planets (fig. 2.5). The Ptolemaic explanation of planetary motions was never completely successful.

The Heliocentric Theory

A breakthrough in the direction of the truth was made by Nicolas Copernicus (1473–1543) who suggested that the Earth and the other planets all revolved around the sun in circular orbits and that the sun was near the centre of the planetary orbits with the moon revolving around the Earth in its own orbit. This was therefore called the *Heliocentric Theory* (fig. 2.6).

The Copernican system gave an explanation of the observed planetary

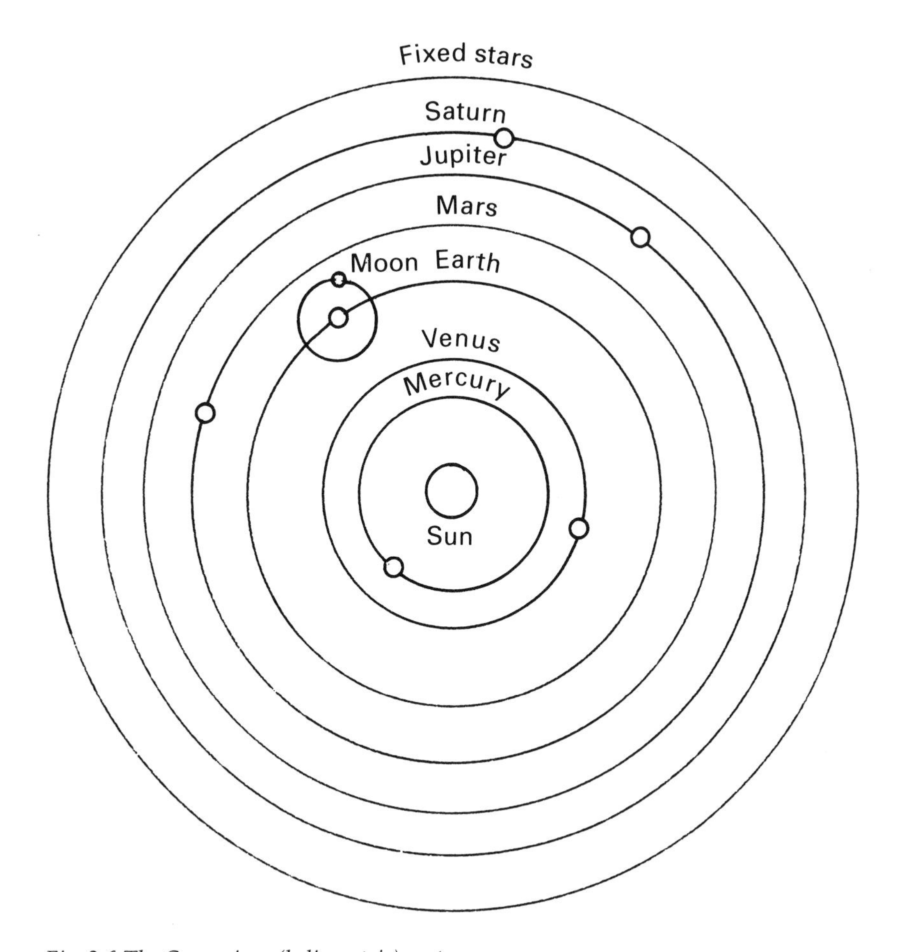

Fig. 2.6 The Copernican (heliocentric) system.

motions as good as that of the Ptolemaic system but it was no better. Its strong point was its greater simplicity and it was a major advance in the direction of truth in that it removed the Earth from its central position in the universe and relegated it to the status of being an ordinary planet.

Finally, as described in chapter 1, it was Kepler who formulated the three laws of planetary motion that bear his name and which accurately describe the motions of the planets.

Further reading

Books

For those interested in the history of the development of ideas about the nature of the Solar System and the motions of the planets, the following books should prove to be interesting. They are only a small selection of the large number of works available on this subject.

The World of Copernicus. Armitage, A., Signet Books, New York (1947).

Changing Views of the Universe. Ronan, C., Macmillan, New York (1961).

A Short History of Astronomy. Berry, A., Dover Publications, New York (1961).

Watchers of the Skies. Ley, W., Viking Press, New York (1963).

3

Telescopes and accessories

Types of telescopes

The primary instrument of astronomical observation is the telescope of which there are three main types – refracting, reflecting and catadioptric.

Refracting telescopes

This type of telescope is the kind familiar to everybody as the common spy glass. In its simplest form it consists of a simple convex lens of fairly long focal length mounted in one end of a cardboard tube. This lens is pointed towards the object being viewed and is called the objective. An image of the object being observed is formed at its focus and is viewed and further magnified by another smaller convex lens, the eyepiece mounted at the other end of the tube. It is usual to mount the eye lens in a separate tube that slides in and out of the tube carrying the objective so that the image seen through the eyepiece can be brought to a focus (fig. 3.1).

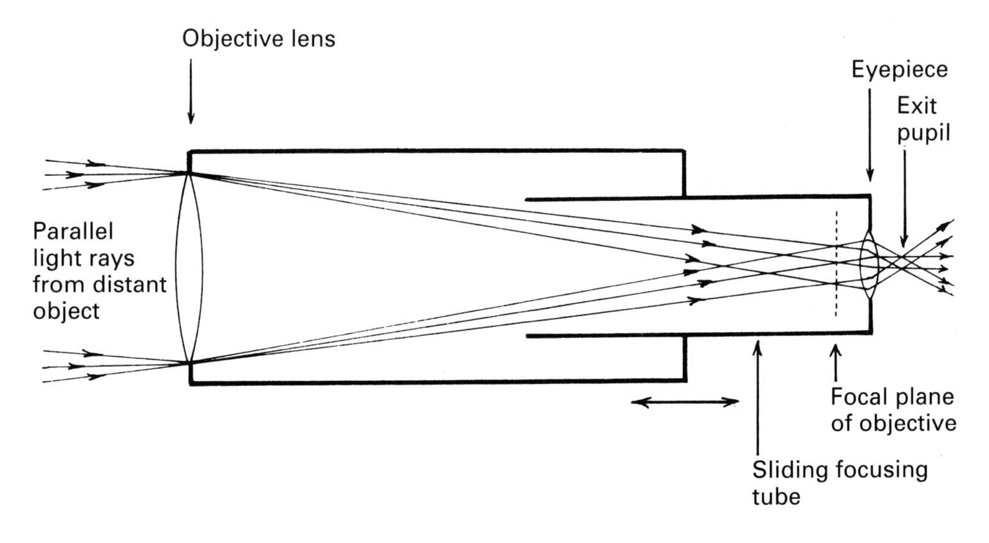

Fig. 3.1 Simple refracting telescope showing the path of the light rays.

This simple design of refractor suffers from two inherent defects or aberrations that make it unfit for astronomical observation: chromatic aberration and spherical aberration. *Chromatic aberration* is caused by the different wavelengths of light being focused at different distances from the objective (fig. 3.2). This causes objectionable colour fringes around the image of the object being viewed. *Spherical aberration* is caused by the light rays refracted from different zones of the objective being focused at different distances from the object (fig.3.3). This makes it impossible to focus the image sharply which consequently appears blurred.

These defects are greatly minimised in the achromatic objective which consists of two lens elements. The component nearest to the object is a biconvex lens of crown glass and the aberrations in the refracted light emerging from the other side of this lens are almost entirely corrected by the biconcave rear lens of flint glass (fig. 3.4). This results in a sharply focusable image that is almost free of spurious colour fringes. Similarly, eyepieces consist of two or more lens elements

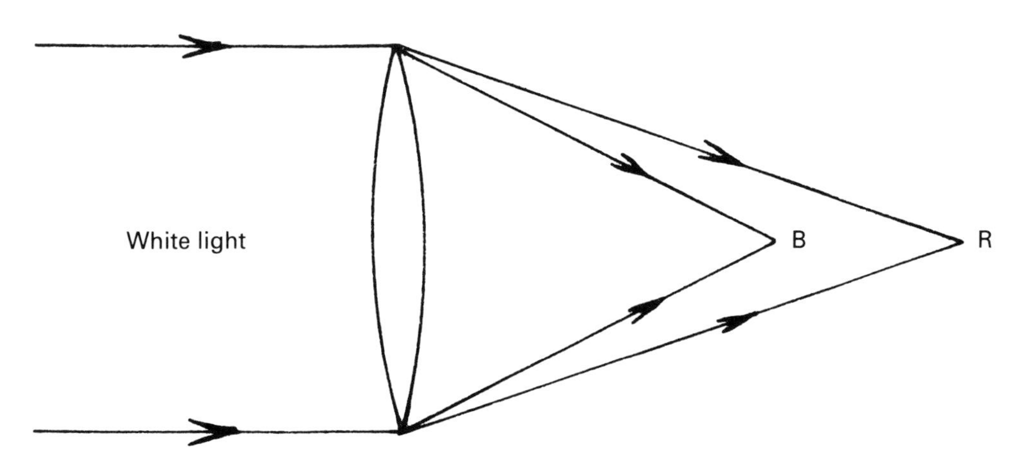

Fig. 3.2 Chromatic aberration in a simple biconvex lens. R is the focus of the red rays and B that of the blue rays.

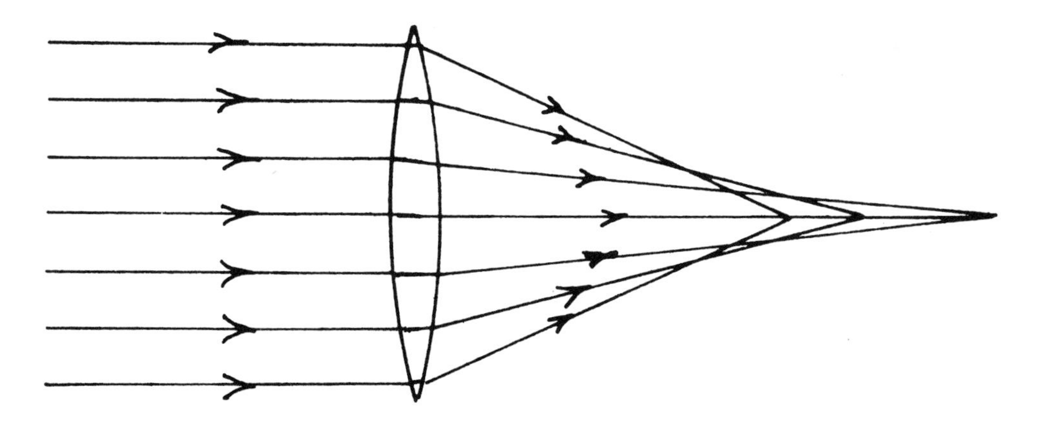

Fig. 3.3 Spherical aberration in a simple biconvex lens.

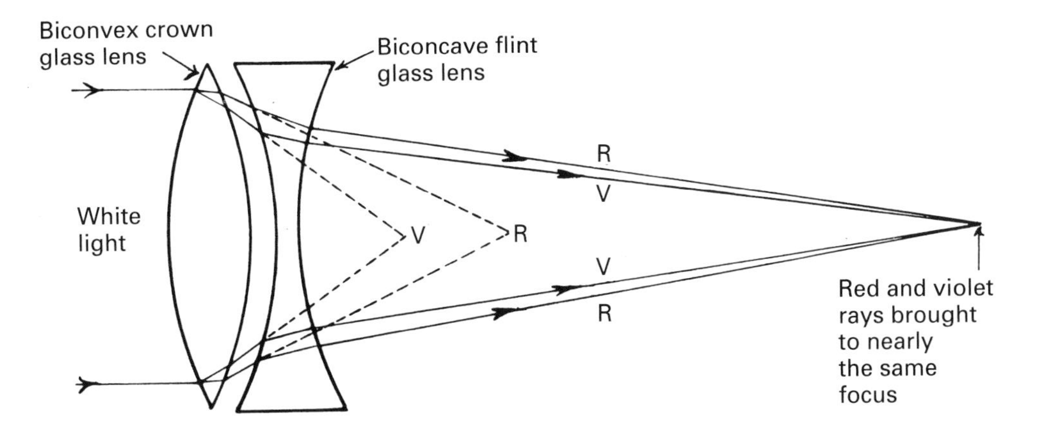

Fig. 3.4 Achromatic objective showing how chromatic aberration is reduced by the double construction. (R: red rays; V: violet rays.)

so that the aberrations inherent in the simple biconvex lens eyepiece are largely neutralised.

An important characteristic of a telescope objective is the focal ratio or *F number*. The focal ratio is defined as the fraction:

$$\frac{F}{D}$$

where *F* is the focal length of the objective, and *D* is its diameter. Hence, a telescope objective 3 inches in diameter – or with 3 inches of aperture as is often said – with a focal length of 45 inches has a focal ratio of $45/3 = 15$ and is said to be an F/15 objective. Astronomical refracting telescopes are usually made with objectives having focal ratios between F/12 and F/20.

Reflecting telescopes

The image of the object being viewed in a reflecting telescope, as the name implies, is formed by reflection of light from a concave mirror instead of by refraction through a lens (fig. 3.5). The image is entirely free of chromatic aberration because the light rays are not refracted. If the reflecting surface of the concave mirror is spheroidal, the image will suffer from spherical aberration unless it is of long focus. In spheroidal mirrors with focal ratios of F/12 and longer this aberration is negligible. If the mirror is paraboloidal the spherical aberration is cured. Reflecting telescopes commonly have focal ratios between F/6 and F/8.

It is impossible to view directly the image formed by ordinary small concave mirrors as the observer's head would completely obstruct the light to the mirror. The image must therefore be diverted to where it can be viewed without light rays impinging on the mirror being completely obstructed. This is achieved in various ways in reflecting telescopes of different designs:

The Newtonian reflector. This is named after its inventor Sir Isaac Newton, has a small plane secondary mirror placed at an angle of 45° on the optical axis of the main mirror and just inside the focus of the light rays coming from the

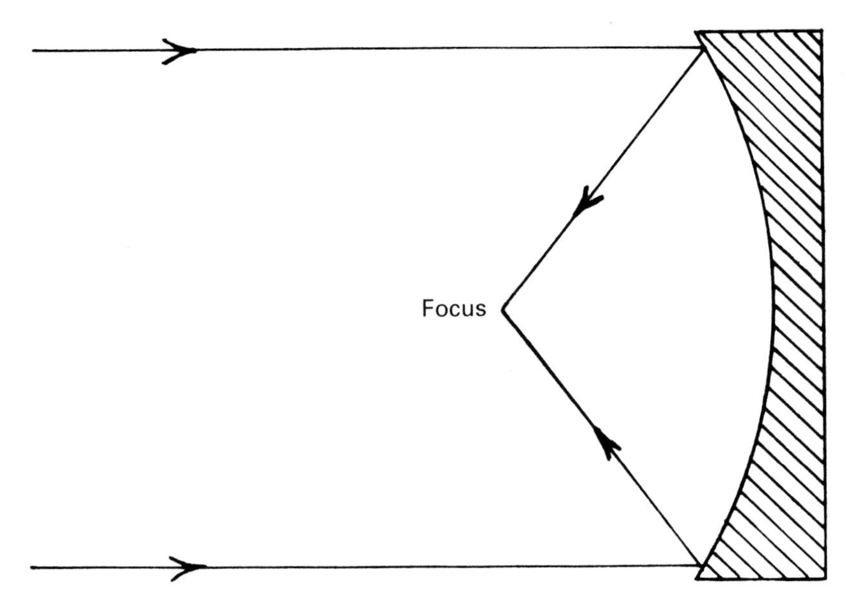

Fig. 3.5 Image formation by spheroidal concave mirror.

main mirror. This diverts the focused rays through a hole in the side of the telescope tube near its upper end (fig. 3.6). The eyepiece is placed here.

The Cassegrainian reflector. This has a small convex mirror placed on the optical axis of the main mirror near the upper end of the telescope tube and just within the focus of the focused light rays from the main mirror (fig. 3.7). The secondary mirror reflects the focused light rays back down the telescope tube through a hole in the main mirror where the image is viewed with an eyepiece.

The Gregorian reflector. This has a design somewhat similar to the Cassegrainian except that the secondary mirror is concave and is placed beyond the focal point of the main mirror (fig. 3.8).

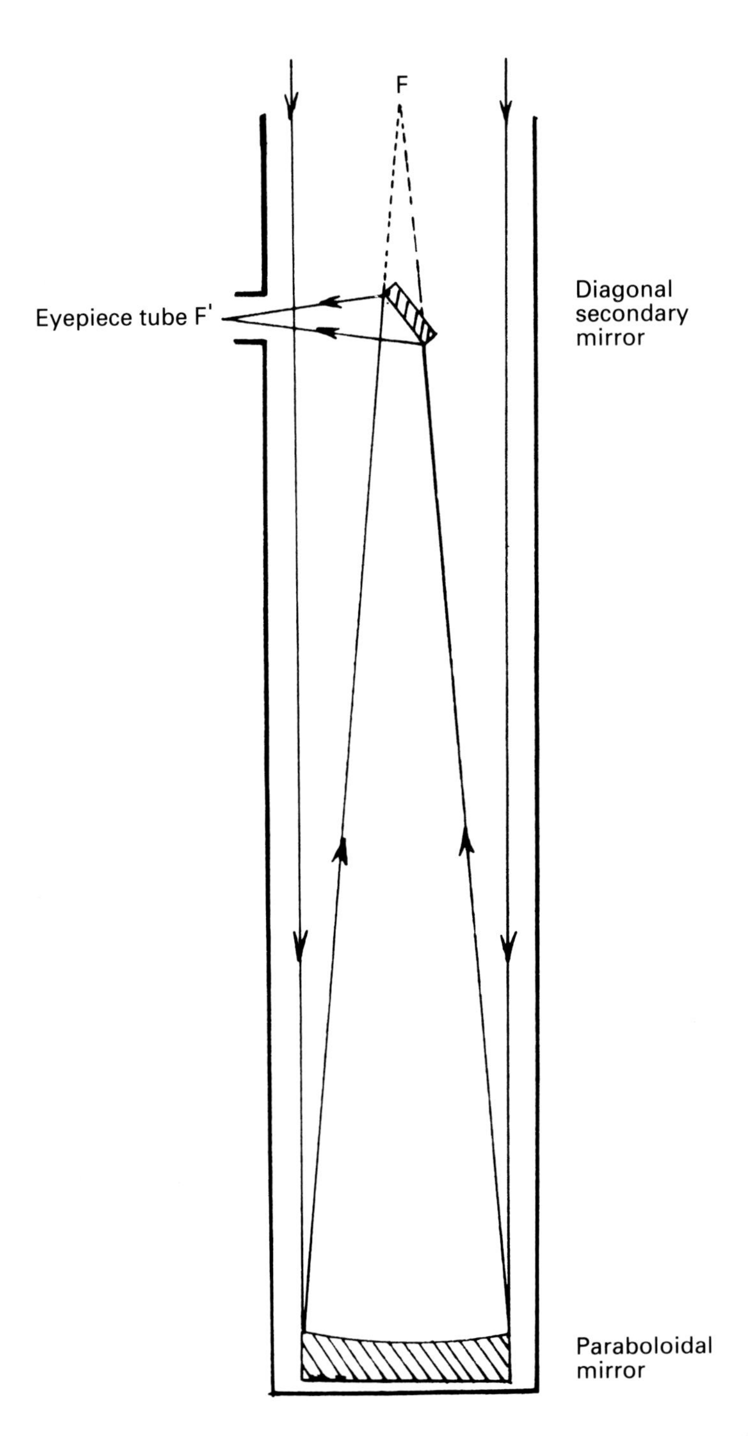

Fig. 3.6 Newtonian reflecting telescope. (F: principal focus: F': Newtonian focus.)

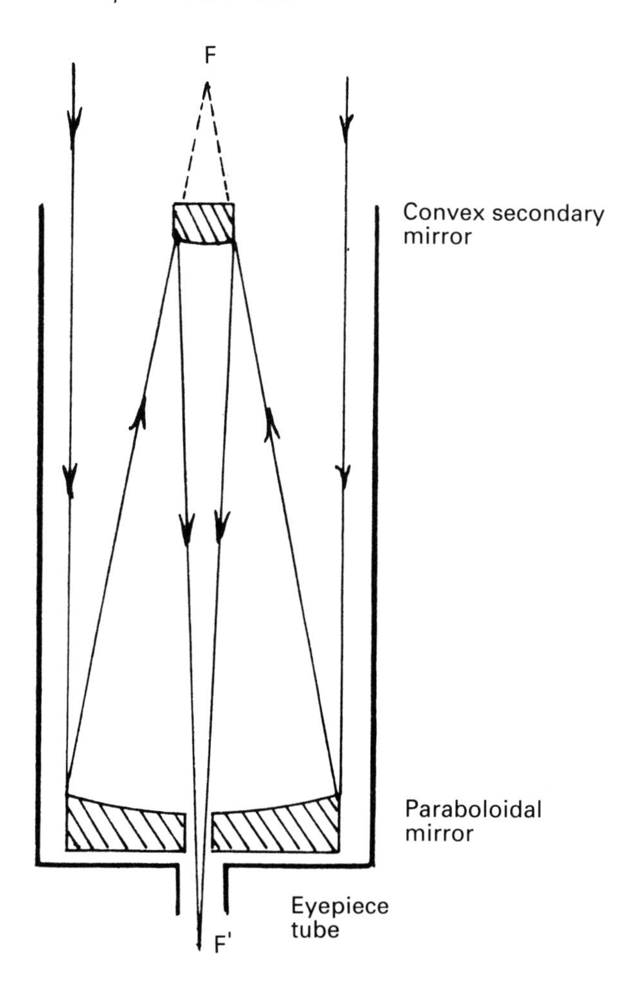

Fig. 3.7 Cassegrainian reflecting telescope. (F: principal focus: F': Cassegrainian focus.)

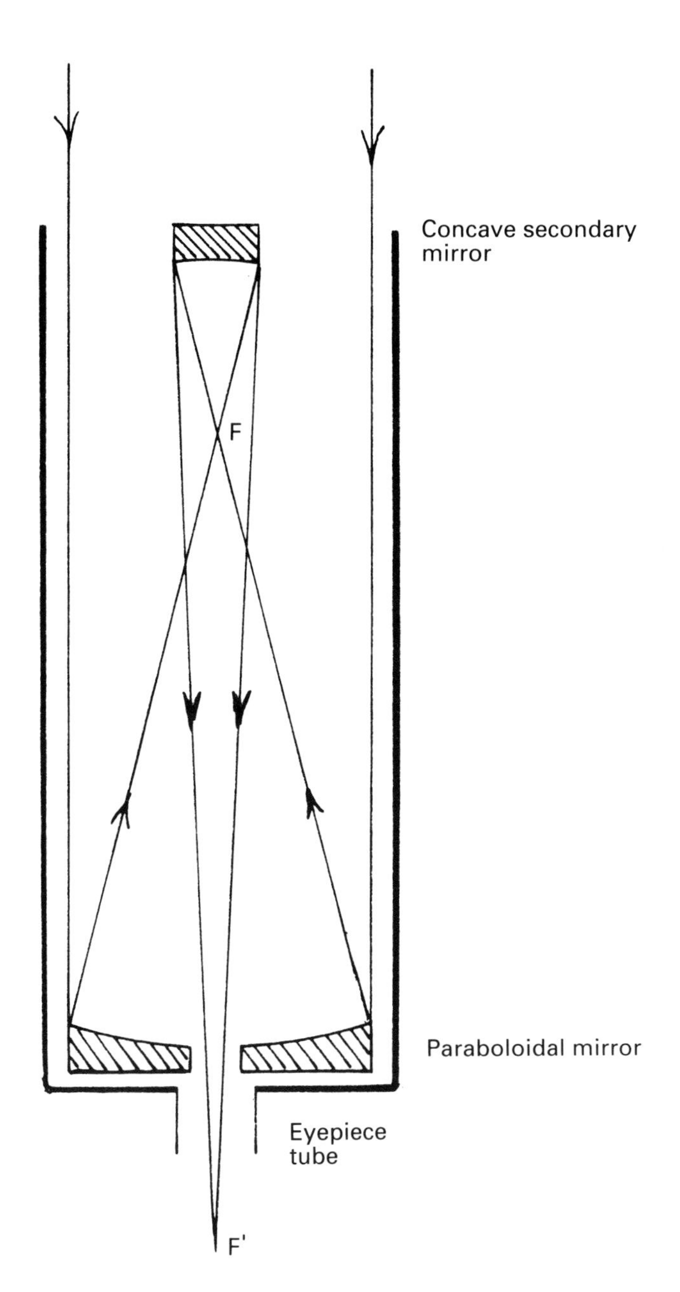

Fig. 3.8 Gregorian reflecting telescope. (F: principal focus: F' Gregorian focus.)

Catadioptric telescopes

These are constructed from both mirrors and lenses. The primary mirror is spheroidal. The light entering the telescope first passes through a lens called a correcting plate situated at the upper end of the telescope tube. This slightly alters the direction of the light rays reaching the main mirror in such a way that the image formed by the spheroidal mirror is free of spherical aberration. There are several different designs of catadioptric telescope:

The Schmidt–Cassegrain. In this, the correcting plate is thin and has a complex aspherical curved outer surface. At the centre of the inner plane side is attached a small convex secondary mirror within the focus of the main mirror. This reflects the focused light rays back down the telescope tube through a hole in the main mirror where the image is viewed with the eyepiece (fig. 3.9).

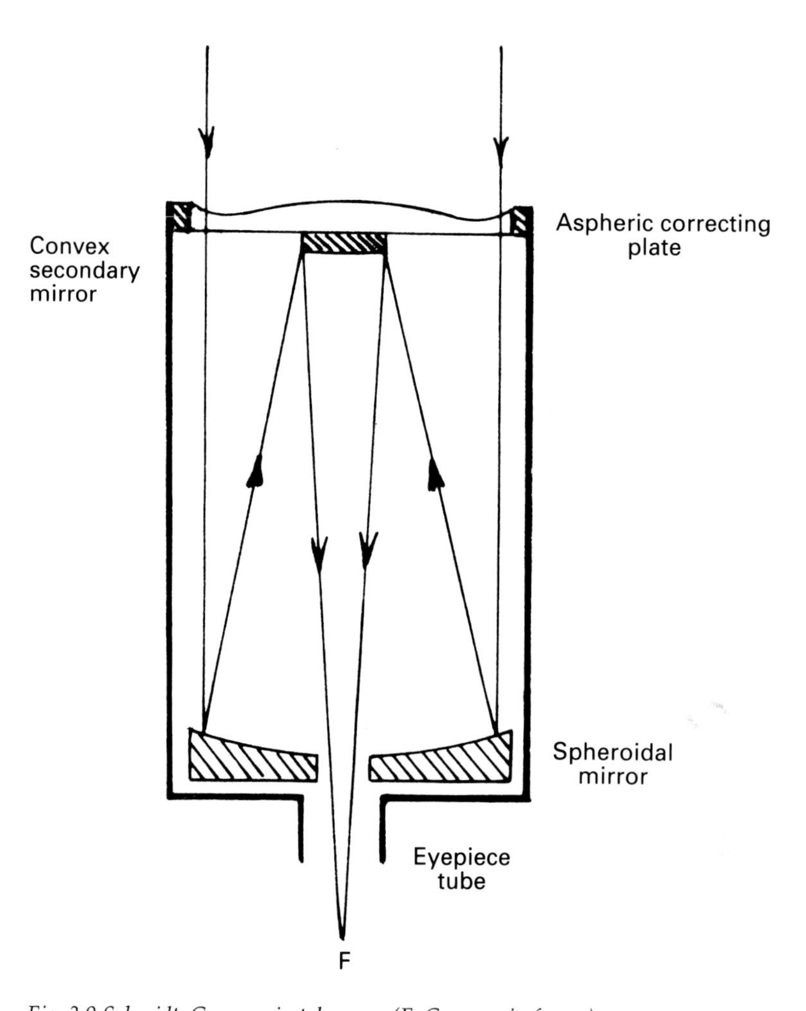

Fig. 3.9 Schmidt–Cassegrain telescope. (F: Cassegrain focus.)

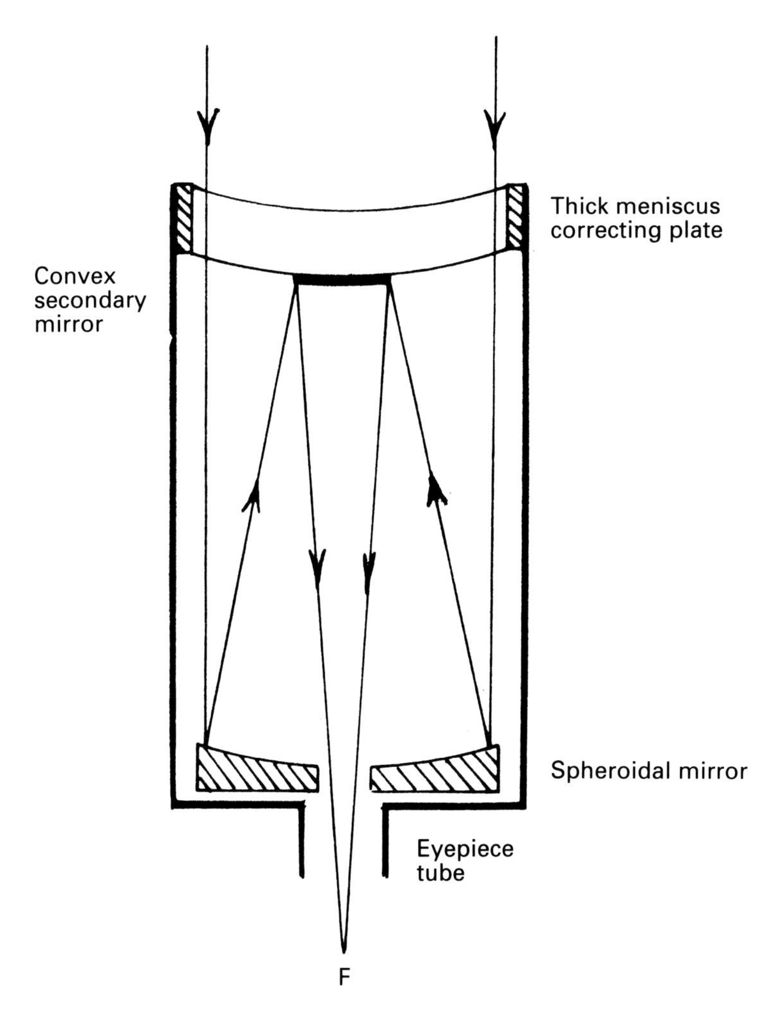

Fig. 3.10 Maksutov–Cassegrain telescope. (F: Cassegrain focus.)

The Maksutov–Cassegrain. This has a thick correcting plate, the concave side facing outwards and the inner convex side facing the main mirror. At the centre of the convex side is a small silvered circular area which performs the same function as the small secondary mirror in the Schmidt–Cassegrain telescope (fig. 3.10).

The Maksutov–Gregorian. Like the Maksutov–Cassegrain, this also has a thick correcting plate but positioned the reverse way from that in the Maksutov–Cassegrain. It has a small silvered circular area at the centre of the inner concave surface beyond the focus of the main mirror which again reflects the focused light from the main mirror back down the telescope tube through a hole in the main mirror and into the eyepiece (fig. 3.11).

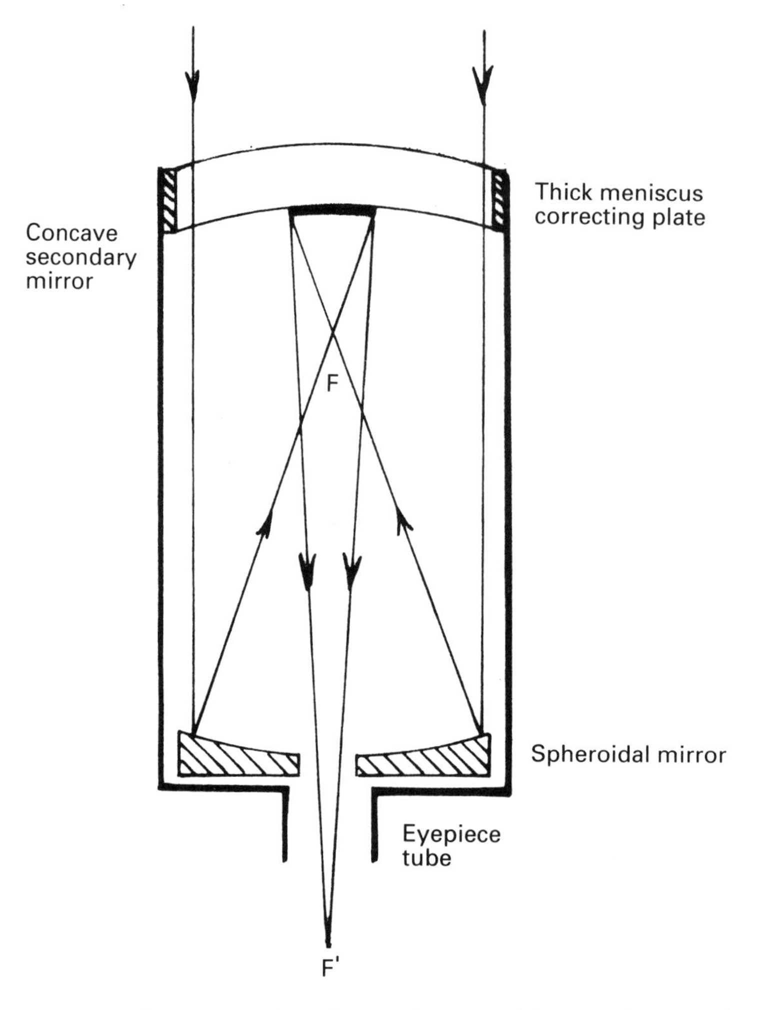

Fig. 3.11 Maksutov–Gregorian telescope. (F: principal focus; F': Gregorian focus.)

The choice of telescope

For planetary observing there is no doubt that the refracting telescope with its unobstructed light path gives the finest planetary images. This is especially so in apochromatic refractors in which the objective is constructed from three components which reduces the chromatic aberration almost to nil. They are superb for planetary observation and give crisp images that bear high magnification. They are very expensive in even 3- and 4-inch sizes but if money is no object go right ahead and buy one! The slight residual chromatic aberration in achromatic refractors becomes obtrusive in the larger sized telescopes. Apart from this, the price of refracting telescopes soars into the stratosphere with each inch of increased aperture over the 3- or 4-inch size. Further, a 5- or 6-inch refractor is not easily portable and is definitely an observatory instrument.

On the other hand, Newtonian reflecting telescopes, apart from their complete freedom from chromatic aberration, can be made with much shorter focal ratios than refractors, usually between F/6 and F/8 so that even an 8-inch reflector is portable and easily affordable by most amateurs. An 8-inch refractor would be financially out of the question for the vast majority of amateurs and would need permanent housing in a large observatory. As a choice for planetary observation, then, there is a lot to be said for the Newtonian reflector in the 6- to 10-inch aperture range.

First, the diagonal secondary mirror and its supports obstruct and diffract some of the light collected by the main mirror and this results in some loss of contrast in the image. However, this is not as bad as some authors imply and can be minimised by using as small a secondary mirror as is possible.

Second, the open tube construction encourages air currents in the telescope tube which can spoil the image. The mirror is not protected from dust and atmospheric pollutants and often gets out of adjustment whereas a refractor objective stays in perfect adjustment indefinitely.

The great strength of the catadioptric telescopes is that they combine long effective focal lengths, due to 'folding' of the light path, with ease of portability. Because of the folded light path a catadioptric telescope is much shorter, more compact and more easily portable than even a Newtonian reflector of the same aperture. Because of its compactness a catadioptric will be much steadier and more easily manoeuvered than a Newtonian or refractor.

The secondary mirror of a catadioptric is usually larger than that of a Newtonian reflector of the same aperture and may be almost one third of the diameter of the main mirror. This has a detrimental effect on planetary images, mainly loss of image contrast. It is not easy to escape the impression that there is some sacrifice of image quality in a catadioptric, at least as they are designed and manufactured today. Catadioptrics appear to be primarily designed to give wide flat visual fields for stellar photography rather than for visual planetary observation. Loss of image contrast is not of much consequence to deep sky enthusiasts who observe faint and ill-defined gaseous nebulae and remote galaxies. Light grasp is more important to them than critical definition. They therefore appreciate the affordability of the larger sized catadioptrics and their portability since they often need to drive for long distances into the country in search of dark skies away from the light pollution of cities.

There are therefore advantages and disadvantages with all the three main types of telescopes. Which type does one choose? On the whole, I would advise planetary observers to keep away from catadioptric telescopes. Assuming that you want to have the largest aperture telescope that you can afford with reasonable portability, in case you don't want or can't afford a permanent observatory, the 6- to 10-inch Newtonian reflector is clearly the telescope of choice. You get the most aperture for your money, it is a simple effective design, well tried and tested and deservedly popular. Most of the great planetary and lunar observers of the past have achieved wonderful results with them.

Eyepieces (oculars)

However good is the image produced by a telescope objective, an eyepiece of inferior design will not bring out all the detail in the image sharply and clearly. The situation is analagous to playing a record album with a blunt needle.

There are many different designs of eyepieces some of which are as follows.

Huyghenian (fig. 3.12) This is constructed from two plano-convex lenses. One, the field lens, is large and is nearer to the objective with its convex surface facing it. The smaller lens, nearer the eye, also has its convex surface turned towards the objective. The focal plane is between the two lenses so that reticles (see later in this chapter) cannot be used with this type of eyepiece, neither can it be used as a magnifying glass. For this reason, the Huyghenian eyepiece is said to be a negative eyepiece.

Ramsden (fig. 3.13) Similar to the Huyghenian except that the flat surface of the field lens faces the objective. In this and all the following types of eyepieces, the focal plane lies outside the eyepiece in front of the field lens so that these eyepieces can be used with reticles. They can also be used as magnifying glasses and so are called positive eyepieces.

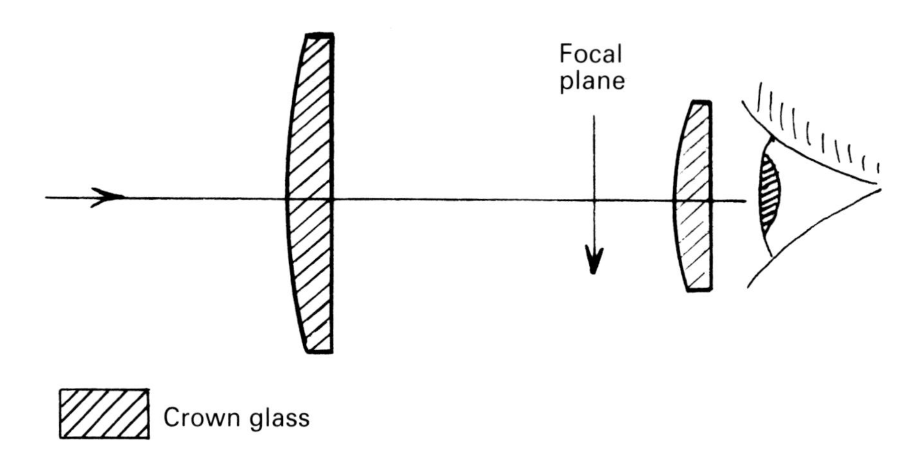

Fig. 3.12 Huyghenian eyepiece.

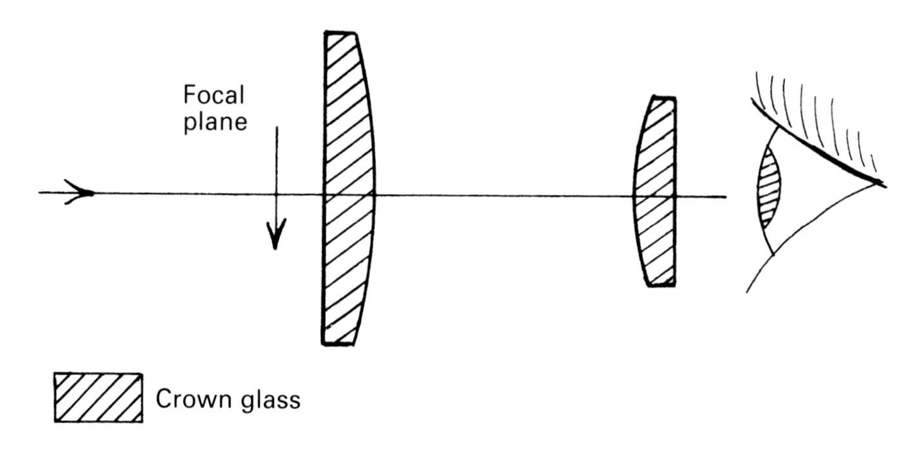

Fig. 3.13 Ramsden eyepiece.

Kellner (fig. 3.14) Similar to but not identical with the achromatic Ramsden and is often confused with it. A disadvantage is that the field lens surface is actually in the focus of the eye lens so that every speck of dust on the former is seen in sharp focus. It also has a tendency to give internal reflections and forms 'ghost images' of bright planets.

Zeiss orthoscopic (fig. 3.15) The large field lens in this design is a triplet and the eye lens a single plano convex lens.

Plössl (fig. 3.16) An achromatic doublet forms the field lens. The eye lens is also an achromatic doublet.

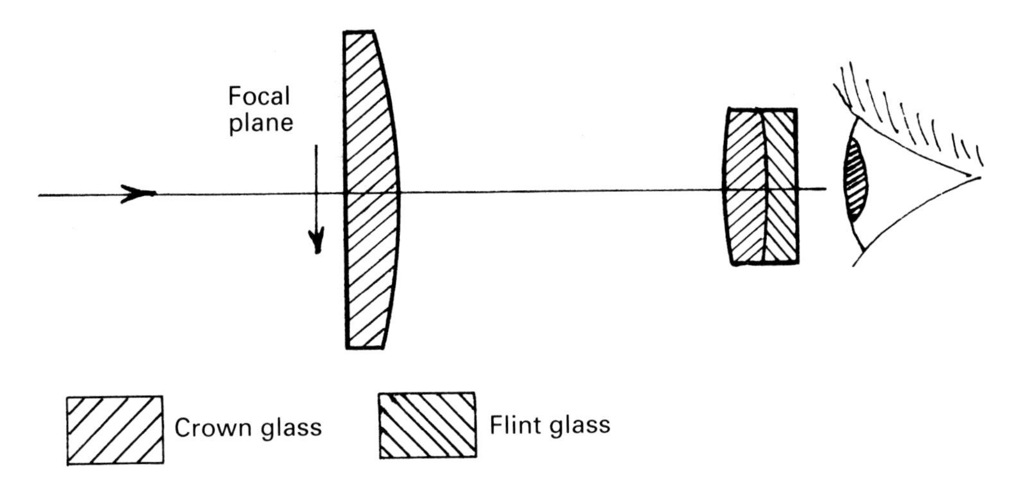

Fig. 3.14 Kellner eyepiece.

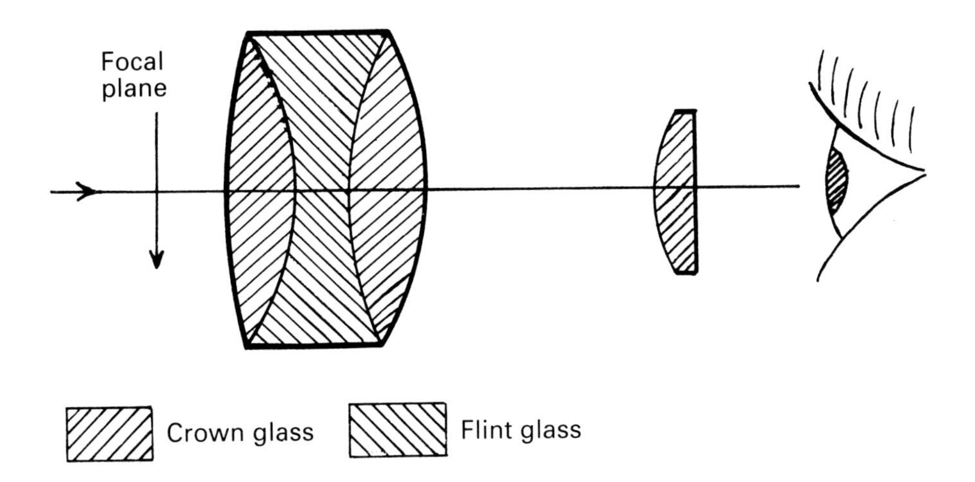

Fig. 3.15 Zeiss orthoscopic eyepiece.

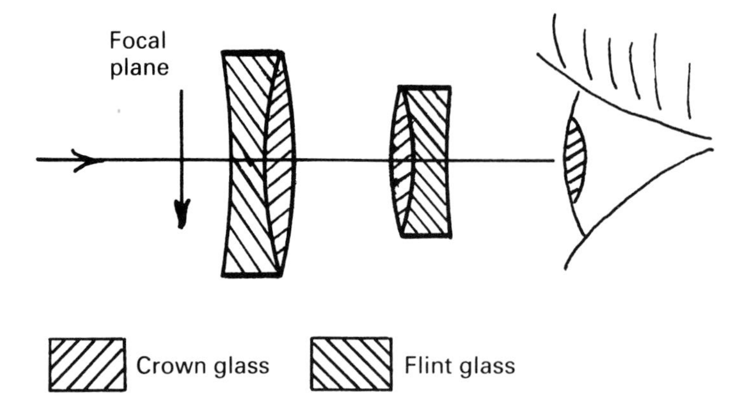

Fig. 3.16 Plössl eyepiece.

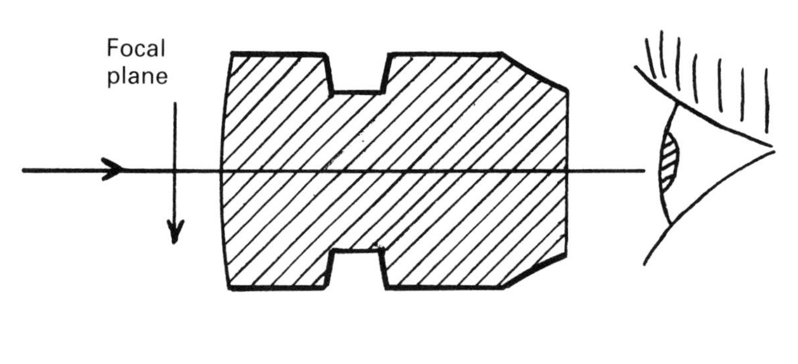

Fig. 3.17 Tolles eyepiece.

Tolles (fig. 3.17) An especially good design for use with reflectors. It is made from a single cylindrical piece of glass, both ends of which are convex. Internal reflections cannot occur.

Monocentric (fig. 3.18) This consists of three lenses cemented together. Internal reflections cannot occur in this design so that this type of eyepiece and the Tolles are free from 'ghosts'.

Of the above eyepieces, although the Huyghenian gives acceptable results with long focus reflectors it is quite unsuitable for use with the common F/6–F/8 Newtonian reflector because it is unable to give a sharp magnified image from the wider-angled cone of rays produced by the mirror. If funds are limited, Ramsdens and Kellners give good results with Newtonian reflectors. Both the Tolles and monocentric eyepieces are employed where critical definition is required in high power observation. The most popular eyepieces used with Newtonian reflectors for planetary work are the orthoscopics and Plössls which although expensive are well worth the money. They have excellent colour cor-

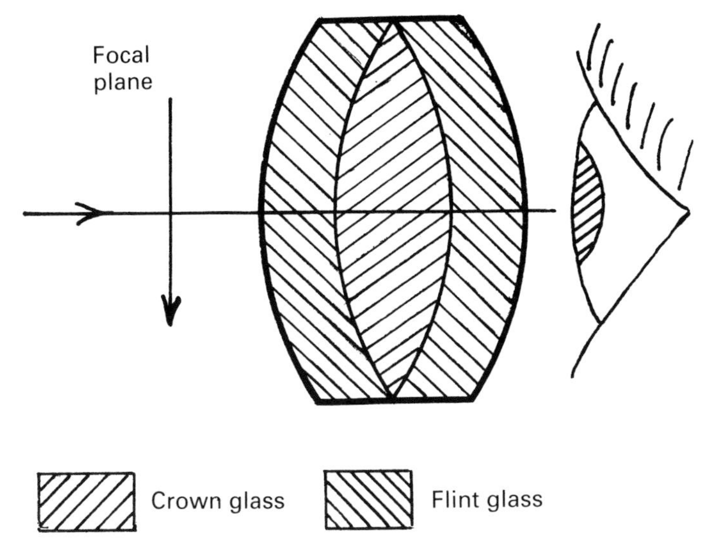

Fig. 3.18 Monocentric eyepiece.

rection are give good sharp images and flat visual fields. In the Ramsden, Kellner, orthoscopic and Plössl eyepieces, the focal planes lie outside of the eyepieces in front of the field lens as previously mentioned. Such eyepieces permit the use of reticles and can be used as magnifying glasses so they are all designated positive eyepieces.

Eyepieces for astronomical telescopes are mounted in cylindrical metal tubes 1.5 inches in diameter and simply slide into the eyepiece holder of the telescope. Some telescopes of Japanese make are designed for eyepieces 0.965 inches in diameter. The size of the eyepiece does not affect its optical performance and either size of eyepiece can be used with any telescope using suitable commercially available adaptors.

The image of a planet or any other celestial object seen in a telescope fitted with an astronomical eyepiece is inverted compared to the naked eye view. This is of no consequence in astronomical work as there is no 'right side up' or 'upside down' for celestial bodies. In telescopes intended for terrestrial use, extra lenses are incorporated into the eyepiece tube that make the image appear 'right way up'. This is never done in astronomical telescopes because the extra lenses absorb more light. This is detrimental in some classes of astronomical work in which faint stars or other objects are being studied.

Every eyepiece has an exit pupil, i.e., the place where the diameter of the converging and diverging beams exiting from the eyepiece is smallest (fig. 3.1). This is where the eye is placed to receive light from the eyepiece. It is actually an image of the telescope objective made by the eyepiece. For a given telescope objective, the higher the power of the eyepiece used, the smaller is the diameter of the exit pupil. If too low a magnification is used the exit pupil will exceed the largest possible diameter of the human eye pupil which is about 0.3 inch and only part of the light collected by the objective will enter the eye. (In my experience there is not the slightest risk of any enthusiastic amateur astronomer using too *low* a power on his telescope!)

Magnification

The overall magnification (M) of the telescope is given simply by dividing the focal length of the objective (F) with the focal length of the eyepiece (f):

$$M = \frac{F}{f}$$

Thus, with an objective whose focal length is 45 inches used with an eyepiece of focal length 0.5 inch the combined magnification will be 45/0.5 = 90 times or 90× as it is usually written. (There is a curious and persistent habit among astronomers of pronouncing magnifications such as 90× as 'ninety ex'. The '×' here is *not* a capital letter x but the sign of multiplication which symbolises magnification and so should be pronounced 'times' because the image in the above example looks 90 *times* bigger than the object. Hence it is quite nonsensical to say 'ninety ex'!)

Most eyepieces range in focal length from 1 inch (24 mm) to $\frac{1}{6}$ inch (4 mm). The higher the power, i.e., the shorter the focal length, the less is the eye relief which means that the eye has to be closer to high power eyepieces, sometimes uncomfortably so, for the highest powers, than with those of low powers.

Theoretically there is no limit to the magnification that can be obtained from a given telescope but in practice there are several factors that limit the highest magnification that is practical. Every increase in magnification also magnifies the effects of bad seeing and reduces the clarity and brightness of the image. If magnification is overpressed, planetary images become dim and fuzzy. The actual field of view in the telescope is reduced with high powers and every little tremor caused by wind or touching the telescope is magnified and the image jumps about.

What magnification range is best for observing the planets? Under average seeing conditions a power of about $20D$–$30D$, where D is the aperture of the telescope in inches, should give optimum results. For an 8-inch telescope this corresponds to powers of 160 to 240. Anything less than the above lower limit of $20D$ will mean that you will not be able to see all the fine detail in the image that the telescope objective is capable of revealing. It is best to use higher magnification than this minimum, however, because this gives a more comfortable view of the detail resolved by the objective. No further detail is revealed by increasing the magnification so any further increase is termed 'empty magnification'. The experienced planetary observer uses only the minimum magnification necessary to reveal the finest planetary features that are being studied. Seeing conditions also affect the maximum magnification that can be usefully employed. Excellent seeing will permit magnifications of $40D$ or even $50D$ but the occasions when this can be done are rare.

In practice you should use only as high a magnification as is compatible with a bright sharp planetary image. If the image is dim or fuzzy at the next higher power, change back to the lower power. A smaller bright crisp image is better than one that is large, fuzzy and dim.

A common defect of the eye is astigmatism which is caused by unequal curvature of the cornea in different directions. This causes objects oriented in different meridians to look unequally sharp. For example, a clock face seen with the naked astigmatic eye will appear to have, say, the figures 12 and 6 looking sharp but the 9 and the 3 will appear unsharp. The use of high powers minimises astigmatic effects as the exit pupil is small with higher powers and so the light beam from the eyepiece traverses a lesser area of the cornea.

The Barlow lens

Users of Newtonian reflectors will appreciate this useful optical accessory. The usual Newtonian reflector has a focal ratio of around F/7. Shorter focal length eyepieces (i.e., more powerful) must therefore be used to obtain sufficient magnification to exploit fully the resolving power of the mirror. The short eye relief (the closeness with which you need to bring the eye in order to see all the field of view) of these powerful eyepieces is a nuisance as the eye has to be uncomfortably close to the eye lens if the entire field of view is to be seen.

The Barlow lens is a negative (diverging) lens that is mounted in a cylindrical metal tube and positioned just within the focus of the telescope mirror. The eyepiece slides into its other end. The Barlow lens reduces the angle of the light cone focused by the main mirror and this greatly increases the effective focal length of the telescope while only adding an inch or two to the actual overall focal length of the mirror (fig. 3.19).

The image produced is thus much larger and therefore lower power eyepieces with their better eye relief may be used and will yield higher final magnifications. Another advantage is that since the light cone transmitted by the Barlow lens is narrower than that directly reflected from the telescope mirror, any eyepiece will perform better with the Barlow lens than without it. Amplifications provided by Barlows are usually 2× to 3×. When selecting a Barlow lens it is best to get one with a non-integral amplification such as 2.5×. By so doing you will not duplicate eyepiece powers that you may already have, e.g., if you possess 12 mm and 24 mm eyepieces, a 2× Barlow used with the 24 mm will merely duplicate the 12

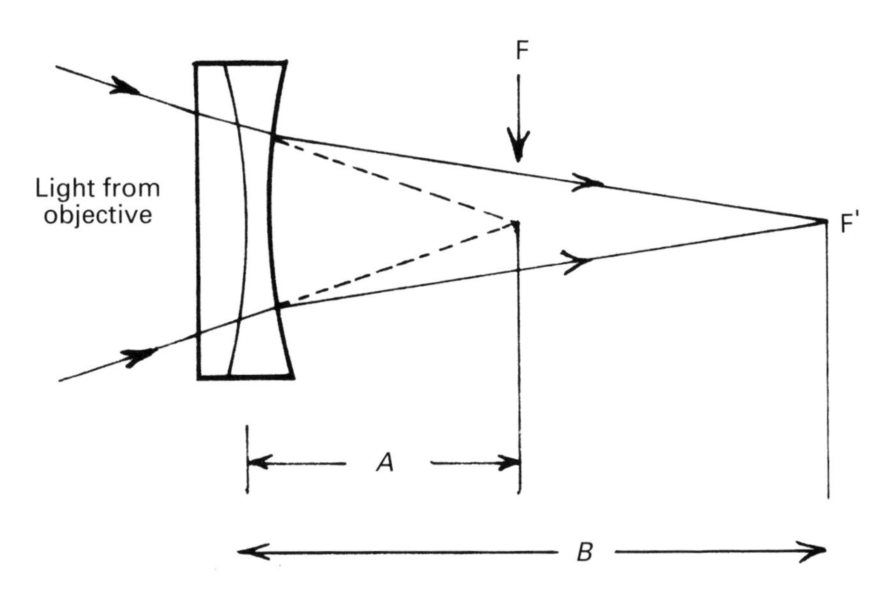

Fig. 3.19 The Barlow lens. (F: focus of the objective; F': focus with the Barlow lens.) Here the distance B *is equal to* 2A, *therefore the amplification factor of the Barlow lens is* 2×.

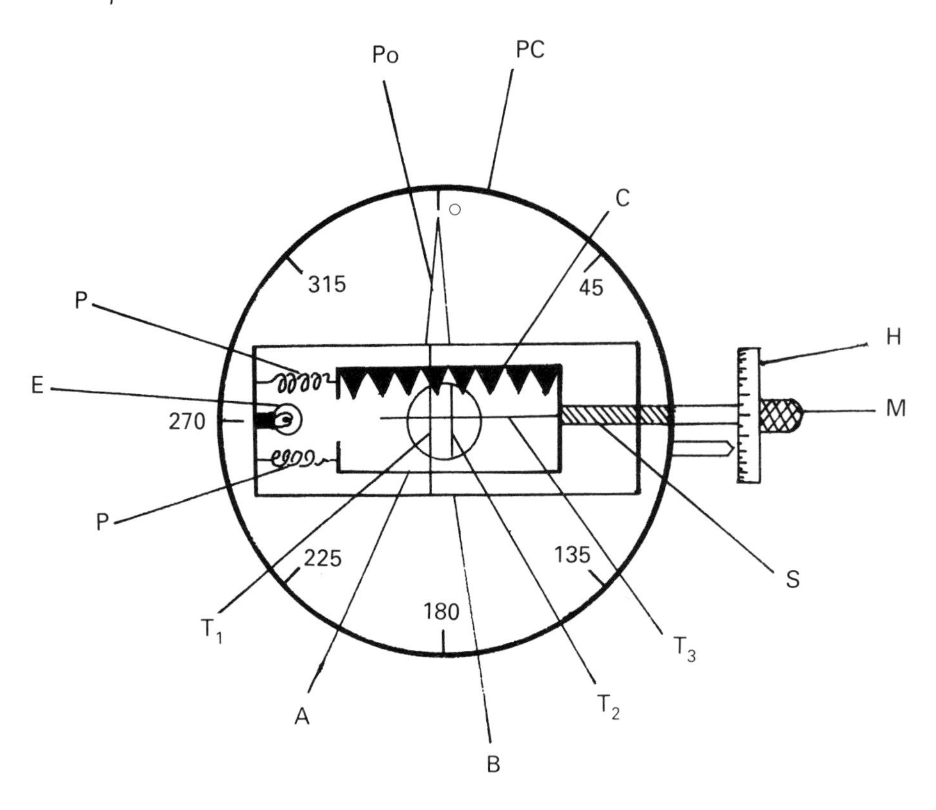

Fig. 3.20 The filar micrometer.

mm. Also, avoid purchasing inexpensive Barlows in which the lens can be moved along the tube to vary the amplification. They may also be non-achromatic. Good Barlow lenses are fairly expensive but are worth the expense.

The filar micrometer

The filar micrometer (fig. 3.20) is a device used for measuring the angular separation between two points seen in the telescope, such as a double star, the angular dimensions of planetary discs or features seen on planetary discs.

In fig. 3.20, A is a rectangular frame which slides over a larger frame B by rotating the fine-threaded screw S which is actuated by the milled head M. The back lash of the screw is absorbed by the springs P,P. Three fine threads of spider web (T_1, T_2, T_3) are stretched across the frames, two in a vertical orientation and one horizontal. They are illuminated by the small electric lamp E. A positive eyepiece is attached to frame A at the position indicated by the small circle and is focused on the spider threads and the circular plate PC is attached to the telescope draw tube with the spider threads lying in the focus of the objective. On looking into the eyepiece the threads will appear like lines drawn upon the sky (fig. 3.21). The diameter of a planetary disc, say, is determined by setting the fixed vertical thread tangent to one edge of the disc, rotating the milled head M until the movable thread touches the opposite side of the disc and then noting the number of turns of the screw corresponding to the distance between the threads. The 'comb'

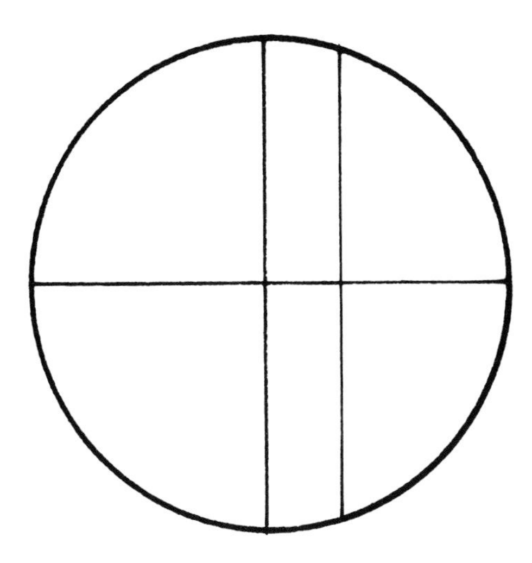

Fig. 3.21 Field of view in a telescope fitted with a filar micrometer.

C visible in the field of view of the eyepiece has teeth, each separated by a distance corresponding to one complete turn of the screw. Thus, complete turns of the screw are easily counted and fractions of a turn are read off on the graduated wheel H.

The instrument has to be calibrated by finding how many turns or fractions of a turn of the screw correspond to the distance between the two vertical threads when they have been adjusted to the same separation as the components of a double star whose angular separation is accurately known. This gives the number of turns of the micrometer screw corresponding to the angular distance between the two stars. It is then a simple matter to calculate what angular distance corresponds to one turn of the screw. If the eyepiece is changed or if another telescope is used, the micrometer must be recalibrated.

For determining the position angle of a double star whose separation is being measured the whole device can be rotated and the position angle read off on the circular plate PC by the pointer Po.

Other designs of filar micrometer are available. In one variant there are two screws, each moving one of the two vertical threads.

Resolution

More important than mere magnification is the resolving power of the objective. This may be simply defined as its ability to reveal fine detail (but see further on). The larger the objective the greater is its resolving power which is directly proportional to its diameter. Hence a 12-inch telescope has double the resolving power of a 6-inch telescope and will reveal detail half the size of the finest detail resolved by the 6-inch. Using the same magnification on a 6-inch and a 12-inch telescope, more detail will be seen in the same size image of the 12-inch because of its larger aperture and resolving power.

Several years ago Dawes put forward a criterion based on actual observation for expressing telescopic resolving power. This is that a very close double star is considered resolved, i.e., seen to be two separate stars, if under high magnification, the diffraction disc images of the two components of the double star overlap to the extent of half their diameter, although this might be difficult to see if the observer was inexperienced or had poor eyesight or if the seeing conditions were poor. (Owing to the wave nature of light, the image of a perfect point source of light, such as a star, produced by a telescope objective, whether a mirror or lens, is never a perfect point of light. It is a tiny bright disc variously known as the diffraction disc, Airy disc or *antipoint*. The larger the aperture of the telescope objective, the smaller will the diffraction disc be and hence the greater will be the telescope's resolving power as shown by its ability to resolve a close double star into its components.) The angular separation of the two components of a double star that are resolvable by the Dawes criterion is considered to be:

$$\frac{4.56}{D} \text{ seconds of arc,}$$

where D is the working aperture of the telescope in inches. A 3-inch telescope will therefore resolve 1.5 seconds of angular separation and a 6-inch 0.76 seconds.

The less rigorous Rayleigh resolution limit based on light wavelength considers that a double star is just resolved when the diffraction disc images of its members just touch. The Rayleigh formula for the angular separation of members of a double star is:

$$\frac{5.5}{D} \text{ seconds of arc.}$$

where D again is the working aperture of the telescope in inches.

Planetary observers will have difficulty in translating the above criteria of telescopic resolution into terms of the resolvability of planetary detail; the separation of two almost touching bright diffraction discs on a black background is a different situation from the resolution of fine irregular detail of varying contrast and brightness on a bright planetary disc. Although usually defined as ability to reveal fine detail, the resolving power and aperture of a telescope objective are alone not sufficient to determine a telescope's ability to reveal detail. Of greater importance is contrast and this has nothing to do with resolving power. Apart from the telescope's aperture, the contrast between fine detail and its immediate neighbourhood on a planetary disc affects the resolvability of the detail. The importance of contrast as opposed to resolution in revealing fine detail in telescopic images has been much neglected in the astronomical literature. The reader is therefore urged to consult R.W. Gordon's paper on this important topic listed in the 'Further reading' section at the end of this chapter.

Light gathering power

The larger the telescope objective the more light it collects from the planet under observation. The light gathering power of an objective is proportional to the square of its diameter; a 12-inch objective collects four times as much light as a 6-inch. The brighter images afforded by larger telescopes enable subtle colourings and tints in planetary images to be more easily and reliably detected than in the dimmer images given by smaller telescopes at the same magnification.

Filters

Coloured optical glass filters that can be screwed into the bottom of eyepieces are important accessories in planetary observation. They are used to reduce glare and to improve contrast of planetary details. Some observers maintain that certain types of planetary detail cannot be seen unless filters are used; others declare that nothing new is brought out by filters, only visibility is improved. Much depends on the colour sensitivity and visual acuity of individuals. The different types of coloured filters, their optical characteristics and their use in planetary observation are fully described in the chapters dealing with observation of individual planets.

The apodising (antidiffraction) screen

This easily constructed accessory reduces the effect of atmospheric turbulence on a planetary image, reduces glare and improves contrast of surface features. It is placed over the upper end of the telescope tube and has the effect of increasingly attenuating the light striking the mirror on its outer and middle zones while allowing the full intensity of the light to fall on the central zone.

The image of a star in a telescope consists of the central diffraction disc surrounded by several much dimmer bright rings that get progressively fainter the further out they are from the centre. The central obstruction in a reflecting telescope causes more light to be diverted into the first bright ring surrounding the diffraction disc than is the case in a refractor. This has the effect of reducing contrast of planetary markings when observing with a reflector and since this first ring is relatively more bright, its shimmering in poor seeing conditions has a more detrimental effect on the image than it does in a refractor.

The light from a star falling on the central zone of a reflector mirror becomes concentrated mainly in the Airy disc while that falling on the middle and outer zones finds its way mostly into the first diffraction ring. Since the apodising screen reduces the intensity of the light falling on the middle and outer zones of the mirror, it follows that the first diffraction ring of a star image will be dimmer – i.e., in brightness relative to the Airy disc it will now approximate to the refractor type of diffraction disc and first diffraction ring. Since the extended telescopic images of the planets can be considered as being made up of an infinite number of Airy discs and ring images of every point in the planet that is being viewed, it follows that contrast and visibility of planetary detail will be improved and the effects of atmospheric turbulence will be lessened.

An efficient apodising screen may be easily constructed from ordinary wire mesh of the sort used to make window screens for excluding insects. First, using strong scissors or shears cut two discs of the wire mesh both equal in diameter to the outside diameter of your telescope tube. Cut a similar third disc but this time leave three tabs spaced at 120° intervals on its edge (fig. 3.22). In this disc cut a circular hole concentric to the circumference of the disc, the diameter of which is 55% of that of the telescope mirror. In one of the other discs cut a larger concentric hole whose diameter is 78% of the telescope mirror. Now lay this disc on top of the first disc concentric with it and with its rectangular wire grid offset at an angle of 30° to the grid of the other disc. Finally, cut a circular concentric hole 90% of the diameter of the telescope mirror in the third disc and lay this on the other two discs with its wire mesh grid offset at an angle of 30° to that of the

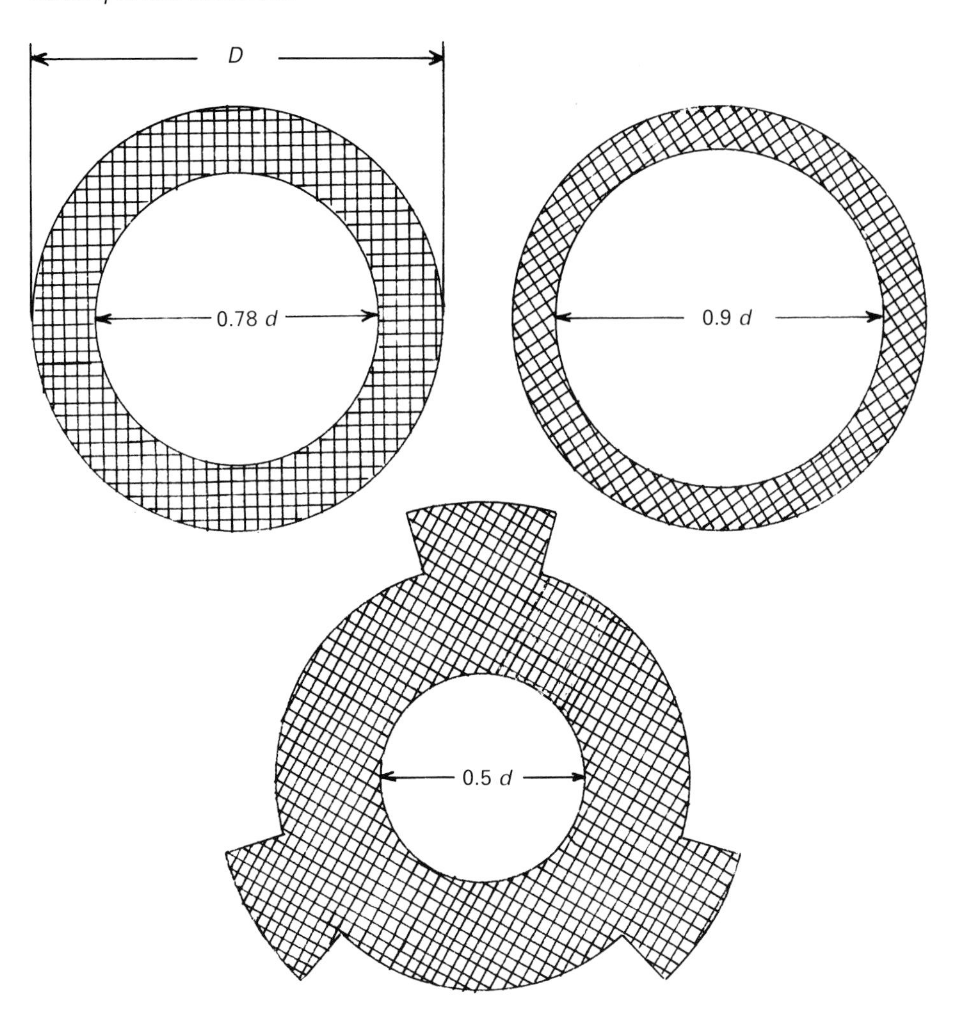

Fig. 3.22 Components of an apodising screen. (D: *outside diameter of telescope tube;* d: *diameter of mirror.)*

second disc. (For a refractor, the hole diameters in each screen layer would be 52%, 76% and 88% respectively of the diameter of the objective.) Now staple all the discs together. Bend the three flaps at the edge of the third disc at right angles to its plane towards the side on which the other two discs are attached. The screen can now be slipped over the open upper end of your telescope tube, the bent-over flaps holding the screen in place. If you prefer, a neater job will result if you attach the bent flaps to the inside of a strip of cardboard bent into a circular shape forming a sleeve that can now be more securely slipped onto the telescope tube. Be careful at every stage of the construction not to puncture your fingers with the cut ends of wires at the edges of the mesh. When a planet is viewed in a Newtonian reflector with this device slipped over the open end of the telescope, glare is reduced, contrast of the planetary image is decidedly improved and any shimmering of the image is lessened. You will see the planetary image at the centre of a small dark circular area from which radiate rows of coloured diffraction spectra caused by the wire mesh of the screen which acts as a coarse diffraction grating.

The great advantage of the apodising screen over 'stopping down' the telescope aperture to improve contrast and reduce image tremors is that the telescope is working at full aperture so that there is no loss of resolution.

Some observers claim that there is no advantage in using apodising screens or that they 'don't believe in them' (!). After years of planetary observing with the help of the apodising screen I disagree totally with such assertions. I have never yet known an evening in which the screen is not of some help in improving planetary images. In fact, I would not think of commencing a planetary observing session without using the apodising screen. I will admit that the planetary image may be a little dim if you also use colour filters in conjunction with it. However, both orange/red filters and the screen used together were better than either used separately and were a tremendous help when I observed Mars during the very favourable opposition of 1989/90.

The apodising screen as constructed from wire mesh cannot be used for solar or lunar observing because the diffraction spectra of the much larger images of the sun and moon overlap and degrade the images.

The eye and its peculiarities

In our preoccupation with the telescope and the way it works, we must not overlook the role of the human eye in perception. The characteristics of the human retina differ among individual people and this is probably of greater importance than the shape of the eye lens. Abnormalities of the lens are largely compensated for in the telescope but retinal characteristics are unaffected.

There are two characteristics of the eye that seem to be related to two types of astronomical vision:

(1) *Sensitivity*, the ability to see faint stars, asteroids and satellites.

(2) *Acuity*, the ability of the eye to resolve fine detail on planetary surfaces.

These two attributes of the eye do not appear to be correlated and may account for discrepancies in what is seen by different observers when viewing the same object with the same telescope. A good example of these vision differences between two observers was on an evening of superb seeing when the Rev. T. W. Webb and the Rev. T. E. Espin were observing Saturn with a 9.3-inch reflector. Mr Webb clearly saw the Encke division of the outer ring but could not see the satellite Enceladus whereas Mr Espin saw Enceladus clearly but was unable to see the Encke ring division. The paper dealing with this topic by W. Sheehan listed at the end of this chapter is recommended to the reader.

Telescope mountings

Serious observational work with a telescope cannot be done unless it is properly mounted on a rigid support and can be moved to point to any part of the sky.

The altazimuth mount (fig. 3.23) is so called because it permits movement of the telescope in altitude and azimuth. It is a good and useful mount, especially well suited to small refractors. However, it is a nuisance to be continually adjusting the telescope in two directions to compensate for the Earth's rotation and to

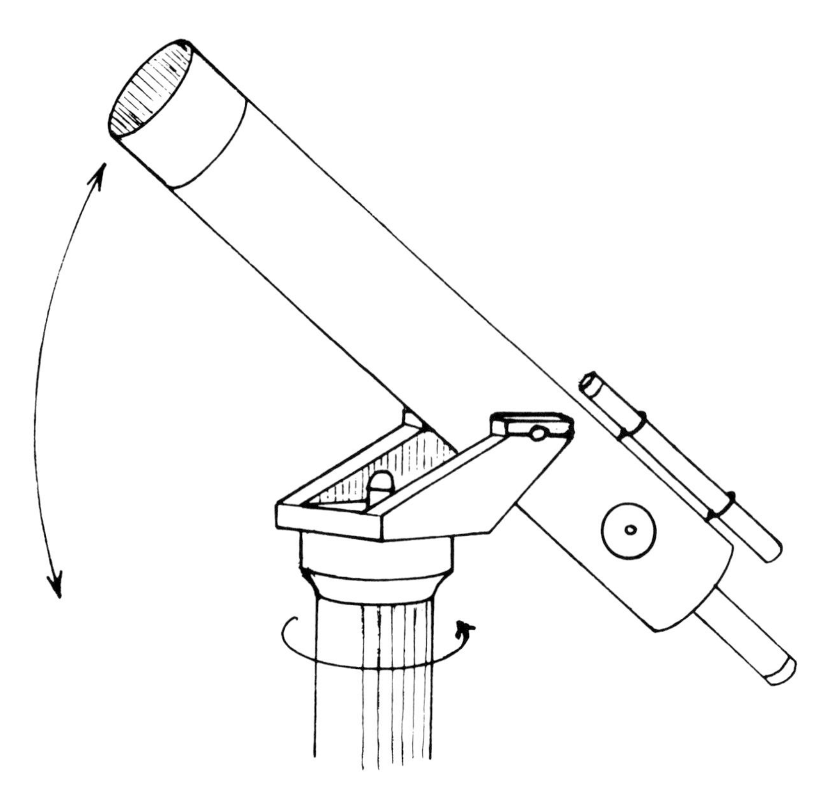

Fig. 3.23 Altazimuth mount for a small refracting telescope.

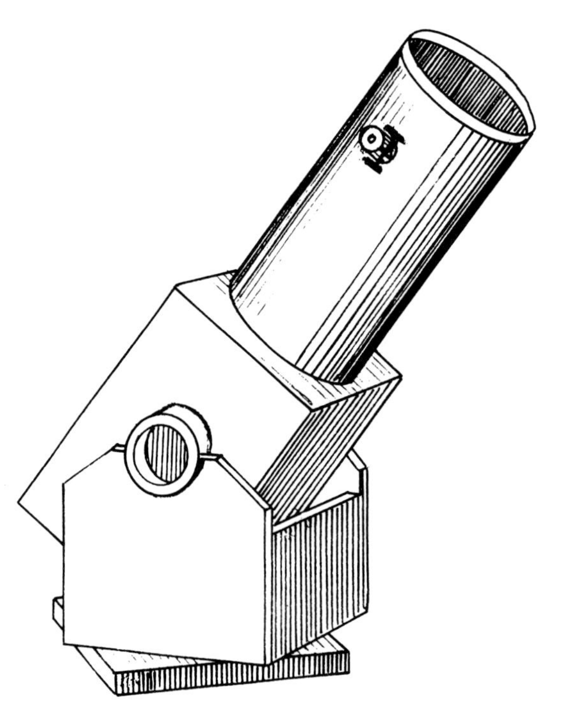

Fig. 3.24 The Dobsonian mount.

keep the planet in the field of view. This is especially irksome when using high powers which necessitate readjustment of the telescope every few seconds.

The Dobsonian mount (fig. 3.24) is a form of the altazimuth used for mounting short focus Newtonian reflectors of large aperture ('light buckets'). Such telescopes are used virtually exclusively by 'deep sky' observers. Planetary observers won't have much use for them.

The equatorial mount is much better and is constructed in many different forms, all of which have one feature in common; the axis about which the telescope turns in an east–west direction is not vertical as in the altazimuth but is

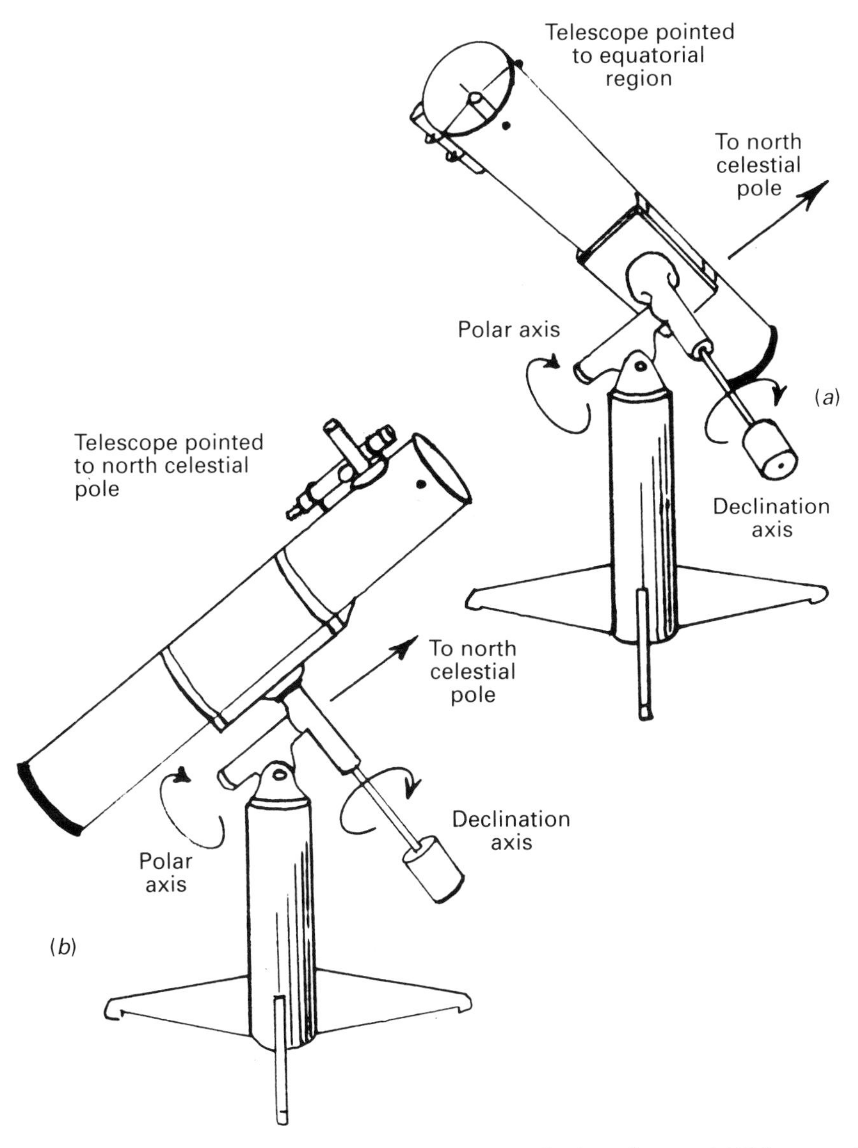

Fig. 3.25 German equatorial mount for a Newtonian reflecting telescope. (a) *Telescope pointed to the equatorial region;* (b) *telescope pointed to the north celestial pole.*

tilted and aligned so as to be parallel to the Earth's axis when oriented exactly in the north–south direction. This is therefore called the polar axis of the telescope. The other axis which permits raising or lowering the telescope is the declination axis. If the telescope is pointed to a planet, rotation about the polar axis alone, which counteracts the Earth's rotation, will keep the planet in the field of view of the eyepiece. In effect, the equatorial is an altazimuth in which the vertical (altitude) axis is tilted so as to be parallel to the Earth's axis and aligned in the north–south direction.

Modern equatorial mounts are fitted with electrically driven motors to keep the polar axis turning in an east–west direction at a rate that just counteracts the Earth's west to east axial rotation. The planet thus stays still in the field of view, provided that the telescope axes have been accurately aligned.

There are two common types of equatorial mount, the German and the fork mount. Commercially made amateur-sized Newtonian reflectors are almost always mounted on the German equatorial type (fig. 3.25) and catadioptrics on a fork mount (fig. 3.26).

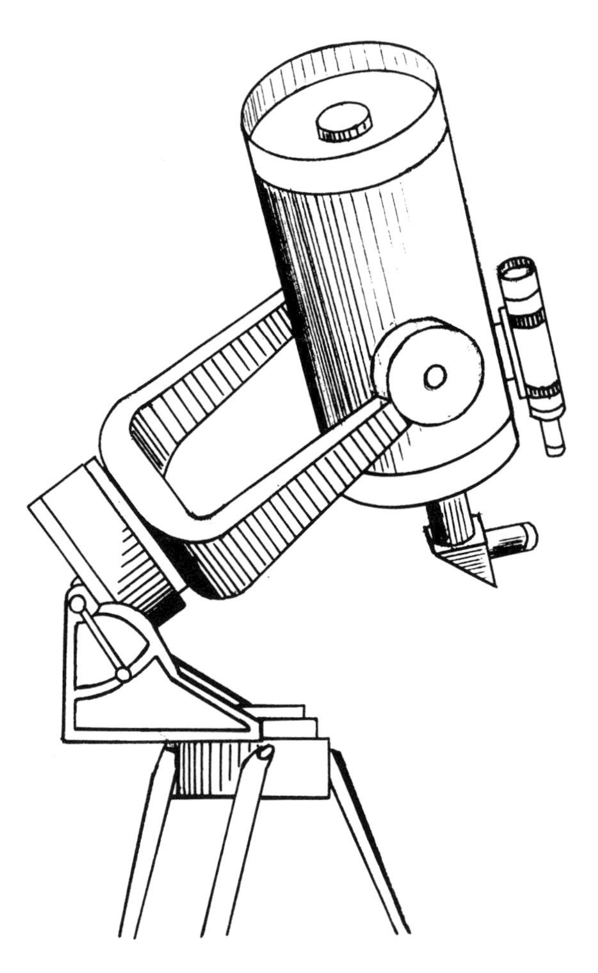

Fig. 3.26 Equatorial fork-mount for a catadioptric telescope. (Celestron International and Meade Instruments Corporation.)

The German equatorial mount is usually stood on a short vertical pillar with three horizontal legs arranged at a 120° angle to each other (fig. 3.25). This makes a rigid vibration-free stand. The equatorial mount itself is often not rigid enough for vibration-free observing. If it were, it would have to be quite massive and heavy which would add considerably to the cost and limit the portability of the telescope.

Setting up and adjusting the equatorial mount

There is no sense in purchasing an electrically driven equatorially mounted reflector if you are not going to set it up and adjust it properly. The following procedure though somewhat crude will be sufficiently accurate for visual planetary observing. First, remove the telescope tube and counterweights from the equatorial head. Then slightly loosen the large nut that permits up and down angular movement of the polar axis about the large horizontal screw that attaches the equatorial mount to the vertical pillar.

Tilt the polar axis at an angle to the ground as nearly as possible equal to your local latitude. Use a protractor to measure the angle. Tighten up the nut sufficiently so that when the counterweights are attached the polar axis will not slip around the horizontal screw and tilt out of the correct angle. This adjustment can

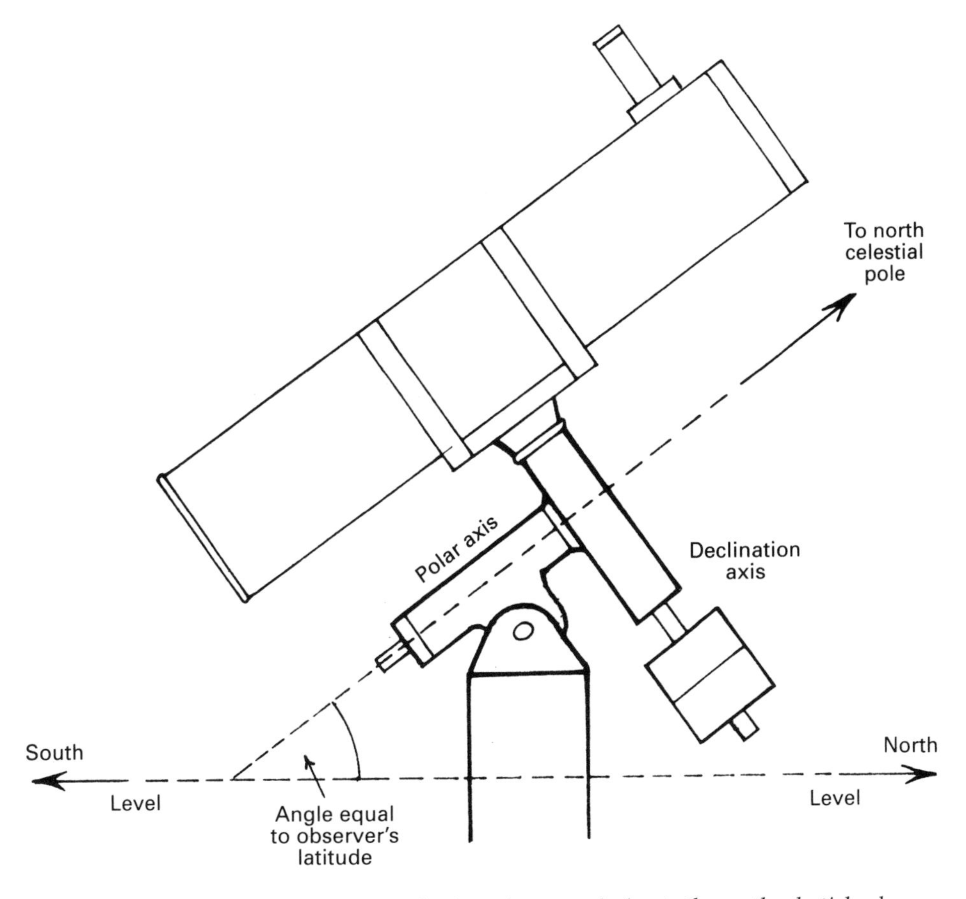

Fig. 3.27 Tube of equatorially mounted reflecting telescope pointing to the north celestial pole.

be checked as follows on the next clear night with the telescope and counterweights attached. Align the polar axis in the north–south direction by reference to the Pole Star (Polaris) which marks the position of the north celestial pole to within about half a degree of arc with the upper end of the polar axis pointing north. Rotate the telescope tube about the declination axis and point the open end of the tube in the general direction of Polaris. Now rotate the polar axis so that the telescope tube is on top of and parallel to the polar axis with the counterweights underneath. The declination axis is now oriented in a vertical plane (fig. 3.27). If the angle that the polar axis makes with the ground is accurate and if the axis is pointing exactly in the north–south direction, then on looking through the eyepiece (preferably low power) and focusing, Polaris should be in the field of view, but it probably won't be perfectly central. If you cannot see Polaris, readjust the angle of the polar axis to the ground. It should be fairly easy to loosen the nut that holds the polar axis at the correct angle and simultaneously look through the telescope eyepiece. With the telescope tube pointing due north, bring Polaris into the field of view by slowly raising and lowering the telescope tube about the horizontal screw. As soon as you see Polaris, centre it in the field and tighten up the nut of the horizontal screw again. When you set up the telescope again with the polar axis pointing due north, start the electric motor drive and point the telescope at a planet. The adjustments that you have made should be sufficiently accurate to keep the planet in the field of view for several minutes of uninterrupted viewing before it is necessary to make the slight adjustment on either axis to bring it into the field again.

Collimating the telescope

For the telescope to yield the best image that it is capable of giving, it is essential that all the optical components are properly lined up or *collimated*. In a Newtonian reflector the eyepiece tube and diagonal mirror are usually accurately squared with respect to each other. The main mirror cell has three spring-loaded screws in its back that enable the mirror to be adjusted with respect to the secondary. Before an observing session, the collimation should be checked as the main mirror is often misaligned and has to be brought into correct adjustment with the diagonal mirror and eyepiece. The diagonal mirror has to be exactly on the optical axis of the main mirror and inclined at an angle of 45° to it, the eyepiece tube must be exactly at right angles to the optical axis of the main mirror and the axis of the eyepiece holder must pass through the exact centre of the diagonal (fig. 3.28). The main mirror is easily put out of alignment by careless knocking of the telescope tube or by just moving it around.

To test the collimation, set up the telescope and pedestal with counterweights attached and point the tube to any bright plain area such as the sky. Remove the eyepiece and look through the eyepiece tube at the reflection of the main mirror in the diagonal mirror. The latter is actually elliptical but will look circular because it is tilted. In it you will see reflected the circular image of the main mirror which will nearly, but not quite, fill the circular outline of the diagonal. In the main mirror reflection you will also see a reflection of the diagonal and its supports with a reflection of your eye in the centre. The diagonal reflection will probably not be exactly central (fig. 3.29), thus indicating that the collimation is not perfect. The main mirror has now to have its tilt adjusted until the reflection of

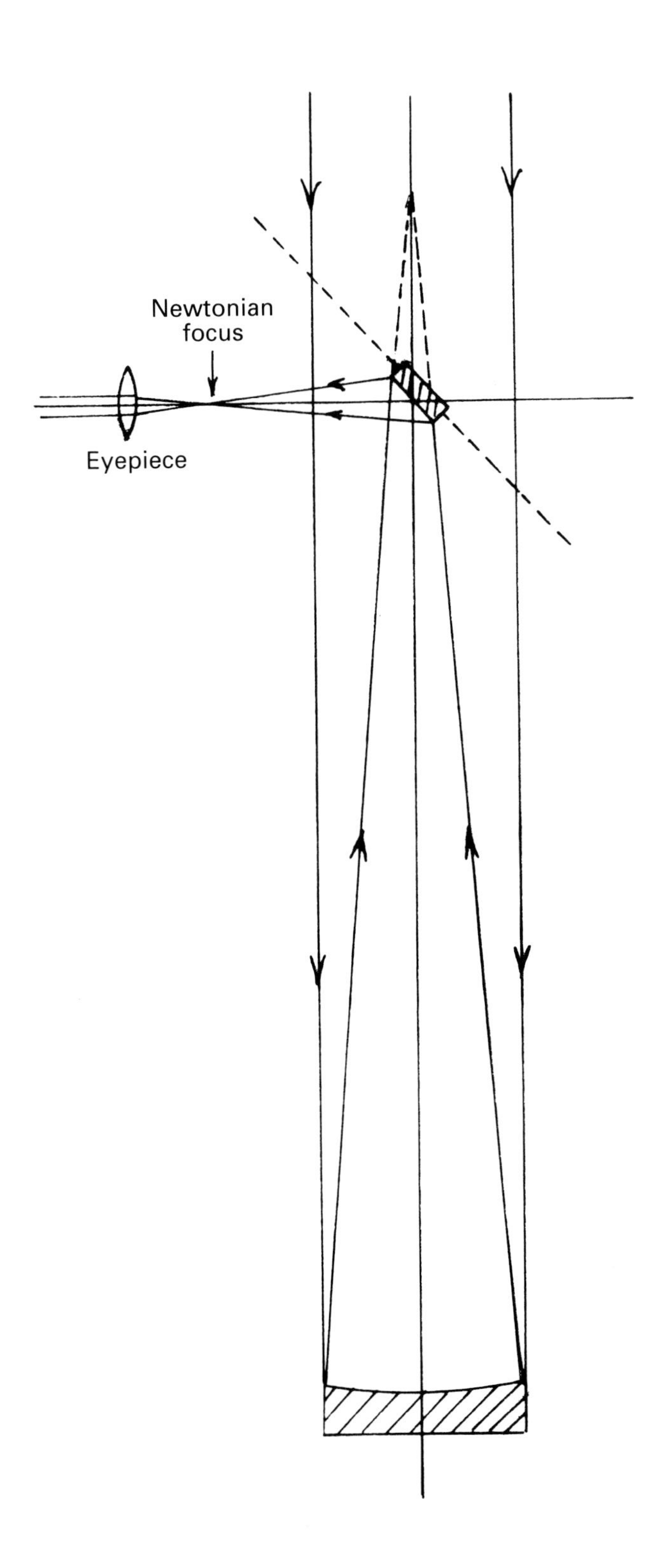

Fig. 3.28 Correctly collimated optical components of a Newtonian reflecting telescope.

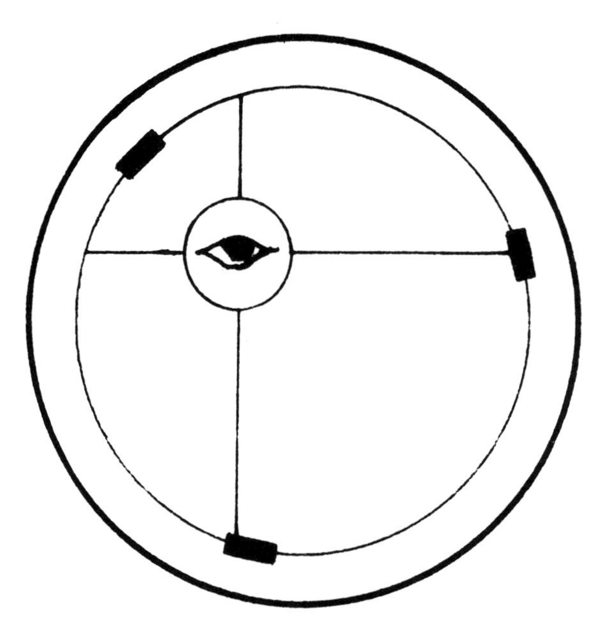

Fig. 3.29 View through the eyepiece holder of an improperly collimated Newtonian reflecting telescope.

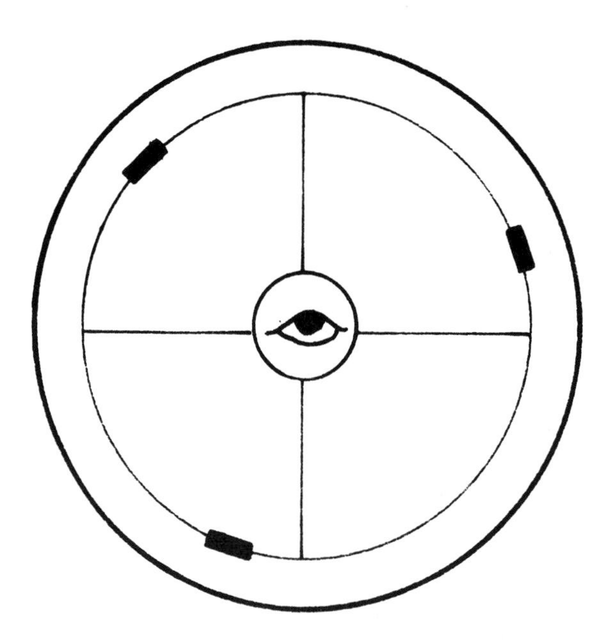

Fig. 3.30 View through the eyepiece holder of a properly collimated Newtonian reflecting telescope.

the diagonal is perfectly central. This is done by locating one of the three spring-loaded adjusting screws at the base of the mirror cell. Give the screw a quarter turn and look through the eyepiece. If the diagonal reflection is more off centre, turn the screw back again. Go to the next screw and repeat the procedure and then to the third and so on until the diagonal mirror reflection is judged to be central (fig. 3.30). Each time try the effect of turning the screws in either direction.

The telescope optics will now be collimated. Rotate the adjusting screws about a quarter turn each time while performing this adjustment. The procedure need not take more than a minute or two. It is a nuisance to have to keep moving away from the eyepiece to step to the other end of the telescope to turn the adjusting screws so it is helpful if you have a companion to do the adjusting for you while you stay at the eyepiece.

For greater accuracy of collimation the centration of the image of the main mirror in the diagonal mirror should be observed through a 'peep sight' that can easily be constructed from an empty 35 mm roll film canister. Saw off the bottom of the canister cleanly and in the snap-on cap pierce a small hole exactly in its centre. Snap it back on to the canister. Slide the cannister into the eyepiece holder with the capped end facing towards your eye. On looking through the hole in the cap you will see the image of the main mirror in the diagonal mirror as before but this time your eye will be looking exactly centrally along the eyepiece-diagonal mirror axis. This will permit greater accuracy as your judgement of accurate centration of the diagonal image in the main mirror image will no longer be affected by the slight off-centredness of the eye with respect to the central axis of the eyepiece tube that can occur if you do not use the peepsight.

The collimation may be even further refined by studying the in- and out-of-focus images of a star and adjusting the main mirror until perfect star diffraction images are seen consisting of perfectly circular and concentric bright rings around the central Airy disc. However, the previously described method will give collimation that is quite good enough for critical planetary work.

Balancing the telescope

The electric drive will not move the telescope around the polar axis unless it is correctly balanced. A simple procedure to achieve this is as follows. After attaching the counterweights unlock the declination axis and place the telescope tube horizontally in its cradle rings with its centre a little way past the midpoint towards the mirror end. Remember that the mirror is quite heavy and so the tube has to be positioned off centre to achieve a balance. If the tube is balanced in the cradle it will remain horizontal and not tilt about the declination axis. Now point the telescope at a planet and slide the counterweights so that the telescope does not rotate around the polar axis under its own weight. This will happen if the weights are too far up on the declination axis. If they are too far down they will pull the telescope around in the opposite direction.

When you have achieved a rough balance, point the telescope at the planet again and focus the image. Most likely you will find it drifting out of the field of view although the telescope drive is running – make sure that it is by putting your ear close to the box enclosing the motor and listen for its sound. If nothing is heard did you switch on the current? If all is well and the drive is working, then the drifting of the planet's image is because the telescope is still not quite accurately balanced about the polar axis by the counterweights. Try shifting the weights a little further up the declination axis, lock them and look through the eyepiece again. If now the planet holds still then everything is balanced. If not try shifting the weights still further up or down the declination axis until balance is achieved. A surprisingly small amount of movement of the weights can make the difference between perfect balance and lack of balance.

When the telescope is perfectly balanced, you will be able to observe uninterruptedly for several minutes. If the planet drifts towards the edge of the field (as it certainly will unless the alignments of the axes are perfect) a touch on the declination slow motion control or – using the declination axis as a convenient lever – a tiny shift of the polar axis will recentre the planet in the field of view.

During a prolonged observing session the planet will alter its position in the sky to such an extent that the telescope may need to be balanced again as it will now be pointing in quite a different direction from that at the start of the observation. A slight readjustment of the weights up or down the declination axis should suffice to correct this. Sometimes you may find that the counterweights are too heavy; however far up the declination axis you move them, the telescope still refuses to turn on the polar axis in response to the electric drive. This is remedied by removing one of the weights and repositioning the other. A balance should now be possible. These manipulations may sound tedious but as you gain familiarity with your telescope mounting and its peculiarities and 'get the feel of things', you soon can make these adjustments quickly and easily.

A few further hints will help in keeping observing trouble-free:

(1) Before you begin observing rotate the telescope tube in its cradle so that the eyepiece holder is on the same side as the counterweights and the declination slow motion control. Adjustments can now be made on both axes while observing without taking your eye from the eyepiece because the controls will be within your reach.

(2) When making an adjustment to the telescope about the polar axis, always rotate the polar axis in a west–east direction. By so doing, the teeth in the gear wheels of the electric drive motor will all be in contact and the telescope will resume following the planet immediately. If you turn the axis from east to west this will disengage some of the motor cog teeth which then have to move through a tiny distance before they again engage with the neighbouring wheels. This usually takes a few seconds during which time the telescope will stop following the planet before resuming again. This may be long enough for the planet to pass out of the field of view if you are using a high power eyepiece.

(3) Suppose that you start observing a planet a while before it transits the meridian with the telescope tube on the west side of the polar axis. If you continue to observe for some time after the meridian passage of the planet you may find it more convenient to swing the telescope around the polar axis so that it is now on the east side of the polar axis. Rebalance and continue observing. If you do this, be sure to remove the eyepiece (and Barlow lens if in use) from the eyepiece holder because during this manoeuvre the telescope tube turns upside down and there is a serious risk of eyepiece and Barlow lens falling to the ground if not first removed.

Adjusting the viewfinder

Owing to the narrow field of view of the main telescope it is difficult to point it accurately to the planet that you want to observe. The viewfinder facilitates this because of its low power and much wider field of view. The viewfinder or 'finderscope' is the small telescope attached to the upper end of the tube of a reflector and is parallel to it and points in the same direction. On looking through its eyepiece you will see two cross hairs at right angles that intersect at the centre of the field of view. The viewfinder must first be adjusted so that its optical axis is exactly parallel to that of the main telescope. With the main telescope accurately set up and with the motor drive running unlock the declination axis. Using a low power, point the main telescope at a planet and position it exactly in the centre of the field of view. Lock the declination axis. Look through the viewfinder and get the planet image exactly at the intersection of the cross hairs by adjusting the screws that hold the viewfinder in position in the mounting rings. You do this by loosening one and tightening another until the adjustment is achieved. It may be a bit tricky to do this without jolting the main telescope. Now look through the main telescope eyepiece to check that the planet is still in the field of view. If so, the viewfinder is adjusted accurately enough for the planet observer. Subsequently, you 'find' a planet by first pointing the main telescope in the direction of the planet. Look through the viewfinder and gently move the main telescope tube until the planet can be seen exactly at the intersection of the cross hairs in the viewfinder. On looking through the eyepiece of the main telescope the planet should be in the field of view.

Setting circles

On the declination and polar axes of your telescope you will see attached metal or plastic discs a few centimetres in diameter, the axes passing through their centres. These are the setting circles. They are graduated around their edges, the declination circle in degrees and the RA circle in hours and minutes. A pointer is fixed on each axis so that the graduations on the setting circles can be read off. Setting circles are so called because they can be used to locate or set the telescope to point at a given faint or small celestial object at night or to planets during the day. Some telescopes are sold with fixed setting circles and movable pointers and others with movable setting circles and fixed pointers. The instructions with a new telescope should make clear what arrangement is used.

If you wish to point the telescope at a celestial object that is difficult to locate with the naked eye, the object's RA and declination are first calculated from the appropriate tables in an ephemeris.

Assuming that your telescope is accurately set up and that the setting circles are correctly adjusted, the object is found by first turning the polar axis until the pointer indicates the RA of the object at that particular time. Then the declination axis is turned until the object's declination is indicated on the circle. The object should now be seen in the telescope's field, preferably using a low power eyepiece, so that if there are slight errors in the adjustments the object will still probably be in the field of view.

Alternatively, if the object happens to be east of your meridian it is a good plan to align your telescope on the meridian and tilt it at the proper declination angle north or south of the celestial equator. Next, note your local sidereal time and wait for the time of the object's hour angle to pass. It should now be seen in the telescope's field of view. It will be at its highest point in the sky since it is on the meridian and so the seeing should be better than if it were an hour or so on either side of the meridian.

It is usually quicker to consult a current astronomical journal or an ephemeris and look up the current position of, say, dim planets like Uranus, Neptune and Pluto with respect to nearby stars and to locate them by 'star hopping'.

Testing the telescope

If you purchased your telescope from a reputable firm you can safely assume that the optical components have been performance tested before being 'passed'. If you made your telescope yourself then you will be already familiar with tests of optical performance. It is when you purchase a second hand telescope, especially if amateur-made, that you should test optical performance before parting with money. Whoever is selling a telescope to you should not object to this.

The telescope should first be perfectly collimated before applying optical tests. A good test object is a moderately bright star on a clear dark night when the seeing is steady. Using a power of about 100× the image of a star when perfectly focused should appear as a perfectly clear sharp point of light without appendages or other irregularities. Now move the focusing tube slowly in and out so that you can study the out-of-focus image on both sides of the true focus. The star image will now appear as a disc with a dark central spot – the silhouette of the diagonal mirror. A series of three or four concentric perfectly circular light rings of equal intensity will be seen surrounding the central disc. The same appearance should be seen on either side of the true focus. Don't jump to the conclusion that your telescope is faulty if you see any departure from these appearances; it is not very often that the state of the atmosphere will permit such perfect results.

The testing should be repeated on several occasions and in different seeing conditions. If something doesn't seem quite right after many nights of careful testing, don't be quick to blame the mirror; remember that the eyepiece or even your own eyes may not be perfect.

Testing for correction of spherical aberration Before testing for correction of spherical aberration the mirror must be allowed to attain temperature equilibrium (as it should before any kind of optical test or planetary observation is carried out) otherwise rapid temperature changes and consequent distortion of the mirror's figure can give appearances resembling spherical aberration.

Examine again the intra- and extra-focal appearances of the star image. If the outermost bright ring appears bright and wide in the extrafocal setting this indicates that the spherical aberration of the mirror is overcorrected, i.e., the curve is too deep and is more hyperboloidal than paraboloidal. If the opposite appearances are seen, then undercorrection is indicated and the curve is too shallow and approximates more to a spheroid than to a paraboloid.

Testing for astigmatism If the out-of-focus star image and bright rings appear elliptical instead of circular, astigmatism is present somewhere. This could be in the mirror or in your own eyes. Try rotating your head one way then the other while looking through the eyepiece. If the elliptical image of the star appears to rotate in the same direction as your head, then your eyes are at fault. Now try rotating the telescope tube while keeping your head still. If this causes an apparent rotation of the star image, then the telescope mirror is at fault.

Another good test for astigmatism that can be done on moonless nights with good seeing and transparency is to use a power of 100× to study the well-known 'double double' star Epsilon Lyrae. Apertures of 4 inches or more resolve the two pairs easily and this should be possible at one setting of the focus. Although both doubles will be resolved at one focal setting, the eyepiece will need to be slightly refocused for each pair if the mirror is astigmatic. The effect could also be caused by a faulty diagonal mirror or by distortion of the main mirror caused by being too firmly held in its cell or occasionally by poor optical alignment or a mediocre eyepiece. These should be checked before blaming the main mirror. If the mirror is at fault, return it to the maker for exchange or correction of the defect.

Tests for resolving power The Dawes criterion for telescopic resolution was defined earlier. In choosing a double star for the test, the components should preferably be yellow in colour and of about the same brightness. If one is much brighter than the other, its diffraction phenomena might swamp the image of the fainter component. Blue double stars, are easier to resolve than yellow doubles and red doubles are the most difficult. Tables of double stars and their angular separations can be found in many texts on practical astronomy.

The following double stars have been recommended as tests of telescopic resolving power:

For a 6- or 8-inch telescope, Zeta Herculis. The components are separated by one second of arc and the magnitudes are 3.0 and 6.5. Use a power of 200×–250×. Eta Orionis is a double in which the magnitudes are 4.0 and 5.0. The separation is also one second of arc.

For 10- to 12-inch telescopes, the blue component of the wide yellow and blue double star Gamma Andromedae, which is itself a double. A power of 250×–300× should resolve the blue component.

Protecting the telescope from dust and atmospheric pollution

The open tube construction of the Newtonian reflector exposes the main mirror to dust accumulation and when not in use the telescope should be protected from it. I remove the tube of my telescope from its cradle and stand it upright, mirror end down, with a cardboard lid over the open end of the tube and with the eyepiece tube and viewfinder objective covered with the plastic caps that were supplied with the telescope. You could stand the telescope with the mirror end uppermost which would lessen the risk of dust accumulation on the mirror but then the telescope tube would be top heavy and likely to fall over if accidentally bumped; it would be a good idea to store the telescope in a tall narrow cupboard if you prefer to stand it mirror end uppermost.

Silvered mirrors tarnish quickly in industrial areas where there is hydrogen sulphide or sulphur dioxide in the air. They therefore require frequent resilvering. A good aluminium coating on the mirror should last for many years and keep its good reflectance properties especially if further coated with one of the many kinds of transparent protective films that are available.

Cleaning the mirror of a Newtonian reflector

Many observers get upset at the sight of even a few specks of dust on the beautiful shiny surface of the telescope main mirror, especially if they have just spent a lot of money on a new telescope, yet overlook the fact that the central diagonal mirror obstruction in the light path is the equivalent of a big blob of opaque matter on the mirror's centre! However carefully dust is excluded from the telescope some always gets onto the mirror. Dust on the mirror is vividly revealed if you point the telescope tube towards a bright light or to the moon and look down the open end of the tube at the mirror. When your head is held at a certain angle at the side of the tube opening, dust on the mirror will be brilliantly revealed as a multitude of bright points speckling its surface. Be reassured that a little dust on the mirror has hardly any detrimental effect on the telescopic image although the light scattering caused by a more than slight sprinkling of dust will affect contrast of delicate planetary detail. Some observers don't seem to be too worried even if the mirror looks like the bottom of a garbage can! Far more damaging than dust are greasy films, smears or fingerprints on the mirror.

To clean the mirror, carefully remove it from its cell. Do not attempt to wipe it with a cloth; gritty particles on its surface would thereby cause scratches on the surface coating. If you have one, direct a jet of air on the surface from a can of compressed air. This removes the coarser debris. Now place a folded towel on the bottom of a sink and fill the sink to a depth sufficient to completely immerse the mirror with room temperature water to which has been added a teaspoonful of good quality mild detergent (I use dish washing liquid). Stand the mirror, reflecting surface uppermost, on the towel well beneath the water surface. Take a cotton wool swab and gently draw it in straight lines across the surface of the mirror still immersed in water from one edge to the other. Do not apply pressure or use a circular motion. Rinse the mirror under the tap and give it a final good rinse with distilled or deionised water. Stand the mirror on its edge to dry (somewhere where pets or children cannot leave nose or finger prints respectively). When dry there should be no streaks or other marks on the mirror. If there are, wet the mirror again with distilled water and try a final rinse with industrial methylated spirit (*not* ordinary methylated spirit which goes cloudy when mixed with water) or 70% isopropyl alcohol which works just as well.

Do not be too anxious to remove the mirror from its cell for cleaning every time you see a speck of dust on it. There is always the risk of scratching or otherwise damaging the reflecting surface. If the mirror looks a bit dusty but seems to be performing well, remember the good common sense rule of 'leave well enough alone'.

Housing and care of the telescope

It is not necessary to have your telescope housed in an expensive observatory with a revolving dome in order to pursue serious planetary observation. On the other hand it would be madness to leave the telescope out of doors unprotected from the elements.

The entire telescope need not be left out of doors. Some observers have the pillar support cemented into the ground with the equatorial head on top of this and covered to protect it from dampness and rain. The telescope tube is kept indoors and brought out and attached when needed. I have never bothered to have even this arrangement but keep the entire telescope in a garden shed and dismantled into three units – the telescope tube, the counterweights and the pedestal/equatorial head. If you decide to do the same set up the telescope for observing by first taking out the pedestal. The pedestal and equatorial head of an 8-inch reflector, even without the counterweights, is quite heavy. When lifting it be very careful not to hurt your lower back. Bend your knees, grasp the pedestal firmly and lift it by straightening your legs. Do not bend your back but keep it straight all the time. Let your legs do the work, they are much stronger than your back. Be equally careful when setting down the pedestal. Keep your back straight and bend your legs to lower the pedestal to the ground. Next, attach the counterweights and finally fix the telescope tube in its cradle and check the collimation. Don't attach the telescope tube before attaching the counterweights or the polar axis will be top heavy – the telescope will swing around and hit the ground as you turn your back to pick up the counterweights. The whole procedure can be done in five minutes or less.

Allow time for the telescope mirror to attain the outdoor temperature if you have brought the telescope outdoors from a warmer or cooler indoor environment. If, say, a warm mirror and telescope tube are cooling down while you are observing, there will be 'tube currents' set up in the air inside the telescope tube. Also, the figure of the mirror may be temporarily distorted due to temperature changes. Both of these effects will result in image unsteadiness and deterioration.

To be able to walk into an observatory, however simple, with the telescope already set up for immediate use is a great convenience. Many observers will be 'clever with their hands' and will want to build an observatory. To give full instructions for such a project is outside the purpose of this book. The reader will find a list of books at the end of this chapter containing information on the building of observatories and various types of shelters for telescopes.

Two easily constructed types of telescope housing are as follows:

(1) A small waterproof shed with hinged roof and sides which are opened when the telescope is used. The shed can be fixed or movable, e.g., on wheels that run on 'railway lines' made of angle iron. Alternatively the shelter can be made in two halves which can be folded back or rolled away to uncover the telescope. Although the telescope is protected from weather it is not shielded from wind when you are observing and this may cause it to vibrate exasperatingly and the planet will 'dance' violently in the field of view.

(2) A shed with a flat run-off roof. As well as protecting the telescope from the weather it will protect it and yourself from wind while observing. The roof rolls away from the walls and exposes the sky. Alternatively it may be in one or two pieces and hinged so that it can be folded back. Such a shed can be larger than a simple cover and have room for a chair and table and some accessories such as a flashlight, books and charts.

Needless to say, observatories however simple or advanced should not be heated in cold weather as the air currents caused by warm air rising will wreck the seeing.

Further reading

Books

Amateur Telescope Making. Books 1, 2 and 3. Ingalls, A. A. G. (ed.), Scientific American Inc., San Francisco, (1949).

How to Make a Telescope. Texerau, J. Interscience, New York, (1957).

Telescopes for Skygazing. Paul, H. E., Chilton Books, Philadelphia, (1965).

Telescopes. How to Make and Use them. Page, T. and Page, L. W. (eds), Macmillan, New York (1966).

Amateur Astronomer's Handbook. (3rd edition) Sidgwick J. B. Faber and Faber, London (1971).

The Amateur Astronomer and His Telescope. Roth, G. D., Faber and Faber, London (1972).

Astronomical Telescopes and Observatories for Amateurs. Moore, P. A., David and Charles, Newton Abbot, South Devon, England (1973).

Telescope Making for Beginners. Worville, R. Kahn and Averill (1974).

Make Your Own Telescope. Spry, R., Sidgwick and Jackson (1978).

How to Use an Astronomical Telescope. Muirden, J., Linden Press. Simon and Schuster (1985).

Papers and articles

Planetary Telescopes. Baker, J. G., *Applied Optics* **2**, 2 (1963).

Get the Best From Your Newtonian. Lees. A. W., *Astronomy Now* **5(1)**, 38–43 (1991).

How to Plan a Newtonian Reflector. Berry, R., *Astronomy* **10(3)**, 47–50 (1982).

How to collimate your telescope. Porcellino, M., *Astronomy* **20(4)**, 61–65 (1992).

Notes on Newtonian Reflector Alignment. Cox, R. E., *Sky and Telescope* **31(3)**, 170–5 (1966).

Easy Steps to Perfect Polar Alignment. Burnham, R., *Astronomy* **10(8)**, 52–4 (1982).

Polar aligning your telescope. Porcellino, M., *Astronomy* **20(5)**, 69–72 (1992).

How to Use a Polefinder. Burke, P. J., *Astronomy* **13(5)**, 51–4 (1985).

Eyepieces: What You Get is What You See! Schwartzenberg, D., *Astronomy* **7(2)**, 42–6 (1979).

A Close-Up Look at Eyepieces. Eicher, D. J., *Astronomy* **16(10)**, 82–7 (1988).

Test Drive Your Telescope. Suiter, D. *Astronomy* **18(5)**, 56–61 (1990).

How to Star Test Your Telescope. Suiter, D., *Astronomy* **11(4)**, 51–4 (1983).

Ten Tips for Improving Your Telescope. Ling, A., *Astronomy* **18(8)**: 66–70 (1990).

Which Cassegrain is Best? Anonymous, *Astronomy* **7(6)**, 54–5 (1979).

Resolution and contrast. Gordon, R.W., *JALPO* **27(9–10)**, 180–90 (1979).

The Art of Seeing. Porcellino, M., *Astronomy* **18(2)**, 66–70 (1990).

The Big Eyes of Small (Tele)Scopes. Chaple, G., *Astronomy* **10(10)**, 51–4 (1982).

Rethinking Your 60 mm Refractor. Underhay, E., *Astronomy* **10(6)**, 48–50 (1982).

Soup Up Your Small (Tele)Scope. Shaffer, R., *Astronomy* **16(9)**, 80–3 (1988).

The Refractor Advantage: Quality Over Size. Dyer, A., *Astronomy* **15(10)**, 66–71 (1987).

The Lure of a Big (Tele)Scope. Clark, T., Astronomy **16(10)**, 76–81 (1988).

How to Buy a Telescope for Your Child. Berry, R., *Astronomy* **8(10)**, 46–9 (1980).

On an Observation of Saturn: the Eye and the Astronomical Observer. Sheehan, W., *JALPO* **28(7–8)**, 150–4 (1980).

4

The atmosphere and seeing

General

We live at the bottom of a vast ocean of air, several miles deep, that envelopes the surface of the Earth. Though essential for the continuation of all animal and human life it is nothing but a nuisance to the practical astronomer. All telescopic observation has to be through this mass of air which rarely, if ever, is quite still. Apart from the obvious obstructions of thick clouds and fog which prevent observation altogether, telescopic planetary images are often ruined by atmospheric turbulence even when the air is perfectly clear. Air currents and differences of temperature at different levels of the upper atmosphere all conspire to cause irregular refraction of light rays reaching us from the planets. This causes shimmering or 'boiling' and telescopic planetary images oscillate and ripple. Fine planetary detail is therefore difficult to see and 'hold'; if the trembling is bad enough little or nothing of disc markings can be seen. These are what is meant by 'bad seeing'.

When this happens there is nothing that you can do but to wait until conditions improve. However, there is no need to stop observing in the usually somewhat poor seeing conditions prevailing most of the year at most observing sites. On a night of tremulous seeing occasional steady intervals occur lasting a second or two during which the definition of the telescopic planetary image can be astonishingly good. Previously unsuspected fine detail will spring into view and previously seen features will be much clearer and show more detailed structure. Most time spent in observing goes into waiting for these steady intervals. Part of the art of observing is making the best use of these steady intervals when they present themselves.

The worst seeing occurs in the summer. The usual cause is the heat radiated by buildings or concrete patios at and after sunset which causes atmospheric turbulence. Conditions will often improve after midnight.

Generally, the best seeing occurs on winter nights when the temperature is close to or below freezing and the sky clear and still. Avoid observing during a thaw – the seeing is usually atrocious.

Apart from steadiness of the air its transparency is important. The clearer the air the better. Obviously, country dwellers enjoy a permanent advantage over urban dwellers in this respect. Unfortunately, good transparency is often accompanied by tremulous seeing. Slight overhead mist or fog is not usually detrimental; under such conditions the seeing is often quite good and the glare of a bright planetary disc is subdued.

Assessing atmospheric seeing conditions

Serious planetary observers keep records in which not only are actual observations entered but also factors that may have an effect on the accuracy or reliability of the observations. One of these is the atmospheric seeing conditions prevailing at the site where the observation was made. It is of the greatest importance to record this but so complex are the causes of atmospheric turbulence in the atmosphere that it is not easy to measure objectively their effect on observation.

Various scales of seeing have been devised in an attempt to express the seeing conditions on at least a semiquantitative basis. One of the best known of these is Antoniadi's in which the seeing is graded from 1 (best seeing) to 5 (worst seeing):

(1) *Very good* Perfect seeing, steady sharp planetary images even with high power.

(2) *Good* Slightly tremulous with steady intervals lasting for several seconds.

(3) *Fair or moderate* Frequent tremors but in the steady intervals the seeing is quite good with medium powers.

(4) *Poor* Almost continuous trembling of the image with only occasional brief glimpses of detail.

(5) *Very poor* Unsteady blurry image even with low power. Impossible to see details or to make a sketch.

The effect of telescope aperture

Assuming that your telescope is optically perfect the quality of the image depends on the size of the objective (aperture) and the magnification when used under similar atmospheric conditions. Magnification augments the effects of tremors in the image and the larger the telescope the less likely will the column of air that it is looking through be completely free of tremors.

On the majority of nights slow-travelling air waves from about 6–12 inches across move at a height of many hundreds of feet above ground level. It is therefore easy to appreciate that if a telescope of larger aperture than the dimensions of these waves is used their effect will be to degrade the image and break it up. If a telescope aperture smaller than these waves is used then the image won't be broken up but will shift bodily. Therefore although the image in both the large and small telescope will tremble, the image will retain a better overall outline in the small telescope than in the larger.

Under the same seeing conditions, then, despite its lesser resolving power, a steadier image and more detail will be seen with a smaller telescope. This has given rise to the mistaken idea that small telescopes perform better than large ones. Whereas this is so under average atmospheric seeing conditions, when perfect seeing conditions prevail the larger aperture always tells and the large telescope outperforms the smaller telescope. When seeing is poor a large telescope can be 'stopped down' to a smaller aperture with a suitable circular diaphragm and this will usually give a steadier planetary image with improved contrast but at the cost of less resolving power because of the reduced aperture. A valuable observing accessory called the apodising or antidiffraction screen described in chapter 3, when attached to the telescope, significantly reduces the effect of atmospheric turbulence without reducing the aperture of the telescope. It is a valuable accessory for planetary observation.

Local effects on seeing

Poor seeing will usually be experienced by observers on the leeward side of a city but this can improve during the early morning hours. The heat from factories and domestic chimneys will usually have ceased by then. Yet anyone actually living in the centre of a city may enjoy quite good seeing conditions. This is because there is usually a 'warm spot' in this type of locale and the air overhead will be fairly homogeneous.

Hills surrounding the observing site can cause atmospheric turbulence which is more pronounced when conditions are windy. At heights of between 5 and 10 miles (about 8–16 km) winds can start up because of large changes in barometric pressure caused by meteorological 'fronts' moving into the area. The deleterious effects on seeing are here caused by layers of air at different temperatures. However, the waves in the air are at a greater distance from the telescope so that their effects on seeing are less than they would be if the observer was at ground level. Since 'high seeing' is a climatic effect, its character varies with the observer's location on the Earth's surface. It is less of a nuisance in the lower latitudes owing to the more uniform atmospheric pressure.

Unlike the effects caused by cities and height above sea level the effects produced on seeing by the immediate surroundings can be largely controlled. Heat waves rising from warm ground near the telescope naturally occur more during the day and in the early evening than at other times. The later evening and early morning hours will therefore give the best seeing. Surrounding the telescope by scrub can help to reduce the effects of this kind of atmospheric turbulence. Avoid observing on a concrete base; a lawn is much better as it retains and radiates much less heat. Also avoid observing over the roofs of houses especially in winter when central heating is in use.

If your telescope is housed in an observatory, open the roof and doors at sunset to allow dissipation of warm air inside that has accumulated during the day. Local effects can also affect the air within a few feet above ground level. You may find yourself standing in a thick layer of warm air up to your shoulders. If this is so, a refractor on a tall stand will perform better than a reflector on a short pillar, especially if the latter has an open framework tube, so that the mirror is buried in the mass of heated air. It may be revealing to take the temperature at different heights above the ground in your observatory locale.

A generally held but mistaken belief is that some areas have clearer skies, i.e., better transparency, than others. It is quite true that the climate in some parts of the world such as North Africa, the Mediterranean and Arizona, is conducive to cloudless night skies. Transparency probably hardly varies from one part of the world to another. Some consider that the English climate favours murky skies but one has only to search through the works of English astronomers like the comet hunter G. Alcock and Sir William Herschel and many others to realise the error in this viewpoint.

Of all the many factors that can affect an observation such as telescope aperture and quality, magnification and collimation, the state of the observer's eyes and even mood, the atmospheric state probably has the most significant influence. No two nights are exactly the same in this respect. It was pointed out by the double star observer Burnham that a 6-inch refractor may show the tiny companion to Sirius on one night yet on the next night when the seeing may seem to be just as good, at least to the naked eye, no trace of the companion is detectable with the largest telescope in existence.

From this it follows that each night must be considered on its merits. Painstaking care must be taken when comparing observations made under various seeing and atmospheric conditions. If similar views of a planet are to be had then obviously the same telescope and magnification must be used. However, unless the atmospheric conditions are exactly the same, discrepancies are to be expected in what is seen from night to night that must be attributed to atmospheric effects.

Further reading

Papers and articles

The study of atmospheric currents by the aid of large telescopes, and the effect of such currents on the quality of the seeing. Douglass, A. E., *Metr.Jour. USA* (March, 1895).

A seeing scale for visual observers. Tombaugh, C. W. and Smith, B. A., *Sky and Telescope* **17(9)**, 449 (1958).

An analysis of the seeing and transparency scales used by amateur observers. Robinson, L. J., *The Strolling Astronomer* **15(11–12)**, 205–12 (1961).

How to improve your image. Pommier, R., *Astronomy* **20(7)**, 60–3 (1992).

5

Mercury

General

Mercury, the first of the two 'inferior' planets, is the nearest planet to the sun and has a diameter of 3010 miles (4843 km) which is slightly more than one third of the Earth's diameter. (fig. 5.1). Its mean distance from the sun is 36.0 million miles (57.9 million km) but it varies from 28.6 million miles (46.0 million km) at perihelion to 43.4 million miles (69.8 million km) at aphelion, owing to the great orbital eccentricity of 0.206 which is exceeded only by the orbital eccentricities of Pluto and several of the asteroids. This means that the sun is 7.5 million miles (11.9 million km) from the centre of Mercury's orbit. Owing to Mercury's nearness to the sun, its orbital speed is greater than that of any other planet, varying from a maximum of about 35 miles (56.3 km) per second when nearest and about 23 miles (37.0 km) per second when farthest from the sun. Mercury has no satellites.

Because it is an inner planet Mercury exhibits phases like our moon (fig. 5.2). It is 'full' at superior conjunction when on the opposite side of the sun to us, roughly half-moon shaped at its elongations, i.e., when at its greatest apparent angular distance from the sun on either side and 'new' when at inferior conjunction when between us and the sun.

The inclination of Mercury's orbit to the plane of the ecliptic is greater than that of the other planets (again with the exception of Pluto and several of the asteroids) and is almost exactly 7°.

Mercury completes one orbital revolution in 88 Earth days, its sidereal period, and its average synodic period, say from one inferior conjunction to the next, is 116 days. Owing again to the ellipticity of the orbit the orbital speed of Mercury varies in such a way that its synodic period has a range of from 111 to 121 days. As seen from Earth, Mercury seems to swing pendulum-like from one side of the sun to the other. At its greatest eastern elongation it will therefore be seen to the left of the sun at sunset as an 'evening star' and at greatest western elongation it will appear to be to the right of the sun at sunrise as a 'morning star' as seen from the northern hemisphere. These favourable times of maximum visibility are called 'apparitions'. The elongations occur very close to the same dates every thirteen years.

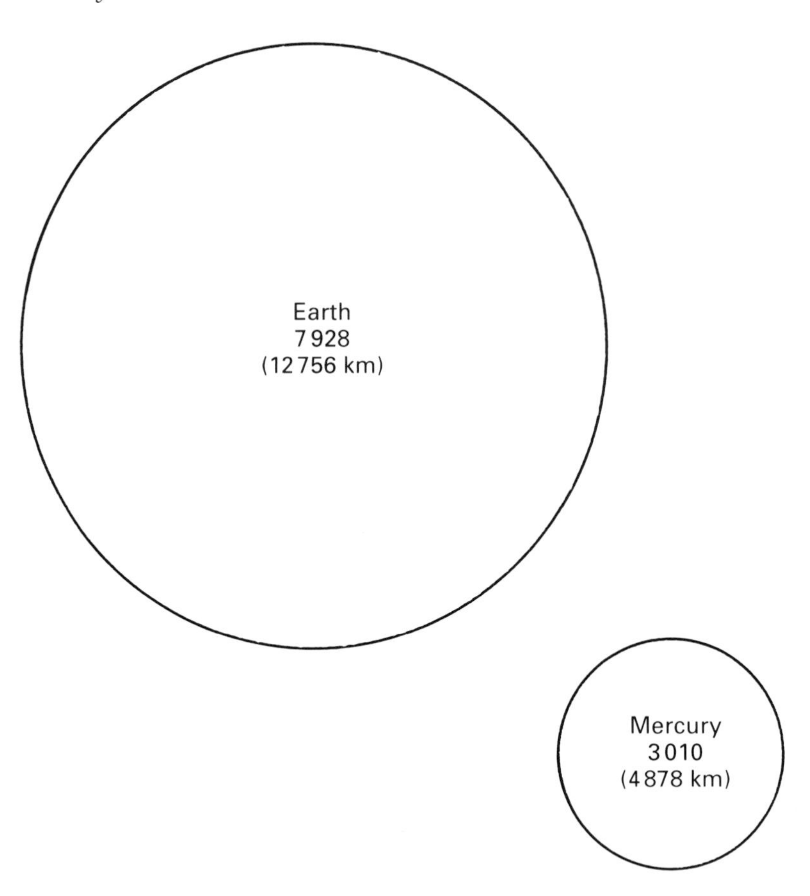

Fig. 5.1 Comparative sizes of Earth and Mercury (equatorial diameters in miles).

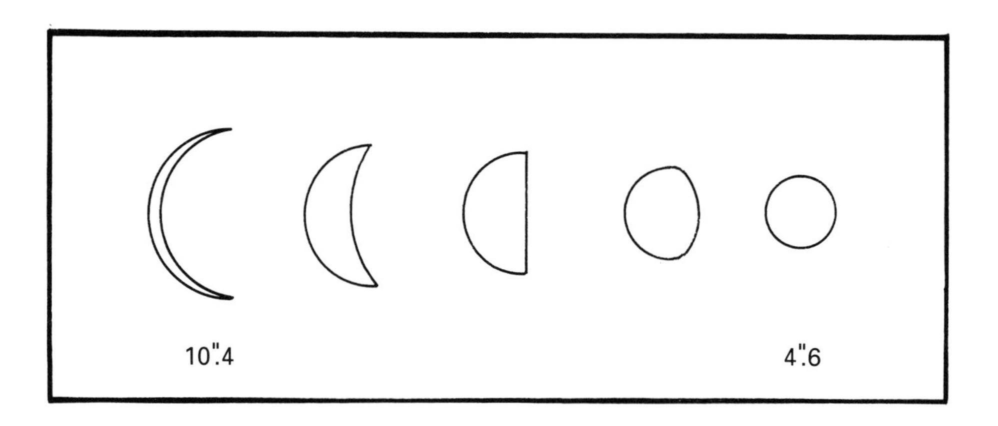

Fig. 5.2 Changes in the apparent angular size of Mercury at different phases.

History of observation

As previously mentioned in chapter 2, the Ancients knew that among the 'fixed stars' that never moved or changed their positions in the constellations there were some star-like objects that did move, the planets, which were five in number. Among them was the planet we now call Mercury. The planet Mercury is never seen far from the sun, is low down in the sky when it becomes dark enough to see it and its apparent movements are fairly rapid; hence it was named after the mythological swift-footed messenger of the Gods. It resembles a star and because it lies so close to the horizon it twinkles. The ancient Greeks therefore called it the Twinkler or Sparkler. It often has a rosy pink hue, an effect due to atmospheric dispersion.

Mercury was probably first observed by shepherds of long ago who had little else to do but watch the skies while guarding their sheep during the long nights. Some unknown shepherd of those far-off days must have noticed from time to time a twinkling star-like point near the sun at twilight that he had not noticed before. Each evening subsequently it would appear farther from the sun before setting. Then on later evenings it would retrace its path getting closer to the sun again and then finally it would be lost in the sun's glare several days later. One fine morning, just before sunrise, a similar twinkling star-like point would be seen low down in the east where no such star was to be seen a few days earlier. These evening and morning 'stars' were thought to be two different objects so that the Twinkler was given two names by the Greeks. The 'morning star' was called Apollo and the 'evening star' Mercury. It was not until much later that it was realised that the morning and evening stars were one and the same object.

The earliest known reference to Mercury is in a report made to Assyria's king by one of Nineveh's chief astronomers. Eudoxus measured Mercury's synodic period circa 400 BC yet some authors say that the first recorded observation of Mercury was made by the Egyptians in 265 BC. That Mercury was a planet was realised and its movements well known before the Classical Period came to a close.

Johannes Hevelius (1611–87) seems to have been the first to notice Mercury's phases although Robert Strom in his *Mercury, the Elusive Planet* gives credit for this to Giovanni Zupus in 1639.

Johann Hieronymus Schröter, (1745–1816) who diligently observed Mercury from his observatory at Lilienthal in Germany between 1780 and 1815, drew the first real map of Mercury. In the year 1800 he observed on many occasions that the southern horn of the crescent phase of Mercury was blunted and this led him to believe that this was due to a mountain 12 miles (19.3 km) high on Mercury. In August 1793, Schröter observed that the terminator of Mercury was slightly concave at a time when it should have been been straight according to calculations. Dichotomy occured about eight days later. Beer and Mädler made many observations in 1836 that confirmed Schröter's and showed that dichotomy occured about 6–8 days later or earlier than the calculated time depending on the direction in which Mercury was moving. This phenomenon is known as the Schröter effect or phase anomaly. Schröter estimated Mercury's axial rotation period to be 24 hours and 4 minutes. Some of Schröter's observations seem doubtful; so thought another observer of Mercury, Sir William Herschel, whose telescopic studies of Mercury never revealed anything interesting.

Giovanni Schiaparelli (1835–1910) observed Mercury from Milan with an 8 ½-inch refractor to which he was able to apply a power of 400× under favourable conditions. Schiaparelli observed Mercury during the daylight hours when it was near the meridian. He said that in 1882 he had been occupied with Mercury and considered it to be easier to observe than Venus and that it resembled Mars more than any other planet. He recorded seeing spots on Mercury that became obscured, sometimes completely, and brilliant white spots that changed their positions. He described the spots and patches appearing as 'extremely delicate, dark streaks' which were 'clear brown in colour against a rosy background'. It was shown sometime later that these appearances were illusory. Although Mercury's disc is smaller at superior conjunction than at dichotomy ('half moon' phase) Schiaparelli found it easier to observe when near superior conjunction because the disc is then fully illuminated. In 1882 an 18-inch refractor was added to Shiaparelli's observatory which greatly encouraged him in his work on Mercury. From his observations of the surface features he concluded that Mercury rotated upon its axis in the same time that it takes to make one orbital revolution around the sun. Schiaparelli's map of Mercury is shown in fig. 5.3.

Other sightings of surface features on Mercury at about this same period are as follows. Prince saw a bright spot a little south of the centre of the disc with faint lines radiating from it on June 11th 1867. Near the east limb of Mercury Birmingham glimpsed a large white spot on March 13th 1870. At Bothkamp, Vogel saw spots on the surface on April 14th and 22nd, 1871.

Percival Lowell (1855–1916), best known in connection with Martian 'canals', observed Mercury with a 24-inch refractor in his private observatory at Flagstaff, Arizona. He made several drawings of Mercury showing linear canal-like features. His description of the surface features of Mercury, however, dwells more on the natural appearance of these 'roughly linear' features that have 'irregularities that are suggestive of cracks'. Lowell's map of Mercury is shown in fig. 5.4.

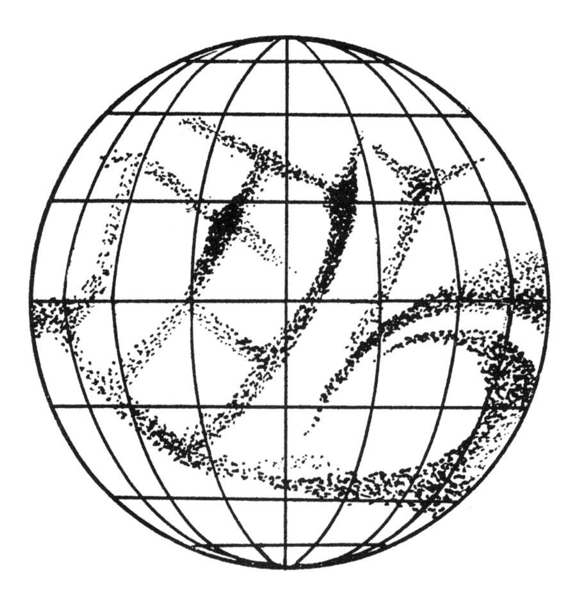

Fig. 5.3 Giovanni Schiaparelli's map of Mercury (188().

Fig. 5.4 Percival Lovell's map of Mercury (1896). (From Mercury the Elusive Planet, *Strom, R.G., Smithsonian Institution Press, Washington, DC, 1987.)*

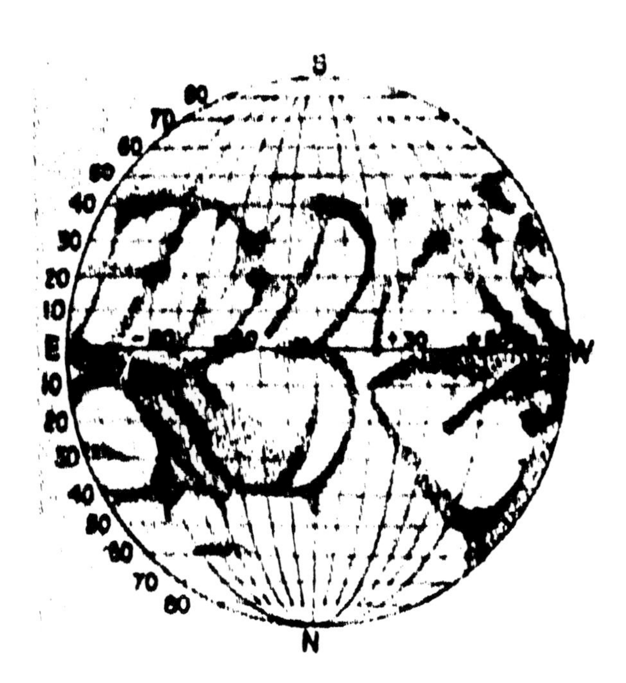

Fig. 5.5 René Jarry-Desloges' map of Mercury (1920). (From Mercury the Elusive Planet, *Strom, R.G., Smithsonian Institution Press, Washington, DC, 1987.)*

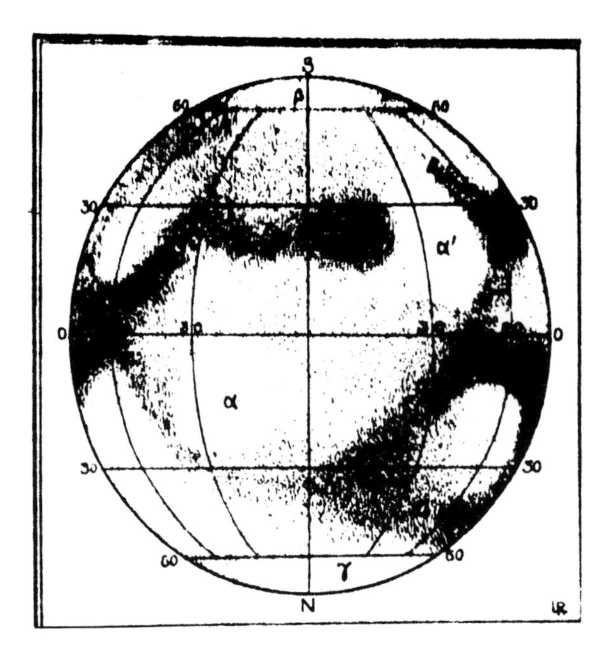

Fig. 5.6 M. Lucien Rudaux's map of Mercury (1927). (From Mercury the Elusive Planet, *Strom, R.G., Smithsonian Institution Press, Washington DC, 1987.)*

The Rene Jarry-Desloges map of Mercury appeared in 1920 (fig. 5.5) and M. Lucien Rudaux's map in 1927 (fig. 5.6).

William F. Denning (1848–1931) was one of the foremost observers of Mercury in England. Observing from Bristol he used a $9\frac{1}{2}$-inch refractor and a 10-inch reflector. He found that the dusky markings on Mercury were easily visible and suggestive of the planet Mars. He concluded that Mercury had an axial rotation period of 25 hours. Two of Denning's drawings are shown in fig. 5.7.

Eugene M. Antoniadi (1870–1943) made extensive observations of Mercury during the late 1920s, using the great 33-inch refractor of the Meudon Observatory just outside Paris. Like Schiaparelli he observed Mercury during the day. He noted dusky hard-to-see greyish patches on the planet. He made drawings and like any good observer recorded only what he saw with certainty. Some of his drawings are shown in fig. 5.8.

Antoniadi constructed maps based on these drawings and gave names to the features on Mercury's disc based on Greek and Egyptian mythology. Antoniadi's nomenclature continued in use until after the Mariner 10 spacecraft mission and one of his maps is shown in fig. 5.9. He thought that he sometimes saw 'local veils' and concluded that Mercury had an atmosphere sufficiently dense to retain clouds of dust in suspension.

Antoniadi observed localised changes in surface brightness of Mercury lasting only about one or two days. Similar transient changes have been described by H. McEwen in the BAA's Mercury reports. Antoniadi attributed these changes to clouds in Mercury's atmosphere but it is now known that what atmosphere Mercury possesses is too tenuous to allow clouds to form or to permit wind-blown dust to remain suspended for long. A possible explanation for these

Fig. 5.7 Drawings of Mercury by William F. Denning (10-inch Newtonian reflector). From Telescopic Work for Starlight Evenings, *Denning, W.F., Taylor and Francis, London, 1891.)*

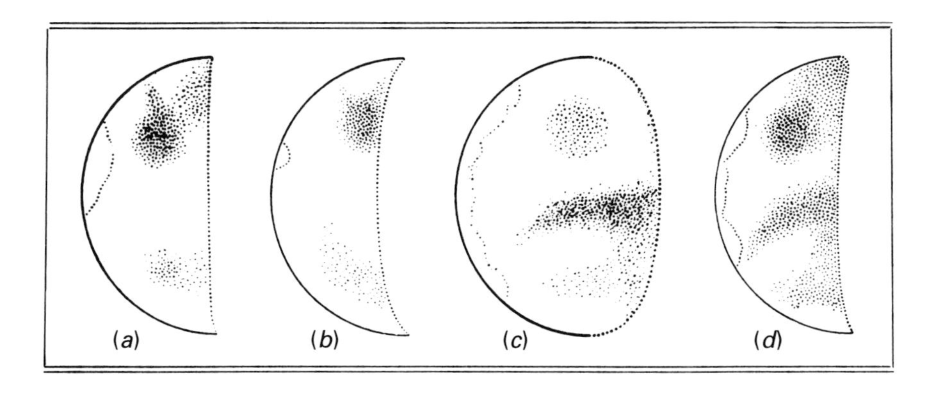

Fig. 5.8 Drawings of Mercury by Eugene E.M. Antoniadi (33-inch refractor). (a) *August 11th, 1924. A very white 'cloud' is shown at the limb.* (b) *June 21st, 1927.* (c) *October 11th, 1927. An enormous irregular 'cloud' of uneven brightness along the limb.* (d) *September 11th, 1929. Large irregular veil at the limb. In* (c) *and* (d) *the dusky surface markings are Atlantis, S. Criophori and S. Aphrodites (see fig. 5.9). (From* The Planet Mercury, *Antoniadi, E.M., English translation, Keith Reid, South Devon, 1974.)*

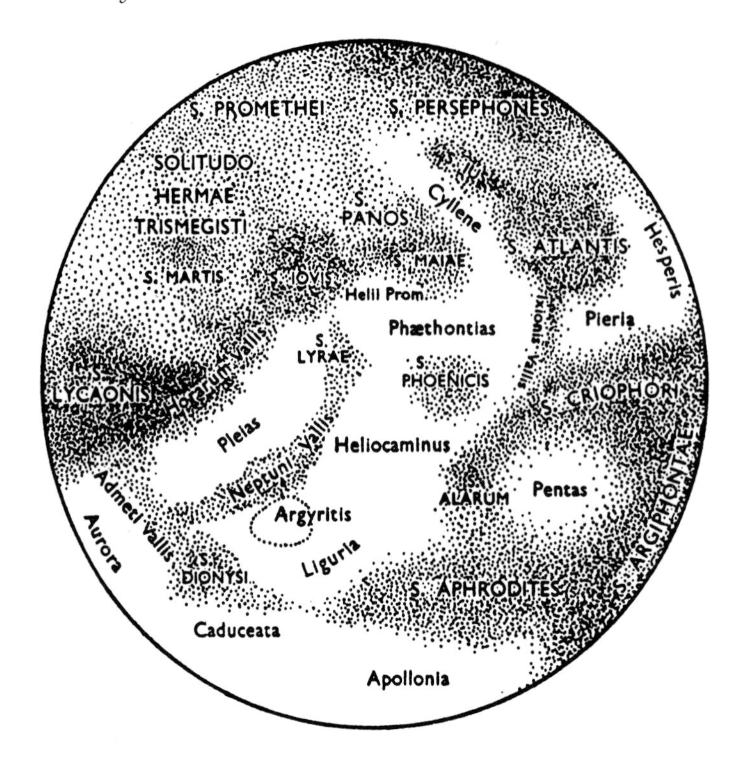

Fig. 5.9 Antoniadi's map of Mercury (1934). (Antoniadi's nomenclature.)

transient brightness changes is that they may be analagous to the luminescent effects sometimes seen on our own moon's surface which are caused by solar radiation. Mercury's surface resembles the moon's in its albedo (light reflectivity) and other photometric properties and also lacks a substantial atmosphere and magnetic field. Mercury's surface is exposed to ten times more solar radiation per unit of surface area than the moon; it would therefore not be surprising if luminescence effects resulted in sufficient localised surface brightening to produce detectable changes in apparent contrast.

Antoniadi's book *'La planete Mercure'* was published in Paris in 1934. Antoniadi, like Schiaparelli, considered that Mercury had a synchronous axial rotation. This belief arose because the same surface features were seen in the same positions every time that Mercury was observed at favourable elongations. Why this mistaken belief was held will be explained later under 'The axial rotation of Mercury'.

The amateur, Gary Wegner, observed Mercury during 1956–62 with a 10-inch reflector and drew the map of Mercury shown in fig 5–10. W. W. Spangenberg constructed the map shown in fig. 5.11 based on drawings made in 1932, 1936 and 1962 using 4- and 5-inch telescopes.

The French astronomers B. Lyot, H. Camichel and the late A. Dolfuss, using visual and photographic methods at the Pic-du-Midi observatory in the French Pyrenees, observed and photographed Mercury under superb seeing conditions. The telescope mainly used in their work was the 23½-inch refractor, which is specially adapted for planet observing. They recorded permanent dark markings on

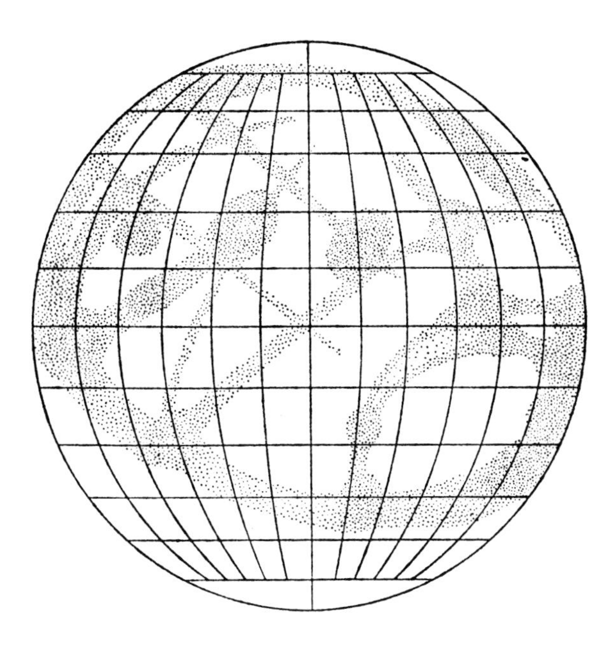

Fig. 5.10 Gary Wegner's map of Mercury (10-in reflector). (From Handbook for Planet Observers, *Roth, G.D., Faber and Faber, London, 1970.)*

Fig. 5.11 W.W. Spangenberg's map of Mercury, based on observations with a 5-inch refractor. Intensity of shading represents frequency of observation. (Redrawn from Handbook for Planet Observers, *Roth, G.D., Faber and Faber, London, 1970.)*

Mercury's surface, the pattern of which was in good agreement with the maps drawn by the old observers. In 1942 they drew a planisphere based on their photographic results.

Lyot deduced that Mercury's albedo was 13% which is somewhat less than our own moon's. He considered that Mercury's surface was rugged, which opinion was later confirmed by the Mariner 10 close-up photography. This revealed

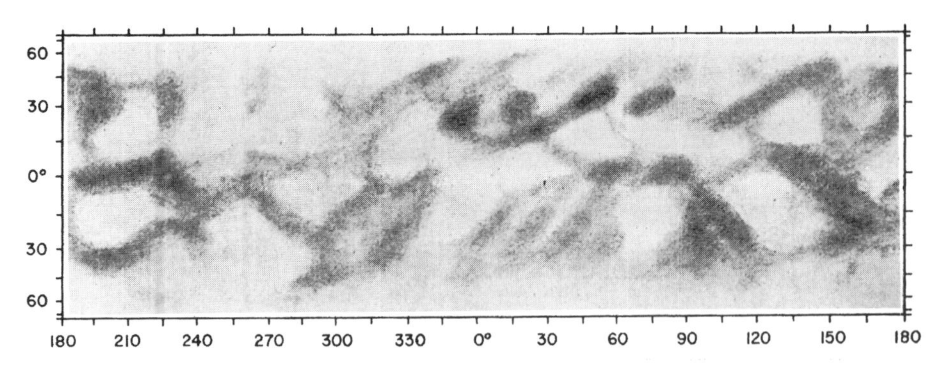

Fig. 5.12 Clark Chapman's map of Mercury: a composite based on 130 visual observations; cylindrical orthographic equal area projection. (From Sky and Telescope **34(1)**, *25, 1967.)*

Mercury's surface to be heavily cratered and strikingly similar to our own moon's.

In 1950 Camichel and Dolfuss drew a new map of Mercury which used Antoniadi's nomenclature slightly modified and extended. The map recorded all surface markings regarded as being clearly defined and shows some agreement with Antoniadi's map.

The Mariner 10 spacecraft confirmed Dolfuss's report of 1950 that he had found Mercury to have a very thin atmosphere although it turned out to be much more tenuous than Dolfuss had believed. In 1963, N. Kozyrev detected hydrogen in the atmosphere and in 1964 V. Morov found indications of carbon dioxide.

The last map of Mercury produced before the Mariner 10 mission was that of Clark Chapman in 1967 which he prepared from a series of drawings and photographs obtained from observers world-wide and the best evidence obtainable (fig. 5.12). Chapman employed the 59-day rotation period of Mercury and used Antoniadi's nomenclature. Oddly, Chapman's map does not agree with others produced contemporaneously and is hard to reconcile with the surface views provided by Mariner 10 photography.

The Mariner 10 spacecraft was launched in November 1973. The distance covered during the 17-month long mission exceeded 1000 million miles and took it three times around the sun.

Mariner encountered Mercury three times but was only able to explore one hemisphere of the planet. Close-up photography revealed Mercury to have a rugged and heavily cratered 'lunar-like' surface as is now well known.

It was soon evident that Antoniadi's nomenclature would have to be dropped and the International Astronomical Union (IAU) devised a new system of names for Mercury's surface features in which six classes of topographical features were recognised: mountains (montes), scarps (dorsa), ridges (rupes), plains and basins (planitia), and, of course, craters.

Visibility of Mercury

In fig. 5.13 are shown the orbits of Mercury and the Earth with lines drawn from the Earth tangent to Mercury's orbit at perihelion and to the sun and lines from

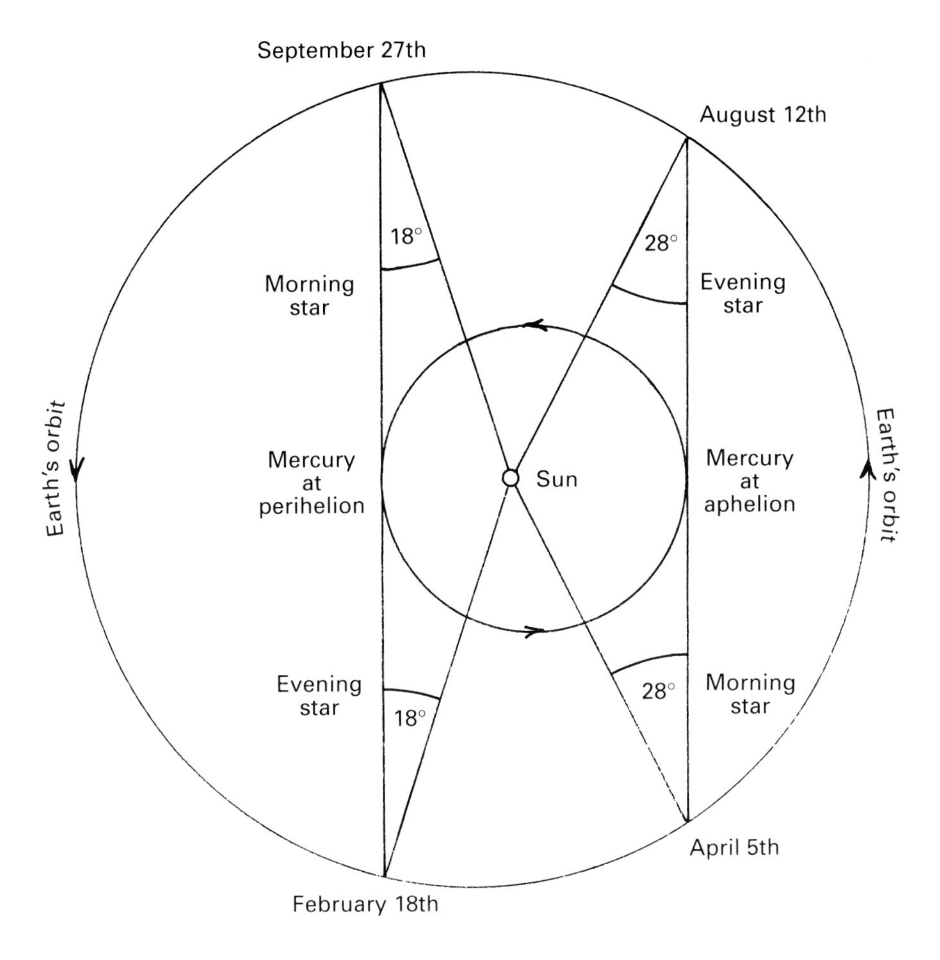

Fig. 5.13 Orbits of Mercury and the Earth: because of its large orbital eccentricity Mercury's greatest elongations range from 18° to 28°.

the Earth tangent to Mercury's orbit at aphelion and to the sun. The angle between the lines from the Earth to the sun and to the perihelion tangent position is 18° and the angle between the lines from the Earth to the sun and to the aphelion tangent position is 28°.

The perihelion line cuts the Earth's orbit at positions where the Earth is found at about February 18th and September 27th so that Mercury's angular separation from the sun is then 18°, the minimum value for greatest elongation. When at perihelion around February 18th, Mercury appears east of the sun and will therefore be an 'evening star'. It appears as a 'morning star' when it is west of the sun near perihelion around September 27th.

The aphelion line cuts the Earth's orbit at positions it occupies about April 5th and August 12th and Mercury's elongation is maximal at these times and is 28°. The planet is west of the sun about April 5th, a 'morning star' and is east of the sun about August 12th where it appears as an 'evening star' when near aphelion.

It doesn't follow from this that the best times to view Mercury are when it is a morning or evening star respectively in April and August; there is something else that influences Mercury's visibility. This is the angle made by the plane of the ecliptic to the plane of the celestial equator. The Earth's orbit and those of the planets all lie nearly in the same plane. If the Earth's axis were perpendicular to its orbital plane then all the planets – and the sun – would appear to lie close to the celestrial equator. However, as explained in chapter 2, the Earth's axis is tilted at an angle of $23\frac{1}{2}°$ to the plane of its orbit and hence the ecliptic – the apparent circular path in the sky traced out by the planets and sun – is tilted at an angle of $23\frac{1}{2}°$ to the celestial equator. It intersects the celestial equator at two points, the March and September equinoxes. At a given Earthly latitude the celestial equator makes a constant angle with the horizon which is equal to 90° minus the latitude. This would be about 39° at the latitude of London, England, which is about 51°–north latitude. Different parts of the ecliptic will make different angles with the horizon as the Earth rotates on its axis.

In fig. 5.14 the western horizon at the latitude of London is shown just after sunset. South is to the left and north to the right. The celestial equator is shown as a dotted line and the ecliptic is shown as it would appear on or about the March 21st equinox or the September 23rd equinox. The sun is therefore at the intersection of the ecliptic and the celestial equator, where it passes from south to north of the ecliptic, i.e., at the ascending node. Therefore, at sunset on or about March 21st the whole of the northern half of the ecliptic is above the horizon and above the celestial equator and is therefore inclined at a relatively steep angle of 39° plus $23\frac{1}{2}°$ or $62\frac{1}{2}°$ to the horizon. Mercury is shown at its greatest eastern elongation position and is nearly vertically above the sun. The dashed line and arrow

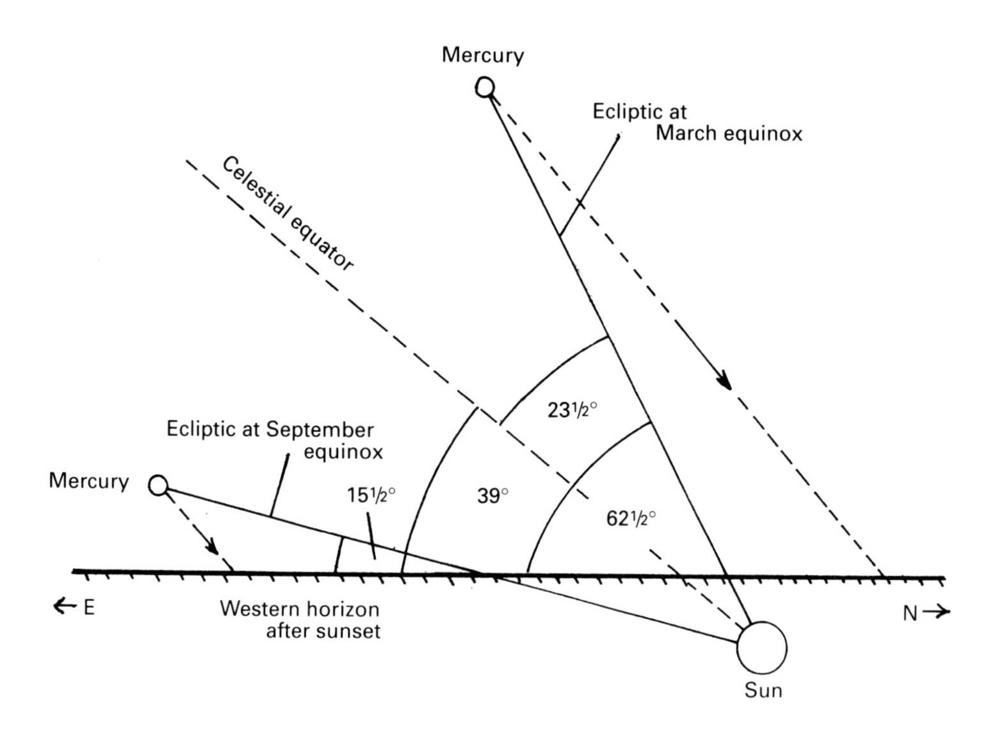

Fig. 5.14 Mercury at greatest eastern elongation after sunset at the March and September equinoxes (latitude 51° north).

leading from Mercury to the horizon is the distance it must travel to its setting position. Now contrast this with the situation if greatest eastern elongation occurred on or about September 23rd, the September equinox. The positions of Mercury and the ecliptic are also shown just after sunset on this date in fig. 5.14. This time the ecliptic makes an angle of 39° minus $23\frac{1}{2}°$ which is $15\frac{1}{2}°$ to the horizon at the latitude of London because the southern half of the ecliptic is now above the horizon and below the celestial equator. Mercury is now much lower in the sky, still to the left of the sun, and takes a much shorter time to set. Now, it is evident that if Mercury was at its least elongation distance – 18° – from the sun on March 21st it would be higher in the sky than it would be even if at its greatest elongation distance of 28° on September 23rd. It follows that the best time to see Mercury as an 'evening star' in the middle latitudes of the northern hemisphere will be on or about March 21st.

When Mercury is at its greatest western elongation it will be seen in the eastern sky as a 'morning star' and will rise a little before the sun. Fig. 5.15 shows the eastern sky just before sunrise and Mercury at greatest western elongation positions around the March 21st and September 23rd equinoxes at London. At sunrise on or about September 23rd when the sun is at the descending node, the whole of the northern part of the ecliptic will be in the sky and north of the celestial equator. Therefore, Mercury will be best seen as a morning star in the northern hemisphere around September 23rd since it will be much higher in the sky than in March.

Even under these best conditions Mercury will never be seen in a really dark sky. Matters are the other way around in the southern hemisphere; the best times

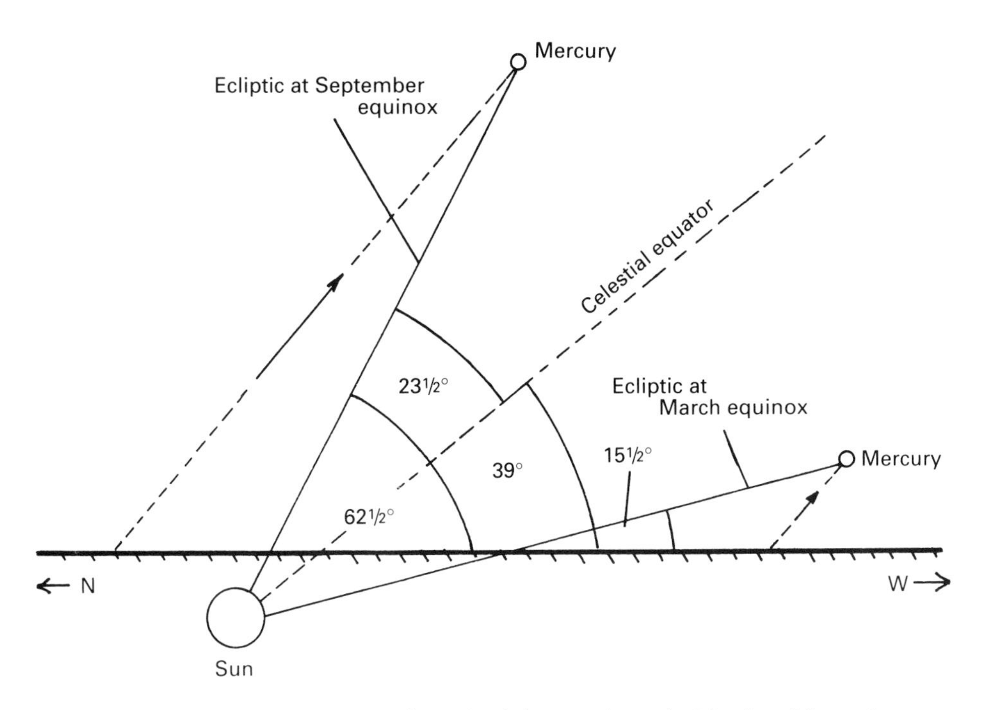

Fig. 5.15 Mercury at greatest western elongation before sunrise at the March and September equinoxes (latitude 51° north).

to see Mercury are as an evening star in September and as a morning star in March. There is an additional factor that makes for better visibility of Mercury at these times that does not obtain in the northern hemisphere; as previously mentioned, the greatest angular distance of Mercury from the sun as an evening star occurs about August 12th which is not far from September 23rd when the ecliptic makes its greatest angle with the western horizon. Thus, these two factors that favour the visibility of Mercury occur quite close together in the southern hemisphere. Mercury can therefore still be seen after dark, unlike in the northern hemisphere when it can only be seen in twilight.

Similarly, these two favourable factors occur close together when Mercury is a morning star. Greatest western elongation occurs on or about April 5th which is close to March 21st so that again, southern hemisphere observers get in addition the benefit of the steep angle of the ecliptic to the eastern horizon. Therefore, in the southern hemisphere, Mercury can be seen in dark skies as it rises in the east before the morning twilight begins.

An additional factor that affects Mercury's apparent position in the sky is the tilt of its orbit to the plane of the ecliptic; because of this, Mercury's declination ranges from ±5°.

To summarise: Mercury is nearly always lost to naked eye viewing in brilliant sunlight and cannot be seen at and near to superior or inferior conjunction. It is always rather low in the sky and can only be seen in twilight or just before dawn in northern latitudes.

In England, even at the best apparitions, Mercury never sets more than slightly over two hours after the sun or rises more than about two hours before the sun. It may be visible about three quarters of an hour after sunset and it fades away about three quarters of an hour before sunrise. It usually looks like a twinkling rose-coloured star but the colour and twinkling are both due to the low altitude. There are advantages to observing Mercury in the daytime as will be shown later.

Mercury can be observed for a period of about five weeks at the elongations and appears to move quite rapidly in the sky relative to the sun. Table 5.1 gives a timetable for viewing Mercury:

Table 5.1. *Visibility and movements of Mercury.*

Days after superior conjunction		Phase
0	Superior conjunction	'Full'
36±9	Greatest eastern elongation	'Last quarter'
47±9	Retrogression begins	
58±10	Inferior conjunction	'New'
69±8	Retrogression ends	
80±5	Greatest western elongation	'First quarter'
116±12	Superior conjunction	'Full'

These data are adapted from *Handbook for planet observers* by G. Roth, Faber and Faber, London, 1970. The ± values are the ranges found by Dr John E. Westfall over six apparitions from April 20th 1988 to March 20th 1990.

The axial rotation of Mercury

When Mercury is best placed for observation – at the half or gibbous phases near the elongations – it shows the same surface features. This understandably led some of the earlier observers to believe that Mercury revolved on its axis in the same time that it took to make one orbital revolution around the sun (88 days, very nearly). Using Doppler radar techniques, G. H. Pettengill and R. B. Dyce discovered in 1965 that the axial rotation period of Mercury is actually 59 days, give or take 5 days. It is now known to be 58.6462 days or exactly two thirds of the orbital period.

The earlier observers were misled into concluding that the axial rotation and orbital revolution periods were the same because of an interesting numerical relationship between Mercury's synodic period, the Earth year and Mercury's orbital revolution and axial rotation periods: Mercury's synodic period (116 days) is about one third of an Earth year, almost exactly four thirds of its orbital revolution time and twice its axial rotation period of 59 days. This means that the distribution and pattern of genuine surface markings seen on Mercury by terrestrial observers agrees with axial rotation periods of 59 days and also 88 days. The explanation is as follows: usually there are three western and three eastern elongations of Mercury in an Earth year, the average time period elapsing between consecutive western or eastern elongations being the synodic period of 116 days. As was explained earlier, the most favourable eastern elongations occur in March and the most favourable western elongations occur in September in the northern hemisphere. Sometimes, there is only one favourable eastern and one favourable western elongation in a year. Usually, in a six-year period, favourable eastern or western elongations follow one another at intervals of three synodic periods. Mercury rotates on its axis three times in the time that it takes to complete two orbital revolutions; this means that 176 Earth days elapse between two successive sunrises at a given point on Mercury. Therefore if an observer sees Mercury at a western (morning) elongation, the next time that the same face is illuminated will be $1\frac{1}{2}$ synodic periods later. Mercury will now be at an evening elongation and we see the opposite hemisphere; but after three synodic periods the same face of Mercury will be seen at the same phase. We saw earlier that a time equal to three synodic periods separates favourable elongations; therefore an observer who studies Mercury only at these elongations will see similar surface markings unchanged in position for six successive years. Thus the incorrect rotation period of 88 days would be deduced.

Maps of Mercury are generally based on a few years of observation with the telescope and so a given observer's individual delineations of Mercury's surface features will all agree with the one map. After a few years, however, the periodicity drifts out of phase. Antoniadi made observations of Mercury between 1924 and 1929 and based one of his maps on them. In his book *La planete Mercure*, there are nine drawings made at eastern elongations and the same face of Mercury is shown in seven of them. Four of his five drawings made at western elongations also show the same face. Antoniadi thus found that most of his drawings agreed with his map and with an axial rotation period of 88 days. Some astronomers observed Mercury at elongations other than the really favourable ones and found disagreements with their own and Antoniadi's observations. Certain of them were probably influenced by the reputations of Schiaparelli and Antoniadi and believed that they should see the same features on Mercury; this may have unconsciously influenced them when they came to make drawings, thinking that their drawings should agree with those of Schiaparelli and Antoniadi. Rationalisations were even attempted to explain obvious discrepancies between what was unmistakably seen on Mercury under the best seeing conditions and the drawings of Antoniadi.

Observing Mercury

Although Mercury can be quite a bright object, attaining a magnitude of –2 at superior conjunction, it is always seen against a bright sky at these times and is virtually invisible. In the northern hemisphere it is never really prominent at its elongations when at magnitude 0 as the sky is not really dark even then. Its comparative lowness in the sky is no help; mistiness, cloud and industrial smog all help to obscure the elusive little planet as it nestles fairly close to the horizon.

Unless, as sometimes happens, Mercury is in conjunction with the moon or a bright planet, the only way to find Mercury is to look up its exact position from an ephemeris or in astronomical magazines which usually have sky calendars in which forthcoming astronomical events are predicted. Usually, observing an evening elongation will be preferred but Mercury should also be observed at morning apparitions although this means rising before daybreak.

Having found out exactly where Mercury will be in the evening sky just after sunset, make sure that your western horizon is unobscured by houses or trees and carefully scan the sky at first without optical aid. With any luck and some patience you should not have too much difficulty in sighting the tiny glimmering point of light that is Mercury. It is a great help if you have binoculars or even opera glasses.

At about 30 minutes after sunset in late summer or early autumn elongations in the northern hemisphere, scan about 3–5° above an unobstructed horizon with binoculars. You should be successful. Satisfaction is found in simply being able to locate Mercury with the unaided eye or with binoculars. It is no mean accomplishment if you can thus locate it several times during an elongation. In fact some observers take a justifiable pride in the number of times they can say that they have spotted Mercury with the unaided eye. It really is fascinating just to hunt and find this elusive planet.

Many years ago, Joseph Anderer of the Chicago Astronomical Society wrote that he observed Mercury on 48 separate days in 1962. The skies of Chicago, a modern industrial city, are certainly no clearer than were the skies near the river Vistula in the Renaissance so that in the light of Anderer's sightings the old story that Copernicus never saw Mercury seems even less credible.

Brightness, colour and phases

As already mentioned, the disc of Mercury never appears large, varying from an apparent diameter of 12.2 seconds of arc at the nearest inferior conjunction to 4.6 seconds of arc at the farthest superior conjunction. At the actual times of superior and inferior conjunction, Mercury is invisible. Mercury's geometric albedo is 0.106 which is slightly less than that of the moon. Its surface brightness is 2.3 times as great at perihelion as at aphelion. The (B–V) colour index is +0.93, very similar to that of the moon (+0.92) and noticeably redder than the sun (+0.62).

The phases may be seen in a 3-inch refractor but a power of at least 100× is needed to discern them; powers of 250×–300× are better. The blunting of the southern cusp of the crescent can be made out with this aperture. The phases are well seen in a 6-inch telescope.

It has often been said that the disc of Mercury has a pinkish tinge. J. Muirden relates that he once viewed Mercury and Venus simultaneously with a 3½-inch refractor when they were close enough in the sky to be seen in the same telescopic field. On this occasion, Venus appeared a brilliant white while in contrast, Mercury was a dull reddish-grey.

Denning made a similar observation on May 12th 1890 when the two planets were evening stars separated by a distance of 2°. With a power of 145× on his 10-inch reflector (which did not permit simultaneous viewing of the planets in the same field of view owing to their angular separation) Denning saw Venus 'like newly polished silver' whereas Mercury was 'of a dull leaden hue'. Denning says that James Nasmyth made a similar observation on September 28th 1878.

If we take the visible variation in the size of Mercury's disc as between 5 and 10 seconds of arc (remembering that it is invisible at the conjunctions), this is a two-fold variation. This means that in going from superior conjunction ('full' phase) to inferior conjunction ('new' phase) the apparent illuminated area of Mercury steadily decreases. The 'full' phase at superior conjunction reflects most light, the other phases reflecting less. The brightness variations of Mercury are represented graphically in fig. 5.16. Mercury's brightness variations suggested a rugged reflecting surface, which has been confirmed by polarimetric studies and, of course, most dramatically by direct close-up photography from the Mariner 10 spacecraft.

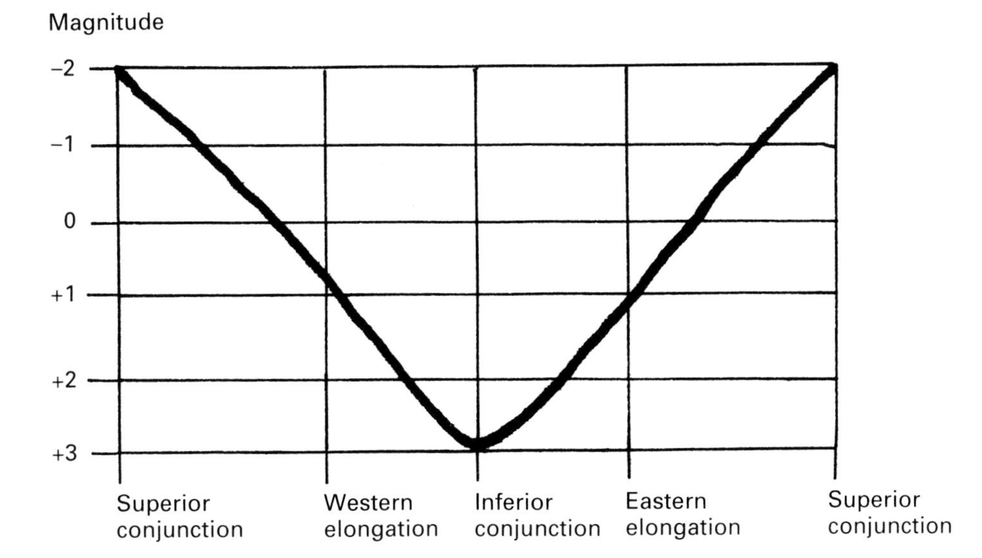

Fig. 5.16 Graph showing the brightness variations of Mercury. (Redrawn from Anderer, J., Rev. Pop. Astron., **LVII(522)**, *12–13, 1963.)*

Surface features

Many observers claim to have seen surface detail on Mercury with moderate apertures yet others, equally competent, are unconvinced of the reality of permanent surface detail. A well-known professional astronomer says that he saw nothing on Mercury with the McDonald Observatory's 82-inch reflector. As previously mentioned, the older observers (Schiaparelli, Antoniadi) were in fair agreement about the existence of dusky streaks on Mercury. The older maps agree well with the permanent detail revealed on Mercury in the results obtained by Camichel, Dollfus and Lyot with visual and photographic methods so that we can safely assume that the surface marking are really there.

If you want to see surface features on Mercury it seems that timing is critical, both with regard to the planet's phase and the time of day that the observation is made. The best phase is around elongation when the planet is at the 'fat' crescent, 'half' phase (dichotomy) or slightly gibbous. The time of day is critical. If you observe in daytime, the bright sky background may be sufficient to 'wash out' pale surface markings of low contrast. The slightest film of dust on objective or mirror may scatter light sufficiently in daylight viewing to obscure the elusive surface markings.

If you wish to observe Mercury in the daytime you should first ascertain its approximate angular distance east or west of the sun from an ephemeris. With an altazimuth mounted telescope try 'sweeping' with a low power to the east or west of the sun, whichever is appropriate and you may be lucky enough to find the planet. Mercury is so close to the sun that there is a grave danger of pointing the telescope at the sun while 'sweeping', so for safety's sake set the telescope up in the shadow of a house so that the disc of the sun is out of sight.

If the telescope is equatorially mounted the telescope may first be pointed at the sun – DO NOT LOOK THROUGH IT – as judged by the small circular shadow it will cast. The telescope should already be in focus and the viewfinder should be covered as a precaution against inadvertently looking through it while the telescope is pointed at the sun. Clamp the declination axis, get the sun out of the field of view, look through the eyepiece and slowly swing the telescope about the polar axis to the east or west with a low power. Mercury is fairly close to the ecliptic – its distance from the ecliptic can vary by $\pm 5°$ – and so you should find it fairly easily by this method.

If the telescope is equatorial and is set up with the polar axis pointing exactly to the north celestial pole, then Mercury should be easily found with the setting circles after first looking up Mercury's RA and declination for the day you are making your observation.

If the sky is nearly dark then the markings may be lost in the planet's glare. Somewhere in between, when the sky background and the glare of the planet are not too great, is the time to choose and it will be short-lived. This brief 'window' should also be a time when the air is clear and free from turbulence. It is hopeless trying to see pale markings on a planetary disc through unsteady air. Observing time may be extended by use of the apodising screen (described in chapter 3) which simultaneously improves contrast, reduces glare and minimises the effects of atmospheric turbulence. Since all these conditions must occur together it is small wonder that many people see nothing on Mercury even with large telescopes. Needless to say the optical quality of the telescope used must be good.

Depending on whether you observe at morning or evening elongations, the pattern of surface markings you may expect to see is shown in fig. 5.17 which represents markings that have been seen and drawn only by experienced observers. Fig. 5.18 is an observational drawing by Richard M. Baum, the noted English planetary observer.

For success, moderate to large apertures, 10 inches and over, are recommended although useful work may be done with lesser instruments. Optical colour filters may be used to improve the quality of visual observations of Mercury. A red filter will make the disc of the planet stand out against a blue sky in daylight observations. The filter could be tried in combination with the apodising screen.

The best time to observe Mercury is just after sunset or just before sunrise; since the Earth's atmospheric distortion of the telescopic image is likely to be

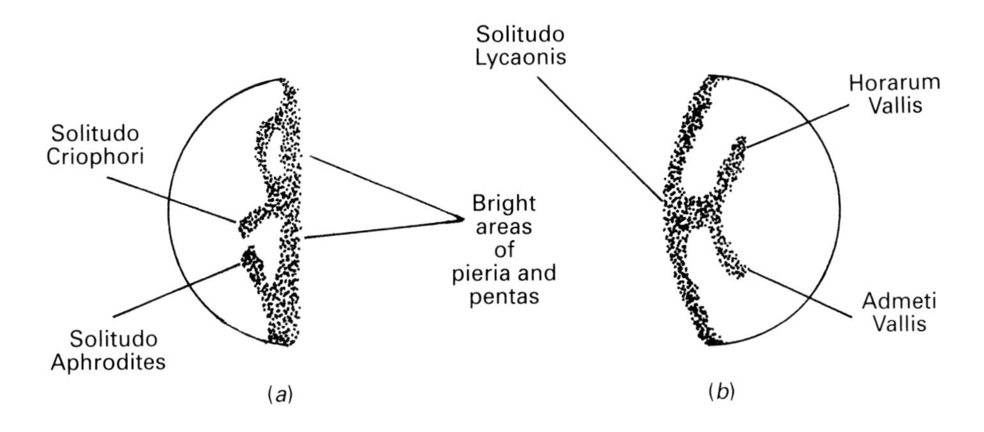

Fig. 5.17 Surface features of Mercury as recorded by experienced observers: (a) *evening apparition;* (b) *morning apparition.*

Fig. 5.18 Mercury: March 15th, 1952, 14.00 UT, 165-millimetre (6.6-inch) refractor. (R.M. Baum, Chester, England.)

serious at these times, an orange filter may be used with high magnifications for observing the phases.

Perhaps the most useful work you can do in observing and recording the markings on Mercury is to attempt to correlate the markings with surface features revealed by the Mariner 10 photography of the planet. In the case of Mars, no correlation has been found between the dark markings and the surface contours. It would be interesting to see if this holds true, or otherwise, in the case of Mercury. It would also be desirable to attempt to map the hemisphere of Mercury that was not imaged by Mariner 10.

When drawing Mercury, some observers like to use different sized circles drawn on their paper to represent the variation in apparent size of the planet at different phases. However, it would seem best to use the same sized circle for all phases, a diameter of about 2 inches being about right.

The phase anomaly or Schröter effect.

Another kind of observation that is worth making is to attempt to determine the exact time of dichotomy or 'half phase', when the terminator is straight, although these are not always equivalent terms. This occurs when the planetocentric angle between Earth and sun is 90°; the planet then looks half illuminated to Earth-based observers. The time of dichotomy and greatest elongation roughly coincide during an apparition, but do not coincide exactly because of the eccentricity of Mercury's orbit. Although the times of both dichotomy and greatest elongation can be exactly predicted, oddly the predicted and observed times of dichotomy often disagree. Practical observers and theoretical astronomers are thus posed an interesting problem – why the disagreement in the predicted and observed times of dichotomy?

It was Schröter who first noticed the discrepancy when he was observing the planet Venus during 1793. It has since become known as the 'Schröter effect'. It has been evident with Venus since that time and Mercury sometimes shows it. Dichotomy occurs late for western elongations and early for eastern elongations. The difference never exceeds five days, if that, for Mercury.

Many different explanations have been put forward for the Schröter effect but none is satisfactory. It is thought that at least some of the effect may be due to physiological or psychological causes in the observer and possibly also to observing conditions and the telescope used. The effect also partly depends on whether a filter is employed or if observation is through thin cloud, both of which will reduce the apparent brightness of the planet. The low sun angle near the terminator may also contribute to the effect.

The available data on the Schröter effect would be greatly improved if more careful observations of the phase were made. The amateur planet observer should not find this too difficult and it provides an opportunity to make a valuable contribution in this area of planetary observation. However, careful records must be made of seeing and transparency conditions, aperture of telescope used, magnification and whether a filter or apodising screen was used; an observation will lose much of its value if any one of the these is not recorded.

Observations of the phase anomaly can be recorded by first drawing a 2-inch diameter disc on white paper. Next, carefully observe Mercury allowing time for your eye to get accustomed to the image and the brightness level. Then, very carefully with a pencil draw the shape of Mercury's terminator on the disc, writing a small letter 'd' on the dark side so that there can be no subsequent confusion over which side was which. Then record the observational data listed above plus the date and time, preferably universal time (UT) when the observation was made. Your results should be sent either to the Mercury Section Director or Recorder respectively of the BAA in Britain or the ALPO in the USA. When these and other observations are sent to the ALPO or BAA, the Recorder or Director will measure your drawings to make a determination of the fraction of Mercury's disc that is apparently illuminated at the times the observations were made. Your data and those sent in by other observers will be used to determine the difference between the observed and calculated times of dichotomy. Of course, you may also wish to sketch in surface details seen on the disc.

An interesting computer-aided method has been used to determine the exact observed time of dichotomy of Venus (chapter 6). Presumably, it would also work with Mercury.

Transits of Mercury

Mercury's orbit is inclined at an angle of 7° to the Earth's orbit so that Mercury sometimes passes in front of or behind the sun as seen from the Earth only rarely. A transit of Mercury occurs when Earthly observers see the planet at inferior conjunction pass across the sun. The first person to observe a transit of Mercury was Gassendi who watched the transit of November 7th 1631 from Paris. The disc of Mercury is so tiny that a telescope is needed to observe a transit. Here I would warn most emphatically that the sun should NEVER EVER be viewed directly with a telescope, however small, unless some foolproof device is used to cut down on the intense light and heat focused by the objective. This can either be a Herschel wedge used with a light absorbing filter of adequate optical density or better still – but more expensively – a full aperture filter placed over the telescope objective. This can be relatively inexpensively made from Mylar film or best of all a reflecting metal-film-on-optical glass type of filter can be used. The full aperture filter reflects most of the sun's light and heat away before it enters the telescope.

Do not use a filter of the type that screws over the eye lens of the eyepiece. There is a risk that the concentrated light and heat of the sun will crack and splinter the glass and the concentrated sunlight will enter the observer's eye with disastrous results before he can pull his head away.

The only really safe way to observe a transit of Mercury is to use the telescope to project an image of the sun onto a matt white surface. Mercury will appear as a tiny round black dot on the solar disc. Whatever optical system is used, a long focal length is desirable to give adequate image size since Mercury's apparent angular size is never very large as we have seen. A power of 100× on a 2.4-inch or 3-inch refractor used with a Herschel wedge and dark filter is adequate for direct viewing. Reflectors should be stopped down to about 3 inches for safety. Projection is safest and permits several people to view the transit simultaneously.

Transits of Mercury occur in pairs about three years apart and always in the months of May and November. Of the transits of Mercury in the years between 1937 and 1999, about twice as many occur in November as in May. The reason for this is that November transits occur at Mercury's perihelion and the May transits occur at aphelion. May transits occur at Mercury's ascending node and November transits at the descending node.

I observed the transits of May 9th 1970 and November 10th 1973 from the Alan Gee Solar Observatory at the Buffalo Museum of Science, New York, USA. The observatory is equipped with a 7-inch fixed vertical solar telescope and a heliostat that reflects its light into the telescope so that the large solar image projected onto a screen in a darkened room does not move. The sun's projected image is so large that Mercury appears as a black disc about 1 centimetre in diameter. At fixed time intervals I took close-up black and white Polaroid photographs of the image of Mercury on the solar disc and combined the pictures freehand into the two drawings shown in figs. 5.19 and 5.20. An interesting phenomenon of the 1970 transit was that Mercury passed across a sunspot. I wondered if Mercury had ever before been seen passing over a sunspot during a transit. The planet's disc was noticeably darker than the sunspot umbra.

The paths of Mercury across the sun's disc in the transits of 1970 and 1973 and in forthcoming transits are shown in fig. 5.21.

When the disc of Mercury has just completed its entry onto the sun's disc – or is just about to leave it – it has sometimes been seen to be slightly elongated towards the sun's limb, an effect due to irradiation or solar limb darkening. It is unlikely to be due to refraction of the sun's light by Mercury's atmosphere which is extremely tenuous.

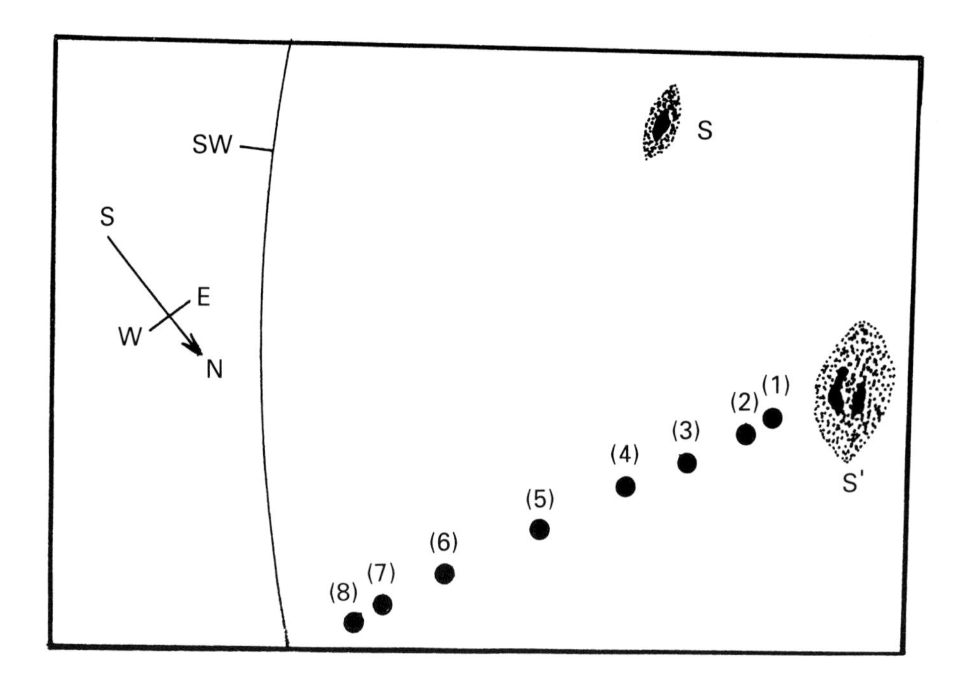

Fig. 5.19 Transit of Mercury, May 9th, 1970. The distances between successive position of Mercury are not strictly proportional to time. This is because the drawing is a freehand composite based on photographs taken with a hand-held camera. The distance of this from the projected solar image and the camera's inclination to the screen varied somewhat from picture to picture.

The times (UT) of the approximate positions of Mercury on the sun's disc are: (1) 11.17; (2) 11.21; (3) 11.27; (4) 11.37; (5) 11.48; (6) 11.55; (7) 12.05; (8) 12.10: S and S' are sunspots.

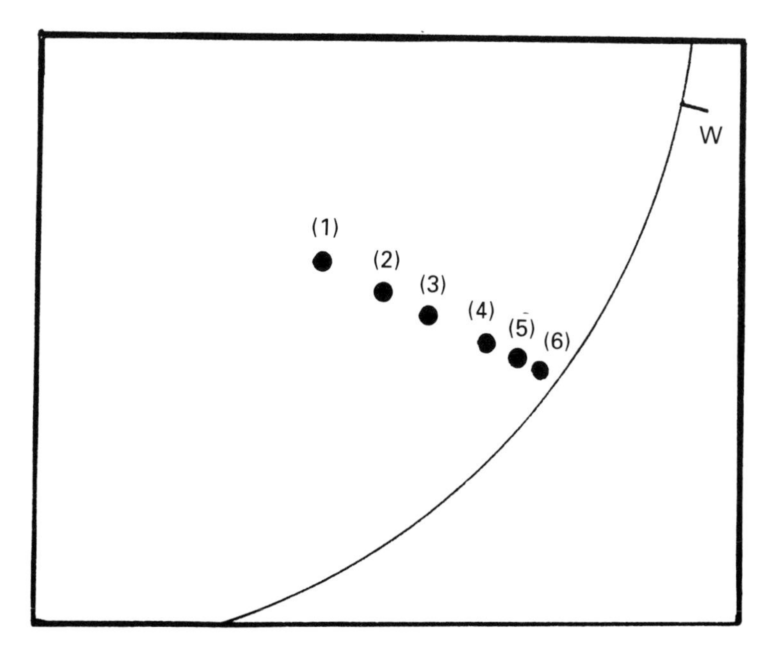

Fig. 5.20 Transit of Mercury, November 10th, 1973. The times (UT) of the approximate positions of Mercury on the sun's disc: (1) 12.35; (2) 12.47; (3) 12.55; (4) 13.05; (5) 13.11; (6) 13.15.

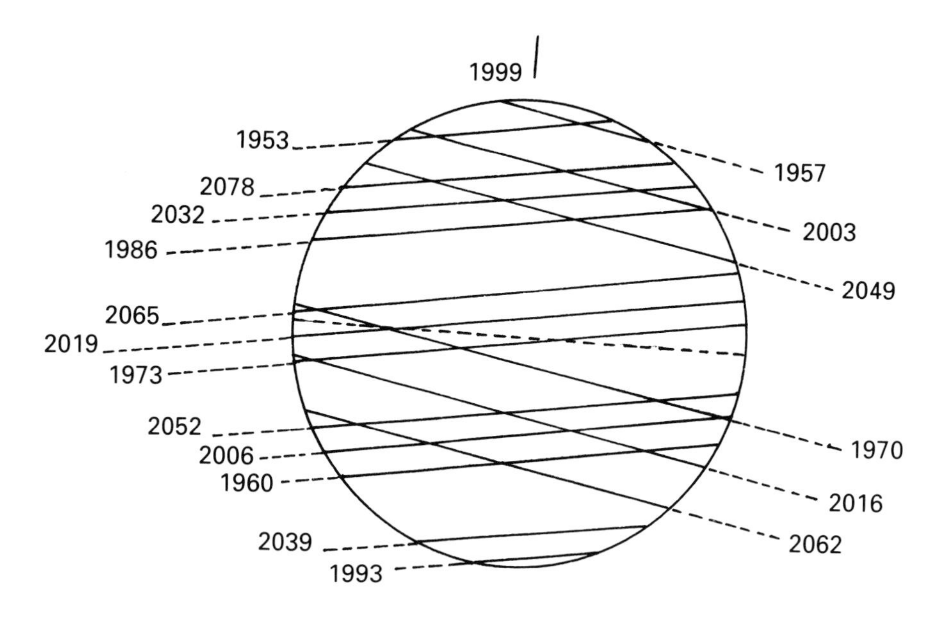

Fig. 5.21 Transits of Mercury, 1920–2080. (J. Meeus.)

When fully on the solar disc, Mercury looks intensely black. Some observers have represented it in drawings with a somewhat dusky border and others with a narrow luminous ring. These are probably optical illusions caused by ocular fatigue resulting from the high contrast between the sun's disc and Mercury.

A similar explanation may account for the appearance of one bright or grey spot and sometimes two, on the dark body of the planet witnessed by Wurzelbauer in 1697, Schröter, Kohler and Harding in 1799, Fritsch and some others in 1802, and Moll and assistants in 1832. It was in 1832 that Harding clearly saw two light spots and Gruithuisen suspected one. Huggins and Elger both saw bright spots in 1868 and again Browning saw two spots. Many English and Belgian observers noted a bright spot in 1878 (two spots were noticed by one observer) although some others did not see anything. Huggins experimented with an artificial disc in simulating a transit but no bright spot or spots were seen. All these appearances of bright spots seem to be due to a fatigued state of the retina. Any appearances seen in astronomical observation, however unlikely, should always be recorded otherwise important new phenomena may escape detection.

The scientifically important data to record during a transit of Mercury are the times of first, second, third and fourth contacts; first and second contacts are when Mercury's disc is at external and internal tangency to the sun's disc at ingress and third and fourth contacts are when Mercury's disc is at internal and external tangency to the sun's disc at egress (fig. 5.22). First contact is impossible to time accurately; fourth contact is difficult. For timing the contacts it is necessary to have an accurate timing device, correct to within one or two seconds. Any good watch or clock that has been checked with short wave radio time signals will be adequate.

If there is a group of observers watching the transit together it is a good idea to have someone to count the seconds out loud before and after the contacts so that the observers can listen and make mental notes of the contact times.

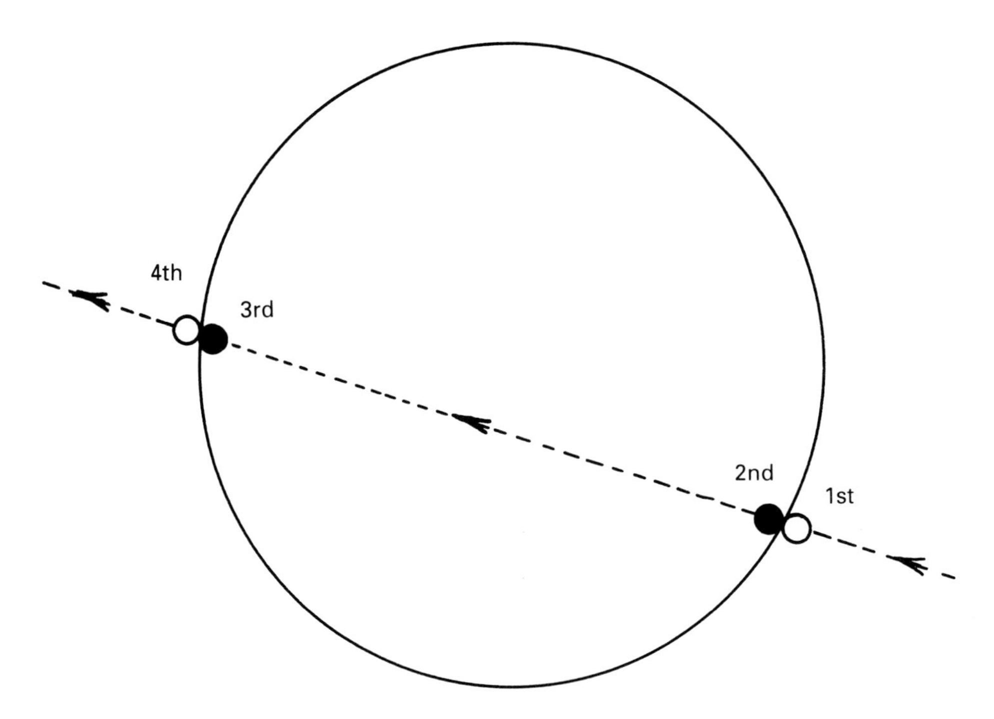

Fig. 5.22 Contacts of a planetary disc with the sun's limb during a transit.

First contact should be watched for about a minute or two before the predicted time so that your eye can adapt itself to the solar image. If the edge of the sun's disc is rippling as it usually will be, you must be even more alert to catch and time the first appearance of a notch at the east edge of the sun's disc which is first contact. As already mentioned, it is not possible to time first contact really accurately.

Just before second contact, as Mercury moves onto the solar disc, the planet will appear to be attached to the sun's edge by a short thick stalk, the so-called 'black drop' effect (fig. 5.23), a purely optical phenomenon. A similar effect is seen in transits of Venus but is due to light refraction by the relatively dense atmosphere of Venus. As soon as a white thread of light is seen cutting though the neck and separating Mercury completely from the sun's edge, this is the moment of second contact.

When the transit draws to a close and Mercury approaches the sun's west limb and nears third contact, keep a close watch on the narrowing band of light between Mercury and the sun's edge. It narrows to a delicate thread and as soon as the thread breaks this marks the instant of third contact. A 'black drop' effect may be seen.

As Mercury moves off the sun it appears as a notch in the limb. The notch becomes more and more shallow, finally disappearing which marks the instant of fourth contact. Contact timing should be recorded to the nearest second. However, poor seeing may make this difficult so that timing may be uncertain by ten seconds or even longer. If so, record the times of the beginning and end of the period of uncertainty.

Contact timings are not of much value unless you know your geographical longitude and latitude to within one minute of arc. These can be ascertained from the Ordnance Survey in Great Britain or in the USA the US Coast and Geodetic Survey or Geological Survey. If you cannot determine your position in this way

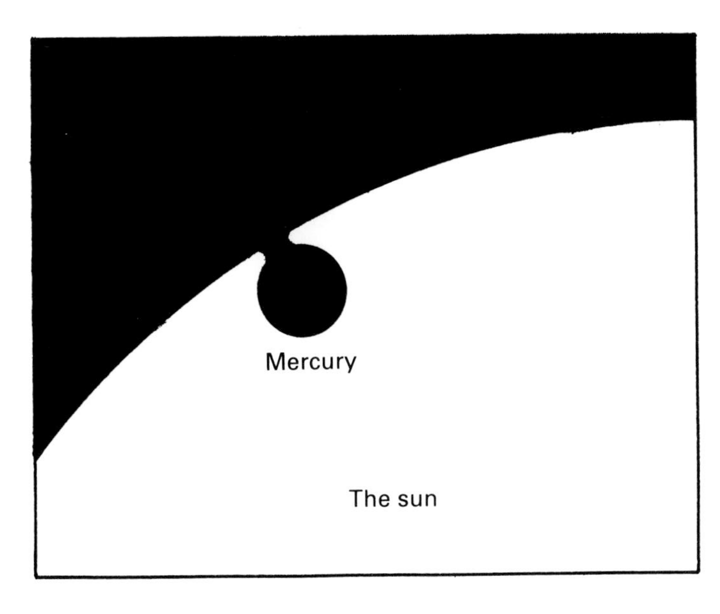

Fig. 5.23 The 'black drop'.

at least record the distance in miles and direction from the centre of the nearest town.

For future transits an interesting project would be to videotape the events, especially ingress and egress and short-wave time signals could be recorded on the sound track.

Your contact timings should be forwarded to astronomical or other organisations requesting them for inclusion with results from other observers so that they can be analysed. The scientific value of the results can lie in refining the orbit of Mercury and determination of the solar diameter with considerable accuracy (the motion of Mercury is about 0″.05 per second of time). The timing data should be accompanied by a note of the following:

(1) Your name, address and geographical position of your observing site.

(2) Aperture of telescope and diameter of aperture stop if used; magnification or diameter of the image of the sun if you used projection to observe the transit.

(3) Filters used, if any.

(4) What your timing device was and how you standardised it.

(5) The observed times of the contacts and information to assist in their interpretation.

(6) Anything like poor seeing or clouds which might affect the reliability of your observations.

Further reading

Books

The Planet Mercury. Antoniadi, E. M., (translated from the French by Moore, P. A). Keith Reid, South Devon, England (1974).

The Planet Mercury. Sandner, W., Faber and Faber, London (1974).

The Atlas of Mercury. Cross, A. C. and Moore, P. A., Crown Publishers Inc., New York, (1977).

Mercury, the Elusive Planet. Strom, R. G., Smithsonian Institution Press, Washington DC, (1987).

Papers and articles

The markings of Mercury. McEwen, H., *JBAA* **46(10)**, 382 (1936).

A suspected phase anomaly of Mercury. Nightingale, H. C., *JBAA* **78(1)**, 45 (1967).

On the phase anomaly of Mercury and Venus. Kirby, G. J., *JBAA* **80(4)**, 293–5 (1970).

The planet Mercury as viewed by Mariner 10. Strom, R. G., *Sky and Telescope* **47(6)**, 360 (1974).

Mercury in 1960. Gaherty, G. Jr, *The Strolling Astronomer* **15(11–12)**, 187–91 (1961).

Observations of Mercury in 1988 and 1989. Schmude, R. W., *JALPO* **34(3)**, 101–3 (1990).

Historical sighting of the craters of Mercury. Baum, R. M., *JALPO* **28(1–2)**, 17–22 (1979).

Findings from Mercury's transit. Ashbrook, J., *Sky and Telescope* **40(1)**, 20–4 (1970).

6

Venus

General

Venus is the second out from the sun of the two 'inferior' planets. It has a diameter at the surface of 7521 miles (12 104 km), only a shade less than that of the Earth (fig. 6.1). Measured at the cloud tops the diameter is 7606 miles (12 240 km). Because of this close similarity in size to the Earth, Venus has often been called 'the Earth's twin sister'. The masses and therefore the densities and gravities are also similar. The resemblances end here, however, because the two planets are vastly different in other respects.

The mean distance of Venus from the sun is 67.2 million miles (108.2 million km) and varies from a perihelion distance of 66.8 million miles (107.5 million km) to the aphelic distance of 67.7 million miles (108.9 million km). The orbital eccentricity is 0.007 so that the orbit is very nearly circular in contrast to Mercury's orbit.

Venus comes closer to the Earth than any other celestial body except for the moon, a comet, an occasional meteor or an asteroid. At its nearest it is only about 100 times the moon's distance from us.

Venus journeys around the sun at a mean orbital velocity of 21.8 miles (35.1 km) per second and completes one circuit of the orbit in 224.7 Earth days, its sidereal period. Its mean synodic period, say from one inferior conjunction to the next, is 584 days.

Venus rotates on its axis once in 243.01 Earth days but it is possibly misleading to call this a 'Venusian day' since its 'year' is only about 225 Earth days long. The combination of orbital motion and axial rotation period means that 117 Earth days elapse between two consecutive sunrises. On Earth, the sidereal day differs from the solar day by only four minutes whereas on Venus the sidereal day is about double the solar day and its 'year' is only two Venusian days long.

During one synodic period, Venus has close to five solar days so that at every close approach to Earth it shows virtually the same hemisphere. (Between the two successive inferior conjunctions of June 13th 1988 and January 18th 1990 there was a 2.8° central meridian longitude difference. W. R. Corliss (*The Moon and the planets*, 'The Source Book Project', Glenarm, USA, 1985, p. 304) gives a $3\frac{1}{2}$-hour difference from resonance.) The curious resonance in the spin cycle of

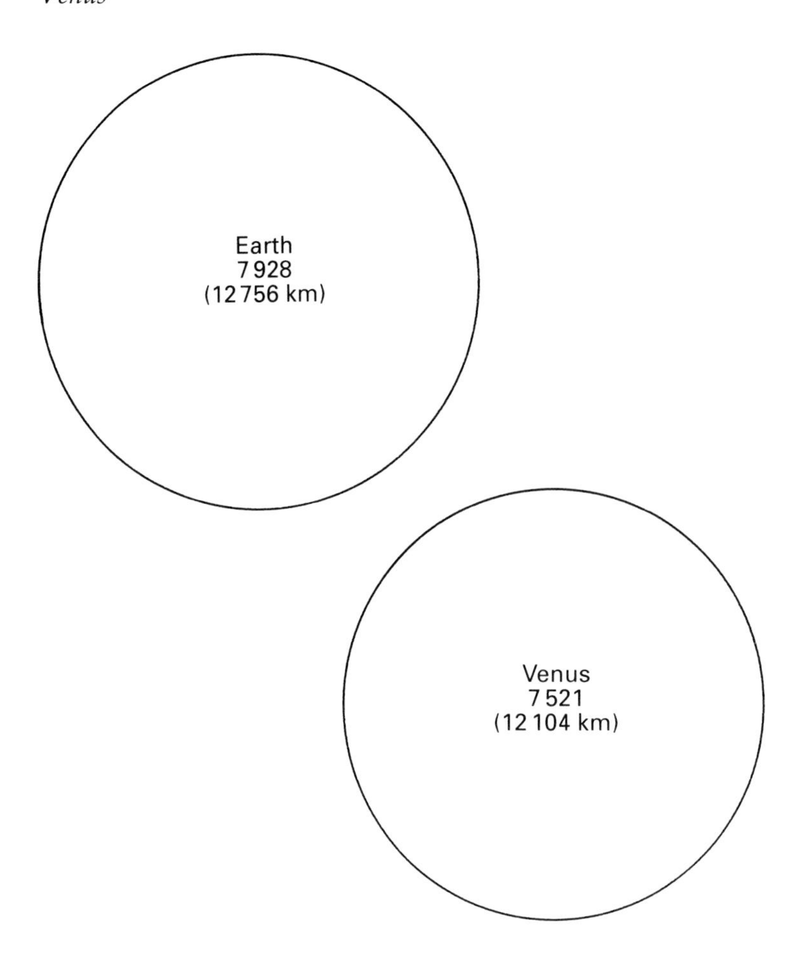

Fig. 6.1 Comparative sizes of the Earth and Venus (Equatorial diameters in miles).

Venus may be coincidence; no completely satisfactory explanation for it has been put forward. One is reminded of the resonance between the Earth and moon which keeps the moon's same face always towards us. This is not a good analogy, however, because over a long time period, Venus presents all longitudes to the Earth.

Like Mercury, Venus is seen first on one side of the sun and then on the other at morning and evening apparitions when it is at its western and eastern elongations respectively, i.e., as a morning or evening 'star' (fig. 6.2). Being an inferior planet it also shows phases similar to those of Mercury but the apparent variations in angular size are much greater and the apparent brightness correlates with the phase in a totally different way from that of Mercury (fig. 6.3); for example, Venus is brightest when it is a crescent 40 seconds of arc in diameter. Next to the sun and moon Venus is the most brilliant object in the sky, outshining the planet Jupiter. A typical telescopic view of Venus at inferior conjunction is shown in the drawing by R. M. Baum in fig. 6.4. The time intervals that separate the conjunctions and elongations of Venus are shown in fig. 6.5.

The orbit of Venus is tilted at an angle of 3°24″ to the plane of the Ecliptic.

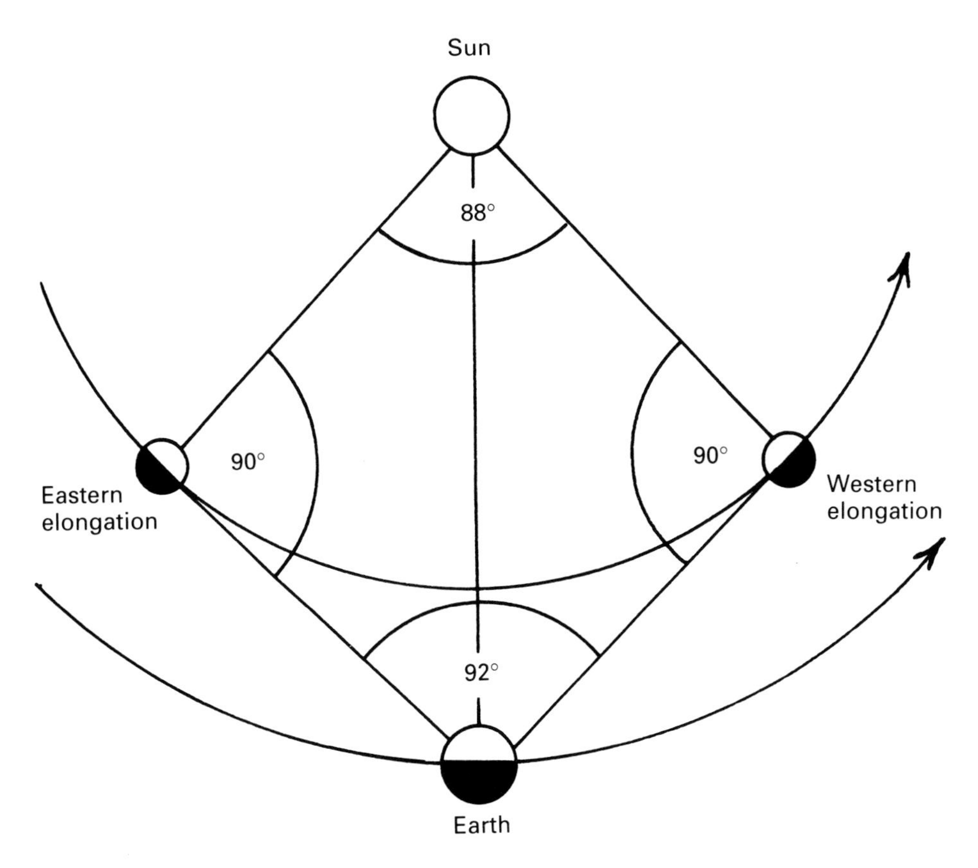

Fig. 6.2 Eastern and western elongations of Venus.

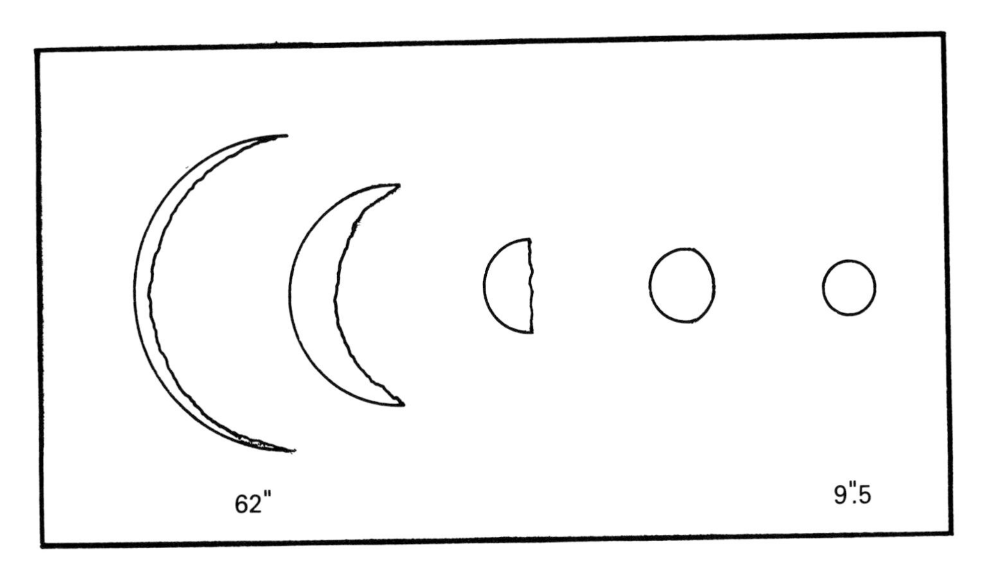

Fig. 6.3 Changes in the apparent angular size of Venus at different phases. The apparent angular size of Venus increases to 63–64" as it approaches inferior conjunction but the sun's glare makes it almost impossible to observe.

Fig. 6.4 Venus prior to inferior conjunction. March 27th, 1977, 18.15 UT, Se 5/10, Tr 5/5, 115 mm OG, ×786 (R.M. Baum.)

Venus, like Mercury, may sometimes be seen passing across the sun's disc at inferior conjunction as seen from Earth but these transits of Venus are much rarer than transits of Mercury.

Unlike Mercury, Venus has a substantial cloud-laden atmosphere. The sunlight reflective power of the cloud layer is what gives Venus its great brilliancy. Like Mercury Venus has no satellites.

The inclination of the rotational axis of Venus is 177° and, as earlier mentioned, the retrograde rotation period on its axis is equal to 243.01 Earth days. A complete rotation of the clouds in its atmosphere, however, takes only about 4 Earth days. The long period of its axial rotation combined with its direction retrograde to the orbital motion makes the solar day on Venus equal to 116.7 Earth days.

History of observation

Venus, like Mercury, was known to the ancients who perceived that it was the brightest of the 'wandering stars'; indeed, at times Venus can be bright enough to cast a perceptible shadow. The brightness of Venus not surprisingly led the ancient watchers of the sky to attach importance to the planet but as in the case of Mercury, they did not at first realise that the morning and evening star appari-

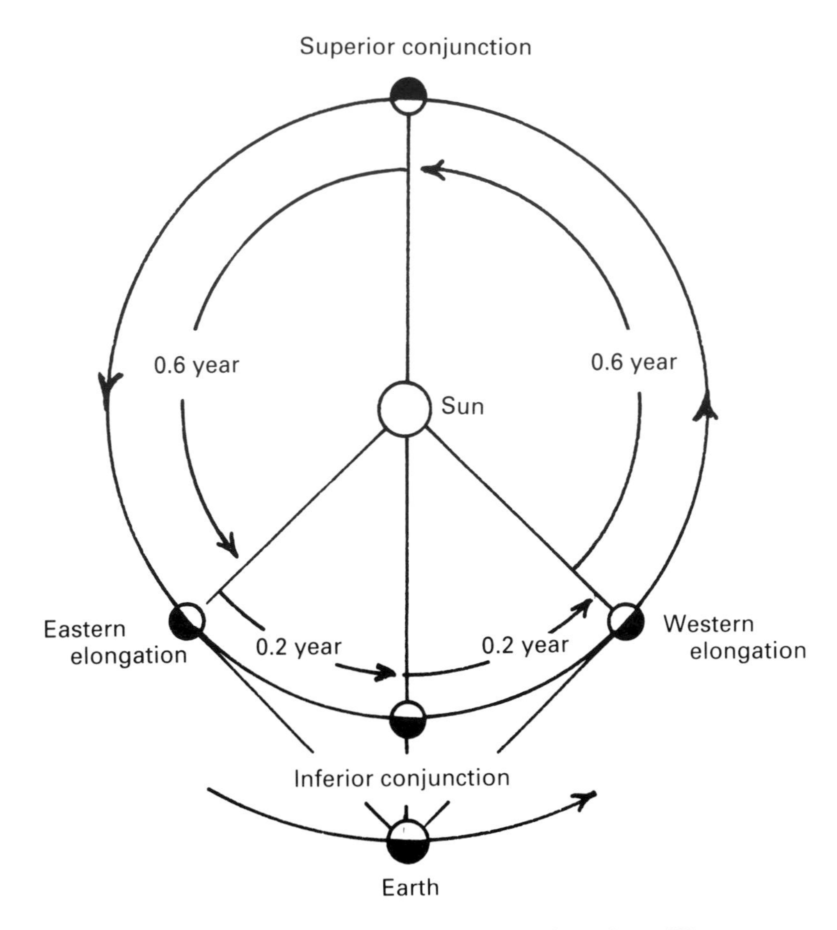

Fig. 6.5 Time intervals separating conjunctions and elongations of Venus.

tions of Venus were one and the same object. (Mercury and Venus being inner planets could not be seen throughout the night as could the other three planets familiar to the peoples of the ancient world – Mars, Jupiter and Saturn).

The evening star was named Vesper and the morning star Lucifer by the Romans; the names used by the Greeks were respectively Hesperus and Phosphorus. It was the Greeks who finally realised that the evening and morning stars were one and the same object and thereafter called it Cytherea after the goddess of love. The Romans subsequently gave it the name Venus which has remained in use up to the present.

The ancient Babylonians watched Venus carefully, noting that it was seen in the eastern sky before sunrise and then vanished for three months and was seen again after sunset in the west. Venus was then noted to stay in the western sky for a little longer than eight months to vanish again for roughly a week before reappearing in the eastern sky for a little over eight months.

There is some evidence that in the clear air of ancient Babylon sky watchers might have perceived the crescent phase of Venus when it was close to its inferior conjunction with the Sun; certain Babylonian cuneiform scripts refer to Venus as being banana-shaped at times.

Galileo was the first to observe the phases of Venus (1610) with the then recently invented telescope.

The planets Mercury and Pluto are unaccompanied by satellites and we know now that Venus also has no attendant body yet for many years there were reports of an alleged satellite. On August 18th, 1686 at 4.15 am, G. D. Cassini reported observing what looked like a satellite of Venus at a distance from it of about three-fifths of the diameter of Venus. He thought he saw it exhibiting the same gibbous phase as Venus and its diameter was about one quarter that of the planet. He earlier records a similar observation on January 25th 1672 from 6.52 am to 7.02 am. Venus was a crescent on this occasion and likewise the satellite. Even earlier than this in 1645 on November 15th, Fontana mentions a similar sighting. James Short the instrument maker saw the satellite on October 23rd 1740 with low power on a reflecting telescope of $16\frac{1}{2}$-inch focus. When the power was increased to 240× he saw the satellite exhibiting the same phase as Venus. He estimated the satellite's diameter to be about one-third that of Venus. He had it in view several times during the hour-long observation but he never saw it on subsequent clear mornings or ever again. Tobias Mayer saw the satellite on May 20th 1759 at 8.45 am. The reality of the satellite appeared to be confirmed two years later by the German observer A. Scheuten. He maintained that during the transit of Venus of June 6th 1761 he saw a tiny black dot attending Venus as it crossed the sun's disc and it remained visible after Venus had passed off the sun. Another observer, Samuel Dunn, specifically looked for a satellite during this transit but saw none. A most convincing set of observations was made by Montaigne who first saw the satellite on May 3rd 1761 in the crescent phase as also was Venus and appearing about one-quarter the diameter of Venus. He saw the satellite again on each clear night subsequently (May 4th, 7th, 11th) in different positions and still showing the same phase. Montaigne stated that he had taken precautions against this being an optical illusion and could still see the attendant of Venus in his telescope even when Venus was out of the field of view. Others who reported seeing the satellite were Roedkiaer on March 3rd and 4th 1764 and Horrebow on March 10th and 11th, both from Copenhagen and Montbaron on March 28th and 29th at Auxerre. Schröter, Herschel and Gruithuisen never saw the satellite. In 1766 Maximillian Hell declared the satellite sightings to be optical illusions and so did Boscovich in 1767. That the satellite might be an asteroid was put forward as a possible explanation by Von Ende in 1811 and this idea was again considered by Bertrand in 1875.

The observations of the 'phantom' satellite were carefully analysed by P. Stroobant in 1887. Some he totally rejected; Montaigne's observations were declared to be ghost reflections in his telescope. On his own admission one of the telescopes used by Wargentin, an eighteenth-century Swedish astronomer, always showed spurious companions to Venus or any other bright celestial object. Other observations were stated to be of faint stars. Roedkiaer may have seen the then undiscovered planet Uranus and mistaken it for a Venusian satellite.

Since it is known that Venus has no satellite the sightings of Cassini and Short must have been due to telescopic ghost images, optical illusions or a nova; it seems odd that such competent observers should have been so easily misled. The mystery is why some observers saw it whereas others did not. One idea put forward by Admiral Smyth was that the satellite was very minute and that parts of its surface may be much less reflective than others. Schorr stated that the satellite varied in brightness and was usually too faint to be seen. Thus the many failures to see it could be accounted for. In 1884 Houzeau suggested that the sightings were not of a satellite but of a planet that he suggested be named Neith, just outside the orbit of Venus.

It is worth mentioning an observation made by E. E. Barnard on August 13th 1892 using the 36-inch Lick refractor. He saw a star-like point of light of about the seventh magnitude in the same field of view as Venus. This does not appear to have been a ghost reflection. Barnard ascertained the position of the object and found that it did not agree with that of any star known. It was suggested by Professor F. J. M. Stratton of the Solar Physics Observatory, Cambridge, England, and later by Joseph Ashbrook, that Barnard may have seen a nova and was the only one who did.

A true satellite of Venus of fair size would most certainly have been detected by space probes yet none has ever been found. We are forced to conclude that Venus, like Mercury, is without an attendant and that no such planet as Neith exists.

It was not until the post-Renaissance period in Europe that serious efforts were made to explain the movements of Venus on a logical and mathematical basis so that future apparitions of the planet could be accurately predicted. The solution of the problem took a long time.

Venus, like Mercury, occasionally transits the sun but the transits are much rarer than those of Mercury. In 1627, Johannes Kepler completed what turned out to be his last great work; this was the compilation of a series of tables of planetary movements that were more accurate than any previous ones. In honour of his benefactor Rudolph II he named them the Rudolphine Tables. He showed that Mercury and Venus would both transit the sun in 1631. Kepler died in 1630 but the French astronomer Pierre Gassendi succeeded in observing the transit of Mercury. Unfortunately he was unable to observe the transit of Venus which occurred at a time when the sun was below the horizon as seen from France during the night of December 6–7th in that year.

According to Kepler no more transits of Venus would occur until 1761. However, new calculations were made by Jeremiah Horrocks, a young English amateur astronomer, which showed that a transit of Venus would occur on November 24th 1639 (December 4th on the Gregorian Calendar). Mr Horrocks was also a clergyman (he was curate of Hoole in Lancashire) and the transit date was on a Sunday. He had set up a telescope in readiness in his observatory to view the transit by projection and in between church services he looked for the tiny disc of Venus on the sun's image but saw none. Clouds had covered the sun for most of the day but cleared at about 2.45 pm. He went to his observatory and to his great joy saw the tiny black disc of Venus on the sun. Horrocks was thus the first ever to record observing a transit of Venus. He made valuable measurements during the transit and in his last year – he died at the tragically early age of 21 in 1641 – also did valuable work on Jupiter, Saturn, comets and the tides. His papers were edited by a Dr Wallace.

Horrocks had a friend, William Crabtree, who lived near Manchester and a brother Jonas who lived near Liverpool, both of whom he notified of the time of the transit. Crabtree saw the transit and confirmed Horrocks's observations for he had set up a telescope in a room over his shop to observe it by projection

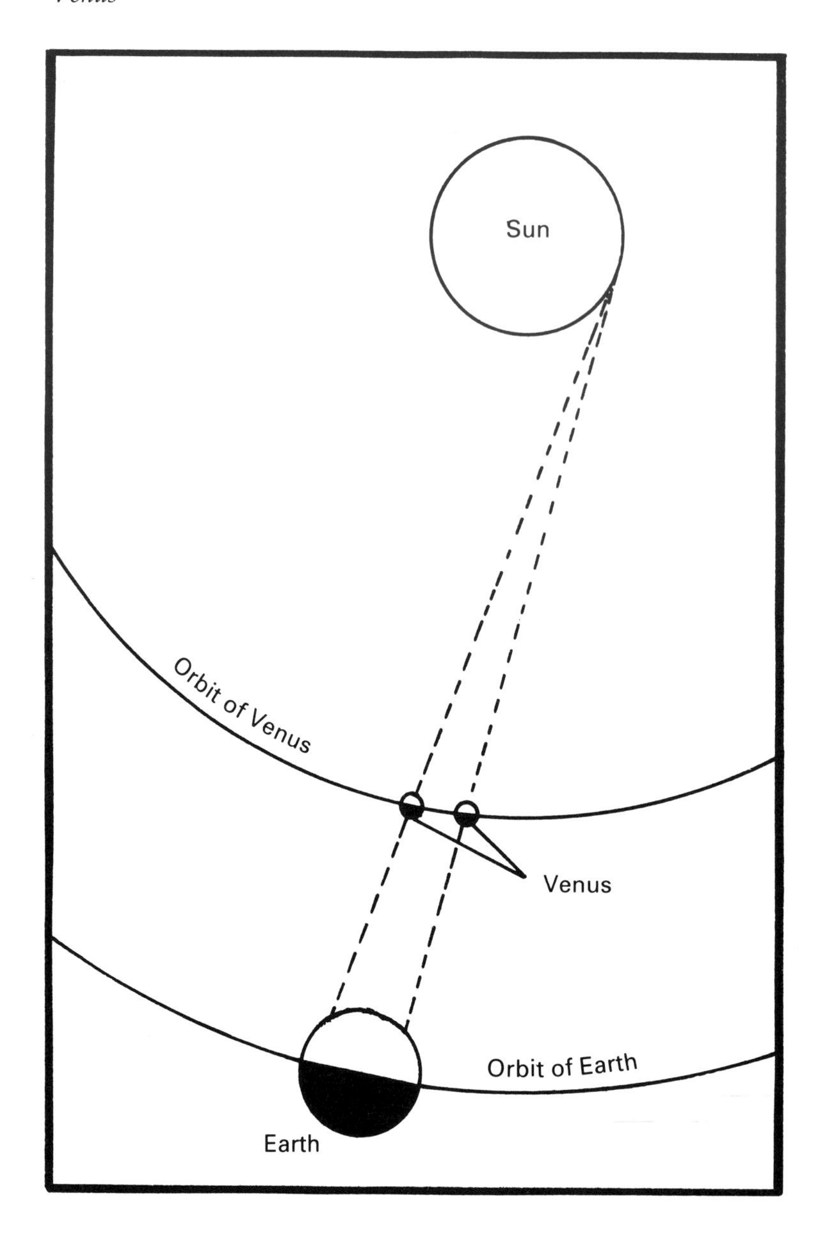

Fig. 6.6 Measuring the Earth's distance from the Sun during a transit of Venus.

which he witnessed at 4 pm. The event is depicted in a charming painting on a wall in Manchester Town Hall. Crabtree is shown staring enthralled at the projected image of the sun with Venus in transit; he is apparently overawed by the spectacle as both his hands are clasped over his heart. Behind him sits his wife also absorbed by the transit, her forgotten knitting in her left hand while at the same time she is apparently pacifying her two excitable children who seem not to realise what it is that is absorbing their parents' attention.

In Westminster Abbey there is a marble scroll on the monument in the nave at its western end which memorialises the event, recording the simultaneous observation of the transit by both Horrocks and Crabtree.

During the next century transits of Venus assumed considerable importance because they afforded astronomers a means by which they felt that they could accurately measure the Earth's distance from the sun. The dimensions of the Earth were known with considerable accuracy and estimates of the solar parallax, the semidiameter of the Earth as seen from the sun, would therefore enable the Earth's distance from the sun to be calculated. Unfortunately transits of Venus are very rare. They occur in close pairs 18 years apart and are separated by more than 100 years. The English astronomer Edmund Halley predicted the pair of transits occurring on June 6th 1761 and June 3rd 1769 which would be only partially visible from Europe. He appealed for careful timings of the transits to be made all over the world. What was required was accurate recording of the exact time, or at least the time difference, from widely separated localities when the disc of Venus was completely on the Sun's disc and the time it touched the sun's disc at third contact, i.e., when the edge of the planet's disc touches the sun's limb as it is about to egress. The time differences would be due to the parallax effect (fig. 6.6) and the distance from the Earth to the sun could be calculated from these timings. It was also necessary to know one's latitude and longitude, which was a problem in those days, and the size of the Earth. However, this and later observations of transits were spoiled by the 'black drop' effect (fig. 6.7) which made accurate timings impossible. This phenomenon is an optical effect caused by the atmosphere of Venus and also by the limited resolution of the telescope employed. We have already noted a similar effect in transits of Mercury but it is not as pronounced.

In spite of this difficulty and the uncertainty of the exact longitude of the stations from where the timings were made, the distance of the Earth from the sun

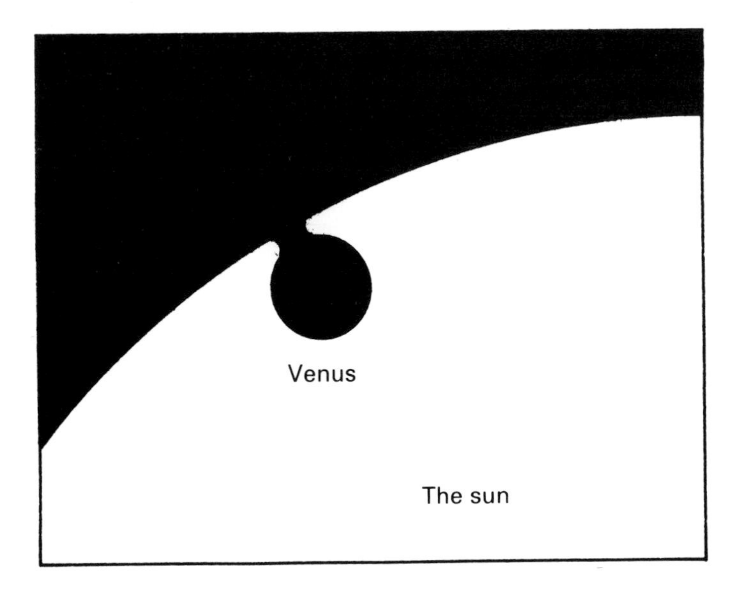

Fig. 6.7 The 'black drop'.

was calculated more accurately than previously although Halley had anticipated and hoped for greater precision.

These same transits provided evidence for the presence of a substantial atmosphere around Venus. M. V. Lomonosov, a Russian astronomer, while observing the transit of 1761 saw that as first contact was about to commence the edge of sun's disc apparently extended out towards Venus. He simultaneously noticed a pale halo around the planet. Lomonosov concluded that an atmosphere surrounding Venus was responsible for these effects and he was later shown to be correct.

Although the black drop phenomenon renders accurate timings of transits of Venus impossible, preparations were nevertheless made for the next two transits, those of 1874 and 1882. More reliance was placed upon micrometric and photographic methods of recording and observing. As was done previously, many observing stations were set up and a large number of sets of measurements were made but once again, the results were disappointing. The photographic results of the 1874 transit were not very good. From the whole mass of contact observations made of this transit three differing values of the solar parallax were calculated by different observers: the official value was 8.76 seconds of arc (Airy), 8.81 seconds was calculated by Tupman and 8.88 seconds by Stone. Less reliance was placed on photography in the 1878 transit and from the results Newcomb deduced a solar parallax value of 8.857 seconds of arc 40.016 seconds. Subsequently transits of Venus were no longer considered of such great importance in measuring the solar parallax as it was anticipated that less costly methods would give better results. There are many literature references to the transits of 1874 and 1882 some of which are listed in the 'Further reading' section at the end of this chapter.

Many years later Schröter noticed that as Venus approached inferior conjunction, the crescent being very thin, it seemed that the horns of the crescent reached much further around the disc of Venus that would be expected from spherical geometry. This too indicates the presence of an atmosphere and supports Lomonosov's conclusion. In 1790 Schröter saw 'a faint gleam' beyond the cusps of the crescent and continuing their curve which was a true twilight effect due to an atmosphere.

Mädler at Dorpat in 1849 observed the horns of the crescent extending to 200° and even to 240°. The horns of the delicate crescent of Venus when very close to inferior conjunction have been seen so extended at times that the dark disc of the planet is completely encircled by a thin circle of light. This can only be caused by refraction and/or scattering of the sun's light by an atmosphere surrounding Venus. This appearance was first noticed by David Rittenhouse who observed the transit of Venus of June 3rd 1769 from Philadelphia. The light ring encircling Venus was seen just after first contact when half of the disc was still off the sun's disc and was seen surrounded by a circle of light.

Then on December 11th 1866, another American observer, Professor C. S. Lyman of Yale College, observed Venus at inferior conjunction with a 9-inch refractor. Venus was unusually close to the line joining the planet to the Earth. Lyman saw the disc of Venus completely surrounded by a circle of light which narrowed to a mere thread on the edge opposite the sun. It seems that this was the only time that this phenomenon, i.e., 360° of illumination, was seen since Rittenhouses's observation of 1769. It has been seen and even photographed many times since then. Nelson calculated from these data that the atmosphere of Venus must approach nearly double the density of the Earth's.

Rittenhouse's observation of the luminous ring at the 1769 transit and Lyman's in 1866 were apparently overlooked because when the phenomenon was seen at the transit of 1874, astronomers were surprised. These appearances at transits of Venus are shown in fig. 6.8.

Another interesting phenomenon exhibited by Venus is the 'ashen light' (fig. 6.9) which when seen at the thin crescent phase of Venus resembles the 'Old moon in the new moon's arms'. The latter is due to the illumination of the dark part of the moon when in the thin crescent phase by sunlight reflected from the Earth, but the cause of the ashen light of Venus is something of a mystery. Schröter reported seeing it in 1806 and that the whole of the dark part of the disc of Venus was seen as an ash-coloured light. The ashen light has also been seen at the gibbous phase of Venus. Some observers are sceptical about the reality of the ashen light because they have been unable to see it in large telescopes while many others using telescopes in the 3–12-inch range have occasionally reported seeing it. It is much too elusive to photograph. Independent confirmation of the phenomenon by several different observers is quite good, however.

The ashen light appears to have first been seen by Giovanni Riccioli in 1643 and was mentioned in 1715 by Derham in his *Astro theology* and then again by an assistant astronomer named Kirch at the Berlin Academy of Sciences on June 7th 1721. It was faintly seen by Herschel on March 8th 1726 and was even seen in daylight by Andreas Meyer on October 20th 1759 a little after noon, and again in daylight and before noon but less distinctly by Winnecke on September 25th 1871. Between January 24th and March 1st of 1806 the phenomenon was seen by

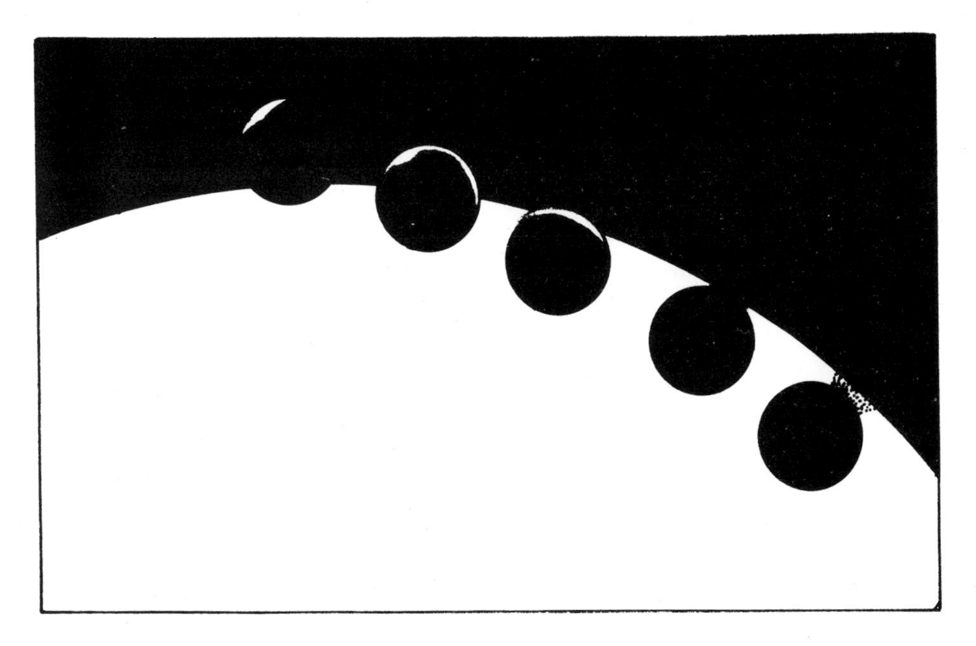

Fig. 6.8 Refraction effects of the atmosphere of Venus as seen during a transit. (Redrawn from Splendour of the Heavens. *Hutchinson, London, 1923.)*

Fig. 6.9 The ashen light: March 29th, 1977, 19.10 UT, Se 6/10, Tr 415 115 mm OG, ×186. (R.M. Baum.)

Harding at Göttingen on three occasions and it was seen many times in the years following by several others. The ashen light was seen by Baron van Ertborn when Venus was close to inferior conjunction in 1876 on July 9th, 13th and 14th.

The colour of the ashen light is variously reported as grey or reddish. Webb reports seeing it in 1878 on January 31st after many unsuccessful attempts in previous years. He used powers of 90× and 212× on his 9.38-inch reflector and saw the ashen light as having a brownish colour. He recommended using a bar in the field of view to obscure the bright part of the disc of Venus.

The noted English planetary observer Richard M. Baum has observed the ashen light phenomenon for over 35 years. He says that it is 'a glow variously described as coppery, deep purple, olive, or even rusty brown.' Often the glow is uniform but occasionally it is patchy. Baum raw the rare copper-coloured hue of the ashen light in 1953 and rendered it in the form of a beautiful realistic drawing published in the August 1988 issue of *Astronomy* magazine.

The ashen light has been attributed to light reflected from the Earth, which would be much too faint. Several authorities think that it may be due to auroral effects as was stated by Arago. In support of Arago's idea was the claim by one astronomer that the ashen light was most pronounced at times when the sun was most active. However, the 'aurora theory' of the ashen light seemed less believable when the Mariner II spacecraft revealed that Venus had no appreciable magnetic field. Many astronomers in the 1930s considered that the ashen light was not real after they had studied observational reports. The origin of the ashen light is still undecided; one theory considers that it might be due to recombination of molecules dissociated by sunlight energy.

The spectrum of the dark side of Venus was studied by Kozyrev in the 1950s and he found indications of ionised gases, so that Venus would have an airglow like the Earth's but 50 times brighter.

Curiously, some have reported seeing the dark side of Venus darker than the dark sky background in seeing conditions when it would be hard to accept it as a contrast effect. This appearance may have been due to the dark part of Venus being silhouetted against the sun's outer corona or against sunlight-scattering dust beyond Venus, something like the zodiacal light.

The surface features of Venus cannot be seen because of the thick atmospheric cloud layer. The centuries of telescopic study of Venus following Galileo's discovery of the phases revealed little about Venus. A dark spot on the planet was reported by Fontana in 1645 and bright and dark spots were seen by G. D. Cassini in 1666 and 1667. Four of Cassini's drawings of Venus are shown in fig. 6.10. Bianchini also made drawings of Venus (fig. 6.11) and constructed a map in 1726–7. Herschel saw dusky markings near the terminator on June 9th 1780

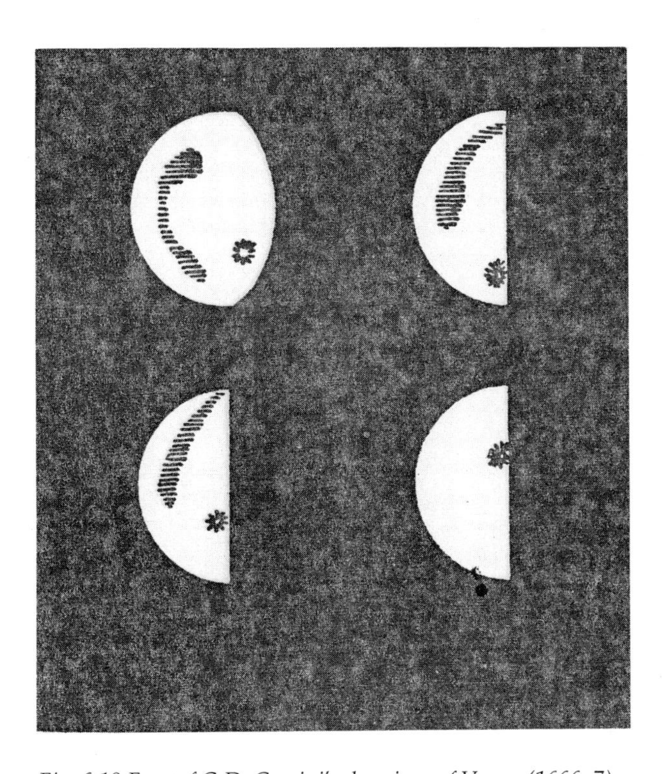

Fig. 6.10 Four of G.D. Cassini's drawings of Venus (1666–7).

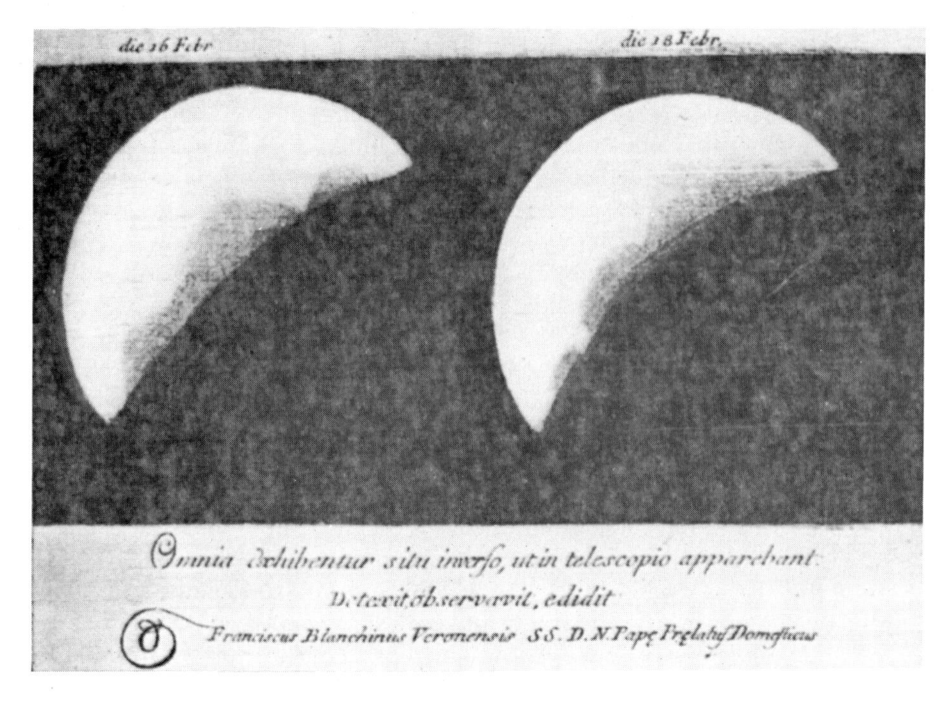

Fig. 6.11 Observational drawings of Venus by Bianchini (1727).

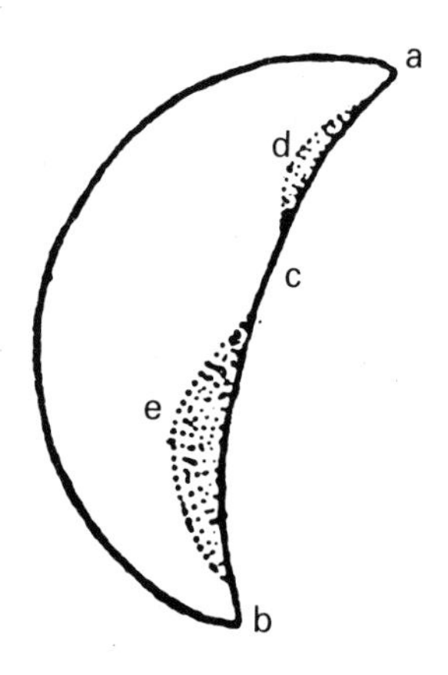

Fig. 6.12 Dusky markings on Venus as seen by Herschel (1780).

which he recorded in a drawing (fig. 6.12). Schröter saw a faint oblique streak on the disc on August 29th 1801 (according to Moore, February 28th 1788) and Gruithuisen several times saw long vertical shadow-like markings. On April 6th 1868 With saw a curious appearance of the entire disc which appeared to be marbled or veined. At the same time he noticed a small bright protuberance on the

Fig. 6.13 Dusky markings on Venus seen by H. McEwen, June 7th, 1892. (From Observational Astronomy, *Mee, A., D. Owen and Co. Ltd., Cardiff, 1893.)*

limb (not terminator); he saw this with an unsilvered 12¼-inch mirror and the observation was confirmed six days later by Key. Dusky markings as seen by H. McEwen are shown in his sketch of June 7th, 1892 (fig. 6.13). Webb mentions many other observations by different observers but these are too numerous to describe here.

In 1813 Gruithuisen reported seeing white patches at the poles of Venus and subsequently they were seen by others. Successful photography of white pole spots was claimed by M. Quenisset in 1908. There is some doubt as to whether these 'polar spots' are actually at the poles.

Some astronomers who observed Venus through yellow filters claimed to have seen faint markings of low contrast and thought that these might be short-lived sightings of the surface through transient clearings in the clouds. However, it is now known that Venus is covered with multiple layers of dense cloud so that the surface can never be seen. Another interpretation was that these markings were indications of groupings of clouds or other meteorological phenomena induced by the influence of high mountains on Venus. Since the clouds extend up to 68 km and because the maximum surface relief is only 13 km, this explanation is unlikely.

Percival Lowell reported seeing linear features on Venus, in particular a pattern of linear dusky features radiating from the subsolar point like the spokes of a wheel and was first to sketch them in 1896 (fig. 6.14). His observations were verified by Foulkes in 1897 observing from Malta. Griffiths saw and sketched the spoke system on February 17th 1897 (fig. 6.15).

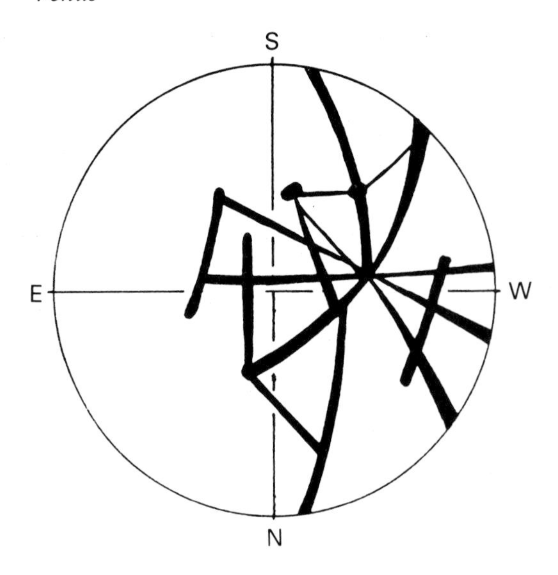

Fig. 6.14 Spoke-like dusky markings on Venus as seen and drawn by Percival Lowell in 1896. (From The Moon and the Planets, *Corliss, W.R., The Sourcebook Project, 1985.)*

Fig. 6.15 The spoke system of Venus sketched by Griffith, February 17th, 1897. (From The Moon and the Planets, *Corliss, W.R., The Sourcebook Project, 1985.)*

Further observations of the spoke system made at the Lowell Observatory in 1898 are shown in fig. 6.16. Antoniadi never saw the spokes and denied their existence. No one else seems to have studied the spokes or even remembered them until Baum 'rediscovered' them in 1951. He saw them as Lowell did and detected a dark spot at the subsolar point (fig. 6.17). What is interesting is that Baum had not read any of Schiaparelli's or Lowell's writings on Venus which argues strongly for the reliability of his observation. He was able to keep the

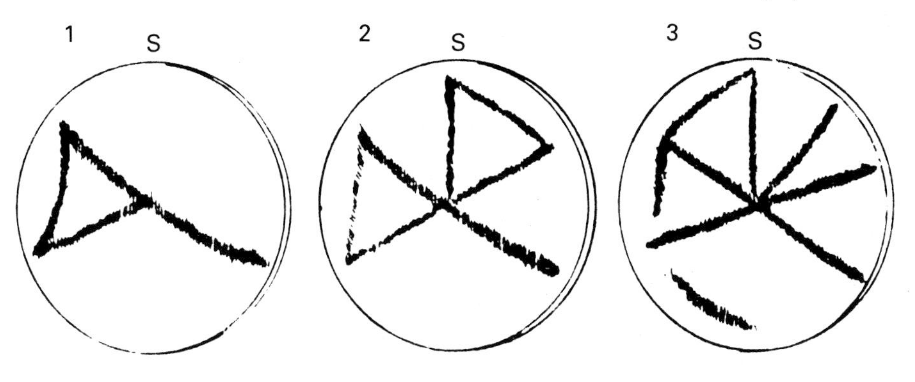

Fig. 6.16 Observations of the spoke system of Venus made at the Lowell observatory in 1898. (From The Moon and the Planets, *Corliss, W.R., The Sourcebook Project, 1985.)*

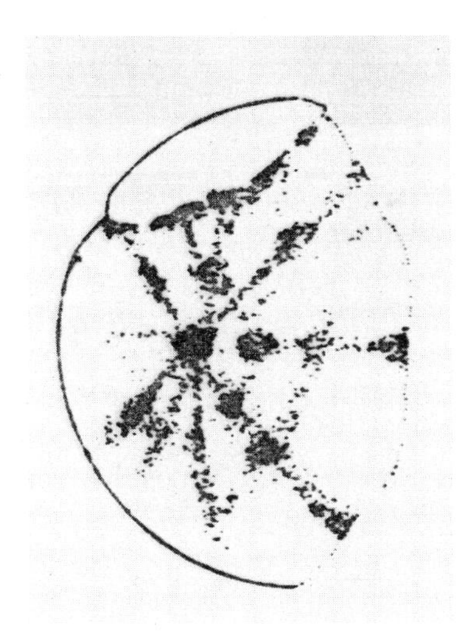

Fig. 6.17 Radiating spokes on Venus drawn by R.M. Baum, April 26th, 1951. (From The Moon and the Planets, *Corliss, W.R., The Sourcebook Project, 1985.)*

spokes under observation for about two months and at times perceived that they had a yellowish-grey colour.

Not every observer is able to see the spokes and there has been difference of opinion about whether they are real or merely an optical illusion. The objectivity of the radiating spokes was demonstrated by ultraviolet photography of Venus in 1927 with the Mount Wilson 60-inch reflector which revealed dusky bands somewhat radially arranged to and originating from the subsolar point. The streaks showed much similarity to the drawings of Schiaparelli, Lowell and several others.

Astronomers have viewed Venus telescopically through different kinds of colour filters. Many observers have seen markings especially with violet or blue filters as recorded in the Venus Section reports of both the ALPO and the BAA. Usually, nothing is revealed when the planet is photographed in visible or infrared light but diffuse markings become visible in ultraviolet which must be due to ultraviolet absorbing material in the clouds. Yellow light photography by Camichel in 1943 revealed vague features.

A highly significant observation in 1957 was Boyer's detection of markings in the cloud envelope of Venus when he observed the planet in ultraviolet light. Further study revealed that the markings changed shape and apparently make a complete rotation around Venus in a time equal to 4 Earth days. Oddly, this was a retrograde motion meaning that it was the opposite of what we see in the Earth's pattern and planetary rotation in general. Peculiar C-, Y- and psi-shaped markings have subsequently been seen in ultraviolet light photographs of Venus.

In 1789 Schröter said that he had observed that the south horn of the crescent of Venus was blunted and he saw a tiny luminous point in the dark sky beyond it which he considered was due to a mountain of prodigious size near the south pole illuminated by the sun (fig. 6.18). He also stated that he saw terminator irregularities. In 1833 and 1836 Beer and Mädler fully confirmed the accuracy of these observations of terminator irregularities but Herschel doubted Schröter's observations.

Webb mentions numerous other observations of terminator and cusp irregularities between the years 1700 and 1876. Such phenomena have been said to indicate that the surface of Venus is rugged to an extent that can interfere with the cloud patterns but this is unlikely as it is now known that the clouds are too far above the surface.

Like Mercury, Venus exhibits a phase anomaly, the calculated and observed times of dichotomy often being discrepant by several days. This was first noticed by Schröter in 1793 and has become known as the Schröter effect. In 1692 Cassini noted that the crescent was apparently too broad when it should have been thinner.

In 1932, Dunham and Adams searched for water vapour in the atmosphere of Venus by studying the spectrum of the planet near 8000 angstroms. Although

Fig. 6.18 Schröter's 'enlightened mountain' (1789–91º.

they found no indications of water vapour there was evidence of carbon dioxide as shown by absorption lines at 7820, 7883 and 8689 angstroms. Kuiper found additional spectroscopic evidence for the presence of carbon dioxide in 1962. A tiny amount of water vapour was detected by Bottema, Plummer, Strong and Zander in 1964.

Carbon monoxide was found by Sinton in 1961 in trace amounts but oxygen was not certainly identified from studies carried out from high altitude balloons in the early 1960s although traces of water vapour were detected.

In 1956 radio waves of wavelength 0.32 cm were detected coming from Venus and indicated a surface temperature of 300° C. Only in 1968 was there general recognition that the surface temperature was at least 480° C as shown by radio and radar observations. The atmospheric pressure at the surface was believed to be at least 90 times that of our own at sea level.

In 1970 radar astronomy was used to chart a large part of the surface of Venus. This was achieved with the 210-foot dish antenna of the Goldstone Tracking Station by astronomers from the Jet Propulsion Laboratory of the California Institute of Technology. About one-sixth of the total surface was covered by one of their maps extending from about 90° W and 30° E longitude and from 45° S and 35° N latitude. The resolution was about double that of the first map made in 1968. Radar studies continued in 1972 and revealed large crater-like depressions in the equatorial region and a huge valley.

By 1975 radar images with even better resolution were obtained by the National Astronomy and Ionosphere Center at Arecibo in Puerto Rico with the 1000-foot radio telescope dish. Among the surface markings recorded was a large bright feature named Maxwell.

Space probe exploration of Venus

Several space craft for exploration of Venus have been launched. The US Mariner 2 spacecraft was launched in August, 1962. Data from the radiometers on board indicated no significant difference between the temperatures of the day and night sides of Venus. The temperature of the tops of the clouds was found to be 250 °C and about 234 °C deeper down. The surface temperature appeared to be 400 K. No magnetic field was detected.

The Soviets launched a series of 20 spacecraft including the Veneras 1–16 between 1961 and 1983. They were designed to soft-land on Venus. Except for Veneras 1–3 the entire series were almost all successful, especially the Venera 9 and 10 landings in 1975, in sending back important data relating to the atmosphere, clouds and surface of Venus.

The US Mariner 5 spacecraft was launched on June 14th 1967. The Soviet space probe Venera 4 was launched two days earlier on June 12th. Venera 4 reached Venus one day sooner than the Mariner 5 craft on October 18th 1967. Signals sent from Venera 4 were received by the large radio telescope at Jodrell Bank and the data indicated that cloud temperatures of Venus varied from 40° C to 275° C and that the surface atmospheric pressure was at least 15 times that of Earth's The atmosphere contained at least 98.5% carbon dioxide. Mariner 5 flew above the surface of Venus on October 19th 1967 at a height of about 2480 miles. In general, the data obtained from Mariner 5 confirmed those of Venera 4 although there were some discrepancies. One thing it revealed was that the shape of Venus is not as oblate as was one thought; in fact it is more nearly spherical than the Earth.

On November 3rd 1973 the Mariner 10 spacecraft was launched. Its primary mission was to study Mercury but it would also closely fly past Venus. Previous US Venus space probes had not recorded images of the surface of Venus, but Mariner 10 had a camera on board. On February 5th 1974, the date of the flyby, Mariner 10 took pictures of the sunlit side of Venus in ultraviolet light that revealed much fine detail of the cloud patterns of Venus and the polar hood. These pictures confirmed the existence of the so-called C-, Y- and psi-shaped markings that are barely visible in Earth-based photographs. Further pictures taken during the days following the close flyby also confirmed the 4-day rotation period indicated by Earth-based observation. Owing to tight budgets, the US Venus exploration programme had to be limited to two flights, a multiprobe, consisting of the basic bus, a large probe and three similar small probes and an orbiter in 1978 (Pioneer Venus 1). The probes analysed the major and minor gas components of the upper and lower atmosphere of Venus, temperature, pressure and density all over Venus and vertically from the surface up through the atmosphere. Two Pioneer photographs of Venus are shown in figs. 6.19 and 6.20. The runaway 'greenhouse effect' was confirmed. Details of cloud structure both globally and vertically were determined and what appeared to be lightning was detected in Venus's atmosphere.

The orbiter measured global winds at the cloud tops and radar pictures of surface topography were obtained. The radar mapping suggested that the surface topography of Venus might resemble the Earth's; at any rate it seemed different from the rugged cratered surfaces of the moon, Mars and Mercury. Several geological formations were noted – mountains, large crater-like structures, plains and continents. A sketch map of Venus on Mercator's projection is shown in fig. 6.21.

A beautiful colour contour map of Venus was published in *Astronomy* magazine in the March 1981 issue on pages 12–13 and in many other publications.

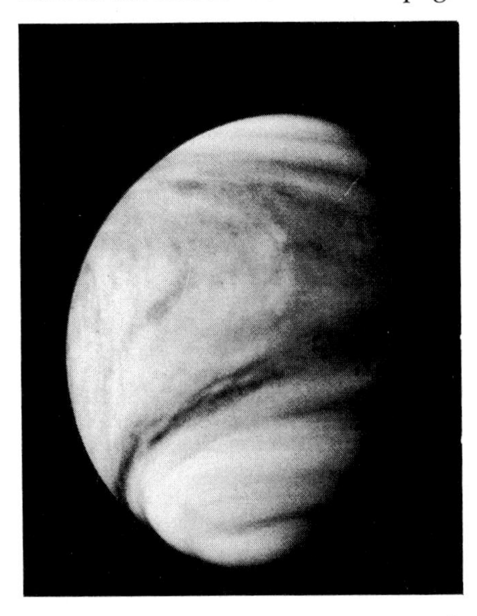

Fig. 6.19 Pioneer Venus Orbiter photograph showing ultraviolet markings.

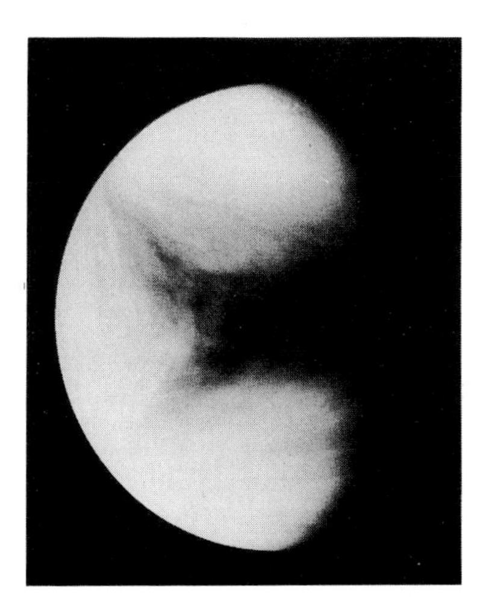

Fig. 6.20 Pioneer Venus Orbiter photograph showing ultraviolet markings.

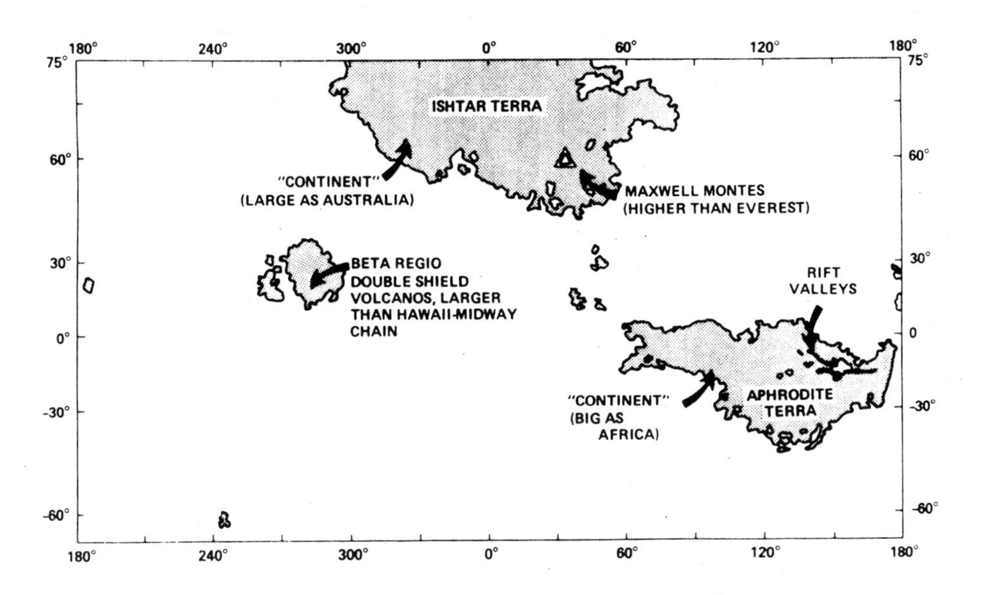

Fig. 6.21 A map of Venus on Mercator's projection, showing the major highland regions (shaded). Ishtar Terra (as large as Australia) looks larger than Aphrodite Terra (as large as Africa) owing to the distorting effect of the map projection of high latitudes. (From Venus – An Errant Twin, *Burgess, E. Colombia University Press, New York, 1985.)*

An atmosphere that rotates faster than the planet even at the equator is known as a superrotating atmosphere. Our Earth, on the other hand, has a subrotating atmosphere. The probes revealed that the rotational speed of Venus's atmosphere falls off rapidly the closer the surface is approached.

The ultraviolet markings are thought to be related to some kind of ultraviolet absorbing medium in the clouds, the nature of which is unknown, although carbon monoxide has been suspected, and nothing is known about its extent.

Observing Venus

Venus is much easier to locate with the naked eye than Mercury owing to its great brilliancy and the much larger angular distance it attains from the sun. Next to the sun and moon it can be the brightest object in the sky, usually even outshining Jupiter. On such occasions it can cast a perceptible shadow.

In contrast to Mercury, Venus becomes brighter as it moves from superior conjunction through dichotomy to about a month before inferior conjunction. It becomes dim at inferior conjunction. This is because the apparent diameter variations of Venus as it goes through its phase cycle are much greater than those of Mercury as previously mentioned and shown in fig. 6.3. Hence the apparent size of the illuminated area of Venus is much greater when it is a crescent than when it is nearly 'full'.

The worst time to observe Venus is when it is brilliantly visible in a dark sky some time after sunset or before sunrise. The disc is so brilliant that false colour effects and flare spoil the telescopic image seen in the most optically perfect telescopes. Also, the low altitude of Venus at such times results in atmospheric turbulence further degrading the image.

Very small telescopes are not much use for observing Venus except for phase estimates; an aperture of at least 4–6 inches should be considered minimal for useful observations to be made.

The best time to observe Venus is while the sun is still above the horizon or if it has just set or is about to rise. The sky is considered too dark if the planet can be seen with the unaided eye.

Whatever kind of telescope is used, only certain kinds of eyepiece should be employed. Those without achromatic corrections are useless and Barlow lenses should be avoided as they may introduce false colour. Perhaps the best types of eyepiece are the Tolles or monocentric. Do not overpress power. Something between about 150× and 250× should be about right for 6- to 10-inch telescopes. For successful observations of Venus it is most important to obtain a cleanly defined sharp image. Higher powers could be used when the disc of Venus looks small when near superior conjunction. It is a pity that the planet has been observationally neglected at this phase.

When making a comparative series of observations of Venus over days or weeks, you should use the same telescope aperture throughout. If two or more telescopes of greatly differing aperture are employed, the drawings or other results obtained with them cannot be compared as reliably as if a single aperture had been used. For similar reasons, the same eyepiece should be used throughout. Further, it is best to be consistent with respect to time of day or brightness of the sky background. If the regular time or sky brightness is deviated from, a note should be made of this to accompany observational drawing or description.

Venus is at its brightest about 40 days after eastern elongation and about 40 days before western elongation. At these times its magnitude increases to anything between about 3.7 and 4.7 at different elongations and it is then the brightest object in the sky next to the sun and moon.

Observers armed only with binoculars ought to be able to pick up Venus when it is at just before or after conjunction and about 5–10° from the sun. Some experienced observers have claimed to be able to see Venus with the naked eye at these times. At inferior conjunction binoculars of good optical quality and mounted on a tripod will sometimes reveal Venus as a very thin crescent with a breadth of only about one second of arc.

Previously it was mentioned that the Babylonians may have seen the crescentic shape of Venus with the naked eye. In at least one instance the shape of Venus has also been described as being like a stretched triangle. It might be interesting to try this test of sight for yourself. The phases of Venus are easily visible in quite small telescopes using only low or moderate powers.

The Venus observer needs similar basic information to that needed for observing Mercury all of which can be found in almanacs and astronomical year books. You need to know the following: RA, declination, the apparent magnitude, apparent diameter (varying from 10 to 64 seconds of arc), the phase angle (or better still, the phase coefficient, K) and the elongation.

Daylight observation of Venus

Observing Venus in the day time is not difficult and is well worth doing. The advantage of daylight observation is that Venus is high in the sky and not close to the horizon as it is at sunset or just after. The effects of atmospheric turbulence and the glare of the planet are thus considerably alleviated.

Whether you use binoculars or an altazimuth mounted telescope, Venus should be fairly easily found in the day time as it will lie a little ahead of or behind the sun on the ecliptic. If you can judge the curve and angle of the ecliptic, finding Venus should be easy simply by 'sweeping' in the area where Venus will be with binoculars or telescope. This can be dangerous as there is always the risk of accidentally pointing your glass at the sun with disastrous results. It is best to stand in the shadow of a building when sweeping for Venus near the sun during the day.

If your telescope is equatorially mounted you can point it at Venus directly by using your setting circles if you know its RA and declination (see the article by J. H. Palmer in the 'Further reading' at the end of this chapter.)

An altazimuth mounted telescope can be pointed at Venus as follows. First, select a star in the night sky with the same declination as Venus; it doesn't have to be exactly the same as long as it is close enough to be seen in the same telescopic field as Venus. The telescope is pointed to this star. Note the time and clamp the telescope so that it cannot move. Now determine the RA of the star from an almanac. You will now be able to calculate the time when Venus will be seen in the telescopic field by using the formula:

$$TV \ 5\ T + (\text{RA of star} - \text{RA of Venus})$$

where T is the local mean time when you pointed the telescope at the star, and TV is the local mean time for observing Venus. (The multiplying factor 0.997 should be placed in front of the bracket if one is calculating mean solar time.) Note that this formula works only if the interval is less than 24 hours. Also, if the RA difference is negative, add 24 hours to it. This formula gives the time when you can expect to see Venus in the field of the clamped telescope.

Daylight observation of Venus can be improved by the use of yellow and red eyepiece filters. These combat the effect of the bright blue sky light in reducing contrast between the planetary disc and the sky. The apodising screen should also be tried with and without filters.

The dusky markings

It is difficult to see the dusky markings on Venus but observers have claimed to have seen them in even a 2- or 3-inch telescope. Some people assert that these markings have no real existence and are illusory. However, a few people possess some visual sensitivity to ultraviolet light and it is probably these who can see the so called 'ultraviolet markings' whose presence has been confirmed by ultraviolet photography. Visual observation with and without filters and photography reveal pronounced time differences in the appearance of the dusky markings which makes it very probable that they are atmospheric phenomena. A typical view of the dusky markings of Venus is shown in Baum's drawings in figs 6–22 and 6–23 of Venus at the crescent phase.

Dusky markings are named and classified as follows:

(1) Banded. These are dusky parallel streaks roughly perpendicular to the line joining the cusps.
(2) Radial. A spoke-like pattern radiating out from the subsolar point.
(3) Irregular. May be elongated or roughly linear dusky streaks. There is no definite pattern.
(4) Amorphous. These have no definite form or pattern.

Observational experience, atmospheric conditions and where the planet is situated in the sky are all important in determining success or failure in observing Venus.

The dusky markings are extremely elusive and it is easy to deceive oneself into thinking that markings have been seen on Venus merely because one expects to see them. There is therefore no imputation of dishonesty when the reliability is questioned of observational drawings of amateurs who use small telescopes and always record markings on Venus whenever they observe the planet. Schröter, whose observational skill may not have been of the best, was nevertheless

Fig. 6.22 Dusky markings seen on Venus, September 29th, 1959, 06.40–07.00 UT, Se 6/10, Tr 415 115mm OG, ×188. (R.M. Baum, Chester.)

Fig. 6.23 Venus, 1954, crescent phase (morning) 115 mm OG, ×186. Left: December 15th, 08.20–08.40 UT, Se 3/10, Tr 5/5. Right: December 17th, 08.30 UT, Se 3/10, Tr 3/5. Note the belt-like features either side of the supposed equator; these apparently equate with ultraviolet features. Also note the fine extensions on December 17th, taking the cusps beyond the geometric norm. (R.M. Baum, Chester.)

scrupulously honest and patient; he observed Venus for 9 years before he was honestly convinced that he had seen a marking on the planet. So don't give up hope if your early attempts are unsuccessful.

The disc of Venus usually appears to be blank apart from a perceptible dimming of light near the terminator. The majority of observations made at elongations of Venus usually reveal a dusky marking or two that appear to be real. The apparitions of Venus last much longer than those of Mercury so that we can follow Venus for weeks, not just days.

If you are lucky enough to see a dusky marking or two, watch them as often and for as long as you can and see if you can detect movement. A blue filter is helpful in these observations but others should be tried as well.

It should be mentioned here that the results of observations of Venus with filters have resulted in controversy. Some observers maintain that filters can render visible otherwise invisible features. Others claim that filters make no difference. The majority of observers find that filters improve contrast but nothing more; no new details are 'brought out'. Since it has been shown that people's eyes differ in sensitivity to different wavelengths of light it is not surprising that opinions differ as to the value of filter observations of Venus.

Most observers of the dusky markings of Venus appear to fall into three groups; those who see large diffuse markings, those that see them mainly as a pattern of streaks and others who lie somewhere in between insofar as they see the two types of marking sometimes in combination and at other times see one or the other type only. Although the spoke-like arrangement of the linear markings is not visible to every observer, too many observers using different kinds of telescopes have seen them for anyone to deny the reliability of the observations. There are plenty of reliable data on the spokes but it is the interpretation of the data that leads to argument.

There still remains the mystery of why some see the spokes, others do not and yet others see them in part. In this connection it is worth mentioning an experiment described by Patrick Moore in the first edition of his *Guide to the planets* (1955). On August 24th,1953 Moore, C. D. Reid and J. B. Hutchings tested their sensitivity to ultraviolet light with a spectrometer. Moore was found to have the lowest wavelength detection limit of 4000 ± 100 Å, Hutchings and Reid 3700 ± 100 Å and Baum 3200 ± 200 Å. Further, on April 21st 1954, Moore, Reid and Baum observed Venus with the same telescope, a 9-inch reflector, with a power of 240×. The results were interesting; Moore saw only a vague greyness and two condensations, Reid saw a Y-shaped marking lying on its side and Baum saw the spoke system. Interestingly, the Y-shaped marking seen by Reid coincided with a part of the spoke system seen by Baum. This clearly demonstrates that the greater the sensitivity of an observer's eyes to the shorter ultraviolet wavelengths, the more will be seen of the dusky ultraviolet markings.

The best way of revealing the dusky markings is by ultraviolet photography. Photographs of Venus taken in ultraviolet light always show details; less frequently details are seen in blue light photographs but with greatly reduced contrast. At longer wavelengths no details are registered. Images obtained in the daytime are better than those secured in a dark sky but bright twilight yields the best photographic images of all. If the sky is bright a high contrast ultraviolet film should be used such as Contrast Process Ortho. Good ultraviolet photographs of Venus have been taken by T. Rackham who used a Chance-Pilkington OX9A filter.

If you have a video camera it is worth attempting to study the elusive dusky markings of Venus using the camera with an ultraviolet filter such as the Kodak 18A. Video photography (videography) is described in detail in chapter 15. The advantage of videography over ordinary photography is that you can actually see the image which makes aiming and focusing so much easier even when using the ultraviolet filter.

The bright markings

The most frequently seen bright markings are the so-called cusp caps or polar spots. These are areas of brightness situated near the apparent poles of Venus. On the whole, they appear to be real. They may remain visible for many days or weeks and they appear and fade out quite slowly. They are more common on the south than the north. Cusp caps are usually brightest when the phase coefficient (k), the fraction of the disc that is lit by the sun, is between 0.8 and 0.1. Occasionally the caps are seen bordered by dark, often diffuse, cusp bands.

Other bright phenomena that should be looked for are light clouds, lines and bands crossing the terminator, a spotted or mottled appearance of the disc and the presence and extent of a light 'limb band'.

In making drawings of the dusky and light markings, the circle representing the disc of Venus should be about $1\frac{1}{2}$–2 inches in diameter. The phase should be accurately delineated. Carefully draw in pencil all shadings or light features. You will probably have to exaggerate the contrast between the dusky markings and the whiteness of the disc. Say so if you have. The north and south poles of Venus should be indicated. When you have finished the drawing, do not add or erase anything or 'touch it up' in any way after leaving the telescope.

The phases

In studying the phases of Venus, there are two lines of profitable investigation: the Schröter effect and terminator irregularities.

The times of theoretical dichotomy ('half moon' phase) are calculable with considerable precision yet it invariably happens that at eastern elongation, when Venus is at the waning part of the phase cycle, dichotomy is 'early' and occurs before the theoretically predicted time. Conversely, at western elongations when Venus is waxing, dichotomy is always late. As mentioned earlier, the discrepancies amount to several days and different observers have given different estimates of the discrepancy, from 6 days by Beer and Madler to as much as 14 days by McEwen and the Rev. J. A. Lees (reported in the *Journal of the British Astronomical Association*, **37**, 345 (1927)). It would obviously be interesting to determine if there is any pattern to these discrepancies. There are, in fact, consistent differences between the observed and predicted phase times over a wide range of phases, not just at dichotomy.

There seems to be no explanation of the phase anomaly. None of the theories put forward is adequate, in particular for explaining the asymmetry of the effect. Atmospheric refraction and irradiation effects have been proposed as explanations but do not account for the leading and lagging phases.

Most likely an atmospheric or subjective effect or both will explain the observations or a psychological effect due to contrast. Dr John E. Westfall suggests that it would be interesting to carry out a programme of naked eye lunar dichotomy estimates in order to see if there are psychological–physiological effects when the optical and atmospheric effects have been eliminated.

The observed time of dichotomy should be noted as accurately as you can so that the extent of the Schröter effect can be measured. With practice, estimates of exact dichotomy can be made by eye with sufficient accuracy for them to be reliable and useful. It has not yet been shown whether the atmosphere of Venus has any effect on the apparent phase. Contrast definitely does so because the phase appears larger when seen through a red filter and much smaller if viewed through a blue filter.

Most observers appear content to estimate the date of apparent dichotomy and define the phase discrepancy as being the difference between the date of observed and theoretical dichotomy. The value obtained is uncertain as it is based on a single observation. Dr John E. Westfall used a more accurate method based on estimating the date of apparent dichotomy from several observations, applying the least square statistical technique to the data. He first used this in 1964 and describes it in a paper listed in the 'Further reading' section at the end of this chapter.

Using this same approach a computer-aided method of determining the exact time of dichotomy was proposed by J. McCue and J. McNichol of the BAA. Essentially this involves observing Venus daily for several days on either side of dichotomy, estimating the percentage of the disc illuminated and constructing a graph of the percentage of Venus illuminated against the date. From this, the date and time corresponding to 50% illumination (dichotomy) can be determined from where the curve drawn through the observational data points intersects the line on the graph corresponding to 50% illumination.

The method was employed by three Canadian observers, Jean-Francois Viens, Todd W. Lohvinenko and Marc A. Gelinas. Between August and November of 1989 they accumulated 34 observations of Venus over a 100-day period using telescopes with between 4 and 10 inches of aperture. They observed in both white and yellow light. Estimates of the percentage of the disc of Venus illuminated were made by visually comparing Venus with calibrated ready-drawn discs and by measuring drawings made at the telescope. The observations were plotted with a computer and the best curve through the observational points was determined with a curve-fitting program. The point where the curve crossed the 50% illumination axis was thus accurately found. In this case the Canadian observers found that dichotomy of Venus occurred on November 4.74, 1989 UT. This differed by 3.02 days before the predicted date of November 7.76 UT. They computed the margin of error of each observation and estimated their result to be accurate to plus or minus 0.12 days.

This kind of project is well worth doing involving as it does only simple equipment. Perseverence and self-discipline are needed and it is ideally suited for amateur observers. Successful and systematic observation of Venus demands much perseverence.

Another thing to look for when observing Venus at the crescent phase are extensions of one or both cusps. Extensions of both cusps are shown in one of the drawings by Baum shown in Fig. 6.23 which also show bright belt-like features on either side of the apparent equator.

Irregularities on the terminator of Venus have definitely been seen on many occasions and there is no doubt about their reality. However, some of the reported sightings may not be authentic because poor seeing conditions and atmospheric ripples will distort the limb or terminator of a planet. An inexperienced or unreliable observer may easily record these as terminator irregularities. Venus seems to be particularly prone to exhibit non-existent phenomena as a result of poor seeing.

Terminator irregularities usually take the form of 'bumps' and 'dents' in the otherwise smooth elliptical curve of the terminator. These should be studied with different coloured filters because the apparent shape of the deformations differ according to which colour filter is used and these differences should be carefully recorded.

Some terminator deformities might be illusory and due to contrast effects. A bright spot close to the terminator may appear to extend beyond it as a 'bump', an effect due to irradiation. A dark spot close to the terminator may look like a 'dent'.

Unmistakable deformations have often been seen such as that recorded by W. Sandner in 1959 (fig. 6.24) and confirmed by several other observers. G. D. Roth has observed terminator deformities using a 4-inch off-axis reflector. Terminator irregularities, various appearances of the cusps and brightening near the cusps are shown in Baum's drawings of Venus near dichotomy in fig. 6.25. Other variations in cusp forms are shown in Baum's drawings of fig. 6.26.

A potentially fruitful and valuable line of investigation that can correlate with results obtained by space probes is opened up by observations of terminator deformations, now that the surface contours of Venus have been mapped and a meridian of zero longitude has been established. Suppose that a definite localised terminator irregularity is seen. Its precise position should be noted as accurately as possible and also the time (UT). If now the longitude of the central meridian is obtained from an almanac it will be a simple matter to determine from a relief map of Venus whether or not a high relief feature on the surface of Venus is close to or beneath the terminator irregularity. If a protracted series of such observations could be made, it would be interesting to note whether there is a correlation between the terminator irregularities and the underlying relief. Such studies might well give indications of the effect, if any, that high relief features have in causing terminator irregularities or the cloud patterns of Venus in general.

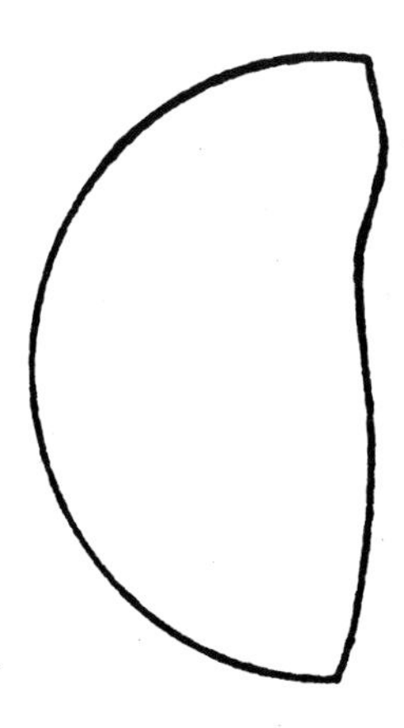

Fig. 6.24 Terminator deformation of Venus observed by W. Sandner (1959). (From Handbook for Planet Observers, *Roth, G.D. Faber And Faber, London 1970.)*

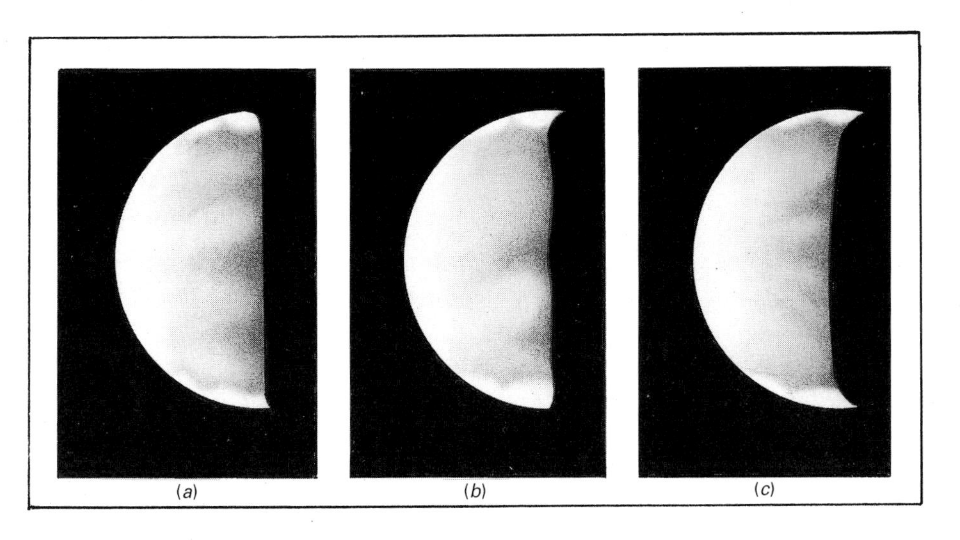

Fig. 6.25 Venus 1959: (a) *June 8th, 20.20 UT, Se 4/10, Tr 5/5;* (b) *June 10th, 20.20 UT, Se, 6/10, Tr 4/5;* (c) *June 17th, 20.10 UT, Se 4/10, Tr 315. (R.M. Baum, Chester.)*

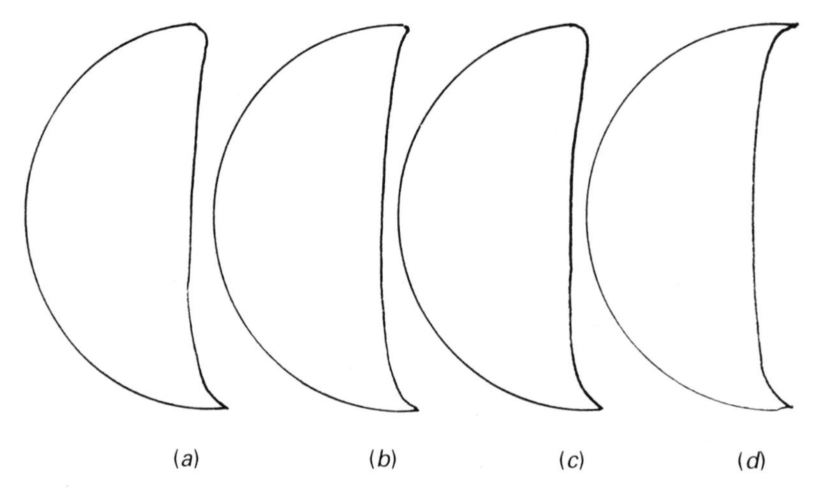

Fig. 6.26 Variations in the cusp forms of Venus, 1975 (115 mm OG ×186): (a) *June 20th, 22.00 UT;* (b) *June 22nd, 21.00 UT;* (c) *June 25th, 22.00 UT;* (d) *June 30th, 22.00 UT. (R.M. Baum, Chester.)*

The ashen light

Reports of sightings of the ashen light have been almost always at eastern (evening) apparitions of Venus but this may be because the majority of observations of Venus have been at evening apparitions.

For successful observation of the ashen light it is necessary to view Venus as the sky is becoming dark and to block out the bright light of the crescent by using some kind of occulting bar or similar device in the telescope eyepiece.

However, you must realise that the scattered light produced by the 'spider' supports of the secondary mirror of a Newtonian reflector are not blocked out by the occulting device. The mirror of a reflector should have a fresh surface and also the diagonal. On the whole, it may be best to use a refractor in observations of the ashen light.

Try using different coloured filters when looking for the ashen light. David L. Graham observed the ashen light with a 6-inch refractor on May 16th 1988 and saw that the light had a mottled appearance. With a red filter (Wratten 25) the light was 'very strongly suspected' and suspected with green (Wratten 58) and orange (Wratten 15) filters but was not suspected with integrated light. Wratten 35 (purple or magenta) filters with 4100 and 6600 angstrom pass bands are used also by BAA members. If the ashen light is seen, make a note of its intensity and colour. Sometimes it has a mottled appearance as mentioned above.

Since 1978 the Pioneer Venus Orbiter (PVO) has been in a close to polar orbit around Venus. It carries instruments for observing atmospheric features of Venus, sampling the interplanetary space in the planet's vicinity and studying Sun–Venus interactions. The PVO is unable to make direct observations of the ashen light in visible light; PVO studies should therefore be complemented by Earth-based telescopic work; there has, in fact, been intensive work carried out worldwide by astronomical associations in studying the ashen light cooperatively with PVO scientists.

At the time of writing, analyses of PVO data and Earth-based visual and photographic data are being performed with the purpose of correlating the Earth-based observations with PVO measurements. Results so far are tentative but the PVO scientists appear to be pleased but cautious about them.

Transits of Venus

The next four transits of Venus will occur on June 7th 2004 and June 5th 2012 at the planet's descending node; December 10th 2117 and December 8th 2125 at the ascending node.

The same general methods as were described for observing transits of Mercury apply to transits of Venus. Attempts at accurate timing of the four contacts may be expected to be more difficult than in the case of Mercury since the 'black drop' effect is more pronounced due to the denser atmosphere of Venus.

The first two transits may be seen at the following locations:

2004. Entire transit: Asia excluding the eastern part, Europe excluding the west; east, south-east and north-east Africa.

Beginning only: eastern Asia, Australasia.

End only: Western Europe, west and north-west Africa, east and south America and north-east and north America.

2012. Entire transit: Pacific Ocean, New Zealand.

Beginning only: North and South America.

End only: Australia and eastern Asia.

Further reading

Books

The Planet Venus. Hunt, G. E. and Moore, P. A., Faber and Faber. London (1982).

An Introduction to Observing Venus. Benton, J., Review Publishing Co., Savannah, Georgia (1973).

Visual Observations of Venus; Theory and Methods. (The ALPO Venus Handbook). Benton, J., Review Publishing Co., Savannah, Georgia (1987).

Venus. Hunten, D. M. *et al.*, University of Arizona Press. Tucson, Arizona (1983).

Venus. An Errant Twin. Burgess, E., Columbia University Press. New York (1985).

Papers and articles

Venus. Oberg, J. E., *Astronomy* **4(8)**, 9–21 (1976).

Venus. Cordell, B. M., *Astronomy* **10(9)**, 7–22 (1982).

Why observe Venus? Robinson, J. H., *JBAA* **90(1)**, 36 (1979/80).

Some naked eye observations of Venus. Ashbrook, J., *Sky and Telescope* **53(1)**, 12 (1977).

The observation of detail on the planet Venus. Warner, B., *JBAA* **71(5)**, 202–5 (1961).

The markings and rotation of the planet Venus. Antoniadi, E. M., *JBAA* **44(9)**, 341–7 (1934).

Bands and belts on Venus. Devadas, P., *JBAA* **73(4)**, 165–9 (1963).

Following the shades of Venus. Eicher, D. J., *Astronomy* **19(11)**, 89 (1991).

How to observe planets during the day. Palmer, J. H., *Astronomy* **17(3)**, 86–7 (1989).

On the phase anomaly of the inner planets. Heggie, D. C., *JBAA* **80(4)**, 288–92 (1970).
The phase anomaly of Venus. Warner, B., *JBAA* **73(2)**, 65–70 (1963).
An investigation into the phase anomaly of Venus. Chambers, R. H., and Taylor, J., *JBAA* **76(5)**, 310–15 (1966).
On the phase anomaly of Mercury and Venus. Kirby, G. J., *JBAA* **80(4)**, 293–5 (1970).
Ashen light on Venus (Radar Reports). *Astronomy* **16(8)**, 86 (1988).
Ultraviolet observations of Venus in 1969. Hiscott, J., *JBAA* **82(3)**, 198–9 (1972).
Photography of Venus (ultraviolet). Rackham, T., *JBAA* **67(5)**, 160–6 (1957).
Venus observed by Mariner. *Sky and Telescope* **47(4)**, 235–40 (1974).
Pioneer Venus: First results. *Astronomy* **7(4)**, 18–23 (1979).
The surface of Venus from Pioneer. *Astronomy* **8(8)**, 58–61 (1980).
A world revealed: Venus by radar. Nozette, S. and Ford, P., *Astronomy* **9(3)**, 6–15 (1981).

Transits of Venus

To list all of the extensive literature pertaining to the 1874 and 1882 transits of Venus would be difficult. Two good sources are the books *Transits of Venus* and *Studies of Venus transits in 1874 and 1882* by R. A. Proctor.

Notes may be found in the *Monthly Notices of the Royal Astronomical Society:*

Volume	Page
24	173 (by Airy)
28	255
29	33, 43, 45–8, 210–11, 249–50, 305–6, 332 and in other places.
35	345 (by Tennant)

Further literature will be found listed in the reference section of *The Planet Venus* by G. E. Hunt and P. A. Moore (see above in the list of books).

7

Mars

General

Mars is the first of the so-called 'Superior' planets, i.e., those whose orbits lie outside the Earth's. It has an equatorial diameter of 4200 miles (6791 km) which is a little greater than one half of the Earth's diameter and intermediate between that of the Earth and the moon (fig. 7.1). The polar compression or oblateness of Mars is 0.005 compared with the Earth's 0.003. Mars has two very small satellites, Phobos at a distance of 5800 miles (9332.2 km) and Deimos at a distance of 14 600 miles (23 491.4 km) from the centre of Mars. Prior to the Mariner 9 mission, the Martian satellites were thought to be spherical in shape. Phobos was estimated to have a diameter of about 10 miles (16 km) and Deimos a diameter of about 8 miles (13 km). Views on the shape and size of the satellites had to be revised after the Mariner 9 mission results.

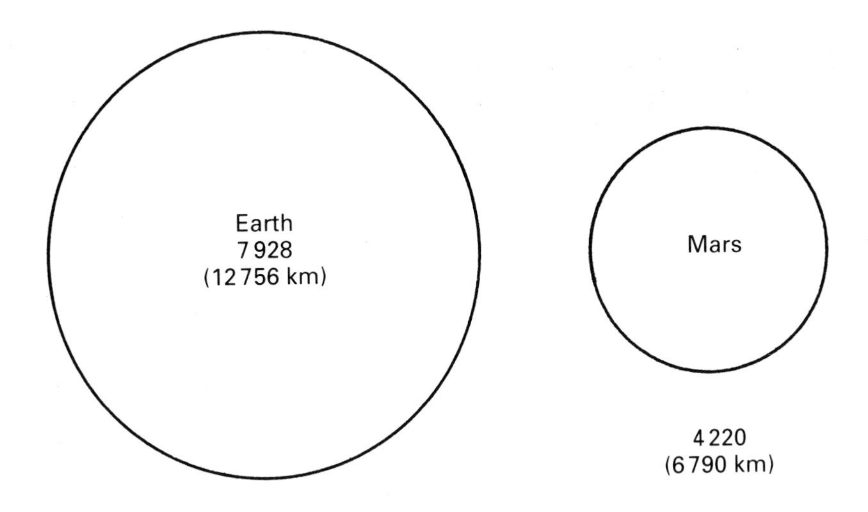

Fig. 7.1 Comparative sizes of the Earth and Mars (equatorial diameters in miles).

Orbital characteristics

The mean distance of Mars from the sun is 141.6 million miles (227.94 million km). Owing to the pronounced eccentricity of the orbit (0.093) which is the largest next to Pluto and Mercury, this distance varies from 154.8 million miles (249.2 million km) at aphelion to 128.4 million miles (206.6 million km) at perihelion. The orbital revolution period of Mars, the sidereal period, is 687.0 Earth days or a little less than two Earth years. Its mean synodic period is 779.9 Earth days, varying from about 25 to 27 months.

Mars moves in its orbit with a mean velocity of 15.0 miles (24.14 km) per second and the orbit is tilted at an angle of 1° degree 51′ to the plane of the ecliptic. Mars rotates on its axis once in 24 hours 37 minutes and 23 seconds and the axis is inclined at an angle of 23° 57′ to the plane of the Ecliptic and 25° 11′ to the plane of its orbit.

The sidereal periods of Phobos and Deimos are 7 hours 39 minutes and 30 hours 18 minutes respectively. Phobos is unique as a Solar System satellite in that it completes one revolution around Mars in a period much less than the Martian 'day' so that as seen from the surface of Mars, Phobos would rise in the west and set in the east $4\frac{1}{2}$ hours later and would exhibit more than one half of its phase cycle from new to full. Successive risings of Phobos would be a little more than 11 hours apart. Deimos with its 30 hours and 18 minutes period would be seen

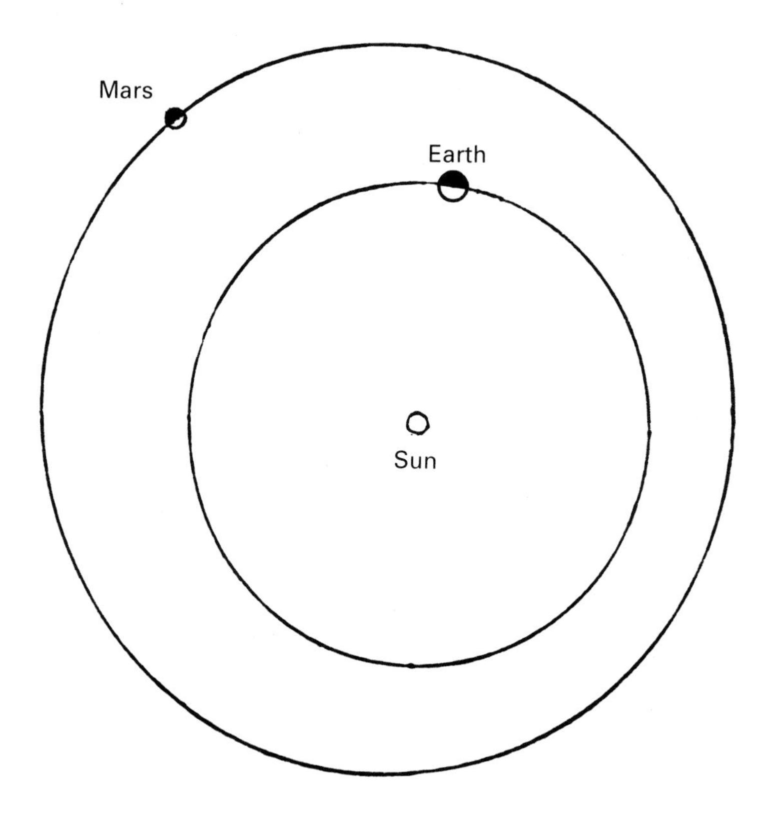

Fig. 7.2 The orbits of the Earth and Mars around the Sun (to scale).

above the horizon of Mars to an observer on its surface for 2½ Martian 'days' at a time.

Since the orbit of Mars lies outside the Earth's (fig. 7.2) it follows that periodically the sun, Earth and Mars are in a straight line with the Earth in the centre; Mars is then said to be at opposition and will be at its closest to the Earth at or near that time (closest approach and opposition may differ by up to 8½ days). Likewise, Mars will be at its farthest from the Earth when on the opposite side of the sun from the Earth and is then said to be in conjunction with the sun.

If the orbits of the Earth and Mars were perfectly circular and concentric, then these maximum and minimum distances would be the same at every opposition and conjunction. Because of the ellipticity of the orbit of Mars its distance from the Earth at opposition will actually vary from a minimum of 34.6 million miles (55.76 million km) at a perihelic opposition to a maximum of 63.0 million miles (101.37 million km) at an aphelic opposition. If the major axes of the elliptical orbits of the Earth and Mars coincided, then the minimum possible distance between the two planets would be realised when opposition occurred with both the Earth and Mars at perihelion. However, although all the planets travel in the same direction round the sun and all the orbits lie nearly in the same plane as the Earth's, there is no law of nature governing the positions of the major axes of planetary orbits relative to one another. Mars comes closest to the Earth at perihelic oppositions but the word 'perihelic' refers to the perihelion of Mars only. Whether or not the Earth is also at perihelion at the same time makes hardly any difference to the distance between the two planets at perihelic oppositions; owing to the relatively large eccentricity of the orbit of Mars it is this that really counts much more than the Earth's. The perihelion and aphelion distances of Mars from the sun are respectively 128 million miles (206.6 million km) and 154.8 million miles (249.2 km), i.e., a difference of 26.8 million miles (43.12 million km) whereas the difference between the perihelic and aphelic distances of the Earth from the sun is much less and is 3 million miles (4.8 million km). Therefore, it is the orbital position of Mars that counts most in determining the distance of Mars from the Earth at perihelic and aphelic oppositions.

Since Mars makes a complete orbital revolution in a little less than two Earth years it follows that oppositions occur at intervals of a little more than two Earth years, for when the Earth has made two complete orbital revolutions after an opposition Mars will have completed somewhat more than one complete revolution. The mean opposition interval (synodic period) as previously noted is therefore 779.9 days i.e., about 2 Earth years and 49 days.

It should be mentioned that it is not quite true that the closest approach of the Earth to Mars coincides exactly with the time of opposition; this would be so if the orbits of the two planets were both in the same plane. Actually, they are titled against each other as already mentioned. The relative motions of Mars and the Earth are rather like two cars moving on sloping ramps so that the moment when the two planets are closest is not necessarily the same moment when they form a straight line with the sun. However, the moments of opposition and of closest approach usually differ by only a small amount, the maximum possible being about 8½ days as occurred during the opposition of 1969. The difference is usually much smaller. At the opposition of 1948 the two events occurred on February 17th, the closest approach occurring just 4 hours after opposition.

Predicting oppositions

Given that an opposition of Mars occurs in a certain year; when will the next opposition occur? The method used to calculate this is as follows: the Earth makes a complete orbital revolution in 365.25 days. Mars travelling much more slowly takes 687 Earth days to make a complete orbital revolutions. This is 321.75 days longer than an Earth year, so that the Earth has made a large part of another complete revolution by the time that Mars has completed one circuit of its orbit. When the Earth reaches the original starting point 43.5 days later, Mars will have moved a bit further on and the Earth will catch up with it after 49.5 days. Now, two Earth years equals 730.5 days so that the entire process takes 730.5 + 49.5 = 780 days or 2 years and 7 weeks. Note that the next opposition will take place at a position where the line drawn through the two planets and the sun will now make an angle of about 49° with the line drawn through the two planets and the sun in the previous opposition.

We find, however, that the actual measured time between two consecutive oppositions differs from the 780 days we have just calculated. This is because in making these calculations we assumed that the Earth and Mars were moving with uniform average velocities in their orbits; actually the orbital velocities are continually varying and are greatest at perihelion and least at aphelion. Thus, at an aphelic opposition, Mars is moving much more slowly than the averaged figure used and Earth simultaneously at its perihelion position will be travelling a little faster under these conditions. Conversely, with Mars at perihelion and Earth at aphelion during an opposition, the Earth will take a week longer to catch up with Mars.

Another interesting point is that because the second opposition that we have been discussing does not occur on the same line as the first but 49° from it, we may ask how many years or oppositions must elapse until another opposition occurs in an identical position. If we express Martian years in terms of Earth years, then Mars returns to a given position in its orbit after 1.8808 Earth years, so that about 15 Earth years are required for Mars to make 8 complete orbital revolutions.

The actual exact time differs from 15 years by 0.047 Earth years which is 17 days and so the two planets and the sun would not lie in the same straight line but would form an angle of about 16°. However, a better correspondence occurs after 151 orbital revolutions of Mars which equals 284.016 Earth years, very nearly, when the angle is only 0.016°. We can therefore say with some precision that a given opposition will repeat itself after 284 years. Using this result we can predict oppositions of Mars reasonably accurately if we have a list of earlier oppositions. For example, the favourable opposition that Maraldi made use of in 1719 will be repeated in 2003. In practice, of course, oppositions are computed individually.

Another consequence that arises from Mars lying outside the orbit of the Earth is that it does not show a full range of phases like Mercury or Venus. It will be 'full' at opposition, and when at quadrature (fig. 1.9), i.e., when the lines joining the Earth and the sun with Mars form a right angle, the planet will appear gibbous like our Moon 3 or 4 days before or after full. This is the maximum phase effect that Mars shows. The fraction of the visible disc illuminated may decrease to 84%.

The retrograde motion of Mars

As seen from the Earth, the planets on the average exhibit a slow eastward drift against the background of the fixed stars of the Zodiac due to their own orbital motion and the Earth's; this is, of course, distinct and quite different from the apparent rapid diurnal east to west drift exhibited by both stars, planets, sun and moon due to the Earth's rotation on its axis.

If one follows the eastward movement of Mars among the stars, at certain times Mars will appear to stop, drift backwards in a westerly direction for many days then stop and resume its easterly drift. Mars thus appears to trace out a loop in the sky (fig. 7.3). The apparent backward movement of Mars is called retro-

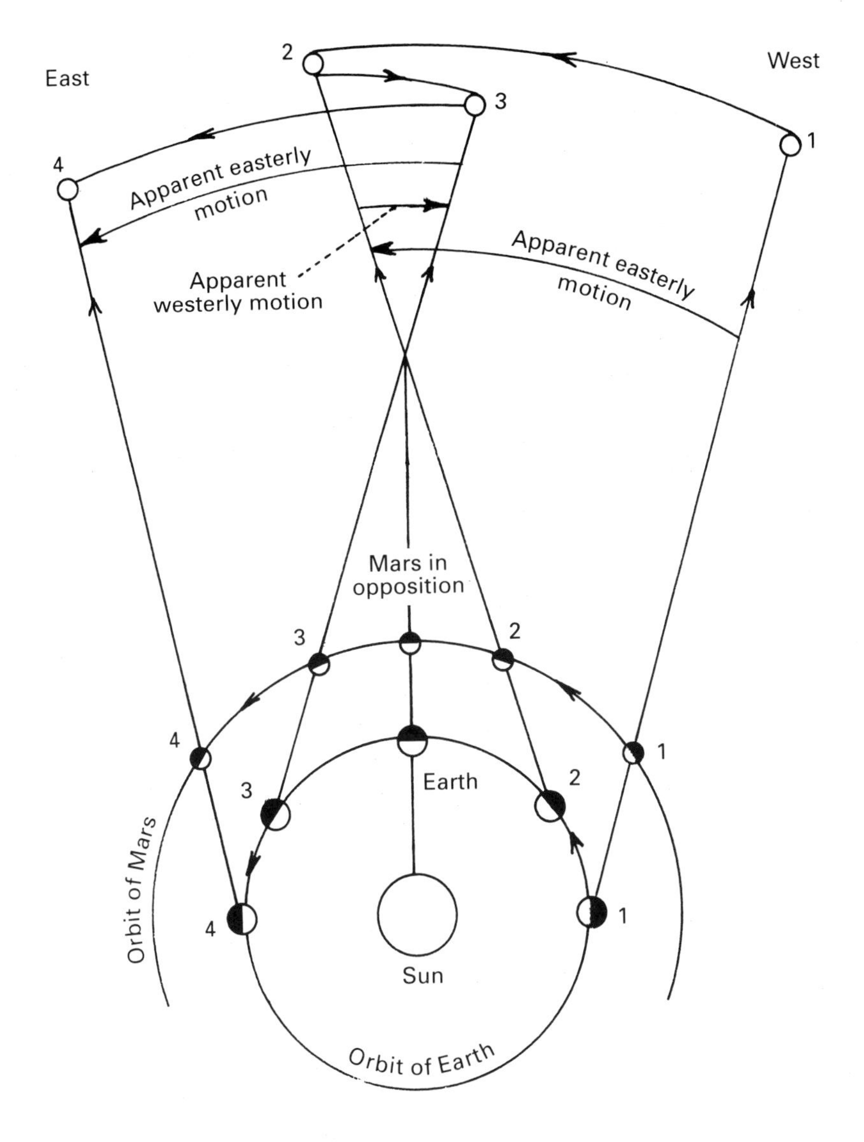

Fig. 7.3 The retrograde motion of Mars.

grade motion and is an illusion caused by the different orbital speeds of the Earth and Mars, the Earth being the faster of the two and thus overtaking Mars; all the superior planets exhibit this. The explanation of the apparent retrograde motion of Mars is also shown in fig. 7.3.

At position 1, Mars is three months from opposition and is ahead of the Earth in its orbit. The line passing through the two planets indicates the apparent direction of Mars as seen from the Earth. The Earth steadily gains on Mars owing to its greater orbital speed with the result that the apparent eastward drift of Mars against the star background seems to slow down and finally stops for a short while just before opposition (position 2). It then commences to retrograde as the Earth catches up with Mars and overtakes it, and stops again (position 3), then recommences its eastward drift and appears in position 4 three months after opposition. During this period of six months the Earth has traversed one half of its orbit whereas Mars has traversed only one fourth of its path around the sun.

Although it was not mentioned in the chapters on Mercury and Venus, these two planets and all the others exhibit retrograde motion. If the orbits of the planets were all in the same plane, then their forward and retrograde motions would appear simply as forward and backward linear motions along the ecliptic. Since the planetary orbits are all slightly tilted to the plane of the ecliptic the planets instead trace out loops when retrograding.

The relative amounts of retrograde motion exhibited by the planets differ considerably. Approximate average values for the number of days a planet spends in retrograde motion and the angular distance it appears to move in a westerly direction can be worked out using the simplifying assumptions that the planets all move at uniform speeds in perfectly circular orbits all of which are coplanar. These assumptions do not greatly affect the accuracy of the values so derived. In the case of Mercury and Venus, the time required for them to gain a lap on the Earth or for the Earth itself to gain a lap on a superior planet is the same as the planet's synodic period, i.e. the time elapsing between two consecutive oppositions (superior planets). Table 7.1 shows the synodic periods of the planets, the time spent in retrograding, the percentage of the synodic period spent in retro-

Table 7.1. *Retrograde motions of the planets.*

Planet	Synodic period (days)	Retrograde interval (days)	% of synodic period spent retrogressing	Westward retrograde movement (degrees)
Mercury	116	23	20	13.8
Venus	584	42	7	16.2
Mars	780	73	9	15.9
Jupiter	399	121	30	9.9
Saturn	378	138	36	6.8
Uranus	370	152	41	4.0
Neptune	367	158	43	2.8
Pluto	367	162	44	2.2

Data from *The beginner's guide to the skies*, Cleminshaw, C. H., Thomas, Y. Crowell Co, New York, USA (1977).

grade motion and the angular distance in degrees that the planet appears to move westward when retrograding. Notice that the planets spend a greater percentage of the synodic period in retrograding the further they are away from the Earth.

Martian seasons

The principal cause of a planet's seasons, assuming that its orbit is almost circular, is the position of its axis. On our own Earth Summer in the northern hemisphere occurs at aphelion. It is the tilting of the northern hemisphere towards the sun, so that the sun's rays hit the globe more vertically, that is much more important than the fact that the Earth is a little further away from the sun than at perihelion. In the same way, the fact that the Earth is nearer to the sun by 3 million miles (5.0 million km) in January when it is at perihelion is completely overshadowed by the northern hemisphere being tilted away from the sun so that the rays now strike the globe at a very oblique angle. The 3 million miles difference between the Earth's perihelic and aphelic distances has a small effect on the seasons amounting to about 7% variation in insolation with perhaps an 8° difference in the temperature.

The situation is a bit different with Mars. Its distance from the sun varies by 26.4 million miles (42.6 million km) during one Martian year so that the the hemisphere which has the summer at perihelion must experience hotter summers than the hemisphere whose summer occurs at aphelion. The hemisphere that has winter at aphelion must have colder winters than the other. It is the southern hemisphere of Mars that is subject to this more extreme range of seasonal temperatures. This also causes the Martian seasons to differ in length to a much greater extent than Earthly seasons. The terrestrial northern summer lasts $93\frac{1}{2}$ Earth days while the southern summer is 89 Earth days long. The Martian northern (aphelion) summer lasts 182 Earth days and southern (perihelion) summer lasts for 160 Earth days. The Martian day is about 37 minutes longer than the Earth day so that the year on Mars lasts for 668.6 Martian days.

A comparison of the lengths of the seasons on Earth and Mars is shown in table 7.2. The southern hemisphere summers are shorter and the southern winters are longer on both the Earth and Mars but the differences are greater on Mars and the seasons last for almost double the time that they do on Earth.

Table 7.2. *Comparative lengths of seasons on Earth and Mars.*

Southern hemisphere (days)	Northern hemisphere (days)	Earth (Earth days)	Mars (Martian days)
Summer	Winter	89.1	155.8
Spring	Autumn	89.7	141.8
Winter	Summer	93.6	176.8
Autumn	Spring	92.9	194.2
		365.3	668.6

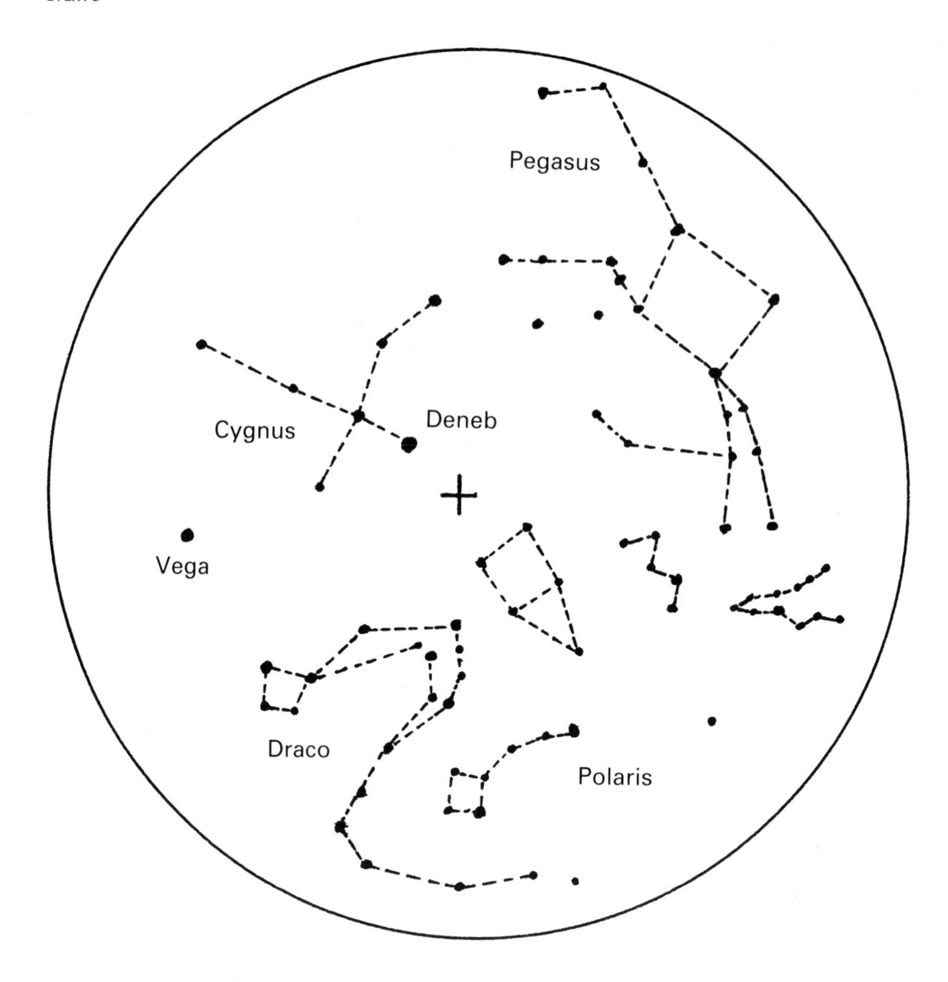

Fig. 7.4 Position of the celestial north pole of Mars (+).

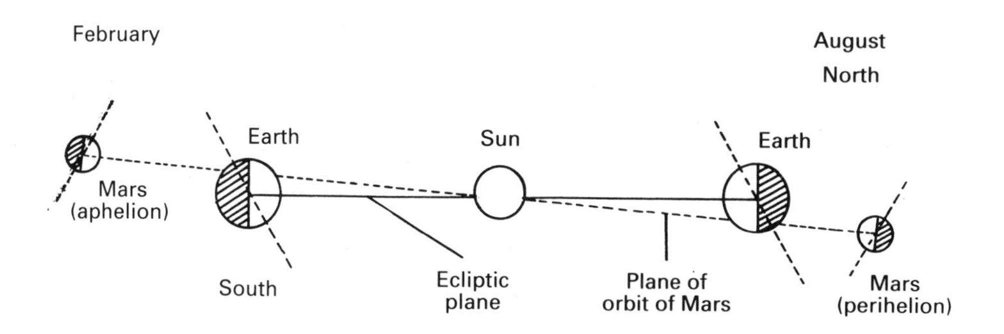

Fig. 7.5 Seasons on Earth and Mars.

The Earth's axis is tilted at an angle of 23° 27′ to the plane of the ecliptic and points almost exactly to Polaris, the pole star. The axis of Mars is tilted to the plane of its orbit to very nearly the same angle, 25° 11′, but points in quite a different direction to a spot in the celestial sphere rather devoid of bright stars but about 10° from the star Deneb in the constellation of Cygnus (fig. 7.4) with a position defined by RA 317.62° (21 hrs 10.5 min) and declination + 52.85° for the epoch 1991. The result of the different position of the Martian axis is that at oppositions Mars is one season ahead of Earth, e.g., at the opposition of November 27th 1990, it was Autumn in the Earth's northern hemisphere but winter in the northern hemisphere of Mars. Seasons on the Earth and Mars are shown in fig. 7.5.

Surface features

Mars is the only one of the solar planets that favours us with a good view of its surface which exhibits prominent permanent markings (fig. 7.6).

To the naked eye Mars is noted for its distinct orange-red colour. This is due to the fact that about two thirds of the surface of Mars is covered by so-called 'desert' areas which have a warm ochre colour when the planet is viewed telescopically. The rest of the surface consists of darker areas of irregular shape possessed of a greyish-green tint, apparently a contrast effect; photometry shows that the dark areas have roughly the same hue as the 'deserts'. These darker regions undergo interesting seasonal and secular changes in shape, intensity and tone which has led to much discussion and controversy as to their nature.

The extended dark areas of Mars are called 'maria' (seas) although we know that they are not bodies of water. The dark areas contain a good deal of 'structure', mostly dark spots, streaks and lighter areas; this alone precludes the maria

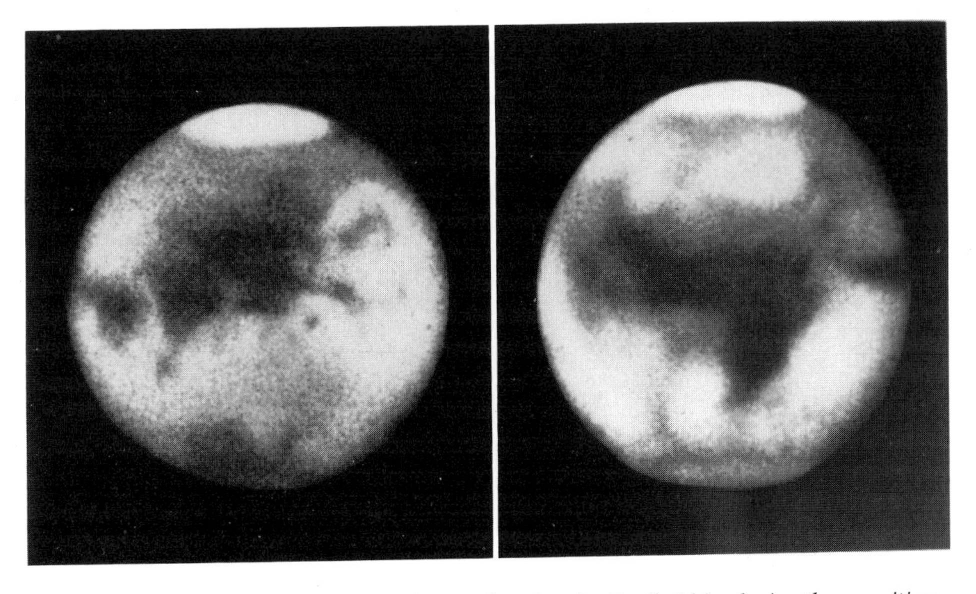

Fig. 7.6 Mars photographed by E.C. Slipher at Bloemfontein, South Africa during the opposition of 1939. Left: July 27th , right: August 9th.

from being bodies of water. The maria extend over a large part of the southern hemisphere. In the northern hemisphere there are more or less isolated dark areas. The maria often have sharp boundaries but they may also be diffuse.

Because of the seasonal cycle of changes of intensity, colour and growth exhibited by the dark areas, the idea has long been held that they may consist of a lowly form of plant life. The most likely candidates were considered to be the lichens, a symbiotic alliance between an alga and a fungus, that can survive under severer conditions, such as extreme cold, than many other forms of living thing. The Orbiter and Lander spacecraft have found no evidence of lichens or other vegetation.

Another idea is the hygroscopic salt theory of Arrhenius put forward in 1912 and revived much later by A. Dauvillier. The dark areas are supposed to be covered with hygroscopic salts, i.e., salts that absorb moisture and in so doing change colour and darken. The dark areas become darker when the polar ice cap melts and the salts supposedly pick up moisture from this which causes their darkening. However, we now know that Mars does not have liquid water on its surface which makes this theory untenable.

A third idea is that dark areas are tracts of volcanic dust. This idea was put forward in 1954 by D. B. McLaughlin of the University of Michigan. Since many of the dark areas like the Syrtis Major are triangular, it could be that there is a volcano at the point that belches out dust that is then spread out fan-wise by the prevailing winds. Apart from the fact that there appears to be no evidence for the existence of active volcanoes on Mars, there are other weaknesses in this hypothesis. One of the these is its inability to explain the seasonal cycle or the 'wave of darkening'. However, this does not exclude the possibility that the dark areas may consist of areas of wind-blown dust, whose changes reflect changes in the wind patterns and atmospheric currents of Mars rather than the activities of volcanoes.

The term 'desert' applied to the ochre-coloured areas of Mars should not be taken to mean that they are exactly like, say, the Sahara. Far from being hot they are quite cool, cooler in fact than the dark areas. At one time it was thought that the ochre or reddish colour was due to iron oxide formed by oxidation of iron compounds by oxygen in the Martian atmosphere. G. P. Kuiper considered that the colour of the desert areas might be due to minerals like felsite which is a rock formed from orthoclase aluminium and potassium silicate – with occluded quartz grains. Dolfuss believes that the desert coating may be of limonite, a sedimentary deposit consisting of hydrated iron (ferrous) oxide.

The third type of surface feature is the polar caps. These appear as brilliant white patches at or close to the poles of Mars and are usually the first features to catch the eye when Mars is viewed in a telescope. They increase and decrease in extent and intensity with the seasons in a way that strongly suggests that they are composed of a solid substance that melts to a liquid or sublimes during the summer in that same hemisphere. Water ice and solid carbon dioxide have been suggested as possible candidates. The polar caps show great differences in size. At the spring equinox the southern cap is 70–80° in breadth while the northern cap is usually only 53° at its greatest breadth, although it did attain a breadth of 65° in 1966–67. This difference is due, of course, to the fact that the southern cap forms during the planet's aphelic Autumn–Winter season lasting 371 days while the northern cap forms in the 298-day perihelic Autumn–Winter season. The southern cap shrinks much faster than the northern cap because of the shorter hotter summer experienced near perihelion; it may even disappear as it did in 1894. The northern cap shrinks more slowly in the longer cooler summer near aphelion and never gets smaller than 6° in breadth.

When the polar caps recede, their outlines do not remain perfectly circular but exhibit irregularities and do not appear of equal intensity overall. Dark bays and bright projections are seen at their edges and rifts are seen dividing the caps. Parts of the caps may become separated and stay as isolated bright spots for many weeks. The caps themselves are variegated by bright and dark markings. Local conditions must affect the shrinkage of the polar caps because the bright areas within the caps, the bright projections and the isolated bright spots external to the caps are observed in the same positions at successive oppositions. The most often observed and most prominent of these irregularities is one that has been seen many times in the shrinking south polar cap, the so-called Mountains of Mitchel (Novus Mons), discovered by Mitchel in 1854 at the Cincinnati observatory.

A dark band or 'collar' ('Lowell's Band') lies closely against the borders of the receding polar caps. There is no doubt as to its real existence; that it is not a contrast effect between the bright polar cap and the much darker surrounding terrain is proved by photography and by its still being seen even when observed through a red filter that makes the polar caps and surrounding surface look equally bright. The dark band has been interpreted as a zone of moisture left behind by the melting of the ice of the polar cap because it is not a permanent feature and closely follows the edge of the receding polar cap.

An interesting fact is that the last remaining remnant of the receding south polar cap is not situated exactly at the pole but is about $6\frac{1}{2}°$ away at latitude 86° (Carr's *The surface of Mars*) and longitude 30°. The northern cap is only about one degree away from the pole at longitude 290°.

There are really two types of polar caps. The cap seen during the Spring–Summer season is quite plainly a surface deposit but the cap seen in Autumn–Winter is a mantle of cloud. Near the time of the Autumn equinox a great hood of dull white colour appears around the poles, the appearance of which varies greatly from one day to the next. These whitish veils, although variable, remain for the duration of the Martian Autumn–Winter but begin to disappear at the Spring equinox at the end of the cold season. The cloud layer then lifts exposing to view the true polar cap beneath as large, white and bright but much smaller than the cloud veil that covered it, being only 70° in width. The clouds disappear completely after the Spring equinox. At about mid-Spring rifts appear in the polar cap that progressively slice it up into sections of unequal brightness. These rifts and sections appear in the caps at the same positions during each Martian year. The whole cycle of appearances is reversed as the Summer draws to a close. Correlated with the Summer shrinkage of the polar caps is the 'wave of darkening' of the Martian surface that spreads from the polar cap towards the equator. When a polar cap begins to shrink the dark areas intensify as though they were 'coming to life' and this 'wave of darkening' spreads from the poles. It is as though moisture being released from the polar caps is causing the lowly vegetation – if that is what the dark areas are – to grow and the ground in general to darken. As the wave of darkening spreads changes occur in the shapes of some of the more prominent dark areas and these regularly occur at the same time of the Martian year. These changes are so regular that they can be predicted and are quite obviously correlated with the shrinkage of the polar caps. Many scientists have questioned the existence of this wave.

Atmospheric phenomena

Mars has an appreciable atmosphere but it is much less dense than the Earth's – only 5–10 millibars at the surface – and has a molecular weight averaging about

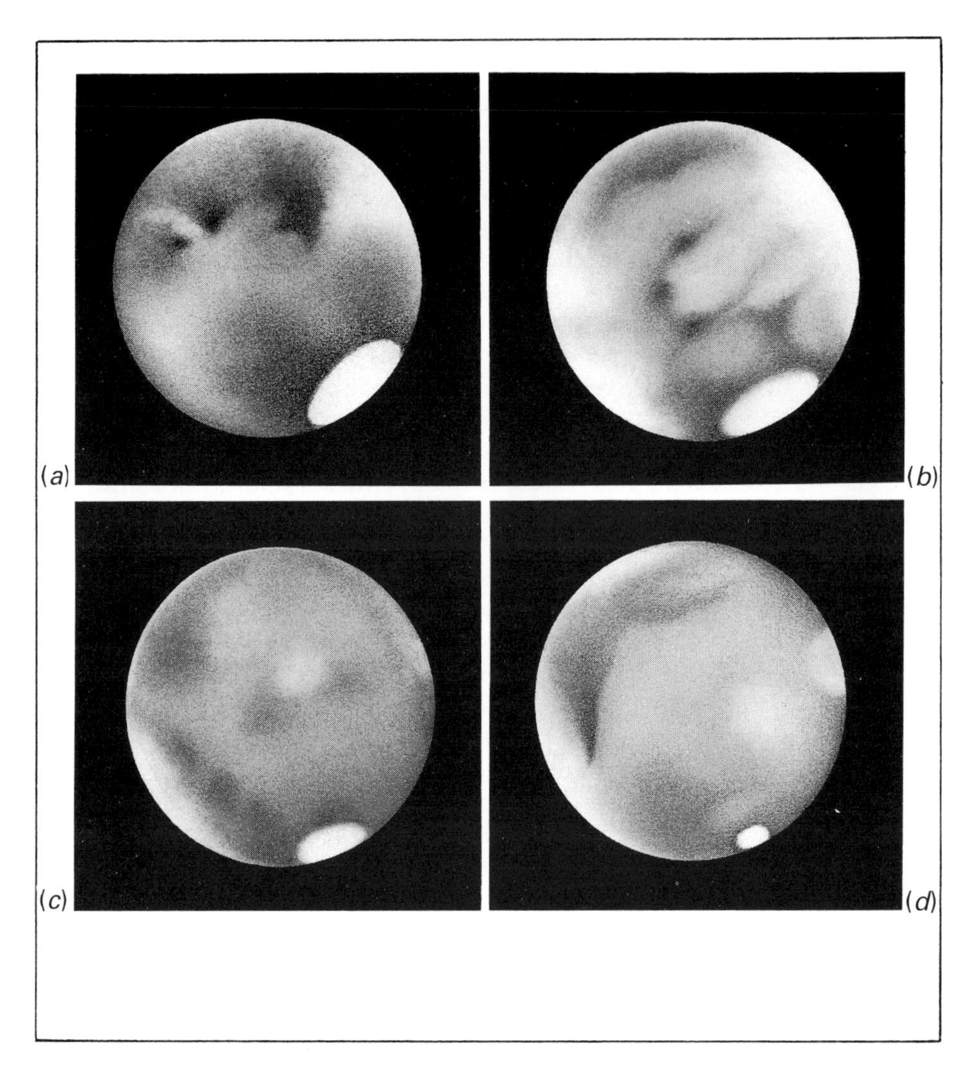

Fig. 7.7 Mars 1980 – cloud and mist. (a) *February 15th, 22.00–22.20 UT, Se II (Anton), CM 268.7°, ×186, Syrtis Major;* (b) *February 23rd, 22.00 UT, Se II (A), CM 199.6, ×186, Trivium Charontis;* (c) *March 3rd, 19.30 UT, Se III (A), CM 84.7°, ×186 Lunae L. and M. Acidalium disappearing;* (d) *March 21st, 23.00 UT, Se II (A), CM 337.6°, ×186, Syrtis Major disappearing. Cloud on terminator. (R.M. Baum, Chester.)*

40 which is consistent with an atmosphere composed mostly of carbon dioxide. The existence of an atmosphere is certainly established by the seasonal variations observed in the polar caps, by the existence of clouds and their motions (fig. 7.7), formation and disappearance, the thick haze that shrouds Mars when viewed in ultraviolet or blue light, the existence of the twilight arc and the rapid changes in intensity and colour of the dark surface markings.

The surface of Mars is nearly always clearly visible in spite of the existence of an atmosphere but clouds do appear from time to time. There are three main types of clouds: 'blue clouds' (high level), 'white clouds' (intermediate level) and 'yellow clouds' (lowest level). There is some disagreement regarding whether blue and white clouds are fundamentally similar or not but the classification of the clouds into the above three types is now well established.

The blue clouds are so-called not so much because of their colour but because they are observable visually only in blue or violet light. The white or whitish-blue clouds are easily seen in blue light but not in red and can usually be photographed. Yellow clouds, best seen in yellow light (Wratten filters 8, 12 and 15), are the most prominent of all. They do, in fact, look yellowish to the eye. It is generally considered that they are dust clouds blown up by winds on Mars. Most of the examples that have been best observed appear to be quite high level phenomena – about 18–20 miles above the surface, at least if they become major dust storms – rather than a few miles high as a few observers have thought.
In addition to the above are the curious orographic clouds that are shaped like a letter 'W'. They form in the neighbourhood of volcanic peaks. They are usually quite large and are best seen in the medium blue to violet wavelengths. They move fairly rapidly.

The yellow clouds tend to be infrequent and sometimes cover but a small fraction of the Martian disc. However, they occasionally can cover large areas of the Martian surface and are most likely due to dust. A great 'dust storm' occurred in 1956, the like of which had not been seen in 60 years. It began on August 20th and lasted about a month reaching its maximum on September 7th. There was a planet-wide 'dust storm' in 1971.

There is evidence indicating that the yellow clouds sometimes are composed not only of dust but of water also. This conclusion is backed up by the observation of transient dark zones that are frequently found adjacent to several of the dense yellow clouds. Sometimes, these dark spots last no more than a day, are very dark and are found in both the deserts and the darker regions. In contrast to the permanent dark regions, these short-live dark patches do not exhibit structure such as the lines and dots seen in the true dark patches, but look more like strongly shadowed areas.

An interesting and puzzling feature of the Martian atmosphere is the fact that although the atmosphere is so thin it appears to be strangely opaque to ultraviolet light. It is as though there is something like a screen in the atmosphere and, whatever it is, the 'screen' is variable. Sometimes it clears away but it always returns. This screen-like layer in the atmosphere has become known as the violet layer, not because of its colour but because of its ability to block violet and blue light. Many theories have been advanced regarding its nature and composition. One of these is that it is a haze of microscopic ice crystals or fine dust in the upper atmosphere but there is evidence against this idea. None of the theories are entirely satisfactory; we just don't know what the violet layer is.

The violet layer appears to be situated beneath the zone of the whitish or blue clouds of estimated heights of 6–19 miles, but it may be as high as 60 miles. When Mars is photographed in violet light, usually no surface detail is seen because of the strong scattering effect of the violet layer on short wavelength light. Occasionally, the violet layer clears and surface features can then be photographed in violet light almost as distinctly as they were in red or orange light before the occurrence of the 'violet clearings' as they are called ('blue clearings' in older literature). The violet clearings may affect the entire planet and have been known to last for up to 2 months. At other times they may affect only certain parts of Mars. It has been noted that the greater the area of the planet that is affected by the clearings the greater is the transparency.

History of observation

The reddish colour of Mars and its somewhat sudden but regular appearances in the sky inspired dread in ancient peoples because it reminded them of blood and fire, war and pestilence. In 1719 Mars was so bright and fiery that it precipitated a panic.

The Egyptians named it Harmakhios and the Chaldean sky watchers called it Nergel which was also the name given to the god of battle and of the dead. It was named the Celestial Warrior (Pahlavani Siphir) by the Persians and the Ancient Greeks called it Ares. The Romans named it Mars after their god of war.

It was Aristotle who made the only observation of Mars in ancient time that deserves to be described. In 356 BC (April 4th 357 BC according to Kepler) he once saw Mars pass behind the moon, what we today would call an occultation. This led Aristotle to conclude correctly that Mars must lie beyond the Moon and was farther from the Earth than the Moon.

The Chinese mention an occultation of Mars on February 14th, 69 BC and Tycho Brahe observed a repetition of this occultation on December 30th 1595. Several other occultations of Mars by the moon have been observed in more recent times between the years 1837 and 1878.

In 1610 Galileo saw the gibbous phase of Mars with his telescope in as much as he perceived that the outline of the planet's disc was not perfectly circular.

Surface features

The first telescope observational drawing of Mars was made by Francesco Fontana in 1636 when Mars was at opposition. The drawing was published in his book of 1655. He made another drawing on August 24th 1638 and this time Mars exhibited its gibbous phase as it was in quadrature; Fontana's drawing exaggerates the phase a bit. In both drawings, the disc of Mars is shown with a dark ring just inside the circumference of the planet's disc and a dark spot at the centre. These appearances are artefactual and due to the poor quality of Fontana's telescope, so that his delineations have no scientific value (fig. 7.8).

The first telescope drawing of Mars that shows a genuine surface feature was made by C. Huygens in 1659. It shows a dark triangular marking which is undoubtedly Syrtis Major (fig. 7.9). The drawing of 1672 is held to be the first to show clearly the south polar cap of Mars.

Huygens also played a part in establishing the axial rotation time of Mars

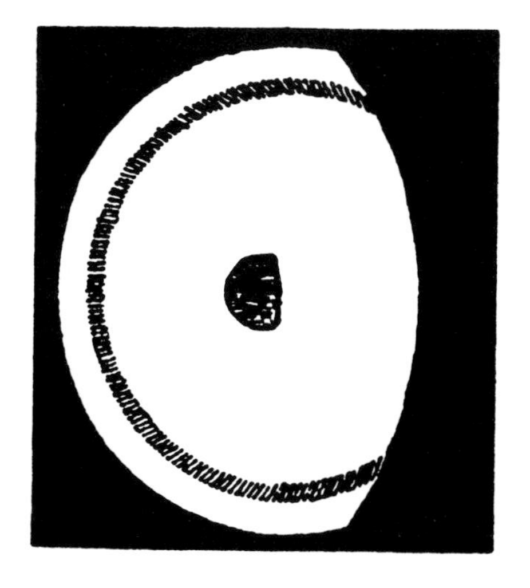

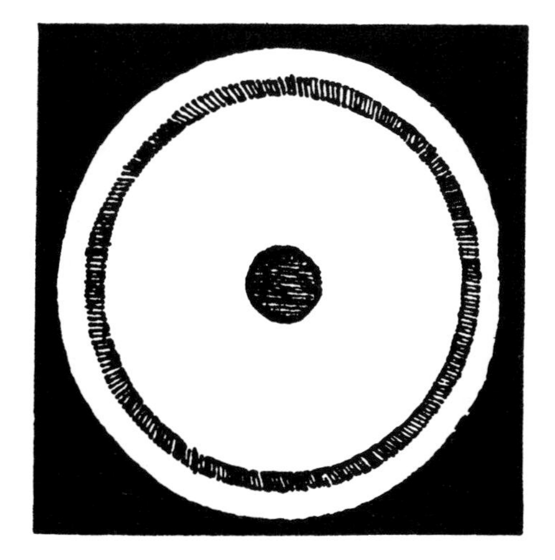

Fig. 7.8 Fontana's telescopic drawings of Mars.

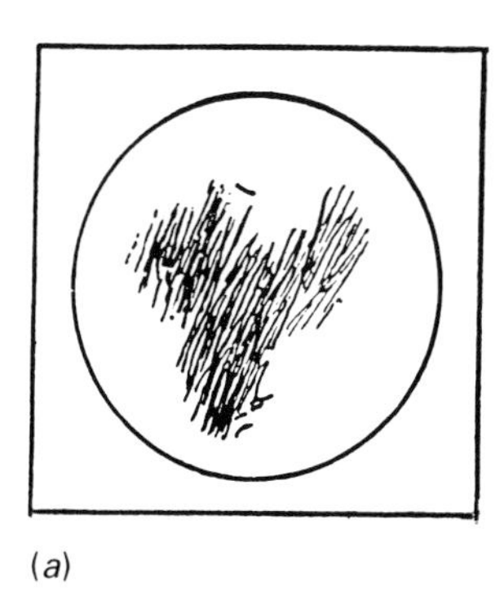

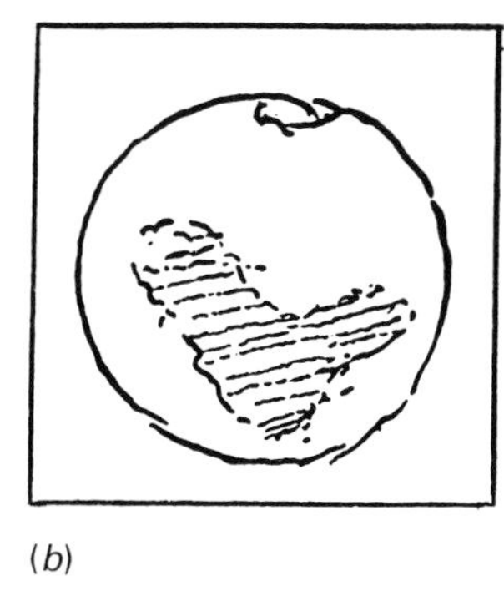

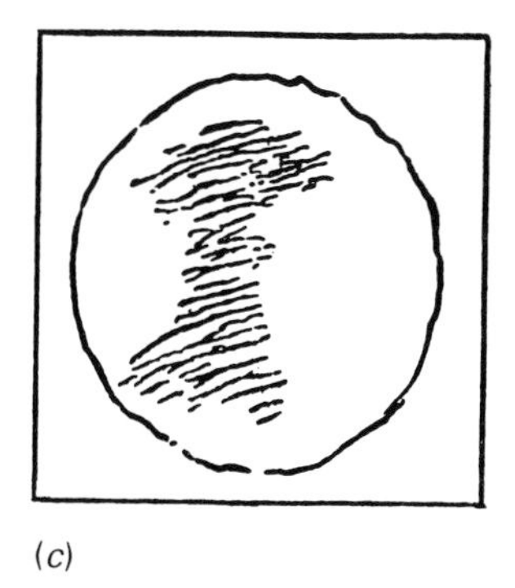

(*a*) (*b*) (*c*)

Fig. 7.9 Telescopic drawings of Mars by Huygens: (a) *November 28th, 1659;* (b) *August 13th, 1672, 10.30 pm;* (c) *May 17th, 1683, 10.03 pm.*

which he gave as '24 terrestrial hours'. This first approximation was refined in 1666 when Mars was in opposition by G. D. Cassini who observed that the dark markings on Mars were seen in the same positions 40 minutes later on successive nights. In this case Cassini discovered the polar caps of Mars. Mars was in opposition in 1672 and G. F. Maraldi observed the planet with the sole purpose of establishing its axial rotation accurately. However, it was not until the 1704 opposition that he actually corrected a previous estimate made by his uncle by shortening the axial rotation time by one minute. Actually, a correction of $2\frac{1}{2}$ minutes would have given a value closer to the actual value of the sidereal period.

In this same year of 1704 Maraldi was convinced that the white patches at the poles and the dark areas underwent changes between oppositions. At the opposition of 1719 he stated that he had also detected changes in the surface markings

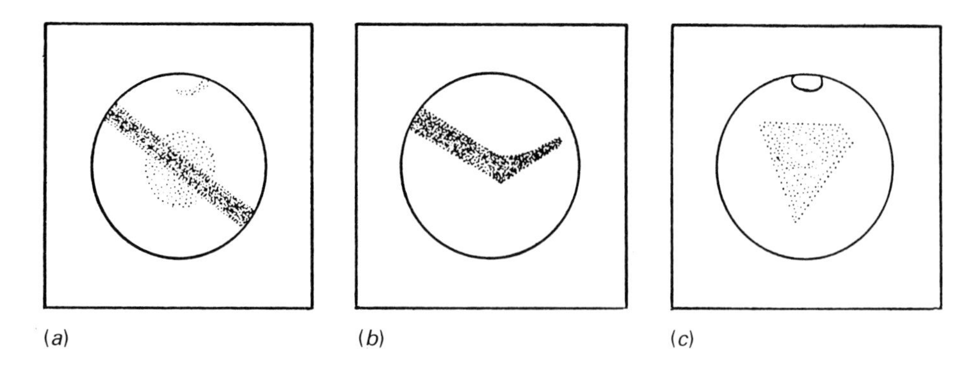

Fig. 7.10 Maraldi's drawings of Mars: (a) *July 13th, 1719;* (b) *August 19th, 1719;* (c) *October 16th, 1719.*

during the opposition. Three of Maraldi's drawings of Mars made at the 1719 opposition are shown in fig. 7.10. He considered that he had witnessed changes in what might have been cloud formations and he noticed that the south polar cap was not placed exactly at the true pole. He did not commit himself as to what the polar cap might consist of.

During the next 50 or so years no further progress seems to have been made in our knowledge of Mars. However, there came a set of favourable oppositions in the years 1777, 1779, 1781, and 1783 and it was Sir William Herschel who then took up the study of Mars with telescopes superior in both size and optical quality. He studied the axial rotation of Mars, which he established as 24 hours 39 minutes 22 seconds (solar) or 24 hours 37 minutes 10 seconds (sidereal). He saw the polar caps and measured the inclination of the axis of Mars which he found to make an angle of 59° and 42′ to the ecliptic. He also measured the oblateness of Mars. Like Maraldi he noticed the eccentric situation of the polar caps with respect to the true poles and he was convinced that the caps were composed of ice and snow. In 1783 he first detected the changing dimensions of the polar caps with the Martian seasons. He was equally convinced of the existence of a Martian atmosphere and observed colour changes on Mars.

Nothing of any real significance was done after Herschel's papers on Mars were published until the 1800s. About two years later after Herschel made his observations, Schröter observed the red planet at his private observatory at Lilienthal, near Bremen, Germany. His observations were carried out during the oppositions from 1785 to 1802. His results were not published until 1887 after his death. Apparently, Schröter coined the term 'areography', the study of Martian surface topography.

Two other German observers more noted for their lunar studies were W. Beer and J. H. von Mädler who observed from Beer's private observatory at Berlin. They observed with a $3\frac{3}{4}$-inch Fraunhofer refractor and produced their famous 'Mappa Selenographica' (map of the moon) in 1834. They also drew the first map of Mars, which although not up to the same standard as their moon map, was the first attempt to bring together in one chart all the then available information on Martian topography (fig. 7.11). In 1840 Beer and Mädler established the arbitrary meridian of zero longitude close to the present zero meridian.

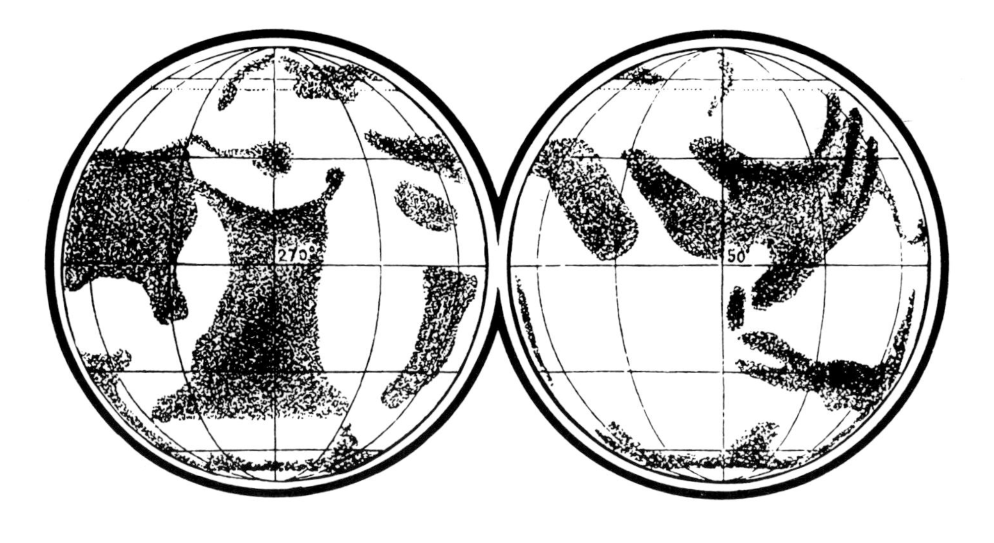

Fig. 7.11 Beer and Mädler's map of Mars.

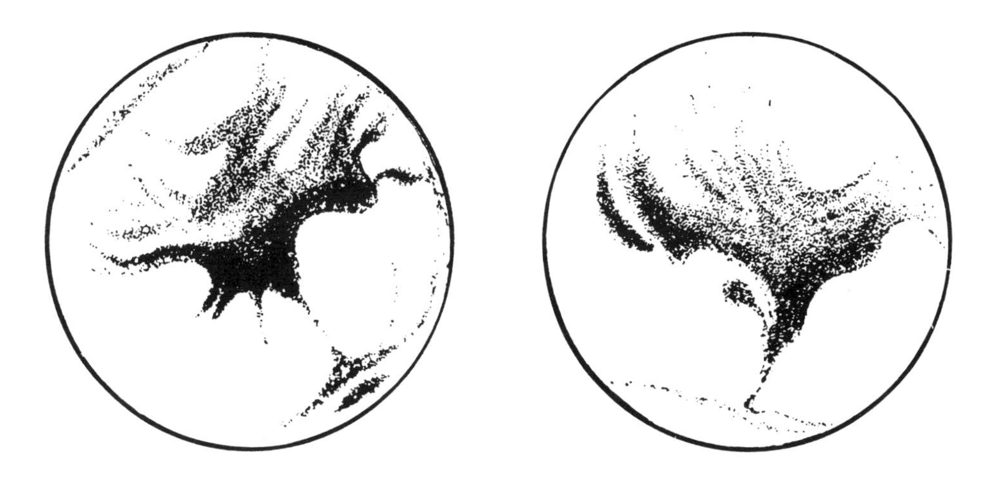

Fig. 7.12 Two drawings of Mars by W.R. Dawes (1864/5).

Several other notable astronomers carried out valuable work whenever Mars presented itself in favourable opposition. Perhaps the best work was done by such observers as Sir J. Norman Lockyer in England and D. F. Arago, F. Kaiser and Father P. Secchi on the Continent. In England, the Rev W. R. Dawes observed Mars during the opposition of November 1864 and executed several beautiful drawings of the planet (fig. 7.12).

A map of Mars was published in 1867 by Richard A. Proctor based on drawings by Dawes which was the first in which names were attached to surface features (fig. 7.13). Proctor named the most important features – 'continents' and 'seas' – mostly after English astronomers, for example: Herschel Continent,

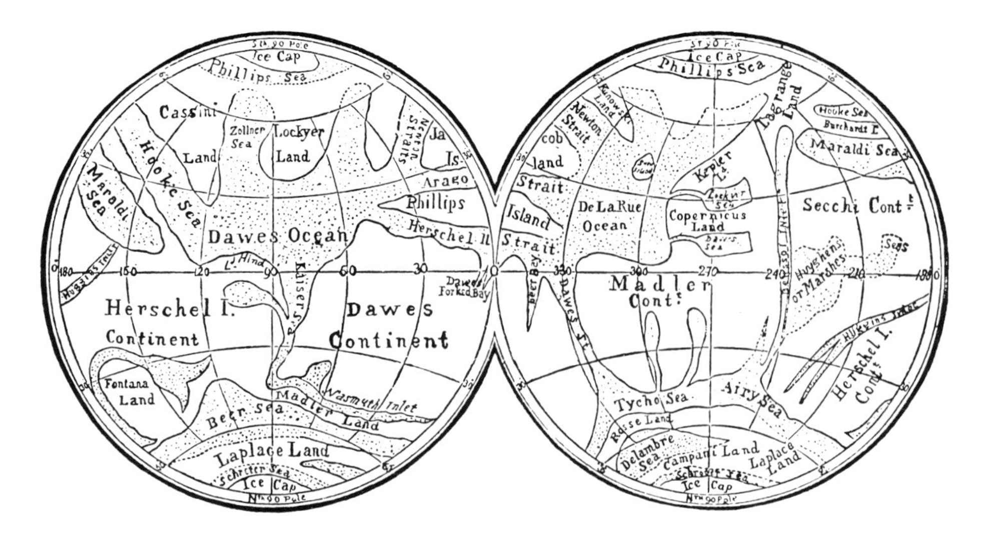

Fig. 7.13 Proctor's map of Mars (1867).

Lockyer Land, Dawes Ocean, Dawes Continent, Delarue Ocean, Kaiser Sea. For the arbitrary meridian of zero longitude he used the location of a dusky feature which he designated 'Dawes Forked Bay', practically the same as that established in 1840 by Beer and Mädler. This location of the zero meridian has remained in use up to the present. It is now more precisely defined by the centre of a small nearly circular crater which lies within a larger crater lying very near the centre of the Dawes Forked Bay (nowadays named 'Sinus Meridiani'). The larger crater is named Airy and the smaller Airy-0 in honour of Sir George Airy who installed the transit telescope at Greenwich Observatory which marks the meridian of zero longitude on Earth.

Wislicenus of Strasburg used the micrometer to establish the positions of the more prominent Martian surface features accurately.

Bright projections on the terminator of Mars in its gibbous phase were first seen by Knobel. On many occasions in 1873 he saw and drew a brilliant white spot on the terminator but did not see it when Mars was at the same phase in 1884.

After the publication of Proctor's map there were no really favourable oppositions of Mars until 1877. In that year Mars was just past perihelion at opposition. G. V. Schiaparelli of Milan observed Mars with his 8¾-inch refractor. He planned to use this favourable opposition of Mars to measure accurately the latitudes and longitudes of various points on the surface of Mars. He was fortunate in having fine seeing conditions and was amazed at the amount and complexity of detail revealed by his telescope on the tiny orange disc of Mars. He discovered hundreds of features and detected inaccuracies in the shapes and positions of several of the features on published maps. He therefore set out to construct an entirely new map and to discard the old names of surface features. In their place he used in part ancient names from the Bible and Classical mythology. Thus, the old Kaiser Sea was renamed Syrtis Major, the Dawes Sea became the Solis Lacus (Lake of the Sun) and Delarue Ocean became Mare Erythraeum (Red Sea).

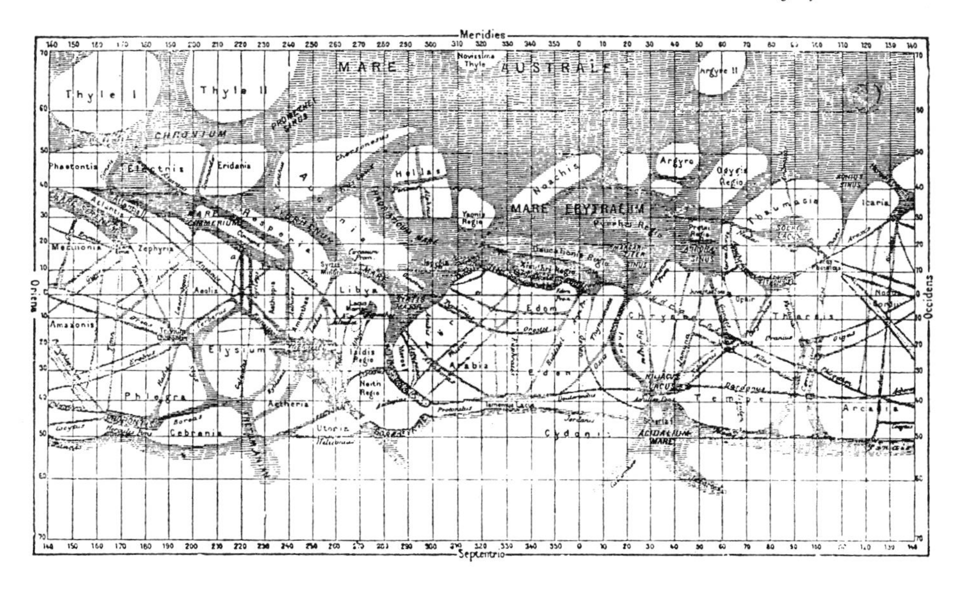

Fig. 7.14 Schiaparelli's map of Mars.

Examples of mythological names are Mare Sirenum (Sea of Sirens), Atlantis and Elysium. Schiaparelli's map (fig. 7.14) is the foundation of modern areography and his names for surface albedo features are still in use today.

At the favourable perihelic oppositions, the southern hemisphere of Mars is turned towards the Earth so its features have become better known than those of the northern hemisphere which is turned Earthwards only during the unfavourable aphelic oppositions. In the 1884 opposition, Knobel set out to combat this inequality and produced several observational drawings of features of the northern hemisphere. He noticed no irregularity in the north polar cap.

The 'canals'

Who has not heard of the 'canals' of Mars? Two important discoveries were made in 1877. First, Asaph Hall announced his discovery of the two very small satellites of Mars, Deimos and Phobos, with the 26-inch refractor of the US Naval Observatory at Washington DC. The second discovery was made by Schiaparelli. He announced the existence of very thin dark lines on the lighter areas of Mars that connected with the darker areas (fig. 7.15). Although he had no logical reason for doing so, he considered these features to be below the general level of the surrounding terrain, i.e., that they were grooves. He therefore named them 'Canali' which simply means 'grooves' or 'channels' in his report of 1878. The translation into English of this word is 'canals' which implies artificially constructed waterways which immediately led to one of the best known controversies in the history of astronomy. Artificial or not, there was serious disagreement over whether the canals really existed since some observers claimed to see them plainly whereas others saw no trace of them.

Such features had been seen before; two canals are clearly shown on one of Beer and Mädler's drawings of 1840 and the drawings of Dawes show many that can be identified with canals shown on later maps. They appear on drawings by

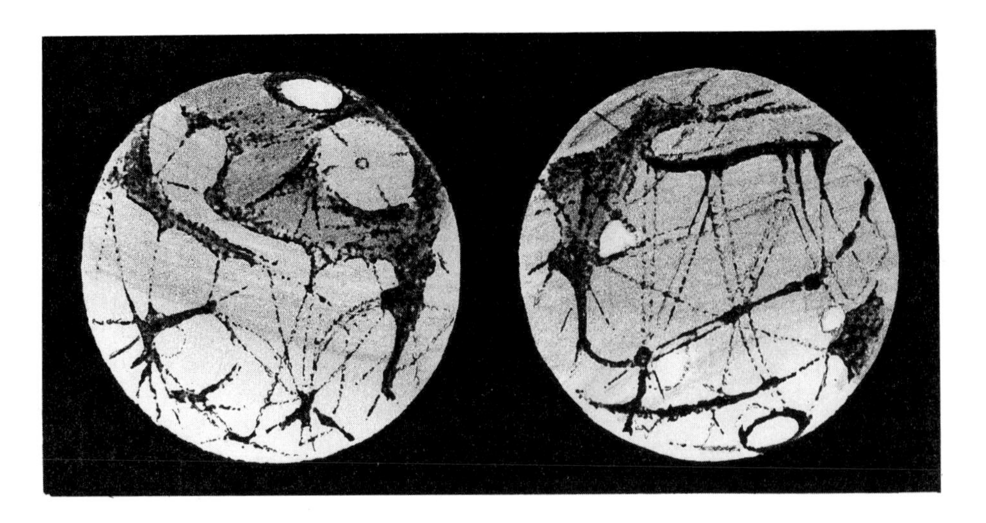

Fig. 7.15 Two views of Mars showing 'canals', 'seas', 'oases', deserts and polar caps, south (left) and north (right).

Sir Norman Lockyer, De La Rue, Kaiser and Lassell, all before 1877. A few canals are indicated on Proctor's map. However, Schiaparelli seems to have seen many more than previous observers. Oddly, no one else seemed to have seen canals on Mars during the 1877 opposition.

Although it is often said that Schiaparelli discovered the canals, he himself never claimed this honour and was well aware of the sightings of these features previous to his own; neither did he originate the term canali because Father Secchi had used the term before him in 1864 although admittedly applying it to a rather short and wide streak. Schiaparelli showed no less than 40 canals on the map of Mars he constructed from many observational drawings that he made in 1877.

In addition to the map based on his 1877 observations Schiaparelli made another based on observations carried out in 1879, again with the $8\frac{3}{4}$-inch refractor. Even today his maps are the clearest and most reliable of those compiled from Earth-based observation and were especially valuable for the detection of secular surface changes but naturally they are exceeded by those constructed from space probe imaging. Some other maps of Mars are the following:

Antoniadi (1924, 1929): visual.

S. Ebisawa: updating of Antoniadi's 1929 chart.

J. A. Roth: colour wall map based on Pic du Midi photographs. (distributed by North American Aviation Inc.)

Gerard de Vacouleurs (1939, 1941): based on combined visual and photographic observations.

G. de Mottoni: updating of Antoniadi's 1929 chart.

Blunck lists 108 maps and globes of Mars. See his *Mars and its Satellites* (2nd edition), pages 188–200.

A larger scale mapping project was initiated at Harvard Observatory in 1958 and has been continued at the University of Texas since 1960. It has been supported by several organisations including the Jet Propulsion Laboratory.

In the winter of 1881, Schiaparelli announced a curious phenomenon displayed by the canals. In place of the single canal he designated Nilus on his 1877 map, there were now two closely parallel canals. This was the first example of the phenomenon of apparent doubling of Martian canals that later was called 'gemination' which he first observed during the apparition of 1879. He noticed doubling of the canals again during the opposition of 1881 and considerable linear structure within the dark areas. Gemination may take place quite rapidly, in a few hours or days at most. The gemination of Martian canals was doubted by astronomers generally for many years but was verified on several occasions and independently by M. Terby, Williams and the Lick Observatory astronomers – although of the last at least one (Barnard) doubted the existence of canals. The phenomenon was most reliably confirmed by Perrotin and Thollon.

During the not too favourable opposition of 1886 which occurred shortly after aphelion, Perrotin and Thollon observing at Nice with a 15-inch equatorial refractor saw several canals, some of them double. They constructed a partial map of Mars that showed more than 20 canals, many of them appearing double. Their map agrees well with Schiaparelli's except that they seem to have used a 'heavier' technique of drawing.

Canals were also seen by observers in England (Burton in 1879 and Maunder in 1881), Belgium and the USA during the 1886 opposition. All the delineations agreed in that the canals always connected up with the dark areas; none ever ended in a light area.

In June 1890, Schiaparelli again observed gemination of the canals using powers of 350 and 500 on an 18-inch Merz refractor (fig. 7.16). A. S. Williams at Brighton saw 43 canals, 7 of them double, with a $6\frac{1}{2}$-inch reflector by Calver using

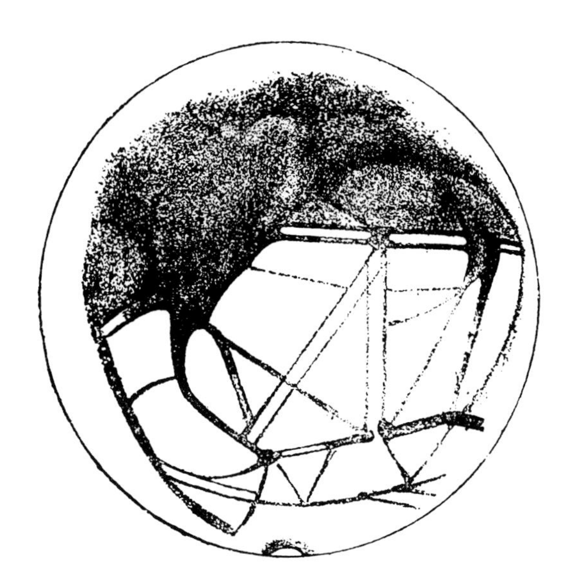

Fig. 7.16 Drawing of Mars by Schiaparelli showing 'gemination' of canals (June 20th, 1890). (From Observational Astronomy, *Mee, A.D. Owen And Co. Ltd., Printers, Cardiff, 1893.)*

powers of 320 and 430. He was the first observer in England to see the canals. Shortly afterwards, Schiaparelli gave up active observing owing to failing eyesight.

Curiously, Asaph Hall, who discovered the tiny and elusive satellites of Mars, never saw a canal. Neither did Antoniadi, a great Mars specialist, until late in life after many years of observation and then only reluctantly did he admit that a few might exist. The visibility of the canals seems to depend on adaptation of the eye to the telescopic image; canals are never seen immediately on looking through the eyepiece. Visibility is also strongly dependent on the moments of perfect seeing. The canals then seem to 'jump out' as fine spider web-like lines.

Two who really believed in the objective existence of the canals were William H. Pickering and Percival Lowell. Lowell devoted many years to the study of Mars from his private observatory at Flagstaff, Arizona, that he equipped with a fine 24-inch refractor. His drawings and maps of Mars are literally criss-crossed with fine spider-web-like lines. He firmly believed that Mars was inhabited by intelligent beings who had constructed the planet-wide system of canals. In 1906 he published a book entitled *Mars and its canals* which started a dispute the like of which has seldom been heard in scientific circles. This book was followed in 1908 by his *Mars as the abode of life* in which he expounded his ideas about intelligent life on Mars in much detail and with great logic.

Around 1892 W. H. Pickering noted that wherever canals intersected there was always a small dark spot. He called these 'oases'. At about this same time, a year in which a nearly perihelic opposition of Mars occurred, a fairly integrated picture of Mars as a planet had been evolved. This was summarized by Schiaparelli in a lengthy contribution to *Natura ed Arte* for February 15th, 1893. This view of Mars did not differ too much from Herschel's even though Schiaparelli had made the greatest contribution to the mapping of Mars than anyone previously.

The reality of the canals seems to be supported by the fact that different observers see the same canals in the same positions on the Martian disc but there are many exceptions (see Sheehan's *Planets and Perception*). This makes it fairly likely that they are not products of imagination or wishful seeing. Antoniadi himself said that the canals were partly illusory but have a basis in fact. Under the finest seeing conditions using the 33-inch Meudon refractor he had seen canals resolved into more or less linear configurations of dark spots and patches. In less than perfect seeing the eye tends to join these together in continuous lines.

Dr Robert S. Richardson of the Mount Wilson Observatory had spent years looking in vain for Martian canals and it wasn't until 1954 that he certainly saw first one and then two others with some degree of probability.

The dark areas

The dark areas undergo regular changes in colour and intensity corresponding to the cycle of Martian seasons. Lowell and Douglas during 1896–7 observed the change of colour of the southern hemisphere maria from green to brown as midsummer approached. The colour of the maria then changed to yellow. This confirmed earlier observations by Liais, Trouvelot and Schiaparelli. The seasonal changes in colour and intensity of the dark areas were established by the year 1900.

In 1924 Antoniadi and Bardet, observing with the 33-inch Meudon refractor, saw most of the greenish areas change to a brown colour as discolouration spread from the south polar region. Even lilac-brown and carmine tints were seen. The brown colouration was short-lived and the Martian date of its appearance varied. Antoniadi was not able to detect the subsequent change to yellow. Fourier and de Vacouleurs observed that the principal seasonal changes seem to be intimately associated with shrinkage of the polar caps; in early spring the circumpolar area darkens considerably, the dark brownish colour moving quickly over the temperate zone. It reaches the equator and extends into the tropics of the other hemisphere. Hence, some areas such as the Mare Erythraeum have a double seasonal cycle.

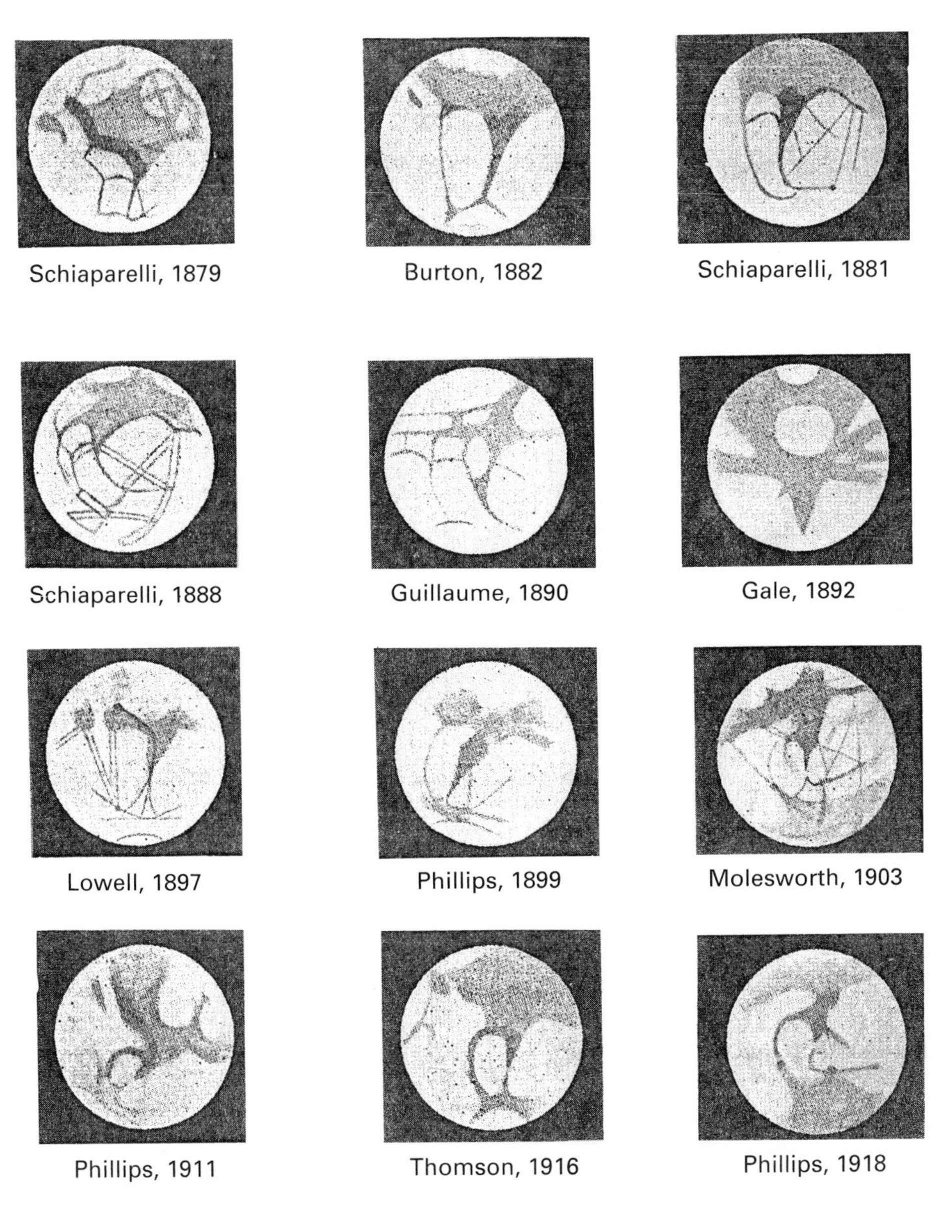

Fig. 7.17 Changes in Syrtis Major, 1879–1918. (From Splendour of the Heavens, *Hutchinson, London, 1923.)*

In summer most dark areas have changed from greenish (probably a contrast effect) to brownish tints. The polar regions become pale once more. The darkening seems to have a tendency to follow the course of large canals like Hellespontus which line up with fissures in the polar caps.

Other seasonal changes observed are changes in intensity and extent of the dark areas. A well-known example is the spreading of Syrtis Major into the light-coloured area called Libya during the Martian Autumn. A well-marked seasonal change in a dark marking is that undergone by Pandorae Fretum. This feature attains its maximum darkness in the Martian Summer and then covers an area of over one million square miles. It is then nearly as large and as dark as Sinus Sabaeus which lies parallel to it and northwards. It fades again in Autumn–Winter and almost disappears and reappears again next Spring. Seasonal and long-term changes have been observed in Hellas. These appearances and disappearances are not caused by atmospheric veils or haze.

Irregular or secular changes of intensity or extent of the dusky regions have also been noted. A remarkable example of this is Solis Lacus. This feature has gone through many changes of size and form since 1892. Oddly, after undergoing a major change it reverts to its original size and shape. Usually Solis Lacus is elliptical in shape measuring about 500 miles (800 km) long and 300 miles (480 km) wide, the major axis being east–west. Maraldi showed it like this in 1704. In 1926 the noted English planetary observer, Rev. T. E. R. Phillips, noted a change in the shape of Solis Lacus. Whereas he had previously seen it as Maraldi had drawn it, the major axis was now in a north–south orientation. Later in the same year Antoniadi observed Solis Lacus with the 33-inch Meudon refractor and made a drawing showing it as consisting of three separate components. It reverted to its pre-1926 appearance in 1930. Then in 1939 it was seen as consisting of

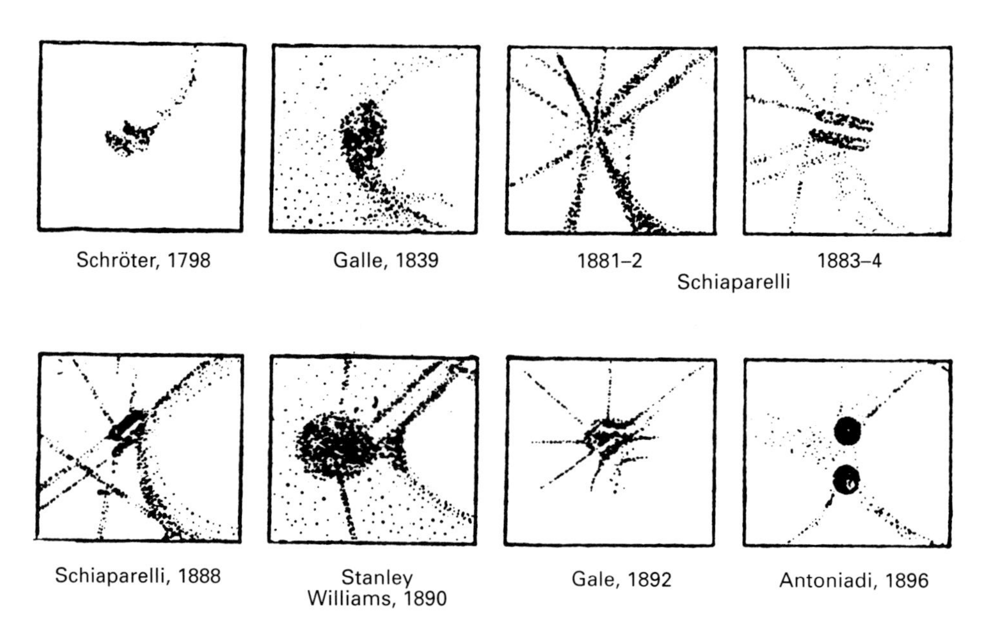

Fig. 7.18 Changes in the Trivium Charontis region of Mars, 1798–1896. (From Splendour of the Heavens, *Hutchinson, London, 1923.)*

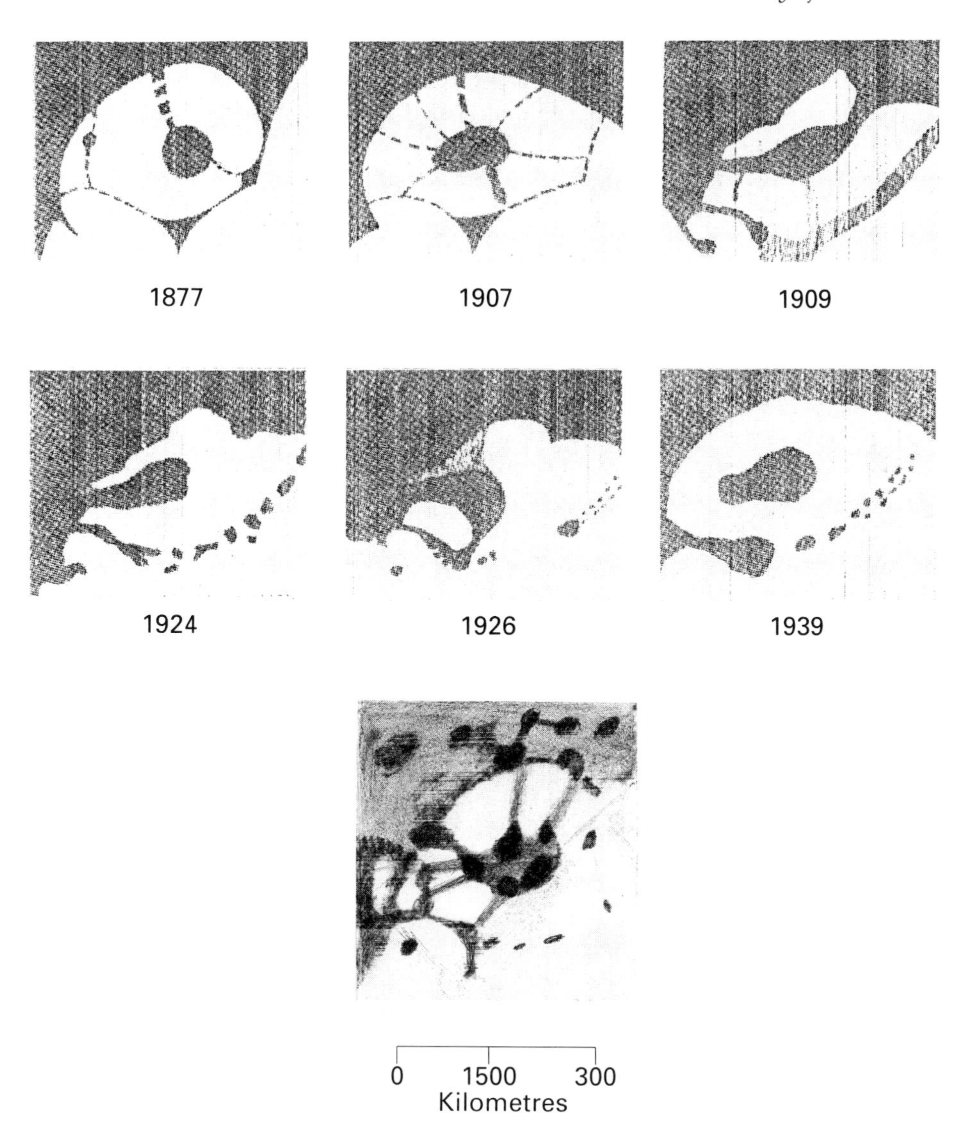

Fig. 7.19 Changes in Solis Lacus: oppositions 1877–1939 (from The Book of Mars, Gladstone, S. National Aeronautics and Space Administration, Washington DC, 1968), and 1988 (a CCD photograph, taken with the 42-inch reflector at Pic-du-Midi).

small dark spots in a dusky area. By 1975 Solis Lacus had reverted to its 1926 appearance.

The darkening undergone by some areas such as Casius (1903), Sinus Gomer (1924) and the dark feature near Nuba Lacus (1954) appears to be permanent.

Another remarkable and unexpected change was the growth since 1909 of the Nepenthes–Thoth system of canals, situated to the east of Syrtis Major. Syrtis Major itself underwent marked changes in form and appearance during the period 1879–1918 (fig. 7.17), Trivium Charontis between 1798 and 1896 (fig. 7.18) and Solis Lacus between 1877 and 1939 (fig. 7.19).

Transient dark spots have been observed from time to time and may be a special kind of feature all sharing a similar genesis. Examples are those that appeared in August 1909, the dark spots that appeared simultaneously with the storm of July 1922, the Aethiopis dark area in October 1926 and the many dark spots that appeared in August–September 1956.

Atmospheric phenomena

Next to the surface features the atmospheric phenomena have claimed almost as much attention. That Mars possesses an atmosphere was made evident by phenomena such as the formation of the polar snows, the changing outlines and distinctness of the dark markings, the way that they disappear at the limb, their easier traceability at the terminator, their greater visibility in summer than in winter (as was noted by Beer, Mädler and Lockyer) and the existence of a twilight arc. The twilight arc is the result of the atmosphere causing the terminator to be indistinct instead of sharply defined. The terminator gradually shades off into the dark part of Mars through 8° of longitude at the equator. This indicates an optically-effective (scattering) atmosphere about 41 miles deep.

Some of the larger dark markings have been seen fringed with something white, reminiscent of snow or clouds on high ground. Schiaparelli saw the feature Hellas (formerly 'Lockyer Land') look nearly as bright as the polar caps. Dawes saw changeable white spots, presumably clouds, on January 21st, 22nd and 23rd in 1865 but these were invisible on November 10th and 12th 1864. In 1867 Browning often saw moving faint white spots which, as they approached the limb, became almost as brilliant as the polar caps. Schiaparelli witnessed similar phenomena in 1877.

Some observers have recorded seeing bright crescents ('menisci') illuminating the east and west edges of the disc. They probably have their origin in the atmosphere. Mädler saw several patches of yellow and bluish light on the limb when observing from Dorpat in 1841. Rev. T. W. Webb mentions seeing a greenish patch on the east limb on October 11th 1862.

Yellow clouds were glimpsed visually by Secchi in 1858, by Dawes in 1864 and were recognised as clouds by Green in 1877. They were first photographed by E. C. Slipher during the oppositions of 1920 and 1922 and were most abundant on the morning side of the disc. Photographs taken on April 12th 1920 revealed unusual very large bright areas on both sides of the disc and mostly over Syrtis Major. Photographs taken about a month later showed the bright area on the morning side of the disc almost gone and the evening side bright area had changed in size and brightness but was still easily visible.

A remarkably conspicuous large white spot appeared over Margaritifer Sinus on July 9th 1922. Subsequently it changed position and shape and became fainter but larger within 24 hours.

Yellow clouds can cover large areas of the Martian surface and are probably composed of dust. Astronomers have observed 40 such 'dust storms' on Mars between 1877 and 1986. In 1956 there occurred the greatest 'dust storm' on Mars for more than 60 years (with the possible exception of the global 'dust storm' of 1924). It began on August 20th, became maximal on September 7th and continued until September 22nd or possibly later. It encompassed the entire planet and appeared most dense in the southern hemisphere. Although usually called 'dust storms', it is not certain whether this designation is entirely appropriate.

A curious W-shaped cloud was seen and photographed in 1954. It appeared each Martian afternoon and corresponded roughly in position with a group of smaller dark markings made up of 'oases' and 'canals'. The 'strokes' of the W corresponded to the canals Ulysses and Fortunae and the 'knots' of the W to the oases named Arsia Silva, Asoraeus Lacus, Tithonius Lacus and Hebes Lacus. This suggests that water may be a possible component of the dark markings. It is now known that these clouds coincide with the Tharsis bulge and volcanoes and are therefore orographic clouds.

As well as blue, white and yellow clouds there are also grey ones. Antoniadi first noted them over Deucaleonis Regio in 1909 and 1911. Several observers in Japan saw grey clouds in 1950 and 1952. These clouds seem to start at a point and rapidly spread out. They have been estimated to be 60–120 miles (100–200 km) high and may have diameters of hundreds of miles. Three Japanese observers including T. Saheki, saw an enormous circular grey cloud about 470 miles (750 km) cross over the Erydania and Electris regions on January 15th 1950. Saheki was observing with an 8-inch reflector. Also on January 15th T. Mitani was observing with the 12-inch refractor of the Kwasan Observatory. He saw something like a cloud at the south-west edge of the Martian disc. Later, on the same evening, T. Osawa observing near Osaka saw a dark grey cloud protruding from the south edge of the disc. A grey cloud was seen by S. Ebisawa with a 6-inch reflector on March 29th 1950. Another was seen in 1952 in the Eridania region and yet another, also in Eridania, by Ebisawa on April 16th 1952.

Brilliant flares have also been reported by Japanese astronomers. The most prominent of these was seen by Saheki on December 8th 1951. It lasted from 21.00 to 21.10 UT and looked as bright as a star of the 6th magnitude. Other flares were observed in 1937 (near Sithonius Lacus, 235/+50), 1951 (on Tithonius Lacus, 83/−4) and 1954 (on Edom Promontorium, 343/−2).

The atmospheric phenomenon known as the 'violet layer' and the occasional 'violet clearings' were described in a previous section.

Exploration of Mars by space probes

The Mariner 4 spacecraft was launched in November 1964 and flew past Mars at a distance of 6117 miles (9844 km) on July 15th, 1965. The television cameras on board sent close-up pictures of the Martian surface back to Earth. The results astonished the astronomical community. Mars was revealed to have a rugged cratered surface with a distinctly 'lunar' appearance; there were craters in both the dark areas and the ochre-coloured regions. Wind and water action have altered the surface which has obviously been eroded. There was no trace of canals. Their existence was thus finally disproved.

Mariner 4 was followed by Mariners 6 and 7 in 1969 and craters were revealed again. Mars was shown to be not like another Earth or like another moon; the Martian landscape is unique to itself.

Virtually no correlation exists between the placement and configurations of the dark areas and the underlying terrain. About the only recognisable correlations are that the short broad canal-like feature called Coprates follows the course of a huge canyon named Valles Marineris and Syrtis Major conforms quite closely to the outline of the somewhat triangular platform on which it lies which is immediately to the west of the large depression called Isidis Planitia.

The dark areas are not necessarily depressed basins as was once thought. Some are high and others are low. Syrtis Major, for example, is an elevated area with sloping sides while the Amazonis desert is a sunken trough.

The Mars-bound Mariner 8 spacecraft unfortunately came to grief. It crashed into the Atlantic Ocean a short time after take-off owing to a malfunctioning rocket.

In 1971 Mariner 9, the first of the orbiting Mariners, set off to Mars and upon arrival went into Martian orbit and continued to function for two years. It sent back 8461 pictures to Earth. The purpose of Mariner 9 was to map the planet but it arrived in time for the great dust storm of 1971 that curtained the surface. When the dust cleared, early Mariner 9 pictures revealed the existence of three large volcanos in the Tharsis region and a fourth enormous one at the location previously known as Nix Olympica. This volcano – the largest known example in the solar system – is now called Olympus Mons. The Mariner 9 cameras revealed a huge system of canyons running along the Martian equator, the Vallis Marineris, corresponding in position with one of the dark surface features, a linear marking known as Coprates near the Solis Lacus region. Channels were found that look very much as if they were carved out by running water. The polar caps were shown to consist of layers of dust and ice.

Two or more spacecraft, Vikings I and II, were sent to Mars five years later. The purpose of the Viking missions was two-fold; to study the geology of Mars with cameras aboard the orbiting spacecraft and to search for evidence of biological activity by setting down landing craft onto the surface from the orbiting vehicles.

The Viking I spacecraft entered Mars orbit on June 19th 1976. It finally circled the planet from a minimum distance of 960 miles (1544.64 km) to a maximum of 20 400 miles (32 823.6 km). The site chosen for the Viking I lander was the area called Chryse Planitia, a relatively smooth plain. The Viking II lander was scheduled to land in Utopia Planitia.

On the night of June 22nd–23rd, the improved camera optics on board the spacecraft revealed more detail than had been recorded by Mariner 9; some of the pictures had resolution better than those obtained from Mariner 9 by a factor of 10. Most exciting was confirmation of the Mariner 9 findings of what seem to be ancient shore lines along channels on the Martian surface. This indicates abundant water that once covered the larger land forms in some regions.

The Viking I spacecraft alighted on the surface of Chryse Planitia on July 20th. Viking II arrived on August 7th and touched down in Utopia Planitia on September 3rd. The Viking II orbiter, which at the time of writing is still orbiting Mars, continued to obtain images and other data for two years and the Viking I orbiter for four. The two orbiters sent back to Earth a total of 53 808 excellent pictures of Mars – Viking I secured 33 100 and Viking II 20 708. The landing craft contained equipment for studying atmospheric composition, the composition of the crust, weather and ground quakes. There was also equipment designed to test for the presence of living organisms. The two landers took about 10 600 images.

Needless to say, the Viking missions have given us a much more detailed and intimate knowledge of Mars than was ever obtained by telescopic observation. The most noteworthy feature of the surface of Mars is its division into two contrasting hemispheres separated by a line at an angle of 35° to the equator. The southern half is made up of highly cratered terrain cut through with channels and other topography suggestive of water action and a once dense atmosphere in the very remote past. The north half consists of smooth plains with volcanos and few craters. Straddled across the dividing line between the two types of terrain is a huge elevation called the Tharsis Bulge, literally an enormous lump on Mars. Here is situated the giant Olympus Mons and other huge volcanos. Hellas, almost diametrically opposite the Tharsis Bulge on the other side of Mars, is the largest of several huge roughly circular depressions. The next largest of these are Argyre and Isidis. The polar caps are permanent and are composed of water ice with generally impermanent solid carbon dioxide caps that grow in the Martian Autumn and Winter and shrivel again in Spring and Summer but some carbon dioxide remains all year round at the south cap. The polar caps are surrounded by broad belt-like zones of dust dunes. Regarding the search for 'life' on Mars, the Viking lander results were inconclusive. Not even organic compounds were detected with certainty in the surface materials.

Since the Viking missions, telescopic observation of Mars has continued but detailed accounts of the many observations made since then would be inappropriate here and would take up too much space. They are recorded in the reports of the Mars Observing Sections of the BAA the ALPO and other astronomical organisations. However, a summary of observations made during the exceptionally favourable Mars opposition of 1988 will be included here. In 1988 Mars was near the celestial equator in the constellation Cetus at opposition so that the planet could be observed in both the northern and southern hemispheres and it was unusually high in the sky for observers in midnorthern latitudes.

During the best observing time – from mid-July to late November – the declination of Mars averaged $-2°$ and was never further south than $-5°$. In the previous opposition (1986) Mars remained at declinations between $-23°$ and $-28°$ and the seeing at best was only fair. In the 1988 opposition Mars could therefore be observed through half of the air mass from about latitude 45°–50° that it was in 1986.

The 1988 opposition was perihelic and the apparent diameter of the disc of Mars swelled to more than 23 seconds of arc. This large image size permitted observers with small telescopes to see a surprising amount of detail on the planet. A piece of good fortune was the transparency of the Martian atmosphere during the best viewing period. Usually, the sun's heat at a perihelic opposition is enough to raise the yellow dust storms that can hide surface detail over large areas of the surface as happened during the oppositions of 1956, 1973 and 1977. Most of the planet's surface was hidden by dust during the 1971 perihelic opposition. There was a major dust storm in 1988 between June 3rd and 27th when the apparent angular diameter of the disc was between 10.4 and 12.6 seconds of arc and another in late November, eight weeks after opposition, when the diameter of the Martian disc had decreased to 14 seconds of arc. The storm was heralded by a brightening of the southern hemisphere of the planet and was detectable with small refractors.

During this opposition observers readily saw the light and dark areas and the shrinking of the south polar cap. More careful scrutiny revealed atmospheric phenomena such as fogs and clouds. Hazes were sometimes detected along the limb of Mars. A series of wonderfully detailed photographs were taken in red light using the 61.2-inch Catalina telescope and a camera at the University of Arizona.

Solis Lacus appeared 'normal' in 1988; during late August–September many observers saw fine detail surrounding this feature in the form of spots and lines. Hellas, the large circular light depression lying due south of Syrtis Major, was unusually bright in 1988. During June, observers saw again the dark features Noctis Lacus, Eosphoros, Gallinaria Silva and Acampsis which had been invisible since the nineteenth century. About a month later observers with small telescopes detected a hitherto unrecorded canal-like linear marking joining these features.

Interesting phenomena were seen in the south polar cap by observers using large telescopes. Small bright spots were occasionally seen with the cap. Between June and September many observers saw small linear rifts that made the polar cap look as if it was breaking up. One of these rifts increased in breadth and became prominent as it did in 1909 when it was observed by Antoniadi and named by him Rima Australis.

The best-known features of the south polar cap are the low areas that stay ice-covered for a long time after the cap has disappeared from the neighbouring plains. These are the 'Mountains of Mitchel' (Novissima Thyle). They look like an isolated detached portion of the polar cap. Novissima Thyle was first seen in late June of 1988 and detached on August 7th and could be seen in small telescopes for many weeks afterwards.

Small white clouds (orographic clouds) were visible near the equator of Mars to observers with especially good eyesight using blue and violet filters. These clouds appear around local noon when moist air rises over the volcanos near the equator of Mars. Clouds over Arsia Mons were especially prominent in the afternoon. The plain Libya, to the east of Syrtis Major, was an area of much cloud activity from late September into early December. The fact that such small features can be seen with small telescopes over these enormous distances should be an encouragement to owners of modest optical equipment.

Sometimes, hazes were noticed along the limb of Mars. These are morning fogs and manifest themselves as a slight brightening along the limb. Some evening hazes were seen on September 12th and 13th in 1988.

The north polar hood was seen by observers with even small telescopes and it was especially pronounced in September.

The violet clearing was observed in June and at times in September and October when surface detail became visible in violet light.

The dust storm of late November was signalled by the appearance on November 23rd of a yellow cloud in the vicinity of Nectar in the Solis Lacus region. Two days later two dust clouds were seen, one over Nectar and the other reaching from the south-east part of Thaumasia in a westerly direction across Bosporus and into Argyre. Much of the southern hemisphere of Mars was covered during the last days of November. The storm extended to Noachis by December 1st and spread westerly into Phaethontis and Daedalia. By this time the dust storm covered an area about 3400 by 900 miles (5500 by 1500 km). The storm subsided as the angular diameter of Mars decreased to 11 seconds of arc in December. The storm had covered most of the southern hemisphere of Mars but did not become planet-wide.

This account of the fascinating observations made during this almost once-in-a-lifetime opposition of Mars should convince owners of even modest telescopes that they can see plenty of interesting surface and atmospheric detail on the red planet. It is hoped that amateur astronomers will be encouraged to observe Mars

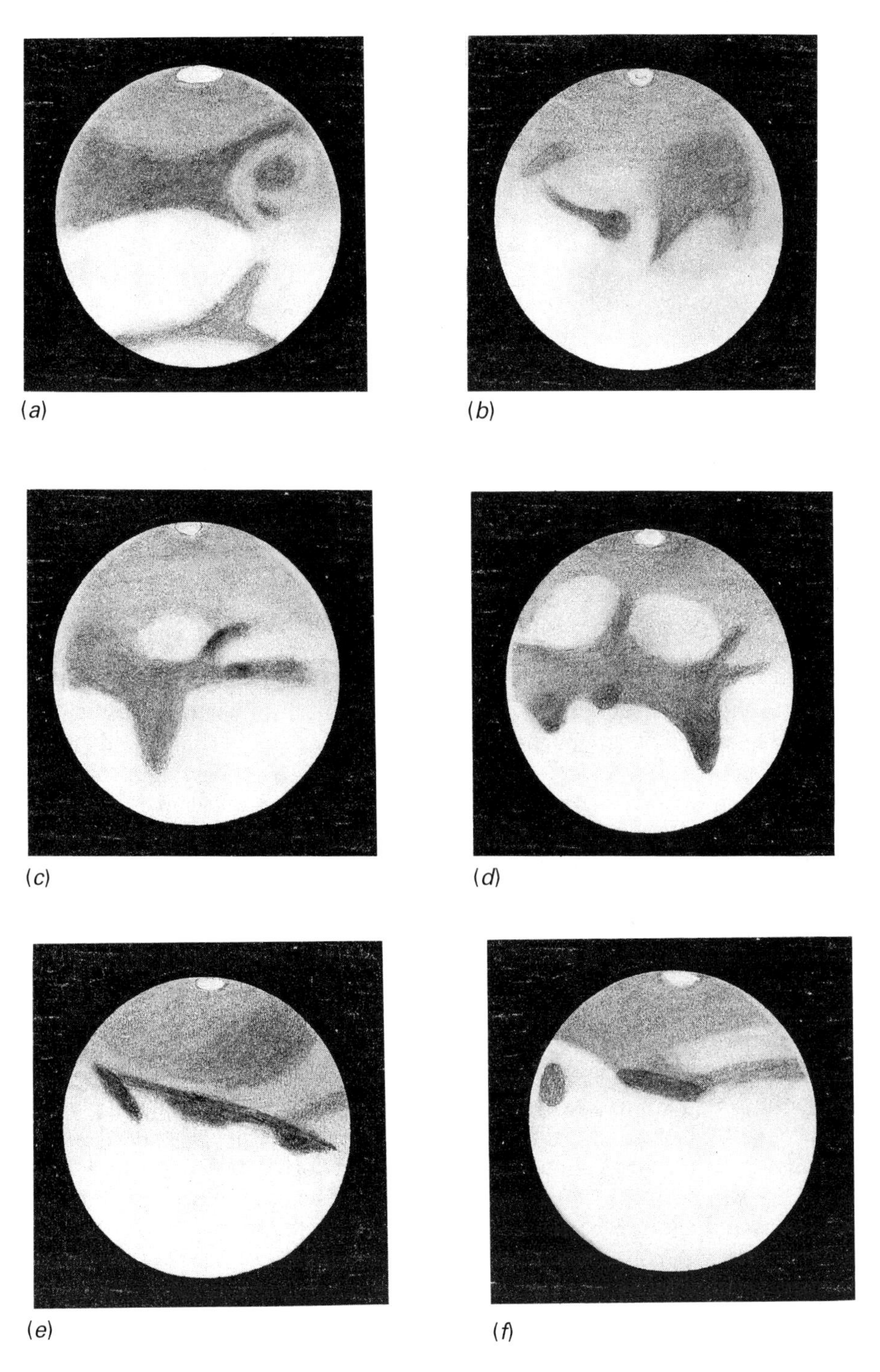

Fig. 7.20 Views of Mars during the 1988 opposition (F.W. Price). 8-inch Newtonian reflector ×225 used with apodising screen vernonscope filter no.23A (orange). Seeing – fair. (a) *September 14th, 1988, 04.15 UT,* D = *23".50;* (b) *September 24th, 1988, 06.50 UT,* D = *23".;* (c) *September 24th, 1988, 03.30 UT,* D = *23".79;* (d) *September 26th, 1988, 02.25 UT,* D = *23".77;* (e) *October 10th, 1988, 04.06 UT,* D = *22".43.;* (f) *October 12th, 1988, 03.17 UT,* D = *22".1.*

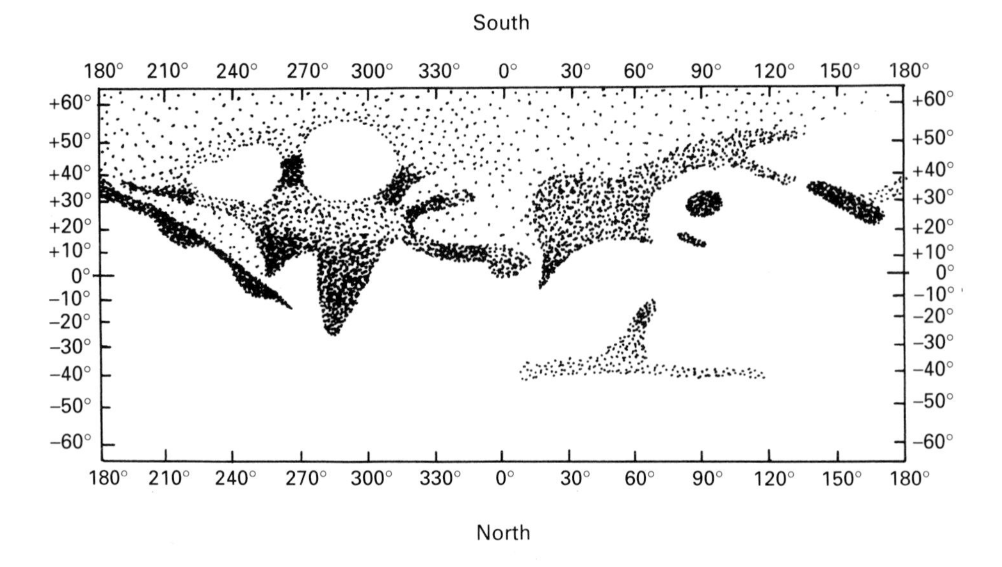

Fig. 7.21 Map of Mars, 1988. Mercator's projection based on observations with 8-inch Newtonian reflector. (F.W. Price.)

as often as possible at favourable oppositions and to study and record the many intriguing and unpredictable changes that are seen.

A selection of some of my own observational drawings of Martian surface features and a map based on them made during the 1988 opposition in average seeing conditions are shown in figs. 7.20 and 7.21. The telescope used was an 8-inch F/7 Newtonian reflector with an apodising screen and a Vernonscope no. 21 orange eyepiece filter. I was not biassed by knowledge of what I ought to see, the drawings and map showing only what I saw with certainty, which cannot be said about some published amateurs' drawings of Martian surface features which have come to my notice.

Visibility of Mars

Mars is easily recognised by its orange-red colour and its greater range of brightness variation as compared to the other planets. Unlike the inner planets Mercury and Venus, which appear relatively close to the sun and are visible in a dark sky for only a short time after sunset or before sunrise, Mars moves all around the ecliptic. It is visible in the night sky for several hours at and around oppositions and culminates on the meridian at midnight at the time of opposition. These apparitions of Mars occur at intervals of 2 years and 50 days. Roughly 50 days after conjunction with the sun, Mars may be seen low down in the morning sky. Gradually it ascends higher each morning until a year after it is first seen in the dawn sky it reaches opposition. It commences retrograde motion 47 days before opposition and continues to retrograde until 47 days after opposition. This is easily observed as Mars appears to trace out a loop-like curve against the background of the fixed stars. Table 7.3 is a timetable describing the cycle of the visibility of Mars and its movements:

Table 7.3. *Visibility and movements of Mars.*

	Mean values	Actual for 1989–1991
Mars seen in the morning sky	54 days after conjunction with sun.	
Beginning of retrograde motion	353 days later	385.7
Opposition	390 days later	424.1
End of retrograde motion	427 days later	458.9
Mars leaves the night sky	726 days later	726.0
Conjunction with sun	780 days later	769.6

Mars is perhaps the most interesting planet to observe with the naked eye. Its apparent movement against the background of fixed stars is so rapid that motion can be detected even after a few days, usually less than a week. Its brightness changes considerably, the difference in its apparent magnitude amounting to as much as 4.6 M. The 1985–91 range was from -2.6 to $+2.0$.

Observing Mars

In the telescope Mars is usually a disappointment to the beginning observer. All that can be seen is a rather small ochre-coloured disc with no definite features. Things are much better when Mars is at or close to a favourable opposition. The disc may attain an angular diameter of 25 seconds of arc if the opposition is perihelic (fig. 7.22). Considerable detail is then visible in telescopes of 4 inches aperture and over. Even a 3-inch refractor will reveal surface features especially if the instrument has apochromatic optics. With such a telescope, changes in the configurations of the dusky markings of the Martian surface can be detected from one opposition to the next.

You may ask 'What is the point of Earth-based observation of Mars since the Mariner 9 and Viking spacecraft gave remarkably detailed photographs of the planet that are better views than any Earth-based telescope can yield?' The answer is that the photographic and other instruments of the spacecraft missions are now turned off. Mars presents an ever-changing spectacle and so if you want to see what is going on now on Mars you must get your telescope and see for yourself. Variable phenomena should be especially studied, particularly meteorology and the polar caps.

Usually, the first Martian surface features to catch the eye are the brilliant white polar caps. When the eye becomes accustomed to the image the dark areas become more easily apparent against their ochre yellow-coloured background, (fig. 7.23).

The Mariner and Viking spacecraft missions and what they revealed have taken away some of the wonder and mystery that were formerly accorded to Mars but the interest and value of Earth-based Martian observation have not been diminished one bit.

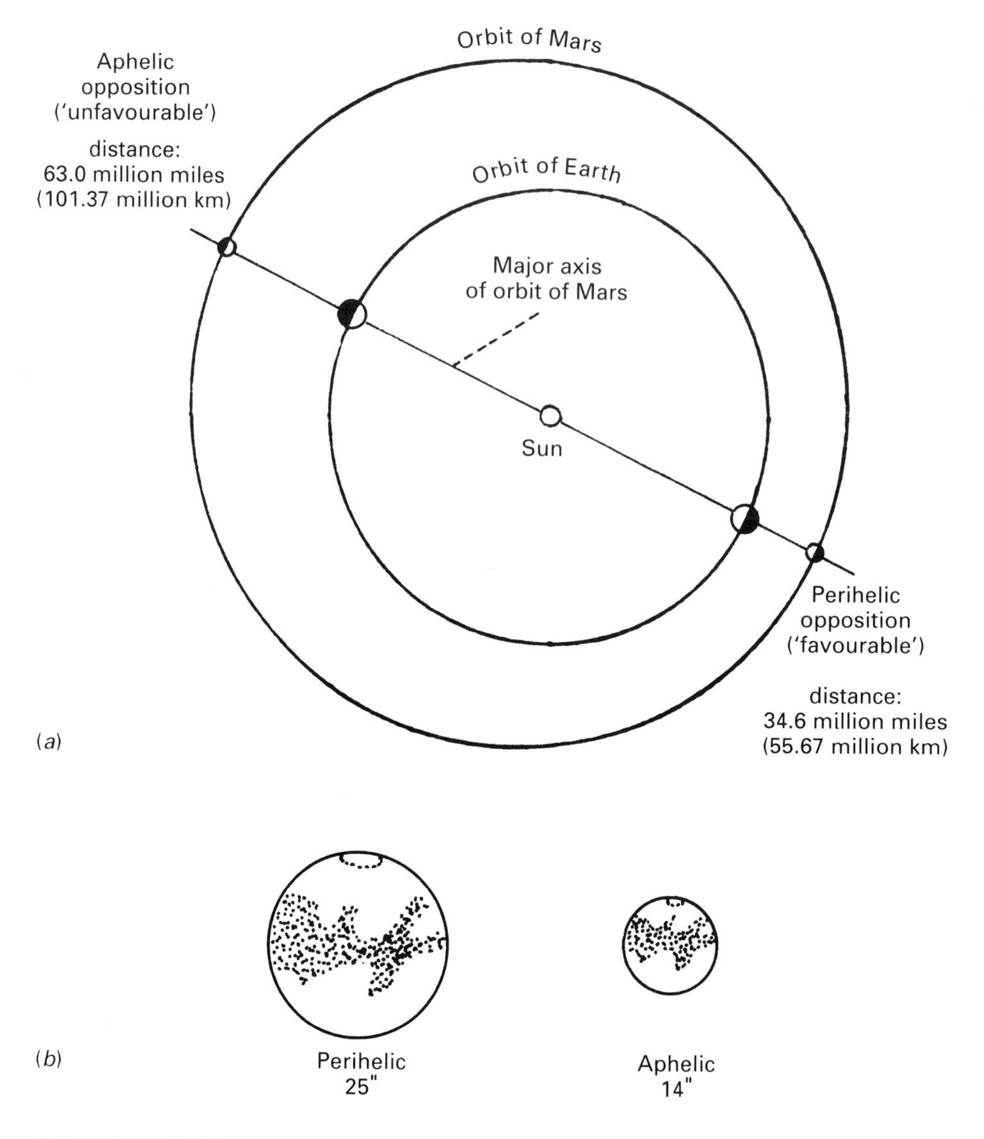

Fig. 7.22 (a) *Perihelic and aphelic oppositions of Mars.* (b) *Comparative apparent angular diameters of Mars at perihelic and aphelic oppositions.*

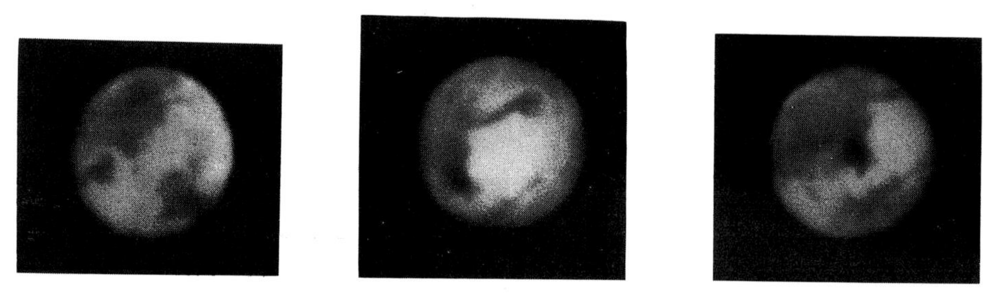

Fig. 7.23 Photographs of Mars during the 1969 apparition.

Telescope, magnification, eyepieces

The type of observing programme that you intend pursuing will indicate the choice of telescope. The aperture should not be less than 6 inches for a reflector or 4 inches for a refractor. Perhaps the optimum instrument from the point of view of resolving power, light grasp and cost would be a Newtonian reflector of 10 inches aperture. It is often said that telescopes with apertures greater than about 12 inches are only rarely more effective than smaller apertures, owing to atmospheric turbulence and tube currents and to a certain extent this is true. However, you should not be discouraged by this from obtaining the largest affordable telescope. The large instrument always tells over the smaller under good seeing conditions. Martian colours will be more truly rendered and readily perceived because the light intensity in the image of a large telescope stimulates the retina to cone cell vision which is responsible for colour perception. Less light intensity stimulates only rod cell vision which is predominantly grey.

Long focal ratios improve the clarity of the image. A refractor or a Cassegrainian may therefore give results better than the usual Newtonian of the same aperture. If you intend using a Newtonian, try to get one whose focal ratio is at least F/8 rather than the commoner F/7 or F/6 types and with a diagonal reduced in size to the minimum possible.

An important attribute of the telescope mounting is that it should have a motor drive and should be capable of tracking Mars fairly precisely. High powers are required, somewhere around 40–50 per inch of aperture, for effective observation of disc detail. Observation of the rather faint diffuse dark areas on the Martian disc is trying and needs patient application, so you won't want to be hampered by having to make frequent adjustments to a non-motor driven telescope.

Adequate magnifications would be as shown in table 7.4. Good quality eyepieces only should be used, the orthoscopic design being best, whether a reflector of refractor is employed. Plössl eyepieces are also good.

Filters

The use of filters in observing Mars is of paramount importance and no observing programme can be considered complete without their use. Filters are used for making surface detail more easily visible and for detecting atmospheric phenomena that might otherwise be invisible. They are useful for reducing glare from the planet's image and also help in objectively evaluating Martian colours.

Gelatin filters are cheap but are easily scratched and damaged, though they are fine if mounted in glass slide holders. Optical glass filters are to be preferred and are worth the extra expense. Manufacturers usually mount filters in threaded

Table 7.4. *Telescope apertures and magnifications for observing Mars.*

Telescope aperture	Magnification
6 inches (15 cm)	300 ×
8 inches (20 cm)	400 ×
10 inches (25 cm)	450 ×
12 inches (30 cm)	500 ×

Table 7.5. *Wratten filters.*

	Colour	Characteristics
25, 25A, 29.	Red	Show great contrast between the dusky areas and the 'desert'. Best with steady seeing and low powers.
15, 21, 23A	Yellow/orange	Perhaps the best filters for surface features. They subdue the brilliance of the reddish areas revealing fine details within them and within the dark marial areas. They also suppress light scattered in the atmospheres of Mars and Earth and tend to make the telescopic image steady.
56, 57, 58	Green	Useful for viewing the dusky band surrounding the polar caps and revealing yellow dust storms. White areas on the reddish surface stand out more clearly.
38, 38A, 80A	Blue	Reveals violet clearings and water clouds in the upper atmosphere.
47	Violet	Reveals violet clearing, high altitude clouds, equatorial cloud bands, limb and polar hazes and polar clouds.

metal holders that screw directly into the base of eyepieces. They should only ever be held by the mount and the finger tips should never be allowed to touch the glass. Because certain Martian features may be see with one filter but not with another it is a good plan to begin an observing session by viewing Mars through a set of filters one after another. This will permit a rapid check to see whether there are features visible with one filter but not another. A good set of filters to acquire is chosen from the Kodak Wratten series. Table 7.5 is a list of the most useful of these filters, their transmission characteristics and the types of Martian features which are 'brought out' by them.

Other accessories

Telescopic studies of Mars may be increased in scope with the aid of the following accessories:

(1) A watch accurate to within one minute of any reliable time signal or WMM radio transmission from short wave radio. With its help you can determine the longitude of Martian surface features accurately by timing central meridian transits (see later).

(2) *The American ephemeris and nautical almanac.* This is obtainable from the US Government Printing Office, Washington, DC. It is published annually and provides useful data for every day in the year such as the longitude of the central meridian of Mars for 00.00 UT, the apparent magnitude of Mars, size of the disc and the planet's phase angle. Also, the *Solar System Ephemeris* published by the ALPO.

(3) A camera. Photography of the planets is described in chapter 15.
(4) A filar micrometer. Filar micrometers at affordable prices are now made commercially. This instrument permits accurate determination of the latitude of Martian features. It may also be used at the recession of the polar caps in the Martian Spring and Summers.

Usually, the micrometer is used to make measurements relative to the apparent size of of the planet's disc. This may be obtained for each day in the year from the *American ephemeris and nautical almanac*: the disc size is given in seconds of arc. In using the micrometer, you first measure the Martian disc size, then the size of the Martian disc for that date.

General hints for observing Mars

The period during which useful observations can be made is usually stated to be limited to about 40 days before and after opposition (see table 7.6). However, observers are encouraged to continue for a longer period because this is necessary to follow the evolution of the polar caps and meteorology. It has also been stated that the amateur observer is better off with visual Martian observation than with photography because the camera cannot reveal delicate detail and subtle contrasts of shading and colour that can be perceived in telescope observation. The advent of CCDs has made this view obsolete.

The apodising screen when attached to the telescope is a great help in reducing the glare of the disc, reducing the effects of atmospheric turbulence and in enhancing contrast. I used it every time I observed Mars during the 1988 apparition.

Accurate recording of detail is possible with a little practice at making disc drawings at the telescope eyepiece. A good quality bond paper should be used attached to a clip board and a medium soft pencil with a soft eraser on the end. This is better than having a separate eraser; you won't want to waste time hunting for the eraser if you drop it. It could bounce almost anywhere. A circle about 2 inches in diameter drawn on the paper will be quite large enough for the Martian disc (The Mars Section of ALPO recommends a disc diameter of 42 mm). Some observers like to use circles of varying size to reflect the changes in apparent angular diameter of the planet before and after opposition but I don't see any special advantage in this.

The telescopic image of Mars is difficult to focus easily because of the glare of

Table 7.6. *Forthcoming oppositions of Mars,*

Year	Date	Angular Diameter (seconds of arc)	Declination at opposition	Magnitude
1993	January 7th	14.9	26.3 N	−1.2
1995	February 12th	13.8	18.2 N	−1.0
1997	March 17th	14.2	4.7 N	−1.1
1999	April 24th	16.2	11.6 S	−1.5
2001	June 13th	20.8	26.5 S	−2.1
2003	August 28th	25.1	15.8 S	−2.7

the limb and because of the indistinct outlines of the dark areas. Particularly avoid the terminator when focusing. The polar caps should be more helpful as objects on which to focus.

Before commencing a drawing, study the telescopic image for a few minutes so as to accustom your eye to it. The polar cap will be most easily visible but little else will be seen on the orange-yellow disc. After a while you will begin to see the maria as faint grey-green diffuse markings. Fig. 7.24 will give you a good idea of the appearance of Mars in a moderate telescope under good seeing conditions and the map of Mars in fig. 7.25 should help you to identify the major features that you see on the disc.

Whether you are studying and drawing surface markings or atmospheric phenomena, draw only what you see. Never record anything that you saw only doubtfully or worse still, what you think ought to be there because you have

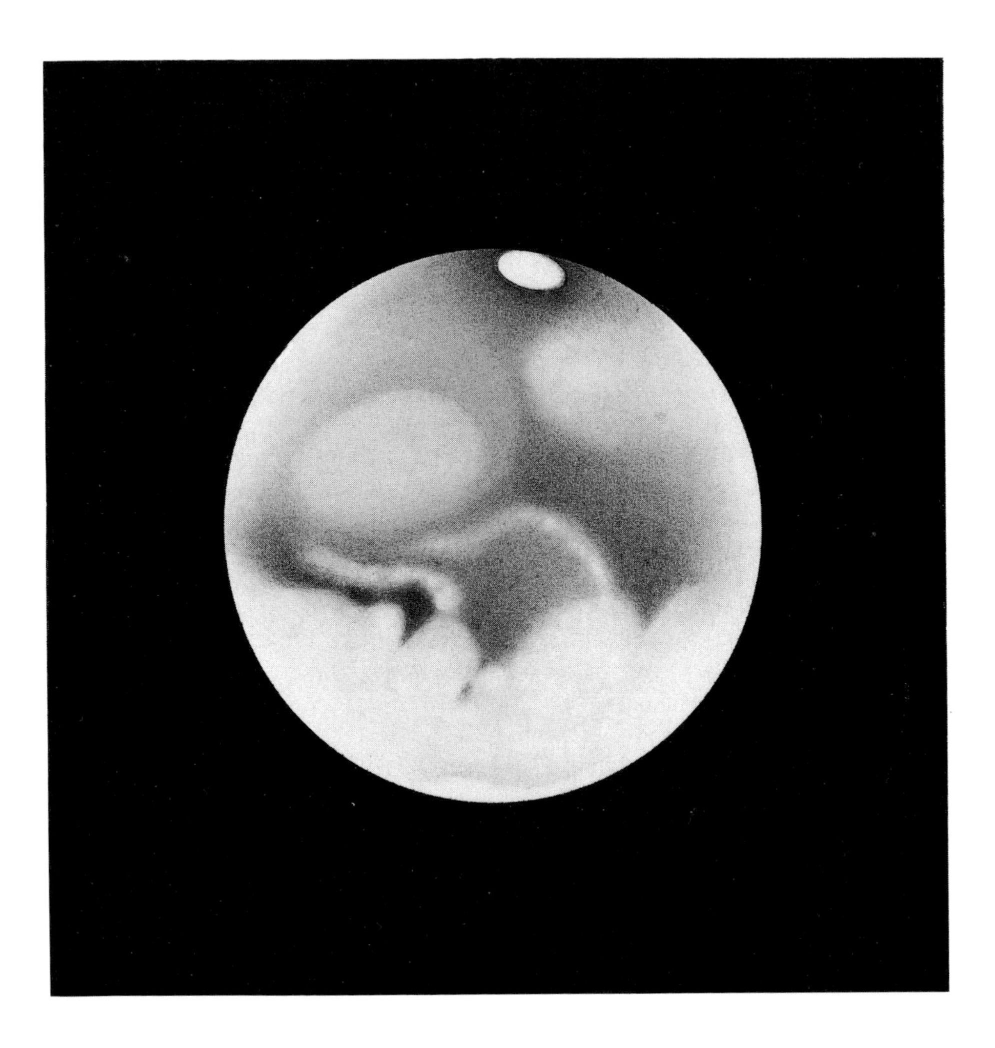

Fig. 7.24 Mars, September 8th, 1988, 22.32 UT, CM 15–32.9°, 115 mm Og, ×186. (R.M. Baum, Chester.)

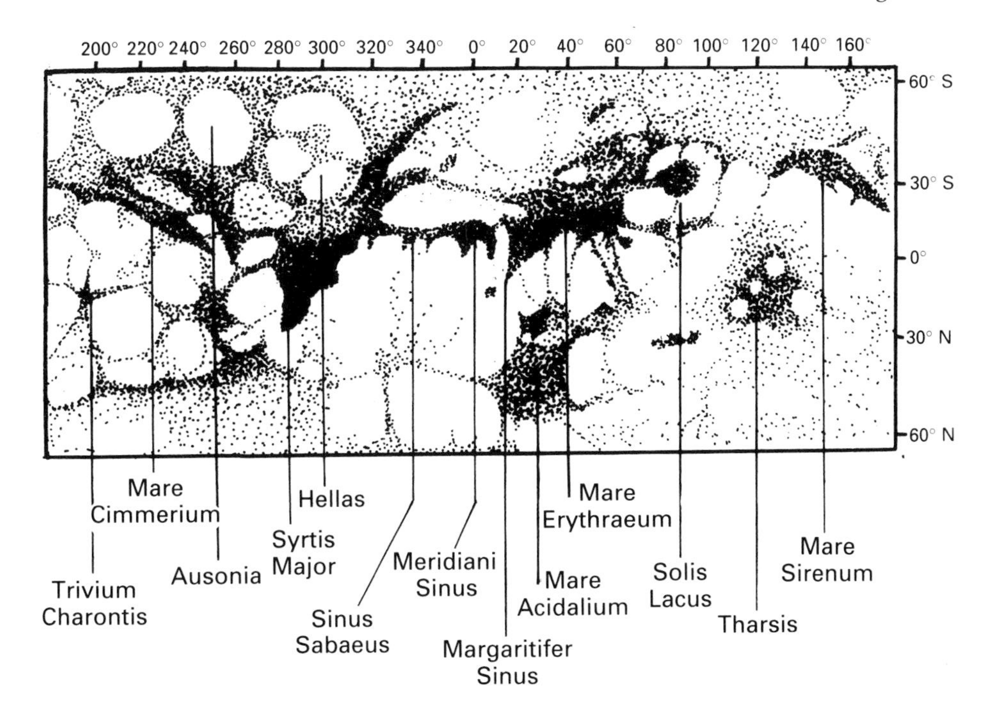

Fig. 7.25 Map of Mars (Mercator's projection) showing principal surface features.

always seen it before. Remember, surface features can be obscured by dust storms.

Try to observe Mars as often as possible when seeing and weather conditions are good. Every second night is good to aim for. If time permits, make two drawings in an evening so as to record a greater range in longitude than a single drawing would afford. If a second drawing is made 2 hours after the first, then a change of about 30° of longitude will have occurred due to the axial rotation of the planet. Remember also that since Mars takes about 37 minutes longer to rotate on its axis than does the Earth, it follows that if you observe Mars on a given night at a certain time, then on the next night the features you saw previously will appear in the same positions 37 minutes later.

When observed on consecutive evenings at the same time Mars will therefore appear to rotate slowly on its axis in a west to east direction. In this way you can witness a complete axial rotation of Mars over a period of about 5–6 weeks. A given feature will be visible from only one Earth hemisphere for an extended period. On the other hand if you observe Mars for several hours on a given night it will be seen to rotate in the opposite direction, its true direction, from west to east.

In making a drawing a good plan is to follow a definite sequence in recording what you see and to keep to that sequence for all observations. I recommend the following:

(1) Delineate the polar cap or caps, if both are visible, and the terminator at an early stage. Note – the polar caps do not indicate the exact north and south

points of the disc because they are eccentrically placed with respect to the true poles.

(2) Prominent features near the centre of the disc should next be accurately drawn. The outlines of large dusky surface markings can be used to position accurately small details that are difficult to see.

(3) Draw features close to the preceding limb (the one on the left of the disc as oriented with south at the top in accordance with the telescopic view.) This is because features near this limb are about to disappear because of the planet's axial rotation. Features near the following limb have recently been brought into view for the same reason.

(4) Look carefully for the fine details such as atmospheric clouds, delicate linear features and dark and bright spots.

Since a drawing of Mars may need about half an hour to complete there is ample time to wait for the moments of extra steady seeing when fine detail becomes visible.

(5) Features seen with filters that were invisible without them should now be added. Include notes detailing which features were seen with what filters.

The following data should be written on your drawing.

(1) Your name, the date and time the observation was made; always use universal time (UT) not your local time.
(2) The type of telescope that you use, its apertures the magnification used and what filters were employed, if any, including the apodising screen.
(3) The transparency of the air. This can be estimated by noting the magnitude of the faintest stars that can be seen in the vicinity of Mars. The figures 1–6 can be used to indicate transparency, these being the magnitude of the faintest stars seen.
(4) Image steadiness on the 0–10 scale (0 = worst, 10 = best).
(5) Note any unusual or short-lived phenomena on the planet.
(6) Visual intensity estimates can be made on a 0–10 scale, or they can be indicated on a separate drawing.
(7) Note the longitude of the central meridian when you commence your drawing.
(8) Be sure to indicate the phase if Mars is gibbous.

Accurate copies should be made of your original drawings. Keep the originals and send the copies to:

The Mars Recorder,
Association of Lunar and Planetary Observers,
12911 Lerida Street,
Coral Gables,
Florida 33156
USA

Features for observation

The polar caps

Perhaps one of the most important and valuable studies that can be made of Martian surface features are observations of the shape changes undergone and the speed of recession of the polar caps during the Martian Spring and Summer in the relevant hemisphere.

The south polar cap is large and can be easily seen at perihelic oppositions with moderate Earth-based telescopes. It undergoes pronounced changes of form and size. As summer approaches at a perihelic opposition we get good views of the south polar cap owing to the strong tilt Earthwards of the axis of Mars. The cap apparently breaks into two sections during the Martian Spring. This is due to the slow melting of the Mountains of Mitchel (Novus Mons), large areas that retain snow. Surrounding the edge of the south polar cap are frequently seen delicately detailed fissures that originate in the thawing cap and radiate out into the adjacent plains. Melting of the south polar cap is shown in fig. 7.26 and in more detail in fig. 7.27.

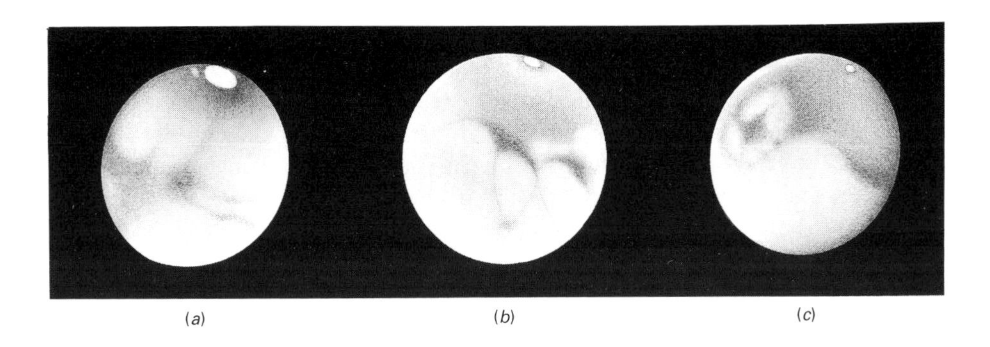

(*a*) (*b*) (*c*)

Fig. 7.26 Mars, 1988, showing shrinkage of the south polar cap: (a) *August 7–8th, 23.00–24.15 UT;* (b) *October 2nd, 21.30–22.30 UT;* (c) *November 11th, 21.05–21.15 UT, 115 mm OG ×186. (R.M. Baum, Chester.)*

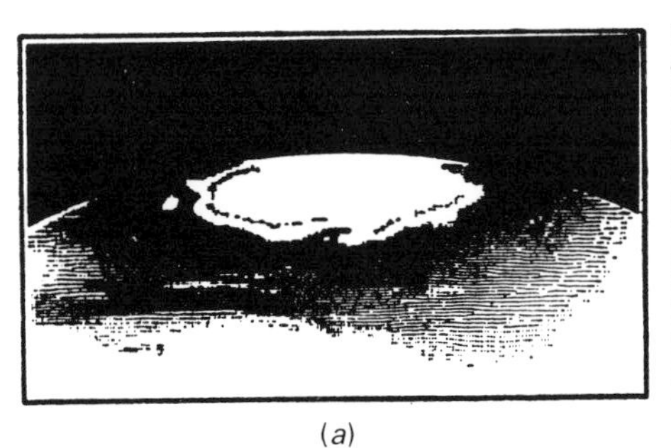

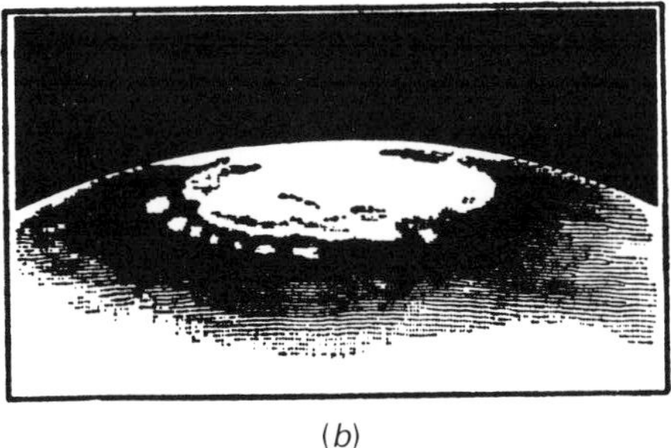

(*a*) (*b*)

Fig. 7.27 Melting of the south polar cap of Mars as observed by Dr Nathaniel E. Green from his private observatory at Madeira: (a) September 1st, 1877; (b) *September 8th, 1877. (From* The Exploration of Mars, *Ley W. and Von Braun, W. Viking Press, New York 1956.)*

The north polar cap does not undergo such pronounced seasonal changes as the southern cap. It never completely disappears even during the summer of an aphelic opposition. It may attain a maximum extent of 72° but never contracts to less than 6°. The behaviour of the north polar cap cannot be predicted which makes it especially interesting and worth observing carefully.

At the approach of Autumn in the Martian northern hemisphere there appears a haze that originates from the polar regions and the decline of the polar cap stops for a while. Strangely, during late summer it sometimes stops receding and begins to grow in size once more. Increase in size of the cap always occurs with the reappearance of the north polar haze. There is also the so-called 'aphelic chill' – the sudden appearance of the north polar haze in late Spring.

Studies of the extent of melting of the cap and the effects and causation of the north polar haze are needed and will be valuable in helping to understand the behaviour of the north polar cap.

Observation of the melting of the polar caps This can be followed by making careful drawings or by measuring the outline and size of the caps on photographs. This should be pursued until Mars is no longer to be seen in the sky.

A filar micrometer may be used to make accurate visual measurements of the polar caps at the telescope. The micrometer is used first to make an accurate measurement of Mars from its north to south pole; useful work can also be done with an eyepiece reticle. Better still, measure from cusp to cusp to avoid the phase. This gives its micrometric dimension for any particular day which is then converted into the degree equivalent of the micrometer thread measurements by looking up the angular diameter of Mars in the *American ephemeris and nautical almanac*. Next, the width of the polar cap is similarly measured and converted. A melting or freezing curve of the polar cap may then be constructed from these measurements (fig. 7.28).

Details within the caps During moments of very steady seeing the polar caps should be scrutinised under high power. Look for any interesting structure in the caps themselves e.g., the 'Mountains of Mitchel' in the south polar cap at about 75° S latitude. These appear on virtually the same Martian date year after year and so provide an exact 'milepost' in the retreat of the polar cap.

Studies of the visibility of polar cap detail give useful data pertaining to the temperature and terrain and how these correlate with atmospheric phenomena.

Meteorological phenomena When the polar caps start to melt look for the appearance of clouds, fogs and mists. Look for these again when the caps begin to freeze once more. These studies should be continued during the Martian local Spring and mid-Summer and also when the cap begins to form again during the Autumn.

During perihelic oppositions the south polar cap initiates meteorological phenomena in the same hemisphere. At aphelic oppositions the north polar cap likewise causes meteorological activity in the northern hemisphere.

Usually the north polar cap haze will be seen during the northern summer and white clouds will appear in the Hellas region when Mars is at a perihelic opposition during our July and August.

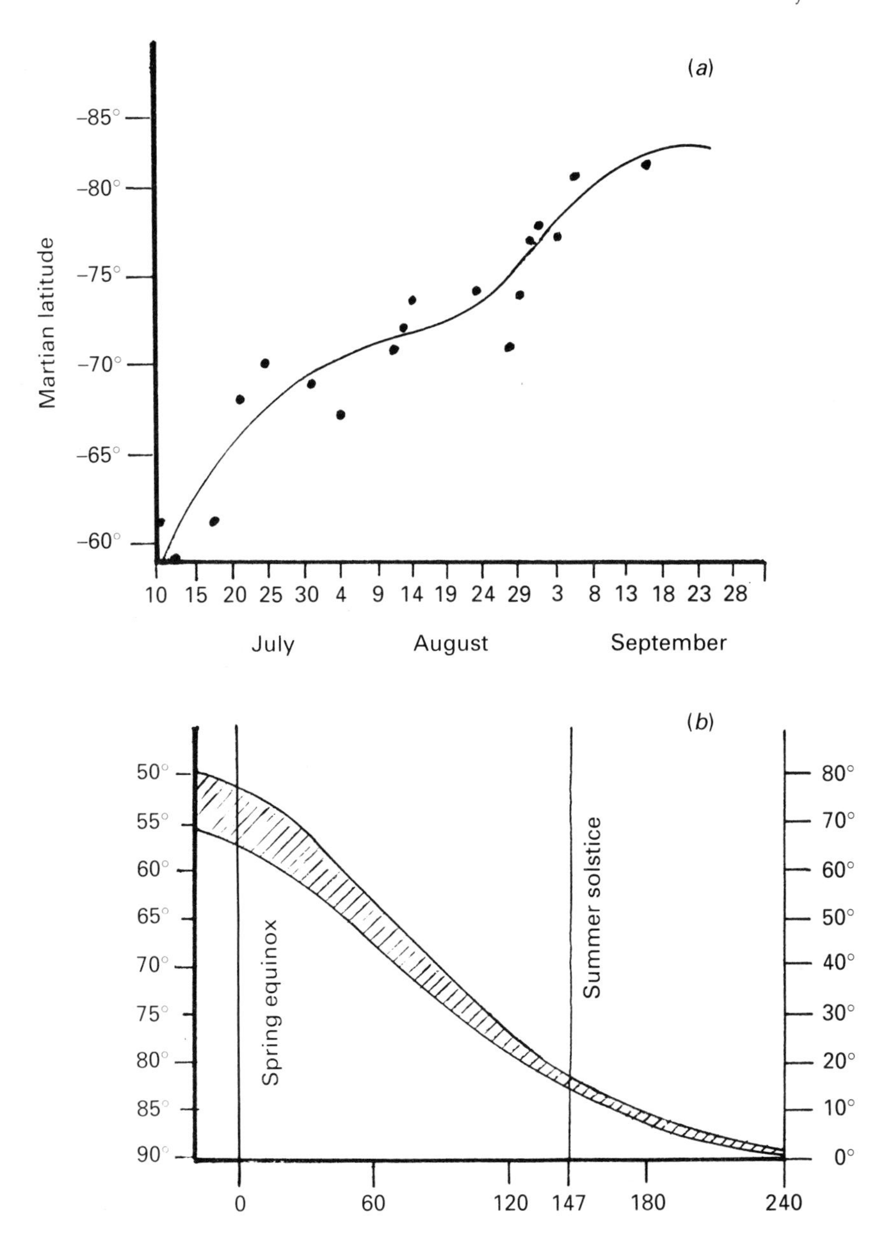

Fig. 7.28 Melting of the south polar cap of Mars: (a) *As recorded in 1956 by W. Sandner.* (b) *Variations of dimensions of the south polar cap during Martian spring and summer recorded by E.E. Slipher.*

Orange, yellow and green filters are useful for studying these phenomena and blue filters for orographic clouds.

The melt line The melt line is the dark band that surrounds the edge of a polar cap that appears as the polar cap melts during Martian early Spring and

Summer. This has often been ascribed to the melting of water ice and the polar cap and the spreading out of liquid water; however, the only place on Mars where surface water can exist, even theoretically, is at the bottom of the Hellas Basin. It seems more likely that the dark polar band or 'collar' is formed by winds blowing off the the polar caps (see Dobbins, Parker and Capen in *Introduction to observing and photographing the Solar System*, page 76). It remains visible until Summer. Comparative studies of the speed of shrinkage of the polar cap and the breadth of the dark melt line are of value. The time when the melt line first becomes visible and when it disappears should be recorded.

The dusky maria and seasonal changes

The seasonal changes on the surface of Mars other than the variations in the polar caps that are most prominent are those that are displayed in the dusky maria. Observation of these should be correlated with polar cap melting.

Watch for the so-called wave of darkening. It starts in the middle of Martian Spring and proceeds until the polar cap has almost disappeared. The darkening spreads rapidly from the polar regions to the equator as the polar caps shrink. It is more noticeable in perihelic than in aphelic oppositions which may be correlated with the different densities of the two polar caps.

The grey-green maria not only darken during the Martian Spring and Summer but they may increase or decrease in size or undergo shape changes. In order to detect these phenomena you will obviously need to know Martian topography fairly well. The darkening of the maria may be roughly quantitated on a scale of 1:10. 1 would indicate the least degree of contrast between a dusky area and its surroundings and 10 the maximum degree of contrast. Some of the features showing pronounced seasonal changes are Pandorae Fretum, Syrtis Major, Solis Lacus Nilokeras/Lunae Lacus and Margaritifer Sinus. Brightening and darkening should be looked for in the maria and in the yellow-orange plains.

The best filters for observing the maria are the Wratten nos. 21 and 25 (orange and red respectively)

Atmospheric phenomena

Not much is being done about Martian atmospheric phenomena from Earth-based observatories apart from amateur astronomical studies. There are five types of atmospheric phenomena that can be effectively studied by the amateur Mars observer:

(1) the violet clearing,
(2) blue-white and white clouds,
(3) yellow clouds,
(4) dust storms,
(5) W-shaped clouds.

It is essential to have a good map of Mars in order to identify and locate accurately short-lived clouds and also a good knowledge of the Martian topography.

The violet clearing When Mars is viewed through a deep blue filter (Wratten 38A) little surface detail will usually be seen apart from a nearly featureless disc and a bright polar cap. Even less is visible if a violet filter is used such as the Wratten no. 47. This is because the blue and violet wavelengths are scattered by the Martian atmosphere so that what we really see with a blue or violet filter is a view of the atmosphere of Mars. Orange and yellow wavelengths are able to penetrate to the surface, except during dust storms and so the view we get with an orange or red filter is of surface detail.

Occasionally, surface features do become visible if Mars is viewed with the blue (Wratten no. 80A) or violet (Wratten no. 47) filters and this 'violet clearing' may persist for days. The phenomenon is of considerable interest to astronomers. A possible explanation of the violet clearing is that the light reflectivity (albedo) of surface features or the degree of polarisation of this light may be partly responsible. At one time it was thought that a layer high in the atmosphere of Mars reflected blue and ultraviolet light and that its occasional dispersal may be the cause of the violet clearing .

In watching for a violet clearing you should first view Mars with the blue filter in your telescope eyepiece and then view the planet without the filter. The best filter for this work is the previously mentioned Wratten blue no. 80A. If you think that there is a violet clearing in progress try the denser no. 47 filter.

Whenever you detect a violet clearing record the date, whether the clearing is widespread – or if localised note its position – and the effects observed when the filters were used. Usually the violet clearing is planet-wide.

Blue-white and white clouds and hazes White clouds are considered to be seasonal, commencing and ending respectively in Martian Spring and Autumn. The really white clouds become more numerous and cover a larger area as the polar caps melt and may occur in both hemispheres of Mars.

Both the blue-white and the white clouds appear to be close to the surface and are probably fogs. They may also be deposits of frosts in sheltered depressed areas on the surface. White clouds may also be day-long phenomena. They start on the morning terminator and are dispersed later by the sun's heat.

Record the appearance of any cloud that you see and where and when the cloud was visible – morning or evening terminator; visible all day. It is of importance to record these times and appearances carefully as it can help to give support to the belief that near-surface fogs and surface frost deposits may occur on a daily basis.

The following regions are associated with seasonal appearance of white clouds: Aram, Elysium. Isidis Regio, Memnonia, Neith Regio, Nix Lux, Nymphaeum, Olympus Mons, Ophir, Tharsis, Zephyria and Hellas Alba Areia.

Observations of white clouds are much enhanced by using Wratten filters no. 58 (green) and nos. 80A, 38 and 38A (blue) for maria.

The majority of white clouds are seen in the equatorial and midtemperate regions. They appear to be more prevalent in the southern than in the northern hemisphere but this is probably an artefact; the southern hemisphere is always better presented than the northern at perihelic oppositions respectively and also the reflecting of sunlight from clouds is at a much more favourable angle for visibility than when the northern hemisphere is presented to us.

Yellow clouds These originate mostly at perihelic oppositions at the time of the southern summer solstice. They are probably caused by the sun's heat raising high winds that stir up the dust, giving rise to clouds and the planet-wide dust storms but most yellow clouds do not develop into dust storms.

Yellow dust clouds have their origin in the Hellas–Noachis–Serpentis Mare area and the Solis Lacus, Chryse, Isidis Regio, Casius Aetheria and Aeria areas. The yellow clouds appear quickly and at times may spread rapidly. A planet-wide storm can obliterate all visible features on the Martian surface from telescope vision. The dust storms have been known to persist for weeks or months. Almost always yellow clouds and dust storms are low atmospheric phenomena and blow over the surface. The short-lived circulation patterns of the Martian atmosphere may be charted during the extended periods when dust storms are active.

The best filters for observing the yellow clouds and dust storms are Wratten nos. 12, 21 and 25 (yellow, orange and red).

W-shaped clouds These odd-shaped clouds appear in the neighbourhood of large volcanos. In 1966 Charles Capen confirmed the original reports of W-shaped clouds by E. E. Slipher and also that they are recurrent.

The W-shaped clouds are often quite large. They appear more frequently in Martian Summer than Autumn and it seems that they are seen more often in the southern hemisphere than in the northern. They make their appearance in the vicinity of the features known as Olympus Mons, Pavonis Mons, Ascraeus Mons and Arsia Mons. These are fairly close to one another and all four are large volcanos. The clouds move quite quickly so as soon as you observe one record the date when it was first seen and accurately note its movements in relation to familiar surface markings. The place where the cloud originated should be reported and also the latitude and longitude of the clouds. W-shaped clouds are visible in yellow light but are brighter in blue.

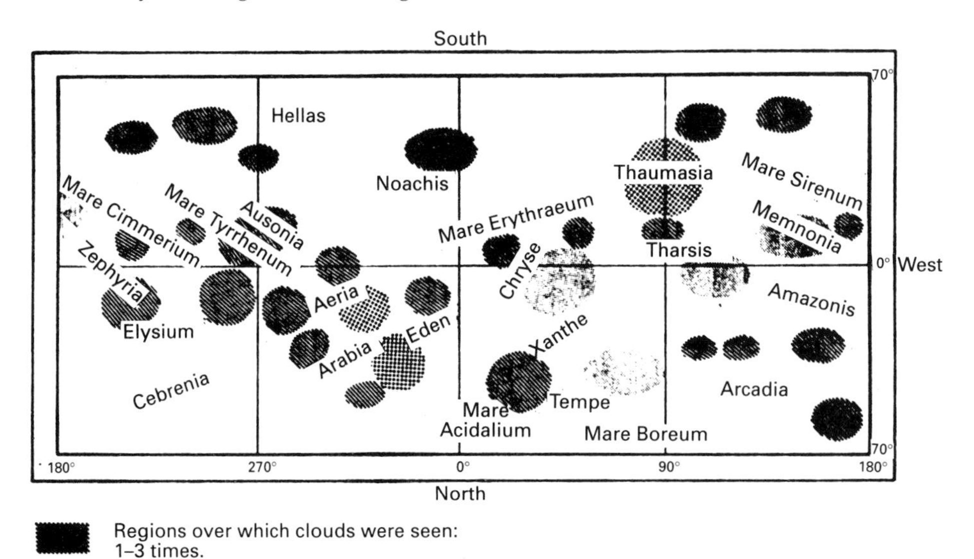

Fig. 7.29 Cloud frequency chart of Mars based on observations from 1950 to 1961 by W. Sandner. (From Handbook for Planet Observers, *Roth, G.D., Faber and Faber, London, 1970.)*

Observational drawings of Martian mists and clouds are shown on fig. 7.7 and a cloud frequency chart of Mars is shown in fig. 7.29.

Longitude determination of Martian features

As mentioned earlier, the meridian of zero longitude on Mars was established by Beer and Mädler in 1840 and actually passes through a small nearly circular crater (Airy-0) near the centre of Sinus Meridiani (formerly known as 'Dawes Forked Bay') although, or course, Beer and Mädler could not possibly have known of the existence of Airy-0! Longitude is measured westward from this point all the way around the Martian globe.

The scientific value of observational drawings of Mars is increased if the longitude of the central meridian (LCM) of Mars is determined with reference to this zero meridian of longitude. The central meridian (CM) is the imaginary straight line passing through the north and south points of the Martian disc which therefore bisects it into halves (fig. 7.30). The longitude of the CM should be recorded when the drawing is made and when it is finished.

The reason for doing this is that it enables you to determine the exact location in longitude of surface features as well as their horizontal angular extent. However, it is important to remember that Mars can have a significant phase defect which could bias CM transit timings. It is therefore preferable to make such timings near to opposition or to make corrections.

To measure the longitude of a Martian surface feature you first note the time to the nearest minute when the feature is exactly on the CM. At this instant the CM bisects the feature.

Having determined the time of CM transit of the feature, the time has to be reduced or converted to longitude.

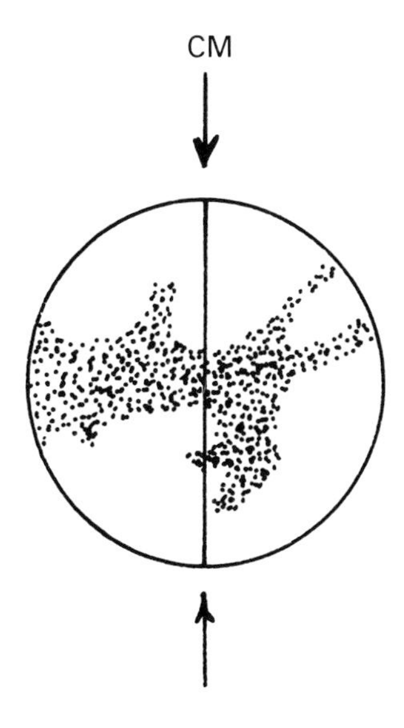

Fig. 7.30 Central meridian (CM) of Mars.

The longitude of the CM of Mars at 00.00 UT on every day in the year will be found in the *Astronomical almanac.* To calculate the longitude of the CM, first determine how many hours and minutes have elapsed since 00.00 UT on the day in question. Since Mars rotates through 14.6° of longitude in one hour, the longitude of the CM increases by this amount every hour.

To convert your local time to UT is simple. Suppose that the time was 22.00 (10.00 pm) EST on a certain date, say March 3rd. Since EST is 5 hours behind UT, then five hours is added to EST to convert it to UT. Thus, 22.00 EST on March 3rd becomes 03.00 UT on March 4th. Note that conversion to UT may alter the date by one day.

Suppose then that we wish to determine the longitude of the CM at a certain time on a given day. First we consult the *Astronomical almanac* for the longitude of the central meridian (LCM) at 00.00 UT on that day. Then from the accompanying table we read off the increase in degrees of Martian longitude of the CM that has occurred during the time that has elapsed since 00.00 UT on the same day. This value is added to the longitude for 00.00 UT and gives the LCM for that date and time.

Example: Calculate the LCM of Mars at 02.15 UT on December 3rd 1990. From the almanac we find that at 00.00 UT on December 3rd the LCM of Mars is 118.38°. (The almanac gives the CM longitude at 0 h 0 mins Dynamic Time. Currently, add +0.23° to compensate for this). From table 7.7 we see that the change in longitude of the CM undergone in 2 hours and 10 minutes is 31.7°. To this we add the change undergone in five minutes from the smaller table at the foot of the main table; this is 1.2°. Therefore, the LCM of Mars at 02.15 UT on December 3rd 1990 is 118.36 + 31.7 + 1.2 = 151.3° (realistically rounded to one decimal place).

Further reading

Books

Mars. Lowell, P., Longmans, Green (1896).

Mars and its Canals. Lowell P., MacMillan, New York (1906).

Mars as the abode of life. Lowell, P., MacMillan, New York (1909).

The planet Mars. De Vaucouleurs, G. (translated from the French by P. A. Moore, Faber and Faber, London (1950).

Exploring Mars. Richardson, R. S., McGraw-Hill, New York (1954).

The exploration of Mars. Ley, W. and Von Braun, W., Viking Press, New York (1956).

Mars: The photographic story. Slipher E. C., Sky Publishing Corporation, Cambridge, Mass. (1962).

Guide to Mars. Moore, P. A., Frederick Muller Ltd, London (1965).

The book of Mars. Glasstone, S., National Aeronautics and Space Administration, Washington, DC. (1968).

Mariner IV to Mars. Ley W., Signet, New York (1966).

The New Mars. The Discoveries of Mariner 9. Hartmann, W. K. and Raper, O., National Aeronautics and Space Administration, Washington, DC (1974).

The planet Mars. Antoniadi, E. M. (translated from the French by P. A. Moore.) Keith Reid, South Devon, England (1975).

Table 7.7. *Change in Martian longitude with Universal Time.*

UT	Add	UT	Add	UT	Add	UT	Add
00 : 00	0.0	06 : 00	87.7	12 : 00	175.4	18 : 00	263.2
10	2.4	10	90.2	10	177.9	10	265.6
20	4.9	20	92.6	20	180.3	20	268.0
30	7.3	30	95.0	30	182.8	30	270.5
40	9.7	40	97.5	40	185.2	40	272.9
50	12.2	50	99.9	50	187.6	50	275.4
01 : 00	14.6	07 : 00	102.3	13 : 00	190.1	19 : 00	277.8
10	17.1	10	104.8	10	192.5	10	280.2
20	19.5	20	107.2	20	194.9	20	282.7
30	21.9	30	109.7	30	197.4	30	285.1
40	24.4	40	112.1	40	199.8	40	287.5
50	26.8	50	114.5	50	202.2	50	290.0
02 : 00	29.2	08 : 00	117.0	14 : 00	204.7	20 : 00	292 .4
10	31.7	10	119.4	10	207.1	10	294.8
20	34.1	20	121.8	20	209.6	20	297.3
30	36.6	30	124.3	30	212.0	30	299.7
40	39.0	40	126.7	40	214.4	40	302.2
50	41.4	50	129.1	50	216.9	50	304.6
03 : 00	43.9	09 : 00	131.6	15 : 00	219.3	21 : 00	307.0
10	46.3	10	134.0	10	221.7	10	309.5
20	48.7	20	136.5	20	224.2	20	311.9
30	51.2	30	138.9	30	226.6	30	314.3
40	53,6	40	141.3	40	229.1	40	316.8
50	56.0	50	143.8	50	231.5	50	319.2
04 : 00	58.5	10 : 00	146.2	16 : 00	233.9	22 : 00	321.6
10	60.9	10	148.6	10	236.4	10	324.1
20	63.4	20	151.1	20	238.8	20	326.5
30	65.8	30	153.5	30	241.2	30	328.0
40	68.2	40	156.0	40	243.7	40	331.4
50	70.7	50	158.4	50	246.1	50	333.8
05 : 00	73.1	11 : 00	160.8	17 : 00	248.5	23 : 00	336.3
10	75.5	10	163.3	10	251.0	10	338.7
20	78.0	20	165.7	20	253.4	20	341.1
30	80.4	30	168.0	30	255.9	30	343.6
40	82.5	40	170.6	40	258.3	40	346.0
50	85.3	50	173.0	50	260.7	50	348.5

Minutes	Add	Minutes	Add	Minutes	Add
01	0.2	04	1.0	07	1.7
02	0.5	05	1.2	08	1.9
03	0.7	06	1.5	09	2.2

(Assuming 350.89 degrees per day)

The Geology of Mars. Mutch, T. A. *et al.*, Princeton University Press, Princeton, NJ, USA (1976).

Mars and its Satellites. Blunck, J., Exposition Press, USA (1977).

To the Red Planet. Burgess, E., Columbia University Press, New York (1978).

The Channels of Mars. Baker, V. R., University of Texas Press, Austin, Texas (1982).

Planets and Perception. Sheehan, W., University of Arizona Press, Tucson (1988).

The Surface of Mars. Carr, M. H., Yale University Press, Cambridge, Mass. (1982).

Introduction to observing and photographing the Solar System. Dobbins, T. A., Parker, D. C. and Capen, C. F., Willman-Bell Inc. Richmond, Virginia (1988).

Papers and articles

Observing the red planet. Capen, C., *Astronomy* **8(1)**, 38–43 (1980).

Ten pointers for Mars observers. Astronomy **16(7)**, 76–77 (1988).

The use of filters for observing fine martian surface detail and atmospheric phenomena. Capen, C. F., *The Strolling Astronomer* **8(5–6)**, 55–7 (1954).

Mars through colour filters, 1958, Heath, A. W. *et al.*, *JBAA* **70(6)**, 270–2 (1960).

Charting the Martian surface. De Vaucouleurs, G., *Sky and Telescope*, **30(4)**, 196–201 (1965).

The polar regions of Mars. Cutts, J., *Astronomy* **5(10),** 10–17 (1977).

Exploring the Martian Arctic. Parker, D. C., Capen, C. F. and Beish, J. D., *Sky and Telescope* **65(3)**, 218–20 (1983).

Photographs of a recent Martian dust storm. Murrell, A. S. and Knuckles, C. F., *Sky and Telescope* **47(3)**, 168–9 (1974).

Dust storms of Mars. Idso, S. B., *Astronomy* **5(3)**, 34–9 (1977).

Dust storm observations from New Mexico. Kirby, T. B. and Robinson J. C., *Sky and Telescope* **42(5)**, 264–5 (1971)

Mars in 1969. Capen, C. F. and Capen V. W., *Sky and Telescope* **37(3)**, 190–4 (1969).

Telescopic observations of Mars in 1971. De Vaucouleurs, G., *Sky and Telescope* **42(3)**, 134–5 (1971).

Telescopic observations of Mars in 1971 (II) De Vaucouleurs, G., *Sky and Telescope* **42(5)**, 263–4 (1971).

Mars at Opposition. Mullaney, J., *Astronomy* **3(12)**, 50–8 (1975).

Some high resolution photographs of Mars. Larson, S. M. and Minton R. B., *Sky and Telescope* **42(5)**, 260–61 (1971).

A portfolio of amateurs' Mars photographs. *Sky and Telescope* **42(5)**, 310–14 (1971).

Memories of Mars. (The 1988 apparition.) Eicher, D. J. and Troiani, D. M., *Astronomy* **17(4)**, 74–9 (1989).

Mars pictures from Mariners 6 and 7., *Sky and Telescope* **38(4)**, 212–21 (1969).

Mars: Five years after Viking. *Astronomy* **9(7)**, 8–17 (1981).

Deimos and Phobos, Oberg, J., *Astronomy* **5(3)**, 9–17 (1977).

8

The minor planets (asteroids)

General

The minor planets – asteroids or planetoids as they are often called – are a swarm of many thousands of small solid bodies most of which revolve around the sun in orbits between those of Mars and Jupiter (fig. 8.1). F. G. Watson estimated that on the basis of the sizes of the larger minor planets and the brightness of the smaller bodies, they altogether would make up a body about 620 miles (1000 km) in diameter which is only about four times the diameter of Juno, one of the larger asteroids, and the total combined mass would be about 1/4000 of the Earth's. Orbital revolution periods of the main belt asteroids range from 3.5 to 6.0 years. The most frequent period is somewhat less than half of Jupiter's.

Of 1568 minor planets whose orbits had been determined by 1953, 97% had mean distances from the sun ranging from 195 to 288 million miles (313.8 to 463.4 million km). Their orbital eccentricities average at 0.15 which is greater than that of the larger planets with the exception of Mercury and Pluto. Some individual asteroids have eccentricities 3–4 times the average. The longer-period asteroids with large orbits except Hidalgo have small eccentricities. Orbital inclinations to the ecliptic of asteroids nos. 1–3000 average 9.4° which is greater than that of the larger planets with the exception of Pluto; a few almost lie in the ecliptic plane and a few others have orbital inclinations exceeding 30°. In spite of this, the asteroids are true planets; every one has direct motion around the sun contrasting with many comets that have retrograde motion. The direction from the sun of most minor planet perihelia is approximately towards Jupiter and the fewest are in the reverse direction which indicates the dominant influence of Jupiter's gravitational pull on them which will be referred to again.

The sizes of the asteroids are difficult to measure because they are so small. Very good results may be obtained by timing stellar occultation by asteroids, but such events are uncommon for a given location owing to the asteroids' small size. Infrared and radar measurements have also been helpful.

The shapes of the asteroids vary from roughly spherical to elongated or irregular. Some of the irregulars are thought to consist of two or more separate

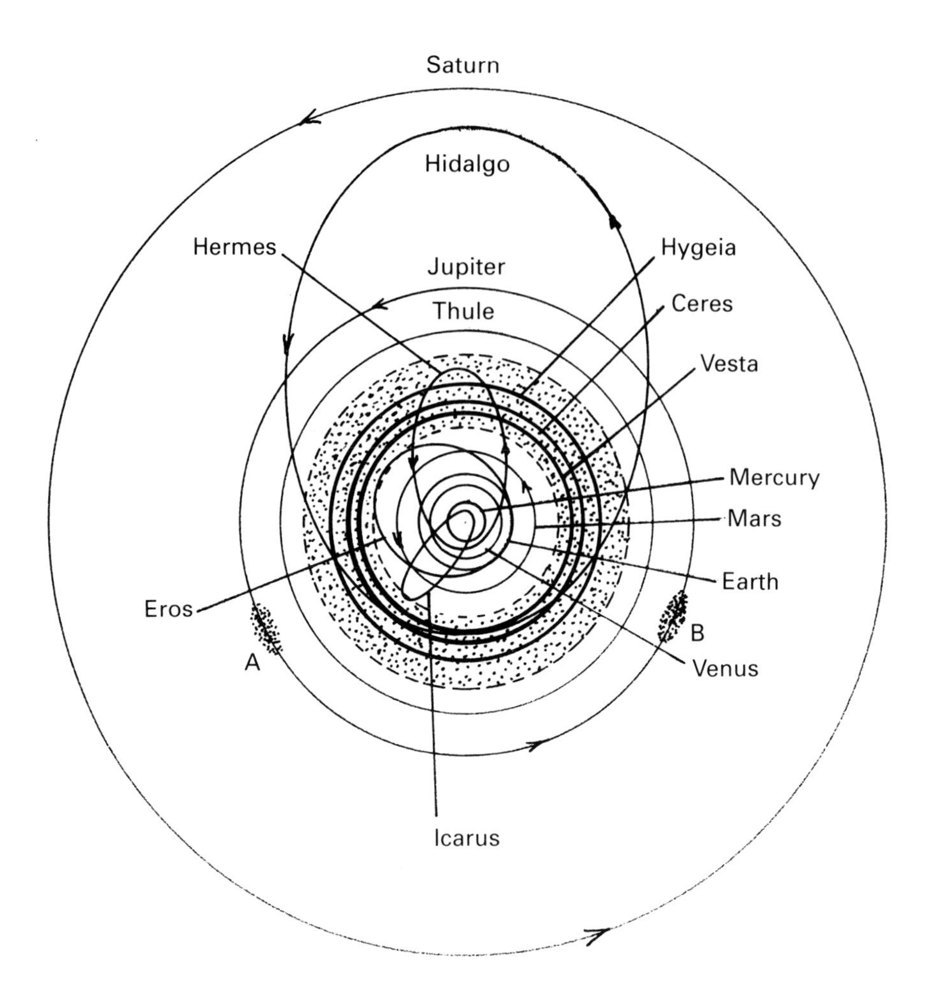

Fig. 8.1 The asteroids (minor planets, planetoids). The main asteroid zone is shown dotted within the two broken circles. Orbits of some individual asteroids are indicated. A: trailing Trojans (Patroclus group); B: leading Trojans (Achilles group).

bodies. There is some evidence that 2 Pallas is attended by a large satellite, as is possibly also 12 Victoria. Data pertaining to some of the largest and brightest asteroids are given in table 8.1.

Studies of asteroid spectra and albedos show that there are several asteroid classes: the C-type (carbonaceous), the most common type (about 75% of the asteroids), that have dark surfaces and albedos ranging from 2% to 5%; the S-type (siliceous or stony) with albedos around 9–24% (Tholen) and the M-types which are moderately bright (albedos 7–21% Tholen). In the small group of unusual asteroids such as Vesta, the majority have relatively high albedos (21–33%).

Table 8.1. *Data of some of the brighter asteroids.*

No.	Name	Max. Mag.	Diameter miles	km	Orbital period (years)	Brightness amplitude (magnitudes)	Fluctuation period (hours)
1	Ceres	7.0	567	913±43	4.6	0.04	9.08
2	Pallas	6.3	325	523±20	4.61	0.03–0.16	7.81
3	Juno	6.9	152	244±12	4.36	0.14–0.22	7.21
4	Vesta	5.08	311	501±24	3.63	0.12	5.34
6	Hebe	7.1	119	192±4	3.78	0.05–0.20	7.27
7	Iris	6.7	126	203±5	3.69	0.04–0.29	7.14
8	Flora	7.8	88	141±3	3.27	0.02–0.10	12.79
9	Metis	8.1	94	151	3.69	0.04–0.36	5.08
15	Eunomia	7.4	169	272±6	4.30	0.4–0.56	6.08
20	Massalia	8.2	94	151±11	3.74	0.17–0.27	8.10

Discovery and history of observation of the minor planets

Before the discovery of asteroids, the unusually large gap between the orbits of Mars and Jupiter had aroused curiosity among astronomers especially as Bode's law (chapter 1) seemed to indicate that a planet should be found there at a distance from the sun of 2.8 AU. Long before the Bode–Titius Law was put forward, Kepler was impressed by the large gap between the orbits of Mars and Jupiter and considered that there must be a planet there. In fact he once wrote 'Between Mars and Jupiter I put a planet'.

In 1800, a group of six astronomers in the town of Lilienthal in Germany set out to search for a planet between the orbits of Mars and Jupiter. Baron Franz von Zach called a meeting of them. They gave themselves the nickname 'Celestial Police' and were headed by J. H. Schröter whose observatory was at Lilienthal. However, their hopes of finding a sizable planet between Mars and Jupiter were frustrated; someone other than one of their number was first to discover a planetary body between Mars and Jupiter. On January 1st 1801, Piazzi at Palermo, who was compiling a star catalogue, noticed a star-like body that yet differed from a star in exhibiting rapid movement even over a few hours. Subsequent studies of its movement established its planetary nature. Piazzi named it Ceres. It had a distance of 2.77 AU from the sun so was a close approximation to the anticipated distance of 2.8 AU but it was unexpectedly small, only 623 miles (1003 km) in diameter. Then in quick succession three other small bodies were discovered between Mars and Jupiter: Pallas by Olbers in 1802, Juno by Harding in 1804 and Vesta by Olbers in 1807. The sizes of these four first-discovered asteroids compared with the Earth's moon are shown in fig. 8.2.

Owing mostly to unavailability of good star charts, no further asteroids were discovered until 1845 when K. J. Hencke discovered another after searching for 15 years. It was named Astraea. When new and better star charts were prepared (to aid in the finding of Neptune) these helped in the discovery of the asteroids Hebe in 1847, Iris and Flora (from London) in 1847, Metis in 1848 and Hygeia in 1849. Asteroid 10 had a diameter of 267 miles (429 km). In 1850 three more were added (nos 11–13). The subsequent rapid rate of discovery of asteroids gave rise

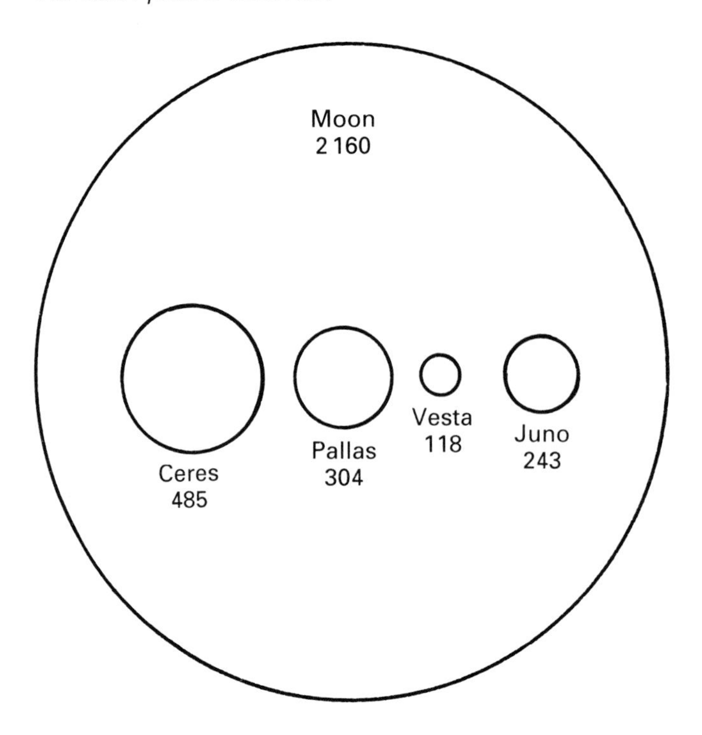

Fig. 8.2 Relative sizes of the largest asteroids and the Earth's moon (diameters in miles).

to difficulties in naming them. Names were chosen from classical mythology, mostly feminine. Masculine names were conferred on asteroids of especial interest. The choice of suitable names gave rise to arguments during the 1850s owing to the rapid rate of discovery of asteroidal bodies.

By 1870, 109 asteroids had been discovered and all had numbers and names and their orbits had been computed. The total rose to 287 by 1890. All were localised between the orbits of Mars and Jupiter.

A guided photographic exposure of the stars of an hour or two through a telescope betrays the presence of an asteroid as a short bright streak on the photographic plate among the tiny discs of the fixed stars. The asteroid image is drawn out into a streak owing to its motion relative to nearby stars. Photography thus enabled Max Wolf (Heidelberg) to discover asteroid no. 323 in 1891. From this time onwards asteroid discoveries increased by leaps and bounds. Wolf found 231 asteroids during the next ten years, Palisa found 83 and C. H. F. Peters more than 50. Later, all these were left behind by K. Reinmuth of Heidelberg who discovered 980 which included rediscoveries and single observations. Of Reinmuth's asteroids only 189 received numbers although 228 of Wolf's final total of 582 were numbered. Not only naming but numbering of asteroids was becoming a problem and had to be discontinued. Those whose orbits have been computed accurately are given numbers.

A photographic search at Mount Wilson was carried out in 1938 to find additional satellites of Jupiter. As well as two new Jovian satellites being found, 31 new asteroids were discovered on the photographic plates. Asteroid discoveries totalled 2799 by 1939; of these, numbers were given to 1489. The year 1948 saw

216 discoveries and another 275 the following year. 1568 asteroid orbits were recorded in the 1950 list.

Some asteroids have been 'lost' and found again. The reason for this is that it is difficult to compute a reliable orbit for an asteroid. Three accurate observations are needed for this, separated by many weeks. The perturbations caused by Jupiter and Saturn cause asteroid movements to be uncertain and difficult to compute. An example of a 'lost' asteroid is 132 Aethra which was missing between 1873 and 1922 even though it was diligently sought because of its highly eccentric orbit.

In 1898 G. Witt of Berlin discovered the asteroid 433 Eros. Its mean distance from the sun of 136 million miles (218.8 million km) was the least known at the time for an asteroid. It can come as close to the Earth as 14 million miles (22.53 million km). F. G. Watson believed Eros to have an irregular shape with a length of 14 miles (22.53 km) and a width of 4 miles (6.44 km). (More modern values are $14.1 \times 14.5 \times 40.5$ km.) It rotates about its minor axis as deduced from its brightness fluctuations. The shape of Eros and its light changes were actually seen by Van den Bos in 1931 with a large refractor. When the Earth passes through the equatorial plane of Eros, end and side views are presented in sequence and produce large light changes. When the pole is presented to the Earth, no light changes are perceptible.

The Albert Group of three asteroids have highly eccentric orbits. They move outwards almost as far as the orbit of Jupiter and inwards almost as far as the Earth's orbit. They are designated 719 Albert ('lost' in 1911), 887 Alinda (orbit variable; 3 : 1 Jupiter resonance) and 1036 Ganymed (not to be confused with Jupiter's satellite Ganymede). Although outside the Earth's orbit their speed is faster than the Earth's when close to it owing to their orbital eccentricities of 0.382 for Ganymed, 0.483 for Albert and currently 0.558 for Alinda. Hence, at opposition they do not appear to retrograde. Some of the asteroids come closer to the Earth than Eros.

2101 Adonis was discovered by E. Delporte of Belgium in 1936. It is 1.5 million miles (2.4 million km) distant from the Earth at its closest approach. The orbit is highly eccentric (0.724) and extends from slightly outside Mercury's orbit to double the distance of Mars from the sun.

1221 Amor, also discovered by Delporte, was about 10 million miles (16.09 million km) distant from the Earth when it was discovered. The orbital eccentricity is high (0.448). It does not cross the Earth's orbit but overtakes and passes the Earth with a speed nearly one third as fast again as the Earth's.

The 'Earth grazing' asteroid Hermes was photographically discovered by Reinmuth on October 28th 1937. It came to within nearly 500 000 miles (800 000 km) of the Earth which is less than double the distance that separates the Earth from the moon. On October 30th 1937 Hermes, which was of the 8th magnitude, sped past the Earth at an angular pace in the sky of 5° per hour, so fast that its distance could not be deduced by using the usual mathematical methods. Instead it had to be calculated from its parallax as seen from two different observatories. It was 'lost' in 1937.

1862 Apollo, also found by Reinmuth, in 1932, came to within about 7 million miles (11.3 million km) of the Earth (Watson gives 3 million km). It was the first asteroid whose orbit was found to extend within the orbits of the Earth and Venus; it also goes outside the orbit of Mars. Hence, this and other asteroids are both exterior and interior. Therefore Apollo, which is best seen at opposition, can also have inferior conjunctions and exhibit solar transits although it is much too small to be visible at such times. These close-approaching asteroids have orbital periods of less than 3 years. Observation was too short for reliable orbits to be computed and so Apollo, Adonis and Hermes soon were 'lost'. Apollo was 'lost' for several years and rediscovered in 1973. It made a close approach to Earth in 1982. Asteroids whose orbits cross the Earth's are collectively known as Apollos and are still being discovered. All are very small and are no more than a few kilometres across.

Those asteroids whose orbits are inside that of the Earth's are known as the Aten Group. There are not many of these known but there may well be many of them. The best known are 2062 Aten and 2100 Ra-Shalom.

There is always a chance that one of the hundreds of the 'flying boulders' that intersect the Earth's orbit may collide with the Earth and wreak great local devastation but that chance is extremely low, probably not as often as once in 100000 years but this value depends on the assumed size.

In 1866, D. Kirkwood demonstrated that at mean solar distances where their orbital periods would be simple fractions of Jupiter's, asteroids were sparse. He concluded that the accumulated effect of Jupiter's gravitational attraction at regularly repeated intervals would compel the asteroids to enter other orbits and

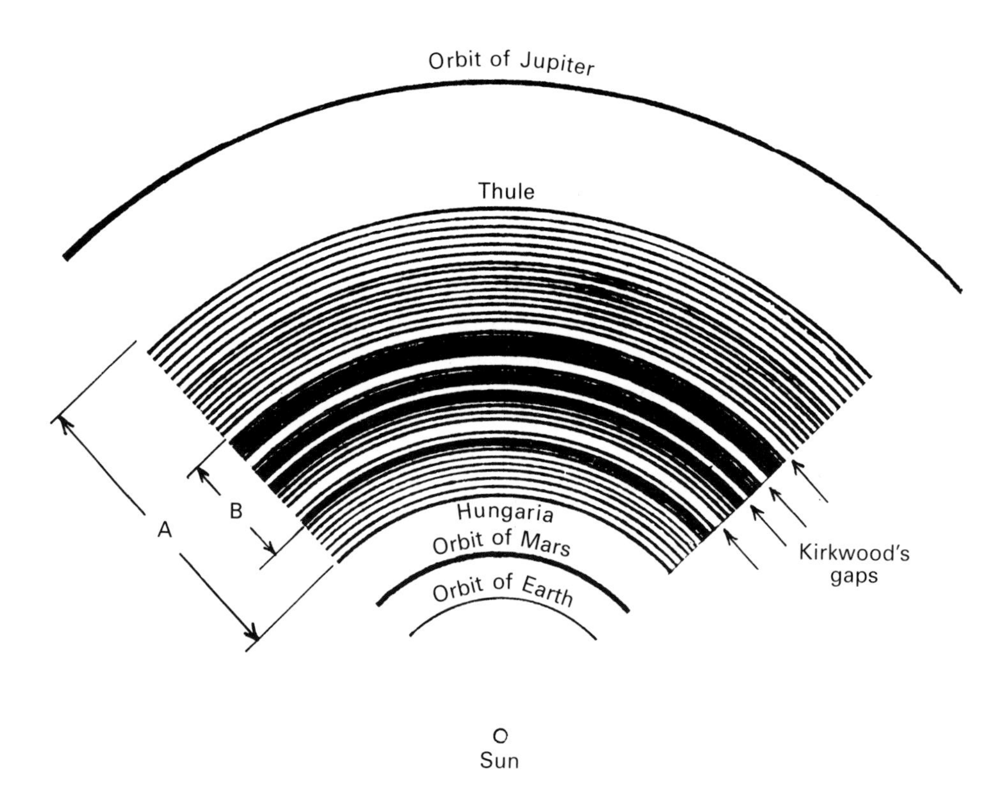

Fig. 8.3 Kirkwood's gaps in the asteroid belt. Most of the asteroids revolve around the Sun between 2 and 4 AU (zone A) with the greatest concentration at 2.2–3.2 AU (zone B). However, some of the asteroids have highly eccentric orbits (see fig. 8.1 and text).

gaps would be formed as in the case of Saturn's ring divisions (chapter 10). Although continuing discovery of asteroids has tended to fill Kirkwood's gaps, as they are called (fig. 8.3), they are still prominent at periods of 1/3, 2/5, 3/7, 1/2 and 3/5 of Jupiter's orbital period. The Hecuba family of asteroids, which is an enormous group of 400, have a period of a little less than half of Jupiter's orbital period. This is reminiscent of the brightness, no doubt due to the high particle concentration, of Saturn's B ring.

There have now been discovered asteroid groupings instead of gaps whose period ratios to Jupiter's are 2/3, 3/4 (occupied only by 270 Thule) and 1/1. To account for these clusterings, E. W. Brown and K. Hirayama have demonstrated that other forces in addition to Jupiter's gravitational field must be involved if both gaps and groupings are to be explained. Mathematical investigation of asteroid orbits enabled Hirayama to characterise five families of asteroids which he named the Flora, Maria, Koronis, Eos and Themis families (after the names of the first discoveries in each). Their orbital periods are close to, respectively, 2/7, 1/3, 2/5, 3/7 and 1/2 of Jupiter's. Because the members of each family had certain orbital features common to them Hirayama was led to believe that the members of these 'Hirayama families', as they came to be called, had a common origin.

Period groups are groups of asteroids with similar orbital periods but not necessarily having common origin. They include not only the Hecuba and Trojan groups but also the large Minerva and Hestia groups and the smaller Hilda group with periods close to 1/3 (Hestia), 2/5 (Minerva) and 2/3 respectively of Jupiter's. The asteroid Thule whose period ratio is 3/4 was thought to be the outermost until the group known as the Trojans (fig. 8.1) began to be discovered in 1906. They have the same orbital period as Jupiter. They provide an example of one of Lagrange's theoretical solutions to what is known as the three-body problem; if a small body, a large one and an infinitesimal body having negligible mass are located at the apices of an equilateral triangle, they will remain in the same relative positions, each of them revolving about the system's centre of gravity. Since they possess mass and are perturbed by Saturn and other planets, the Trojans do not obey this law exactly. Where two of them have similar oscillation periods, each influences the other. The sun, Jupiter and a point of 60° in front of Jupiter and on its orbit form an equilateral triangle. A group of Trojans oscillates about this point. Likewise, the equilateral triangle formed by the sun, Jupiter and a point on Jupiter's orbit 60° behind it has a group of Trojans oscillating at this point.

The orbits of the Trojans are normal and concave to the sun but the eccentricities and inclinations differ from Jupiter's. There is a common centre of oscillation for the slow long-period motions of each group but collisions are not likely for they are very small bodies, probably only a few miles in diameter and they oscillate in different directions. The motions are perfectly stable and the Trojans are likely to remain in their present orbits permanently.

W. Baade discovered the asteroid 944 Hidalgo in 1920. It takes 13.9 years to make a circuit of its orbit which currently extends to 9.68 AU which is just beyond the orbit of Saturn (9.52 AU). The apparent magnitude fluctuates with distance from 12 to 20. The orbital eccentricity is 0.658 and the orbital inclination is 42.4° which is even greater than that of Pallas. It therefore resembles a comet and is protected from close encounters with Jupiter and Saturn.

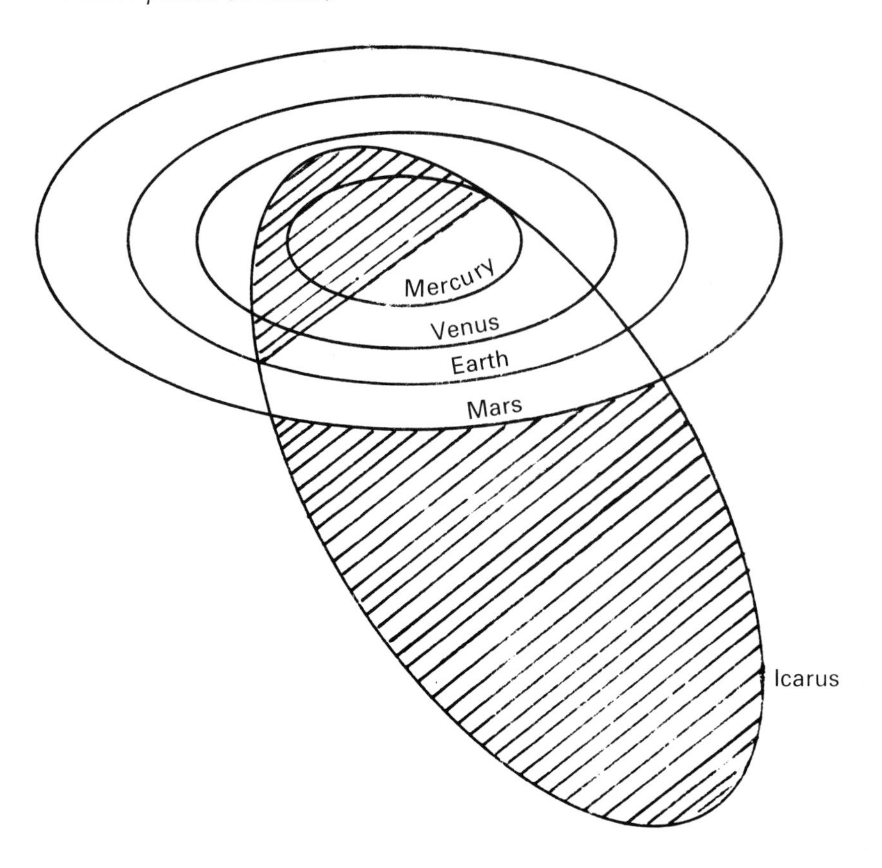

Fig. 8.4 The orbit of Icarus.

In many ways the most remarkable of all the asteroids is Icarus which was discovered by Baade in 1949. Its distance from the Earth was 8 million miles (12.8 million km) when discovered. It is not an Earth grazer as it cannot approach Earth nearer than one half of this distance. Its uniqueness lies in the fact that at its perihelion it is only 18 million miles (29.0 million km) distant from the sun which is closer than Mercury. It can therefore be called a 'sun gazer'. Its aphelion distance is 183 million miles (294.4 million km) which is well beyond the orbit of Uranus. The orbit of Icarus is shown in fig. 8.4. In 1968 Icarus approached the Earth closely at its minimum distance of only 16 times the moon's distance. Small changes of magnitude indicated an axial rotation period of 2.273 hours and its colour seems to be somewhat more blue than most asteroids.

In 1977, from the Mount Palomar Observatory, C. Kowal discovered a very strange object, the orbit of which lies mostly between the orbits of Saturn and Uranus. Its magnitude will be about 15 when it next comes to perihelion in 1997. The object, named Chiron, has been given a minor planet number of 2060. Its diameter is uncertain. In 1988 its brightness inexplicably increased and in the following year Kitt Peak astronomers found that it had a coma which measured about 5 seconds of arc across. Chiron may therefore be a kind of comet if this observation is confirmed. However it seems to be too big for a comet.

The list of numbered Asteroids had risen to 4040 by 1987 and currently (1992) is over 5000. The actual shapes and sizes of asteroids have been determined by the method of timing stellar occultations by them which is the most accurate method. Until 1977 only seven stellar asteroid occultations had been observed but the frequency of observation has been greatly increased since then. Some notable occultations that were well observed and timed were as follows:

(1) The occultation of the star 1 Vulpecula by 2 Pallas in 1983. The shadow path crossed North America and no less than 130 timings of this event were made. About four fifths of the limb of Pallas was observed yielding data that enabled its shape to be defined as an ellipsoid about 325 miles (523± km) across. No large irregularities or asymmetries were recorded. Before 1983 the largest number of timings of a single occultation was 20, the average number being 3 per event. The 1983 Pallas 'event' was therefore a significant one in the history of amateur observational astronomy.
(2) Two more well-observed occultations both occurred in January 1991. On January 3rd, 4 Vesta occulted a magnitude 7.4 star and on January 18th–19th 216 Kleopatra occulted a magnitude 9.3 star. The shadow paths crossed North America.

For a long time astronomers have believed that some of the asteroids are binary, compound in nature or even bifurcated (V-shaped) because of the way that the light changes during stellar occultations by certain asteroids and during rotation. It was not until 1989 that the first really convincing evidence was obtained of the binary structure of an asteroid. On August 9th of that year the asteroid named 1989 PB was discovered by Eleanor Helin and coworkers of the Jet Propulsion Laboratory, on photographs obtained with the 18-inch Schmidt Camera at the Mount Palomar Observatory. Less than two weeks later, on August 22nd 1989 PB passed through the beam of the 1000 foot Arecibo Radio Telescope in Puerto Rico. The images obtained of the asteroid reveal detail down to about 1000 feet across. The asteroid is about 1 mile (1.609 km) in length and 0.5 mile (0.805 km) in width. It has a two-lobed shape made up of two irregular-shaped bodies each one about 0.5 mile in diameter. The two bodies appear to be touching but they may also be separate and both orbiting around a common centre of mass. The orbit of 1989 PB extends beyond the orbit of Mars to inside the orbit of Venus and one complete circuit of the orbit takes 400 Earth days.

Upgrading of the Arecibo telescope making it 20 times more sensitive is due to be completed in 1993. Presently it is hoped that from then on images with a resolution of 150 feet or better will be achieved. It will be interesting to see whether two-lobed objects like 1989 PB are freaks or whether they are fairly common.

Visibility of the minor planets

About 72 out of all asteroids are brighter than magnitude 9.5 when at perihelic opposition so that they should be visible in 2-inch (50 mm) binoculars. A 3-inch refractor will show many hundreds, the limiting magnitude for such a telescope being between 11 and 12. The star-like telescopic appearance of these minor planets is what gave them their name of 'asteroids'. Only the very largest asteroids appear as discs in the telescope and large apertures are required to show them thus.

Table 8.2. *Ten-year table of precession in right ascension.*

	RA (If N., read top; if S., read lower)						
Declination	0,12 h	1,11 h	2,10 h	3,9 h	4,8 h	5,7 h	6 h
degrees	m	m	m	m	m	m	m
80	+0.51	+0.84	+1.14	+1.40	+1.60	+1.73	+1.77
70	0.51	0.67	0.82	0.94	1.04	1.10	1.12
60	0.51	0.61	0.70	0.78	0.84	0.88	0.98
50	0.51	0.58	0.64	0.70	0.74	0.77	0.78
40	0.51	0.56	0.61	0.64	0.67	0.69	0.70
30	0.51	0.54	0.58	0.60	0.62	0.64	0.64
20	0.51	0.53	0.55	0.57	0.58	0.59	0.59
10	0.51	0.52	0.53	0.54	0.55	0.55	0.55
0	0.51	0.51	0.52	0.51	0.51	0.51	0.51
	0,12 h	23,13 h	22,14 h	21,15 h	20,16 h	19,17 h	18 h

	RA (If N., read top; if S., read lower)						
Declination	18 h	19,17 h	20,16 h	21,15 h	22,14 h	23,13 h	
degrees	m	m	m	m	m	m	m
80	−0.75	−0.70	−0.58	−0.38	−0.12	+0.19	
70	−0.10	−0.80	−0.02	+0.08	0.21	0.35	
60	+0.13	+0.14	+0.18	0.24	0.32	0.41	
50	0.25	0.26	0.28	0.32	0.38	0.44	
40	0.33	0.33	0.35	0.38	0.42	0.46	
30	0.38	0.39	0.40	0.42	0.45	0.48	
20	0.43	0.43	0.44	0.45	0.47	0.49	
10	0.47	0.47	0.48	0.48	0.49	0.50	
0	0.51	0.51	0.51	0.51	0.51	0.51	
	6 h	5,7 h	4,8 h	3,9 h	2,10 h	1,11 h	

When converting positions to an earlier epoch reverse the signs.
(Data from J. Muirden.)

Table 8.3. *Ten year table of precession in declination.*

RA	Precession	RA	Precession
h	′	h	′
0,24	+3.3	7,17	−0.9
1,23	3.2	8,16	1.7
2,22	2.9	9,15	2.4
3,21	2.4	10,14	2.9
4,20	1.7	11,13	3.2
5,19	0.9	12	3.3
6,18	0		

When converting numbers to an earlier epoch reverse the signs.
(Data from J. Muirden.)

Vesta is the brightest of the asteroids. It may reach magnitude 5.5 and is then easily visible to the naked eye. At a favourable opposition Pallas brightens to magnitude 6.3. Vesta has a diameter of about 311 miles (501±24 km) and Pallas a diameter of 325 miles (523±20 km). Ceres is the largest of the asteroids and is about 567 miles (913±43 km) across but is never brighter than magnitude 7. It would therefore appear to be composed of darker materials than the others.

Over 500 of the asteroids exhibit brightness variations with periods of about 2.45 hours to 80 hours. Among these are Vesta, Iris, Eunomia, Sirona and Tereidina. Eros, with a period of 5.270 hours, does not always show brightness fluctuations for the reason already explained. Brightness fluctuations are not doubt due to rotation and/or surface reflectivity variations on the asteroid.

Observing the minor planets

General

Over the last decade or so, asteroid astronomy has become an increasingly worthwhile pursuit for both amateur and professional astronomers. The positions of the brighter asteroids that are visible each year are given in ephemerides in the *BAA Handbook* and *Astronomical almanac* (which uses Epoch 2000 coordinates but gives ephemerides for the 'big four' only), the ALPO *Solar System ephemeris* and the RASC *Observer's handbook* (both of these are annual with 26 and 21 asteroids respectively, represented in 1992 and both use Epoch 2000 coordinates so their positions can be plotted on both the Tirion and Uranometria atlases). It would therefore seem that locating a given asteroid would be simple. If you possess a good star map showing stars down to the 9th magnitude such as the *Star atlas* by Webb or the *Atlas eclipticalis*, the position of a given asteroid can be directly marked and the star field compared with what you see in the telescope. If the asteroid is of magnitude 7 or 8 it should be easily seen as an additional 'star' where none is indicated on the chart. *Uranometria 2000* goes to about magnitude +9.5 and the Vehrenberg *Photographic star atlas* goes to magnitude +13.5.

There is one point to be careful about. Owing to the phenomenon of precession, the slow 'wobbling' of the Earth's axis, the positions of the fixed stars are very slowly but continually changing in RA and declination. The published position of the asteroid will therefore have to be adjusted to be consistent with the epoch of the star chart. The precessional tables shown in tables 8.2 and 8.3 will enable you to do this. For example, if an asteroid had its position computed for 1970, you would have to subtract the precessional difference for 20 years to place it accurately into the coordinates of the *Atlas eclipiticalis* drawn for the year 1950.

Norton's star atlas is used is used by many amateur astronomers. It shows stars to only magnitude 6, so more work will be necessary. However, the greater challenge makes the successful identification of the asteroid more satisfying. If you use Norton's atlas, first mark the position of the asteroid in the atlas. Then draw a map of all stars visible in your telescope inside a square about 2° along the sides around the point where the asteroid should be.

Thus, you can be fairly certain that one of the 'stars' is the asteroid but the only foolproof way to identify it certainly is either to draw the telescopic star field again or compare the chart with the telescopic field on the next night. See if one

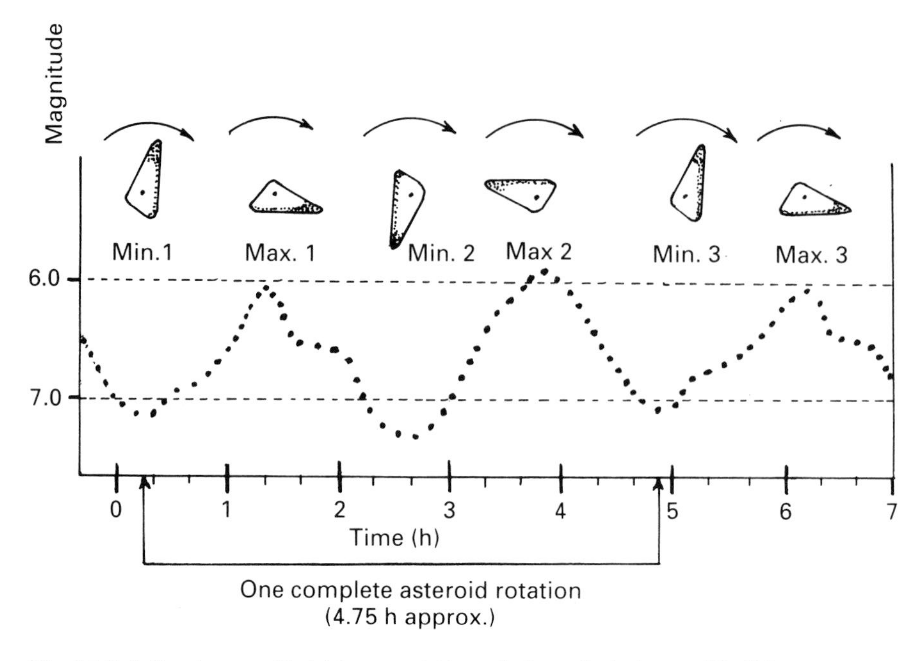

Fig. 8.5 Rotational cause of brightness variations of a hypothetical asteroid's light curve.

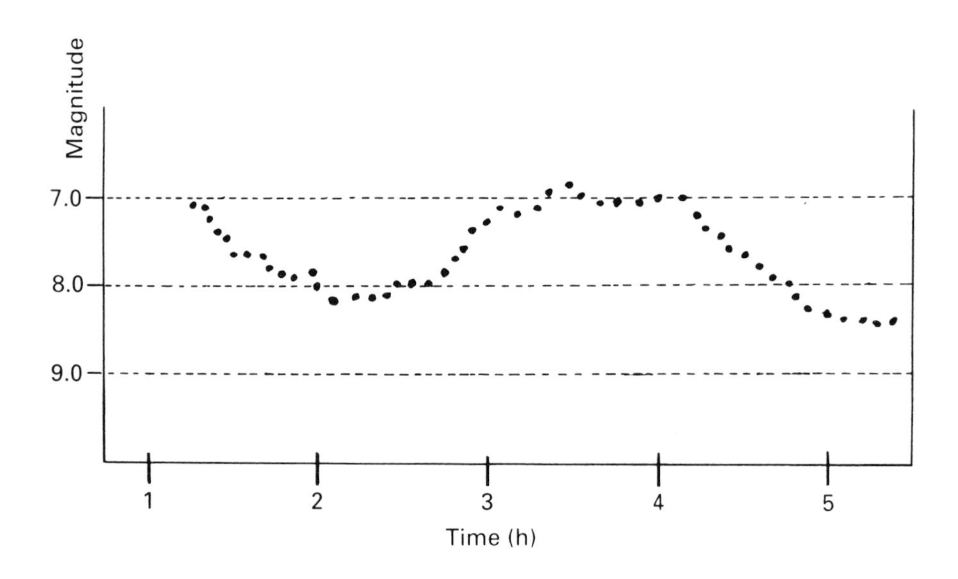

Fig. 8.6 Light variation curve of an asteroid (Eros).

of the 'stars' has altered its position, which indicates its asteroidal nature. Easy though this may seem, a small error in the first map may render the movement inapparent or the asteroid may not even be in the plotted star field.

Photometry of the minor planets

The visual study, measurement and recording of light changes is perhaps the most interesting work that an amateur can do with asteroids. Visual photometry of asteroids can be done with good binoculars, 2- or 3-inch refractors or small reflectors.

Factors that affect the brightness variations of an asteroid are its shape, if irregular, reflectivity (albedo) variations of different parts of its surface and its rotation (fig. 8.5). From the data obtained a light curve can be plotted (fig. 8.6) from which regular brightness variations due to rotation of the body can be distinguished from irregular, but minor and recurrent, variations caused by albedo differences. For a reliable light curve to be constructed, as many observations of the asteroid as possible must be made during the apparition. Rapid brightness changes can occur if the body is very irregular in shape so that measurements of the magnitude should be done every 10 minutes or even every 5 minutes. If no change in magnitude is observed during this interval you should still record the magnitude along with the universal time. If your data are to be unambiguous you should make magnitude measurements for one complete rotation during your observing session. A good plan is to make up a form to make data recording easy, because several brightness determinations will be made at one sitting. Eros, for example, has a rotation period of 5 hours and 16 minutes and more than 30 brightness determinations are needed for each rotation.

The AAVSO *Star atlas* should be used if the asteroid is brighter than magnitude 9.5. Prior to the observation, mark the positional data of the asteroid on the appropriate star chart and choose comparison stars as close as possible to the plotted path of the asteroid. Mark the comparison stars with their magnitudes – consult the accompanying catalogue.

When you start the actual observation select a star that exactly or closely matches the asteroid in brightness. You may have to estimate between two stars, one a little brighter and the other a little fainter than the asteroid. Attempt to find two comparison stars that differ in brightness at little as possible. Direct estimates of magnitude of a faint asteroid are neither feasible nor reliable. Relative changes in brightness should be estimated rather than the magnitude. This may be done by finding two stars that are closely similar in brightness to the asteroid but which differ from each other by only about 0.1 magnitude. You can then estimate the asteroid as 'brightening 0.2' or 'dimming 0.1' and so forth. On your record indicate this as +0.2 and −0.1 with the date and time (UT). The procedure is difficult and is fraught with error even if you are an experienced observer. Provided that your estimates and the instrument that you use are consistent throughout, the aim of the observing programme will be achieved, i.e., to establish the times of greatest and least brightness. So long as you can do this, whatever happens in between, you are enabled to plot the light curve and thus obtain useful data. It is best to observe near opposition so that your light curve extends for as long as possible; you may need to combine curves from more than one night.

To construct a light curve, the hours (UT) are plotted along the horizontal axis from left to right. Separate them by about at least an inch (2.5 cm). The magnitudes are plotted on the vertical axis, brightest at the top, faintest at the bottom, and should be spaced at least half an inch (1.25 cm) apart. If the asteroid has a maximum brightness variation (amplitude) of 0.5 magnitude or less, each magnitude had best be placed at intervals of 2 cm. When your data have been plotted it is a simple matter to derive the amplitude and rotational period of the asteroid (fig. 8.5). Amplitude, which is the maximum range of magnitude variation, is simply determined. It is usual to record it in magnitudes or tenths of magnitudes and is simply the difference between the highest and lowest magnitude in the light curve (fig. 8.7).

The significance of amplitude determinations is that they provide data on the alignment of the axis of rotation of the asteroid and its shifting, if any. This permits a better understanding of the nature of the brightness variations. The amplitude of brightness variations of a given asteroid is found to vary slightly in consecutive apparitions. This is caused by changing orientation of the asteroid to the Earth. If the object is elongated and rotates around its minor axis then equatorial presentations will give large brightness fluctuations but polar presentations will give much smaller brightness fluctuations (fig. 8.8).

You should try to measure amplitude changes as often as you can. Results of these and other observations should be sent to:

ALPO Minor Planets Section,
Department of Physics,
Illinois College,
Jacksonville, IL. 62650

International Amateur–Professional
Photoelectric Photometry (IAPPP),
Dyer Observatory,
Vanderbilt University,
Nashville, TN. 37235

Minor Planet Center,
Smithsonian Astrophysical Observatory,
60 Garden Street,
Cambridge MA. 02138

Axial rotation periods of most of the asteroids have been at least approximately determined. The amateur observer can render valuable service in helping to upgrade the accuracy of these measurements. It is evident from fig. 8.8 that if the asteroid has an elongated irregular figure, two maxima and two minima will be generated in the light curve by one complete axial rotation of 360°. The light curve can therefore be used to determine the axial rotation period. To obtain one full rotation you measure the time interval between every other minimum (minima 1 and 3 in fig. 8.5). Similarly, measure the time interval between every other maximum (maxima 1 and 3 in fig. 8.5) and take the mean value of the two results for the axial rotation period of the asteroid.

As well as direct visual photometry as just described, photography is another useful photometric technique. A good-sized telescope is required – 8 inches of aperture and up – because most asteroids are faint. Photographs of asteroids are made at the prime focus of the telescope. Results are not reliable if the asteroid is fainter than magnitude 8 because this would necessitate a 10-minute exposure during which time the asteroid's brightness could have fluctuated so that the photograph would be worthless. Fast black and white film should be used for photographic photometry of asteroids, such as hypered Kodak 2475. There are many others. Processing is possible to ASDA 1000 or higher.

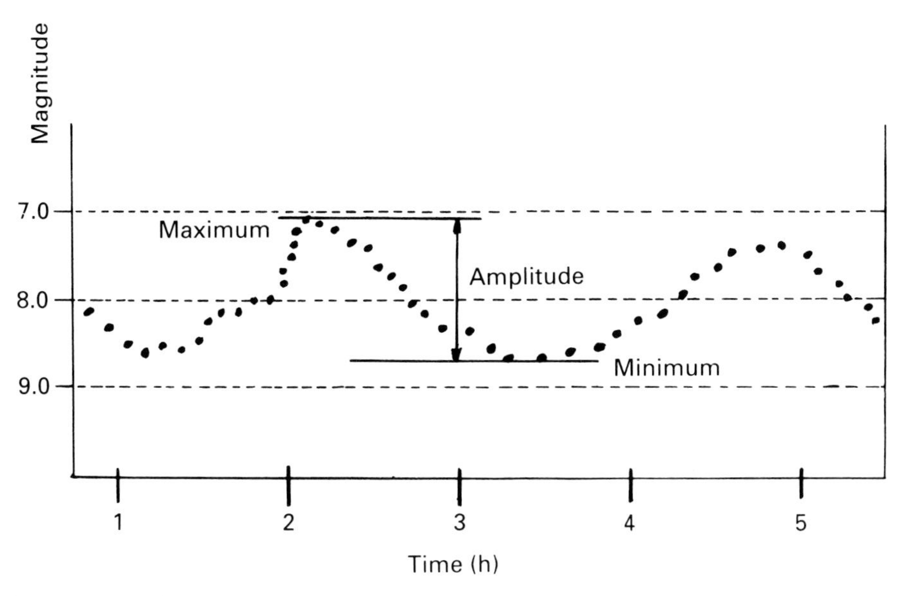

Fig. 8.7 Measuring the amplitude from the light curve.

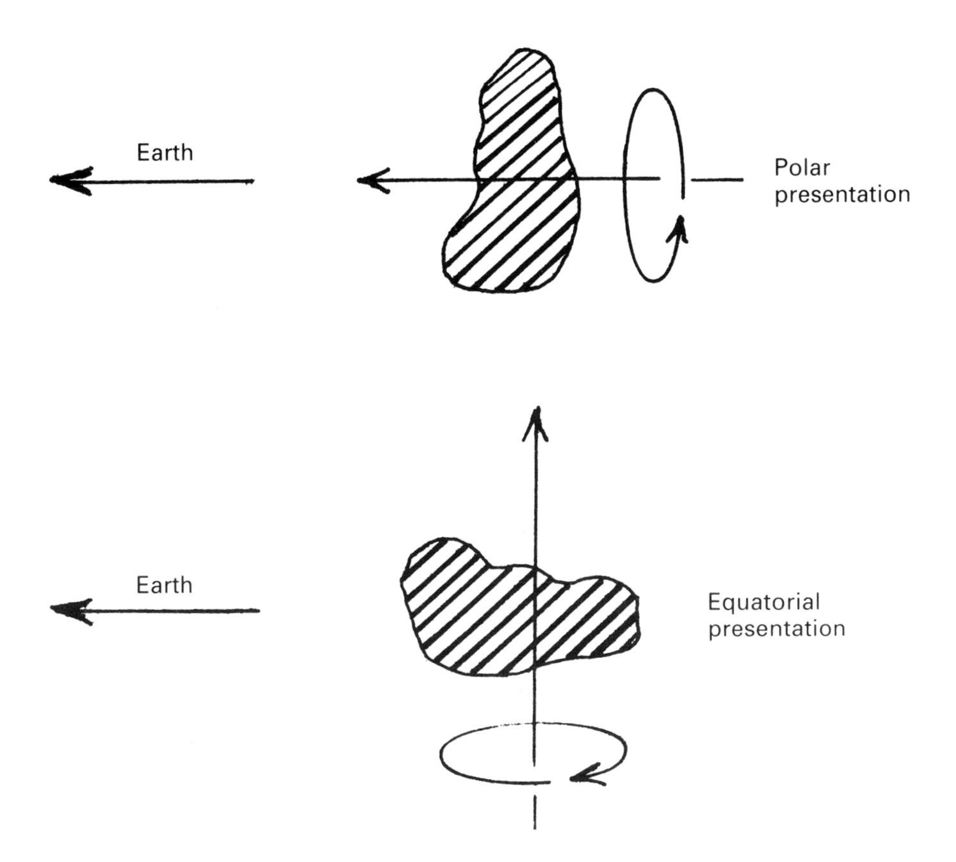

Fig. 8.8 Maximum and minimum amplitudes of a rotating asteroid.

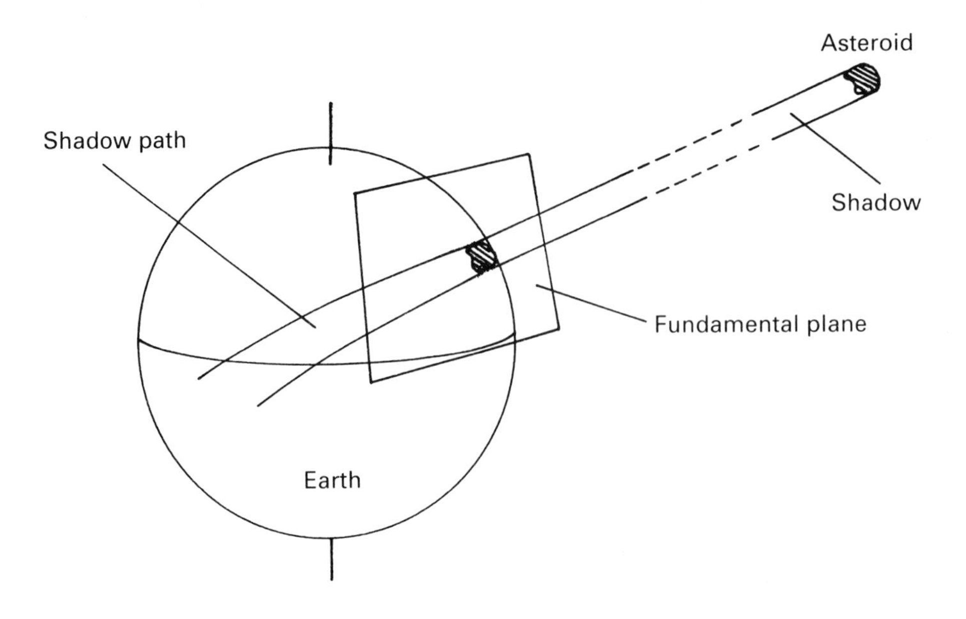

Fig. 8.9 Determination of an asteroid's profile by a shadow measurement during a stellar occultation. (Redrawn from P. Maley, Astronomy **12(2)**, *51, 1984.)*

At the prime focus of a fast optical system, e.g., an F/5 8-inch reflector, you can record magnitude 14 during a 5-minute exposure with such films. Make exposures at 10-minute intervals and keep careful records of each. During processing, be sure to number each print so as to ensure the correct exposure times. The resulting photographs are then used to determine the times of brightness maxima and minima. Image diameters should be measured. You can achieve surprisingly accurate results compiling the light curve from the photographs in conjunction with a reference star chart.

Much more accurate results are obtained with photoelectric photometry in which the eye of the observer and the telescope eyepiece are replaced with an electrometer. This results in the ability to measure brightness to accuracies approaching 0.01 magnitude. Photoelectric measuring instruments are now available at prices affordable by most amateurs who are thus enabled to contribute much valuable research data that can be used by professional astronomers. The construction of the photoelectric photometer and its use in photometric studies of asteroids is described in chapter 16.

Digital CCD cameras can be used for asteroid photometry if equipped with suitable filters. Both conventional photoelectric photometers and digital CCD cameras can be used for colorimetric work in determining colour indices and maybe even classifying asteroids by their spectra.

Observing stellar occultations by minor planets

Really valuable data can be contributed by amateur observation and timing of stellar occultations by asteroids. A real effort should be made to observe these phenomena. Such occultations are quite common but until about 1981 observation of occultations was rarely done because they could not be predicted. Asteroid ephemerides have now been much improved and astronomers who calculate asteroid orbits gladly provide detailed predictions of occultations.

Shadow path maps are now published by the *Occultation Newsletter* of the International Occultation Timing Association (IOTA). Occultation predictions are included in the *BAA Handbook* and the *ALPO Solar System Ephemeris*. Appulses are also predicted. These are close approaches of asteroids to stars. So small are the asteroids that the chance of one actually passing in front of a star is slight at a given location. However, it is interesting to watch an appulse; there is always a chance that the path of the asteroid will be slightly different from that predicted in an ephemeris because of imperfect knowledge of their movements. The error may only be one second of arc but could make the difference between an occultation and a near miss. The parallax effect due to your location though small is significant. It could mean that an observer in Canada would see an occultation while a near miss would be recorded by an observer in Brazil. Don't forget to watch for a second occultation of the star by an invisible satellite or companion of the asteroid. This was observed in 1977 with Hebe and with Melpomene and Herculina the following year.

As already mentioned timing stellar occultations by asteroids is the most accurate method of determining asteroid size and shape. Occultations are therefore well worth observing and timing. All you need to do this is a 4-inch telescope and a stopwatch. Dozens of occultations of stars by asteroids are visible around the Earth each year but from a given fixed location they are rather scarce. Predictions of occultations are given in astronomical journals and ephemerides and up to date information on shadow paths can be obtained by calling the nearest recording of IOTA a few days previous to the event.

Although in the ephemerides the time of an occultation is predicted to within a few seconds it is a good practice to watch for unexpected dimmings both before and after the main occultation, by as much as ± 15 minutes, which may be caused by satellites of the asteroid.

When an occultation is soon to begin the star and asteroid images appear to coalesce into a single point of light. The occultation has not yet occurred; this appearance is due to the eye being unable to resolve two points of light if they are very close together. The occultation is happening when the apparently coincident star and asteroid image suddenly dims. Asteroids have little or no atmosphere so when they pass in front of a star there is a sharp cut-off of the star's light instead of the gradual dimming observed in occultations of stars by planets with substantial atmospheres.

Accurate determination of the times of when the star disappears and reappears, the distance of the asteroid from the Earth and its orbital motion enables the size of the asteroid to be calculated. Many occultation timings must be made from scattered localities whose locations in latitude, longitude and altitude are precisely known across the width of the shadow path. This is necessary to provide sufficient data for determining the silhouette profile of the asteroid.

There is a direct (i.e., almost precisely equal) proportionality between the asteroid's size and its shadow so that measuring the shadow gives information about the dimensions of the asteroid. The different observers will each see a different chord of the asteroid pass across the star, the duration of the occultation as recorded by each observer being determined by the size and shape characteristics of the asteroid. The combined results enable the size and shape of the asteroid to be worked out.

If sufficient timings are made along the path of the shadow on Earth from different stations scattered along it, the dimensions and form of the asteroid can be accurately derived. In order to determine the shape of the shadow and therefore that of the asteroid the coordinates of each observational station are projected onto what is called a fundamental plane. This is a plane perpendicular to a line joining the asteroid and the Earth's centre (fig. 8.9). The observations are then fitted to an ellipse or circle. A good example of this is in the previously described occultation of the star 1 Vulpeculae by 2 Pallas that occurred in 1983 in which 130 timings enabled four fifths of the asteroid's profile to be delineated.

In addition to your telescope you will also need a tape recorder and a short wave radio able to receive stations which broadcast accurate time signals.

The observing and timing procedure for occultations is as follows. While watching the asteroid approaching the star have your tape recorder running with the WWV radio station broadcasting at the same time nearby so that it will be recorded. When the star disappears speak a loud verbal signal such as 'one' or 'gone' and when the star reappears say 'two' or 'back'. Later when you listen to the tape you will be able to time the disappearance and reappearance of the star to within a quarter second accuracy. To allow for reaction time in observing sudden events such as these subtract 1/4–1/2 second from the time recorded on the tape to correct for this.

If you don't have a tape recorder you can still obtain useful data by timing the events with a stopwatch. With this you record the duration of the occultation but not the actual time of disappearance and reappearance of the star. Some electronic stopwatches will allow you to time to the next short wave minute signal allowing UTC to be found by subtraction.

If you don't have WWV radio you can adjust your watch accurately from time given over the telephone by the US Naval Observatory or the Royal Greenwich Observatory. Radio station time checks and commercial recordings may be off by as much as 5 or 10 seconds.

If you have a sensitive video camera and if the asteroid and occulted star are fairly bright the occultation can be recorded if you have a 16-inch telescope. The WWV radio station can be recorded on the audio track. Later you can study the tape and derive event timings accurate to 1/30 second. This is also a very good way to record secondary occultations since the camera can watch the star for many minutes before and after the main event.

The personal equation and accuracy of timing stellar occultations

The personal equation is the small systematic error made by an observer in, say, timing an astronomical event such as stellar occultations by asteroids. The magnitude of the personal equation depends on the observer's characteristics – it is the time needed for the mind to realise that the event has occurred, to know what to do and to make the appropriate response. It is, in fact, the observer's reaction time. Experience and deliberate training can minimise it. For a given observer the actual time when an event occurred is therefore equal to the observed time minus the reaction time.

You can determine your own personal equation in various ways, two of which are as follows:

(1) With your WWV radio turned on, take a stopwatch and start and stop it at exact minutes as broadcast over the radio. Start the watch exactly when you hear the tone of one whole minute and stop it at the tone of the next minute; do not look at the stopwatch. After some practice at this try to do the same without the radio signal. On starting the watch, count seconds mentally just as you did when you heard them on the radio, imagining that you are hearing the beats marking the seconds. Without looking at the watch start it and stop it after one second as closely as you can estimate. When you have acquired some skill at this run the watch for what you estimate is 30 seconds, again mentally counting seconds without looking at the watch. This gives you an idea of how your mental apparatus can subjectively measure time and physically react to the passing of time. Continue with this practice for, say, an hour. After this time you should have gained some proficiency at mental counting of seconds. You can ascertain your personal equation by noting the tenths of a second that you are consistently in error when you judge the lapse of one second.

(2) On the WWV broadcasts, the seconds are marked by a tone but those between 51 and 59 inclusive are interrupted by the voice of the announcer giving out the next minute. As exactly as you can, start the stopwatch on the minute tone and allow it to run and stop it exactly on the tenth tone, i.e., after 10 seconds. Read the watch. It ought to show less than 10 seconds because of your reaction time that caused you to start the watch slightly after the minute tone. The difference between the time shown by the watch and 10 seconds is your personal equation. This difference will be less than 10 seconds but should be not less than 9.7 seconds. Further practice at this should enable you to reduce this difference considerably.

Your accuracy in timing an event bears little or no relation to your personal equation. You may have a very small personal equation but in an occultation timing of a very faint star or in poor seeing conditions your accuracy in timing may be quite poor. However, your timing may be more accurate if your personal equation is smaller than someone else's whose personal equation is large.

When you report the time of an event it should be the time reading minus your personal equation and minus the amount of time that you consider your reading was in error. If you believe that you were 0.5 second late in indicating the occurrence of the event your accuracy is 0.5 second and your personal equation should be subtracted from the value that you got. Your estimate of accuracy should not be subtracted. It only indicates how much the time you recorded could differ. If you timed an event too late by 0.3 second then report this as −0.3 second; if too early by the same amount record it as +0.3 second. In reporting timings always indicate whether you applied your personal equation and, if so, what was its value.

You occultation timing results should be sent to the IOTA. Even if the shadow path of the asteroid just missed you this should be reported. Unless a grazing chord is timed at one site and a miss at another nearby site, the diameter of the asteroid cannot be accurately determined. Further, 'misses' show where possible asteroid satellites are not situated. Asteroid occultation observing differs from observing lunar graze occultation and total solar eclipses which require that you be at a very precisely defined location a few yards away from another observer; but you do need to know your exact position with considerable accuracy.

Discovery and rediscovery of minor planets

Finally, there is always a chance that you may rediscover a lost asteroid or even discover a new one! Usually, such work has been the preserve of observatories equipped with large-aperture Schmidt cameras. However, three happenings in recent years suggest that amateurs may make contributions in this field. First is the manufacture of small Schmidt cameras by *Celestron* that are affordable by most amateurs. Then there was the discovery of the asteroid 2090 Mizuho in 1978 on March 12th by T. Urata, an amateur observer, at 'Yakiimo Station' (see *Sky and Telescope*, August 1979, page 171). Third, the work of large observatory Schmidt cameras is increasingly restricted by budget cuts which place restrictions on their use. If you have equipment that can reach to magnitude 14 or fainter and have a fund of self-discipline, you may well be able to discover a new asteroid. In 1981, R. G. Hodgson suggested that in middle to high latitudes search of the subpolar areas of the sky near the time of the Summer solstice could possibly reveal asteroids within the orbit of the Earth. Since the discovery of the Aten objects whose orbital semimajor axes are less than the Earth's distance from the sun, the possibility of discovering new asteroids may not therefore be as fanciful as it sounds.

Further reading

Books

The system of minor planets. Roth, G. D., Faber and Faber, London (1963).
Physical studies of minor planets. Gehrels, T. (ed.), NASA Special Publication No. 267, National Aeronautics and Space Administration, Washington DC (1971).
Table of minor planets. Pilcher, F. and Meeus, J., private printing (1973).
Asteroids. Gehrels, T. (ed.), University of Arizona Press, Tucson, Arizona (1979).
Asteroids II. Binzel, R. P., Gehrels, T. and Matthews, M. S. (eds.), University of Arizona Press, Tucson, Arizona (1989).
Introduction to asteroids. Cunningham, C. J., Willmann-Bell Inc., Richmond, Virginia (1988).

Papers and articles

The nature of asteroids. Chapman, C. R., *Scientific American* **232(1)**, 24–33 (1975).
Asteroids: Spectral reflectance and color characteristics. McCord, T. B. and Chapman, C. R., *Astrophysical Journal* **197**, 781–90 (1975).
The minor planets – sizes and mineralogy. Chapman, C. R. and Morrison, D., *Sky and Telescope* **47(2)**, 92–5 (1974).
The minor planets: As interesting as ever. Porter, A., *JALPO* **29(3–4)**, 64–9 (1981).
Measuring the shape of an asteroid. Maley, P., *Astronomy* **12(2)**, 51–4 (1984).
Stand in the shadow of an asteroid. Manly, P. L., *Astronomy* **19(1)**, 54–7 (1991).
The minor planet bulletin.

Publishers of almanacs and star atlases.

BAA Handbook British Astronomical Association, c/o Burlington House, Piccadilly, London W.1. England.
RASC Observer's Handbook Royal Astronomical Society of Canada, 136 Dupont Street, Toronto, Ontario, Canada M5R 1V2.

ALPO Solar System Ephemeris Association of Lunar and Planetary Observers, PO Box 16131, San Francisco, California 94116, USA.

Astronomical Almanac United States Government Printing Office, Washington DC, and HMSO London, England.

Norton's *Star Atlas, Atlas Eclipticalis and Photographic Star Atlas (Vehrenberg)* Sky Publishing Corporation, 49 Bay State Road, PO Box 9102, Cambridge, Massachusetts 02238, USA.

AAVSO Star Atlas American Association of Variable Star Observers, 25 Birch Street, Cambridge, Massachusetts 02138, USA.

9

Jupiter

General

Well beyond the zone of the asteroids lies the orbit of the largest of the Solar System planets, Jupiter. Its equatorial diameter is 88 800 miles (142 984.8 km), more than 11 times that of the Earth's (fig. 9.1). Jupiter rotates very rapidly on its axis. At and near the equator the rotation time is about 9 hours 50.5 minutes so that material at its equator is moving at the rate of 28 000 miles (45 052 km) per hour. The rotation time of higher latitudes is near 9 hours 55 minutes. The

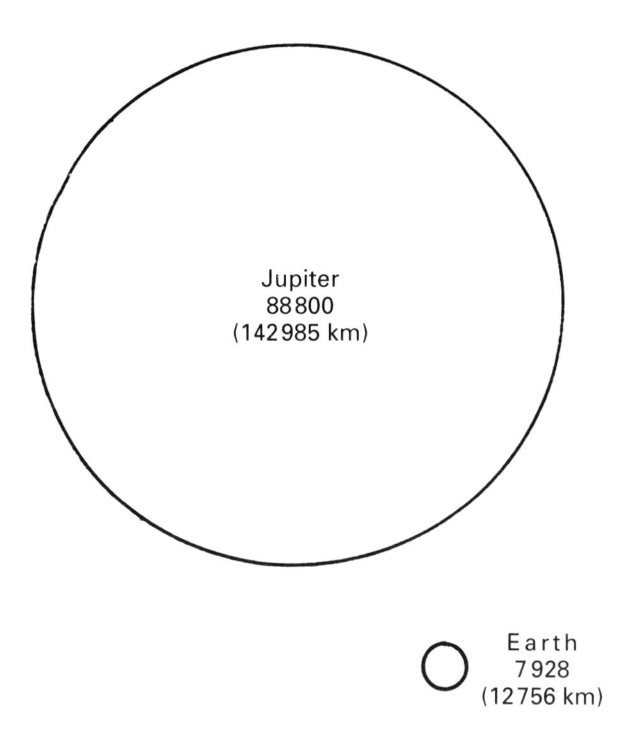

Fig. 9.1 Comparative sizes of the Earth and Jupiter (equatorial diameter in miles).

Fig. 9.2 Jupiter showing cloud belts and bright zones: 8-inch Newtonian reflector. January 6th, 1978, 21.05–21.25 UT. (F.W. Price.)

centrifugal force thus generated causes a pronounced equatorial bulge and flattening at the poles. The polar diameter is 5758 miles (9267.0 km) less than the equatorial diameter. Jupiter cannot therefore be a solid body in the usual meaning of the word. It is certainly known that what we see in the telescope is not a solid or even a liquid surface but a cloud-laden outer layer made up of a series of parallel latitudinally oriented light 'zones' and dark 'cloud belts' (fig. 9.2). See fig. 9.3 for their nomenclature. The ratio of the polar and the equatorial diameter is about 15 : 16 so that Jupiter has a distinctively ellipsoidal shape which is quite noticeable in a telescopic view of the planet.

Jupiter's mean distance from the sun is 483.75 million miles (778.36 million km) and its mean orbital velocity is 8.1 miles (13.07 km) per second. One complete circuit of its orbit takes 11.86 years, its sidereal period, and its mean synodic period is 398.88 days, i.e., 33.63 days more than an Earth year. Jupiter therefore comes to opposition every 13.3 months approximately. Because of its much greater distance from the Earth compared to Mars, its apparent angular diameter throughout an apparition does not vary nearly so much as that of Mars.

During an Earth year Jupiter passes over about one twelfth of the way around the Zodiac, i.e., about 30°. The Earth moves in its orbit about one degree in a day so that it overtakes Jupiter somewhat more than 30 days later in each successive year.

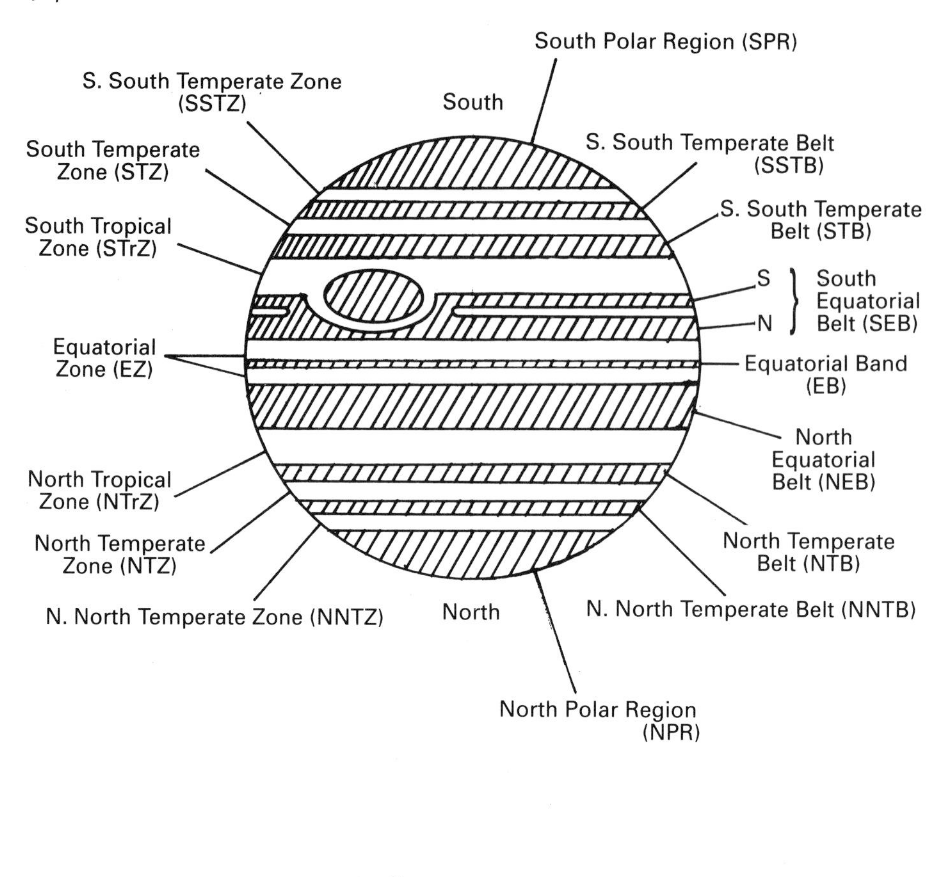

Fig. 9.3 Nomenclature of Jupiter's cloud belts and zones.

Since 11 synodic periods total 4387.68 days and since 12 years total 4383 days, the difference is just 4.68 days so that Jupiter's opposition dates are very nearly the same every 12 years and occur 4 or 5 days later.

Jupiter's orbital eccentricity is 0.048. At perihelion its distance from the sun is 460.5 million miles (740.9 million km) and at aphelion the distance is 507.0 million miles (815.7 million km). Its orbit is inclined at an angle of 1° and 18 minutes of arc to the Earth's and its axial inclination is 3° 7′ of arc relative to its orbit.

Jupiter is attended by 16 satellites. Of these, four of the eight closest to the planet are the largest and were discovered by Galileo in 1609 with his telescope; they were also seen at about the same time by Simon Marius. Whoever it was that first saw them, they are known as the Galilean satellites. They are of planetary dimensions and are named, Io, Europa, Ganymede and Callisto in order of increasing distance from Jupiter. Io and Europa are of about the same size as our own moon whereas Ganymede and Callisto are quite a bit larger. The sizes of the Galilean satellites compared with the Earth's moon, and their orbits are shown in fig. 9.4. The orbits lie in very nearly the same plane.

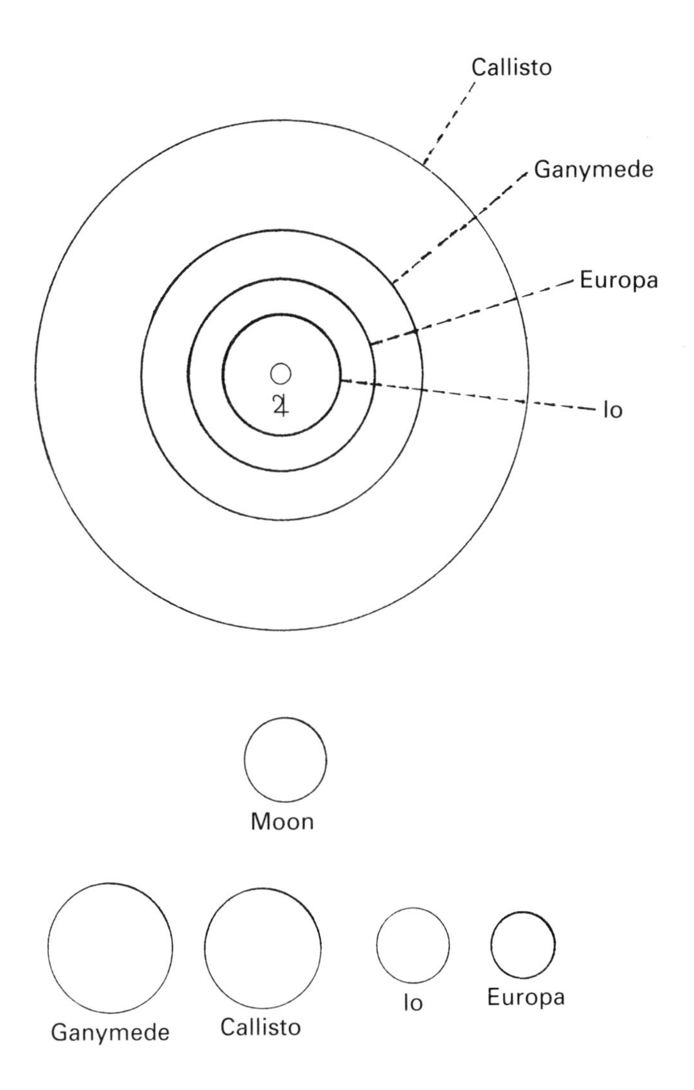

Fig. 9.4 Orbits and comparative sizes of Jupiter's Galilean satellites and the Earth's moon.

Within the orbit of Io are four very small satellites beyond the reach of ordinary amateur-owned telescopes. The closest to Jupiter is Metis followed by Adrastea, Amalthea and Thebe.

Well beyond these 'inner' satellites are two zones of outer satellites, all of them very small and difficult to detect with amateur telescopes; some amateurs have seen Himalia which is theoretically visible with a 15-inch telescope. These satellites are probably asteroid-like in nature. The first group consists of the four named Leda, Himalia, Lysithea and Elara. Another set of four, well beyond these, are named Ananke, Carme, Pasiphae and Sinope. This last group travel around Jupiter in a direction opposite (retrograde) to all the other satellites which go the 'right way' around, i.e., from west to east. Quite possibly they are captured asteroids. The paths of these eight outer satellites deviate considerably from a circle and change with each revolution and their movements are difficult to predict accurately.

Table 9.1. *Satellites of Jupiter*

Name	Diameter		Mean distance from Jupiter	
	miles	kilometres	miles (×1000)	kilometres (×1000)
Metis XVI	24.9	40	79.5	128.0
Adrastea XV	15.5×12.4×9.3	25×20×15	80.2	129.0
Amalthea V	167.8×104.4×93.2	270×168×150	112.7	181.3
Thebe XIV	68.4×?×55.9	110×?×90	137.9	221.9
Io I	2255.6	3630	262.2	422
Europa II	1949.9	3138	416.9	671
Ganymede III	3269.6	5262	665.5	1071
Callisto IV	2982.6	4800	1170.0	1883
Leda XIII	9.9	16	6893.5	11094
Himalia VI	115.6	186	7133.3	11480
Lysithea X	22.4	36	7282.5	11720
Elara VII	47.2	76	7293.0	11737
Ananke XII	18.6	30	13173.0	21200
Carme XI	24.9	40	14043.0	22600
Pasiphae VIII	31.1	50	14602.2	23500
Sinope IX	22.4	36	14726.5	23700
Earth's moon	2160.0	3476.0		

With respect to their distances from Jupiter, the 16 satellites appear to fall into four groups of four satellites each, the second group of four outwards from Jupiter, the Galileans, also being by far the largest.

To avoid cluttering the preceding text with figures table 9.1 summarises the sizes of Jupiter's satellites and their distances from the planet. As well as having names the satellites are also designated by Roman numerals which indicate their chronological order of discovery NOT the order of their distances from Jupiter.

In addition to the satellites Jupiter is surrounded by a thin flat dust ring between about 29 210 miles (47 000 km) and 32 939.7 miles (53 000 km) above the top of its cloud layer (fig. 9.5). The ring has a bright outer zone 497.2 miles (800 km) wide and a darker zone 3107.5 miles (5000 km) wide. The dust ring is invisible from the Earth.

History of observation

Jupiter has always been noted for its brilliance and magnificence among the fixed stars. Its brightness is at times almost enough to cast a shadow.

The planet appears to have been first mentioned by Ptolemy who observed and recorded its conjunction with the star Delta Cancri on September 3rd in the year 240 BC.

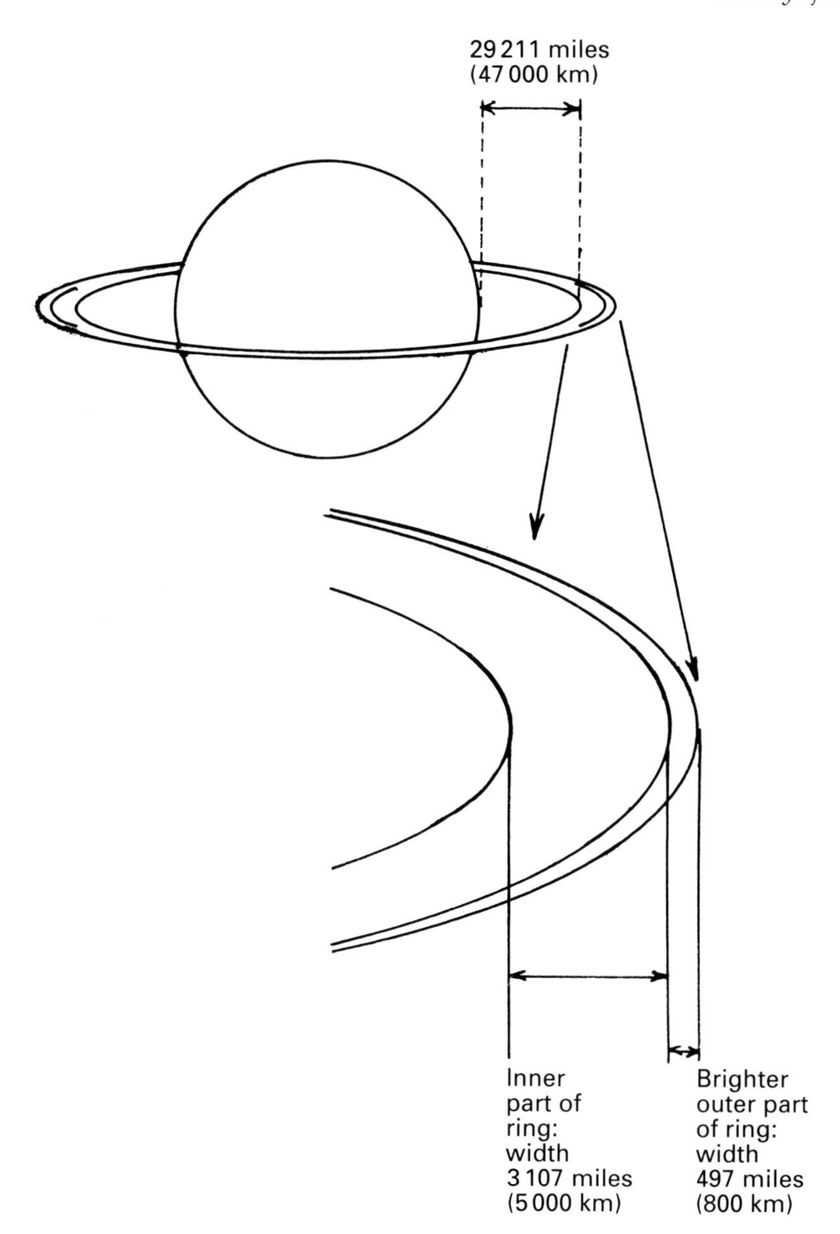

Fig. 9.5 The dust ring of Jupiter.

It wasn't until 1630, nearly a quarter of a century after the invention of the telescope, that the principal cloud belts of Jupiter were discovered. There is some doubt as to who first saw them, the most likely of five possible contenders for the distinction being Campani of Bologna. The remaining four are two other Italians, Niccolo Zucchi and G. D. Cassini, the Englishman Robert Hooke and Father Gilles-Francois Gottingniez, a Belgian.

Cassini measured the axial rotation period of Jupiter by noting the time taken for a dark or light marking on Jupiter's disc to make one complete revolution

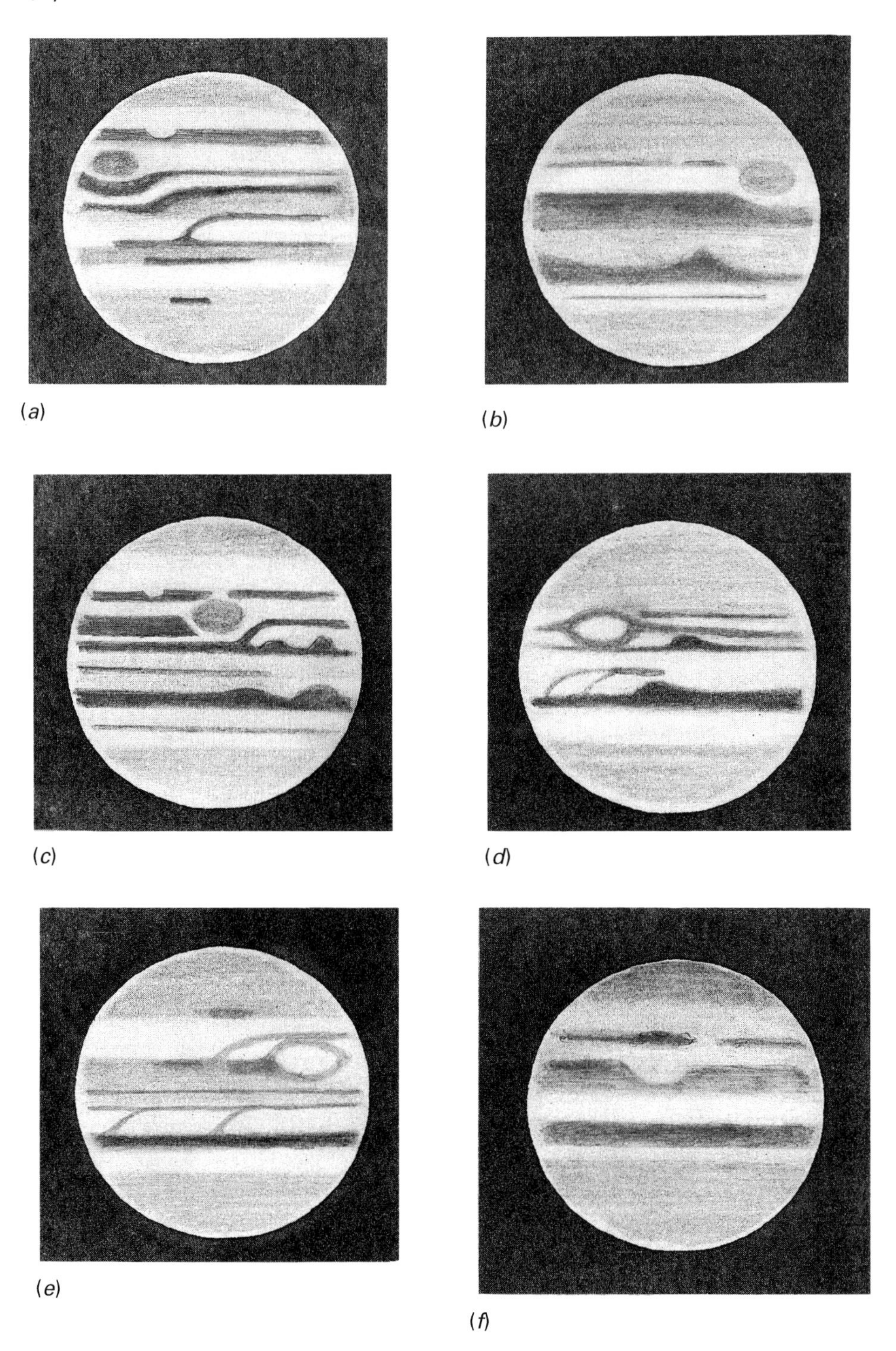

Fig. 9.6 Different aspects of Jupiter's Great Red Spot and the red spot hollow: 8-inch Newtonian reflector ×225. (a) *December 5th, 1977, 03.55 UT;* (b) *January 24th, 1978, 03.22 UT;* (c) *February 15th, 1978, 02.30 UT;* (d) *March 13th, 1978, 01.45 UT;* (e) *April 20th, 1978, 01.50 UT;* (f) *June 7th, 1981, 02.30 UT. (F.W. Price.)*

around the globe. This was done by noting the time between two successive transits of a marking across the central meridian of Jupiter, i.e., the imaginary line joining the north and south points on the disc. This technique is described in the section on observation of Jupiter. In 1665 Cassini noticed that the rotation period derived by observing markings near the equatorial regions were generally about 5 minutes shorter than those given by markings away from the equatorial region, i.e., the features exhibited differential rotation. This clearly indicated that the markings seen on Jupiter were not true surface markings but features of a fluid globe or a cloud-laden atmosphere.

The Great Red Spot

The Great Red Spot is a roughly oval feature situated in a `bay' or hollow as it is usually known in the southern edge of the south equatorial belt (fig. 9.6). Its size is variable and may attain a length of about 30 000 miles (48 270 km) in longitude (major axis) and 7000 miles (11 263 km) in latitude (minor axis) so that its surface area exceeds that of the Earth.

This feature or something like it was possibly discovered by Robert Hooke in the South Tropical Zone of Jupiter in 1664. W. F. Denning suggested that what Hooke saw was, in fact, a very early manifestation of the true Red Spot. Cassini, who was the first director of the Paris Observatory drew the Great Red Spot in 1665, 1672, 1677 and 1691 (fig. 9.7) but it may have been seen even earlier. There is reason to believe that what Hooke saw was not the Red Spot that we know today or even a precursor; writing in the *Philosophical Transactions* Henry Oldenberg (who believed Hooke to be the first observer of the Red Spot) stated that Hooke had seen 'a small spot in the biggest of the three obscurer belts of Jupiter'. This must have been the North Equatorial Belt which was the largest

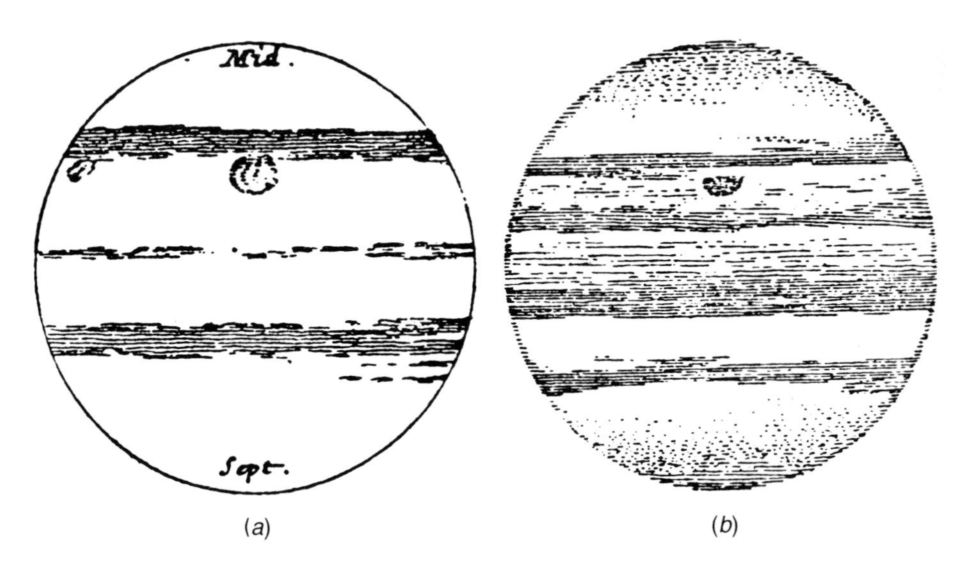

Fig. 9.7 Cassini's drawing of Jupiter's great red spot: (a) *January, 1972;* (b) *July, 1677. The darkness of the equatorial regions in the 1677 drawing may indicate an earlier occurrence of the 1962 disturbance. (From* Le Monde De Jupiter, *Libert, L.)*

and most prominent of the cloud belts in the 1660s, and was so drawn by A. Azout in 1664, by Cassini in 1665 and by Hooke in 1666.

The Red Spot known today is situated in the South Tropical Zone in contact with the south edge of the South Equatorial Belt as was the spot seen by Cassini; also, Cassini's spot was large, about equal to one tenth of Jupiter's diameter. Hooke's spot could not therefore have been the spot seen by Cassini and the Red Spot known today.

The earliest definite record of the Red Spot as it is known today is a drawing by Schwabe made on September 5th, 1831. It shows the Red Spot Hollow and hence establishes that this and the red spot have existed for at least 160 years.

By the year 1878 the Red Spot had grown in size and was elliptical in shape, its dimensions were about 30 000 miles (48 280 km) in longitude and 7000 miles (11 265 km) in latitude. It had also deepened in colour from its formerly pale pink hue to a dark brick-red colour. Since that time it has been systematically observed. From 1891 onwards the members of the Jupiter Section of the BAA played a prominent role in this work.

From 1878 to 1881 the Red Spot remained a prominent object. After 1882 it developed pointed ends and at times became almost invisible followed by returns to greater prominence as it did in 1920 and 1926, though it did not attain the prominence of 1878. However, it did return to its 1878 prominence in 1936. In 1949 the Red Spot appeared sometimes a salmon pink colour and at others it was colourless. Between 1965 and 1971 it was very prominent and of a

Table 9.2 *Visibility and appearance of the Great Red Spot, 1878–1938.*

1878	Usually prominent.
1879–82	Very prominent; striking red colour. Elongated oval shape.
1882–90	Faded steadily.
1891	Revived in visibility and faded again.
1893–94	A fairly easy object.
1896–97	Distinctly reddish (Terby).
1903–4	Faint, difficult to see.
1904–5	Plainly visible.
1905–6	Faint, difficult to see.
1906–7	Faint, difficult to see.
1909–11	'Intensely pink' (Antoniadi).
1914	Distinct well-defined ellipse.
1917–8	Marked revival in visibility.
1919–20	Conspicuous.
1926	Deep red edges (Du Martheray).
1927–8	Outstandingly conspicuous – 'brick red'.
1928–9	Rapid extinction.
1931–2	Moderately visible.
1934	Salmon pink colour.
1936–7	Very prominent, comparable to *ca.* 1880.
1938	Practically invisible.

(Data from B. M. Peek).

strong salmon pink colour. Table 9.2 summarises the visibility and appearance of the Great Red Spot between 1878 and 1938.

Whenever the Spot disappears its position is always indicated by a pronounced indentation or bay in the south edge of the south equatorial belt called the 'Red Spot Hollow'. The red colour of the Spot is real; it appears darkest in photographs taken in ultraviolet or violet light, paler in green or yellow light and it is invisible in the red or infrared.

The Red Spot exhibits a remarkable range of movements in longitude. For example, between 1831 and 1838 its movement in longitude was 1046° which is almost equal to three complete circuits of the planet. Fig. 9.8 shows the longitudinal wandering of the Red Spot from 1831 to 1955. Interestingly, as pointed out by B. M. Peek, the noted English Jupiter observer and one-time director of the BAA's Jupiter Section, the Red Spot is darker when accelerating in longitude, i.e., when its rotation period slows down.

What is the Great Red Spot? Its permanence indicates that it must be something other than a cloud and the remarkable movement in longitude is not

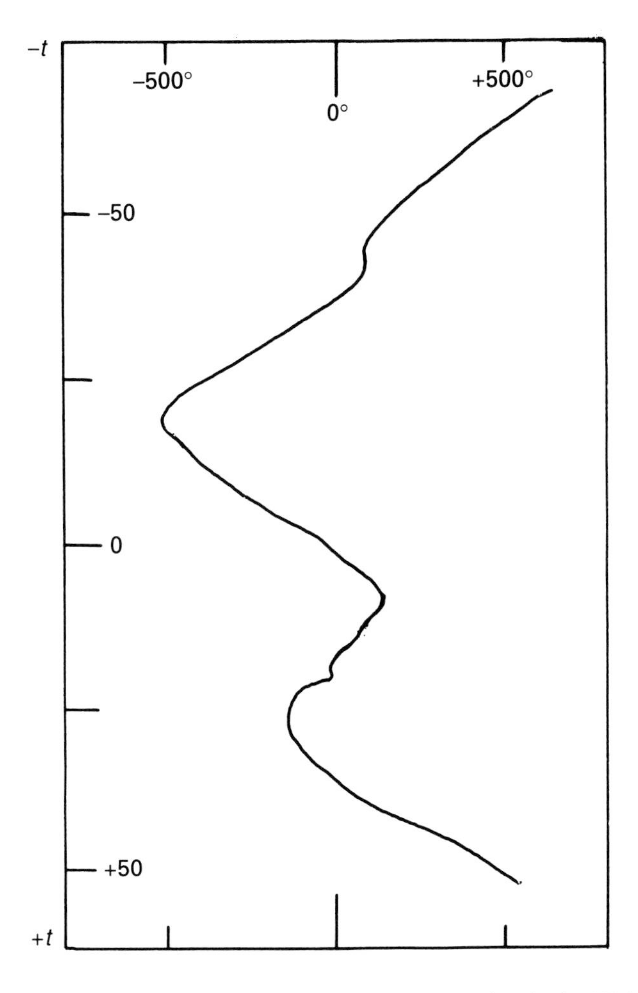

Fig. 9.8 The wandering of the Great Red Spot in longitude, 1831–1955. (After B.M. Peek.)

consistent with it being attached to the surface of Jupiter so it must be floating in the atmosphere. B. M. Peek believed it to be a solid body floating in Jupiter's cloud-laden atmosphere and that relatively slight changes in the depth of the spot within the atmosphere would be sufficient to explain all of its changes in colour intensity and velocity. Small changes in the density of Jupiter's atmosphere in the region of the Red Spot would cause it to sink or rise. When it sinks it would be covered by the clouds and when it rises it would be prominent. Calculations showed that a range of vertical movement of no more than 7 miles (11.26 km) would suffice to cause the observed changes in prominence of the Red Spot.

Another more recent explanation of the nature of the Red Spot is that it is the top of a 'Taylor Column' or a 'standing current' in Jupiter's atmosphere resulting from the planet's rotation. If there is a solid surface somewhere below Jupiter's cloud layer then it may be reasonably expected that there will be strong atmospheric currents there. The presence of a huge obstruction on that surface would disturb the atmospheric currents blowing around it and would result in the formation of a stagnant gas column over the obstruction; this is known as a Taylor Column. That the Red Spot may be the top of such a column was proposed by the British geophysicist Raymond Hyde. He put forward plausible mechanisms to explain variations in the rotation period of the Red Spot.

There is no way of ascertaining which of these two models is correct; perhaps neither is. What is certain is that among Jovian features, the Great Red Spot is unique. Other similar features have often been observed but hardly any as large as the Red Spot and none approaching anywhere near its considerable longevity.

The origin of the colour of the Red Spot has been much discussed; some observations suggest that the reddish colour might be due to hydrogen phosphide (phosphine) but there is as yet no viable model for the chemical make-up of the Spot and the origin of its colour is unknown.

The South Tropical Disturbance

Associated with the Great Red Spot is the South Tropical Disturbance discovered by P. B. Molesworth in 1901 although Dr E. B. Knobel observed something similar in 1890. The Disturbance started as a dark marking stretching across the South Tropical Zone and it quickly extended in longitude. Bright white spots appeared at its preceding and following ends. Since that time it has been seen intermittently and exhibits considerable changes in shape, extent and motions. At one time it was 25° in length. Since 1916, when active, it has always been more than 100° in length. After 1930 its length grew to over 200°. Although the Disturbance itself is often of an indistinct and confused aspect, its ends are clear cut, dark and concave in shape (fig. 9.9).

Initially the rotation period of the extremities of the Disturbance was 20 seconds less than that of the Red Spot so that it would circle around Jupiter and catch up with the Red Spot in about two years. This it did for the first time in 1902 and again in 1904. Encounters of the Disturbance and the Red Spot have since been not so frequent but nine have been recorded in 40 years. When the Disturbance meets the Red Spot it nudges it forward temporarily and at the same time flows below or around it. During the early encounters, the general

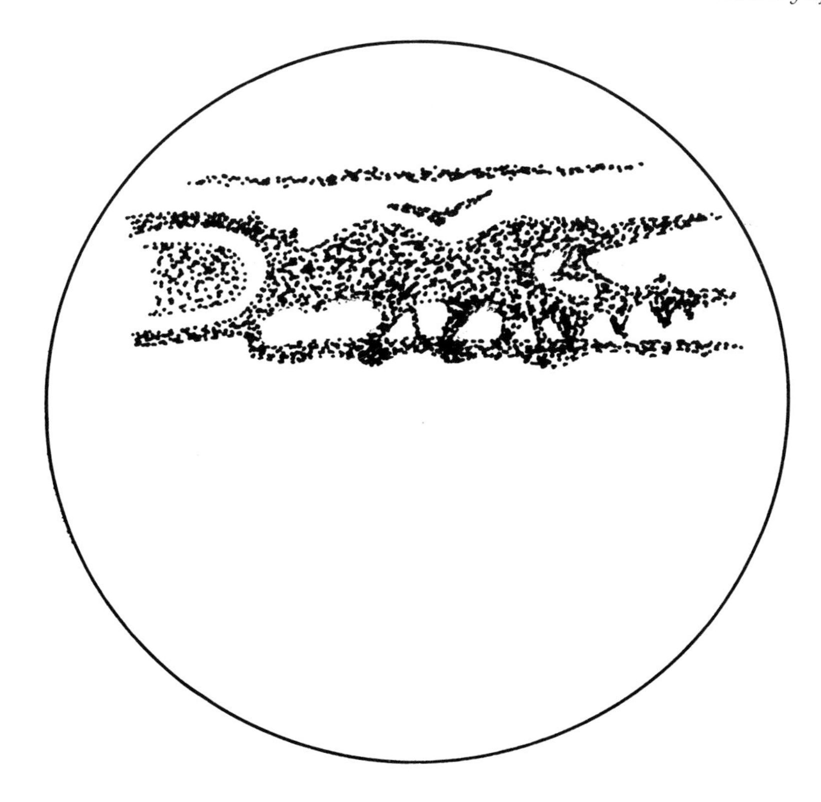

Fig. 9.9 The Great Red Spot in its hollow and to its right part of the South Tropical Disturbance following it. (Redrawn by E.M. Antoniadi, May 21st, 1901.)

effect of the Red Spot on the Disturbance was as if an attractive force was operative; when one or other of the ends of the Disturbance approached the Spot, the motion of that end was slightly accelerated whereas after passing across the Red Spot Hollow the ends of the Disturbance were slowed down. Occasionally, the Disturbance, the Red Spot Hollow and the greater part of the South Equatorial Belt disappear; at these times the Red Spot becomes more prominent.

The Circulating Current

This is one of the most puzzling of the many mysterious phenomena observed in Jupiter's atmosphere. Spots on Jupiter are hardly ever seen to move in latitude so that considerable interest was aroused by the discovery of the Circulating Current by observers of the Jupiter Section of the BAA. At times between the years 1919 and 1934 dark spots would appear at the Red Spot's following end and then drift backwards away from the Spot along the north edge of the South Tropical Disturbance. This had the effect of turning them southward across the Zone after which they reversed and moved forward along the south edge of the Zone (fig. 9.10).

Around 1930–4 the forward motion occurred along a thin horizontal belt resembling a stream of smoke emanating from a vertical dark streak that was nicknamed the 'Smoke Stack', situated at the preceding end of the South

Fig. 9.10 The circulating current between the two ends of the South Tropical Disturbance. The Great Red Spot and Hollow are to the left within the Disturbance. (Redrawn after T.E.R. Philips.)

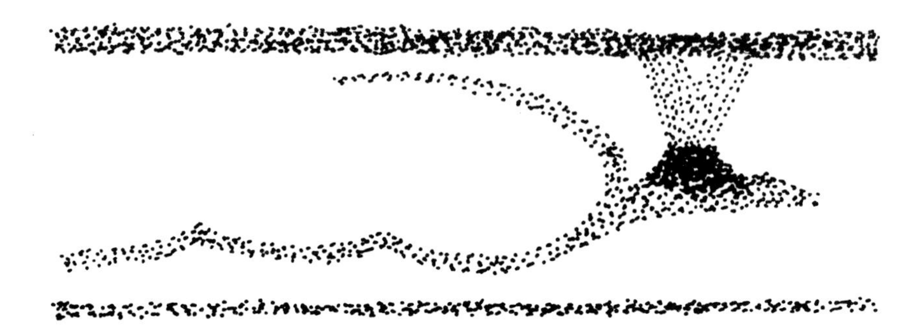

Fig. 9.11 The 'Smoke Stack'. (Redrawn after B.M. Peek.)

Tropical Disturbance (fig. 9.11). The fate of the spots when they reached the following end of the Disturbance and whether or not their circuit was completed by their passage northwards across the Zone to start over again was not observable.

The Circulating Current exists only in the part of the South Tropical Zone not occupied by the Disturbance, i.e., it makes contact with the Disturbance at both ends of the latter.

The dark South Tropical Streak of 1941–2.

An unusual dark feature appeared in the South Tropical Zone in the latter part of October in 1941. It was situated immediately preceding the Red Spot and was first recorded by Ainslee on October 22nd and noted by B. M. Peek on October 23rd. It slowly lengthened in the preceding direction until it appeared as an elongated dark streak and attained a length of 30° of longitude. About mid-November the formerly stationary following end began to move away from the Red Spot at a rate similar to the movement of the preceding end so that the streak maintained a nearly constant length and remained so until the end of March 1942. Detailed drawings of the growth and later behaviour of the streak were made by Hargreaves and B. M. Peek.

T. E. R. Phillips thought that the streak might be merely a partial revival of the South Tropical Disturbance. However, before April 1942 ended the length of the streak began to decrease, apparently from the following end. It was still getting shorter by mid-August and was only 5° of longitude in length by October 5th. It was recorded for the last time by W. Haas on October 6th 1942 and on October 11th, no detail was reported in the South Tropical Zone by H. M. Johnson who was observing the area from the Flower Observatory with an 18-inch refractor.

On January 11th 1946, Hargreaves noticed a dark spot in the South Tropical Zone and again on January 16th and 26th which by this time had increased in length to roughly 12° in longitude. By the end of March it was 22° in length. It now resembled the dark South Tropical Streak of 1941–2 and was continuing to lengthen until it exceeded the length of the 1941–2 streak; in July 1946 at the end of the apparition its length was almost 40° in longitude. It almost looked as though it was a revival of the South Tropical Disturbance. It was not observable again until the beginning of the 1947 apparition and by then it had shortened and was rapidly decreasing in length. The last record of it was on July 9th 1947 when its length was roughly 6°. Careful measurements of the motion in longitude of the ends of the streak indicated that whatever caused it to shorten operated at both its ends. This streak therefore lasted about half as long again as the 1941–2 streak and attained a much greater length.

Eruptions of the South Equatorial Belt

At times, the Red Spot Hollow and the south component of the South Equatorial Belt fade and almost disappear for between one and three years and the South Tropical Disturbance subsides. This is then followed by a great burst of activity on the belt which foreshadows its reappearance. This was seen in 1919–20, 1928–9, 1938, 1943, 1949 and 1952–3. The outbursts appear as two branches, one of them moving towards the following side of the disc along the belt's south edge, the other moving towards the preceding side of the disc along the north edge of the belt. They encircle the planet, meet and pass, accompanied by rapid changes in detail and the appearance of many dark and light spots.

A specially noteworthy disturbance was that of 1949 which included the successive formation of more than 20 bright spots at one locality between the two components of the South Equatorial Belt which then went in procession along the belt looking like a string of pearls! There were also dark spots.

Oscillating spots

At times certain individual spots oscillate in longitude but with a much smaller amplitude than the longitudinal wanderings of the Great Red Spot. Such spots were seen in the two apparitions between 1940 and 1942. B. M. Peek noticed that their oscillations resembled a damped harmonic motion, insofar as the amplitude of the spots' swinging in longitude decreased with time.

Other phenomena

T. E. R. Phillips mentioned what he described as 'the swiftest known current on Jupiter' which was indicated by the occasional appearance of spots on the south edge of the North Equatorial Belt with rotation periods of 9 hours 49 minutes. This is shorter than what is usual for that region by 7 minutes.

B. M. Peek pointed out a peculiar phenomenon in the North Temperate Region where two rapid belt currents were frequently observed to have a much slower current between them. Spots in one current seem to be controlled or at least affected by the proximity of another current.

Variations in the cloud belts

The observational reports of the Jupiter Sections of the BAA Astronomical Associations and the ALPO from 1950 onwards and before are replete with detailed observational information and records of events in the continually changing atmospheric features of Jupiter. All of this would be too much to describe fully up to the present and would make tedious reading so that only a brief summary of changes in the major features – mostly confined to the equatorial cloud belts and the Great Red Spot – will be dealt with here. The two equatorial cloud belts are visible even in quite small telescopes but they exhibit changes including marked variations in intensity and width often from one apparition to the next.

The following is a summary of changes in the cloud belts and Great Red Spot over the 40 years from 1951 to 1991, those from 1976 onwards from my own observational notes.

1951. The Red Spot was inconspicuous and the North Equatorial Belt (NEB) wide and dark. The South Equatorial Belt (SEB) was narrow and fainter while the South Temperate Belt (STB) was prominent.

1952. The planet was fairly quiescent. The NEB was still wide and dark but the SEB had faded somewhat but was now much broader and the STB had become pale.

1953–4. The Equatorial Zone (EZ) was often quite bright. Both the NEB and SEB were wide and double, the NEB the darker of the two equatorial belts.

1955. A fairly quiescent Jupiter showing prominent NEB and thin SEB and STB.

1956. The NEB and SEB were double again. STB thin.

1957. The most prominent belts were the NEB and the STB. The SEB was thin again. The Great Red Spot was becoming more noticeable.

1958. Similar appearance to the previous year with the EZ appearing more active. The Great Red Spot was prominent and late in the apparition the SEB darkened and widened.

1959. The NEB dark and wide. The SEB darker and wider than the STB. The Great Red Spot seemed to have disappeared. EZ a pronounced yellow colour.

1960. The NEB wide and dark again with wavy edges. SEB double and prominent and the STB was also prominent. The Great Red Spot seemed pale.

1961. Some large white spots in the dusky EZ. STB prominent. The Great Red Spot reappeared.

1962. A continuous yellowish band consisting of the fused equatorial belts encircled Jupiter, a very unusual appearance. STB and Great Red Spot visible.

1963. Reappearance of the two equatorial belts although at times they nearly fused again. EZ very active with numerous white ovals. Fairly prominent Great Red Spot. STB prominent.

1964. NEB and SEB both wide and similar in appearance at times. STB inconspicuous. The Great Red Spot was visible.

1965–6. NEB dark and wide, SEB double, dark and wide, STB prominent. Great Red Spot more prominent. Equatorial Band (EB) sometimes visible and much detail in the EZ.

1966–7. NEB prominent. Pale double SEB. Great Red Spot conspicuous. EZ of normal appearance.

1968. A more active-looking planet this year. NEB and SEB conspicuous and double. STB conspicuous, also the Great Red Spot.

1969. NEB most prominent cloud belt, SEB conspicuous but variable in shade. EZ had a yellowish tinge. Prominent Great Red Spot.

1970. The NEB and STB were the most prominent cloud belts with the SEB less conspicuous than the STB. Great Red Spot prominent.

1976 (late). The NEB was the most prominent belt and was narrow and dark with occasional 'bumps' and dents in its edges. The SEB was broad and much paler than the NEB and was sometimes partly fused to the STB, the latter not completely encircling the planet. A light oval was seen indenting the south edge of the STB. The NTB was prominent and the Great Red Spot was pale.

1977 (early). The NEB still the darkest and most prominent belt, narrow with occasional irregularities and dark condensations on its edges. SEB broad and pale, the south edge darker than the rest of the belt and still partly fused with the prominent STB. South STB (SSTB) sometimes seen. NTB prominent, appearing as a dark border to the north polar duskiness.

1977 (late). NEB and SEB now about the same width, the NEB sometimes appearing double. NTB not seen. STB seen in part. SEB appeared double near to the pale Great Red Spot.

1978 (early). NEB and SEB about same width, the NEB often appearing double. SEB appearing double on following side of the Great Red Spot. EB seen. On January 6th the most detailed view I have ever had of Jupiter, all the major cloud belts visible. STB not completely encircling the globe. By March, the NEB was narrower than the SEB.

1979 (early). NEB narrow and dark. Prominent 'rafts' seen at its north edge. SEB broad and pale, narrow near Great Red Spot, Great Red Spot region had a spoon-like appearance, the bright spot area having a dark cloud border. EZ quite dusky at times. STB not completely circling the planet. NTB not seen.

1980. Great Red Spot region, spoon-like appearance. SEB narrow close to the Great Red Spot, broad like the NEB in other parts. NTB not seen in first half of year. STB sometimes seen very broad and touching the SEB at times.

1981. NEB and SEB both narrow and dark, NEB sometimes the darker of the two. The SEB sometimes seen with a darker south edge. Great Red Spot Hollow seen in southern edge of SEB. STB faint, incomplete. NTB very thin.

(There follows a four-year gap in my observations).

1985 (late). NEB very narrow and dark. SEB broader and paler, seen double in some parts. Great Red Spot Hollow seen on southern edge of SEB. STB and NTB fairly easily seen. A large bright oval seen in the EZ in October.

1986 (late). NEB fairly narrow, seen at times with a darker northern edge and 'bumps' on the south edge. The SEB not much broader than the NEB and sometimes seen with a darker northern edge and appearing double on the preceding side of the Great Red Spot. Great Red Spot surrounded by a dark rim. NTB very thin and dark. STB sometimes broad and pale with a light oval inside it, at other times appearing as two thin parallel belts.

1987 (late). NEB narrower and somewhat darker than the SEB. Dark condensations seen locally within the NEB. STB appearing as a darker border to the south polar duskiness. NTB and North NTB (NNTB) both seen as very thin dark lines. Great Red Spot not seen – this does not necessarily mean that the Great Red Spot had disappeared, only that I missed seeing it on the occasions when I observed Jupiter.

1988. NEB and SEB of about the same width and intensity. The NEB once seen with a darker southern edge. Great Red Spot again seen with a dark rim. SEB seen double near Great Red Spot. NTB fairly prominent. STB indistinct.

1989 (early). NEB somewhat broader and paler than the SEB and sometimes dark condensations seen within it. STB appearing as a dark edge to the south polar duskiness. NTB seen on one occasion.

Other observers reported that the NTB, which was prominent in 1986, faded in 1987 and appeared as a thin faint band in 1988 and 1989. In the latter part of February, 1990 it began to increase in prominence. This occurred after a fast-moving white spot in the neighbouring North Temperate Current 'C' had appeared. Dark isolated sections of the NTB were seen in March and April and also dark spots. Finally, the NTB grew to a continuous prominent belt by late May of 1990.

The most spectacular occurrence for many years occurred in 1989. During the summer of that year, the South Equatorial Belt, which was a conspicuous dark feature early in the year, and darker than its North Equatorial counterpart, faded and was practically invisible by August. It was still very faint in December and remained so for several months into 1990. Figs. 9.12 and 9.13 show the appearance of Jupiter during this period.

The disappearance of the South Equatorial Belt was followed by numerous spectacular Jovian atmospheric phenomena during the ensuing months. The Great Red Spot which had grown pale during the previous few years, began to intensify in colour and rare white spots were seen in the North Temperate Belt during February, 1990 (Fig. 9.13).

The North Equatorial Belt was prominent with several condensations, wisps and festoons. The South Temperate Belt was discontinuous but prominent in several parts. The North Tropical Zone was brighter in some areas than in others and there appeared to be a general brightening of the South Polar Region. Around mid-October of 1990 the South Equatorial Belt showed signs of reappearing. Some drawings made of Jupiter in February 1990 showed the South Equatorial Belt to be gaining in prominence and this continued

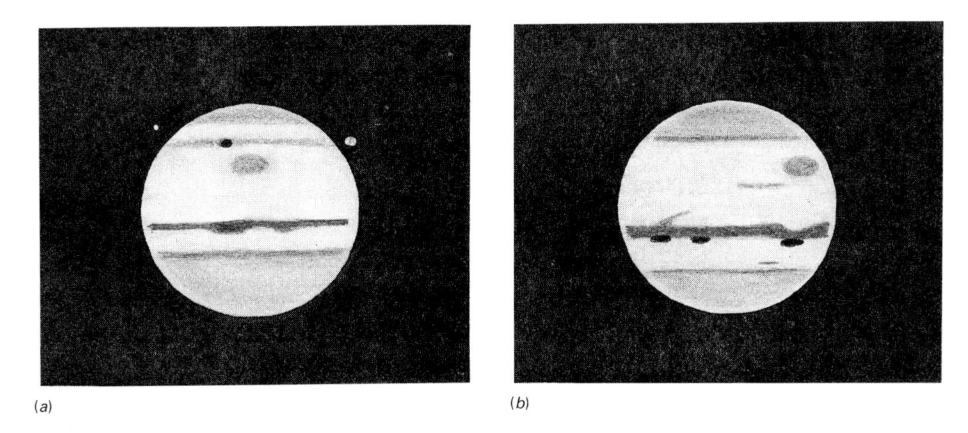

(a) (b)

Fig. 9.12 Jupiter during the disappearance of the South Equatorial Belt in 1989/90: (a) *December 8th, 1989, 04.22 UT;* (b) *February 1st, 1990, 03.45 UT. 8-inch Newtonian reflector, ×225. (F.W. Price.)*

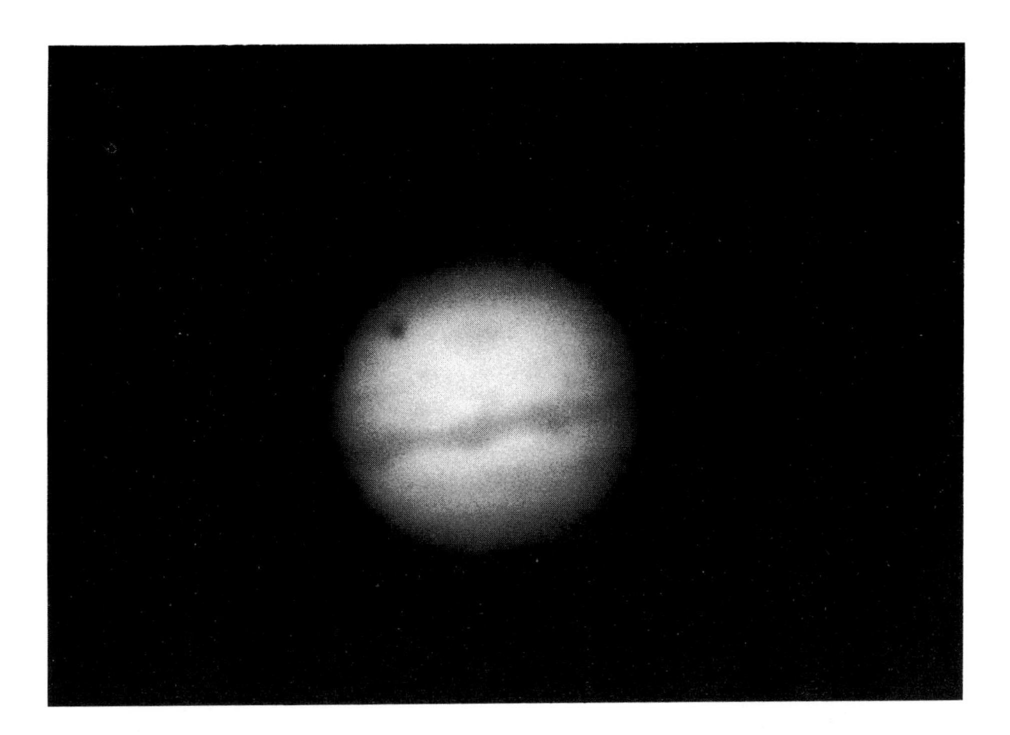

Fig. 9.13 Jupiter, January 20th, 1990, 04.37 UT: Great Red Spot, satellite shadow and absence of South Equatorial Belt. 14.25-inch reflector. (Photograph by E. Witkowski, Buffalo, NY.)

throughout 1990. By the time that I observed Jupiter again in January, 1991, the South Equatorial Belt had completely reappeared. It was broad and paler than the North Equatorial Belt and had a distinct 'warm' dull orange tint which contrasted with the neutral grey of the polar regions. The south edge was sharply defined but the north edge was wavy and very diffuse and extended into the Equatorial Zone. The Great Red Spot appeared like a blister or bubble with a light centre on the south edge of the South Equatorial Belt (fig. 9.14).

The belt dramatically faded on two other occasions, in 1928 and in 1958. It is probably premature to speculate but it looks as though this may be a recurrent phenomenon with a 30 year period. If the belt fades again in 2018 or 2019, then it would seem reasonable to suspect a 30-year cycle.

The fading of the South Equatorial Belt indicates a fantastic change in a weather system as broad as the Earth's diameter. It must surely indicate a considerable alteration in the chemical make-up of the region or that the South Equatorial Belt has sunk beneath a layer of light-coloured zone material. The latter explanation seems more likely in the light of studies by R. Beebe and D. Kuehn at New Mexico State University and astronomers at the Jet Propulsion Laboratory, who photographed Jupiter in light of different wavelengths in February and August–September, 1989. Visible light shows the South Equatorial Belt in February but it is invisible in the August–September images. Pictures taken with filters that isolate wavelengths absorbed by methane in Jupiter's upper atmosphere show no South Equatorial Belt on either of the two dates thus indicating that the disappearance cannot be connected with high-level clouds or the stratosphere of Jupiter. Infrared images, like the visible light

Fig. 9.14 Jupiter, March 9th, 1990, 01.50 UT: reappearance of the South Equatorial Belt in early 1990; 8-inch Newtonian reflector ×225. (F.W. Price.)

images, showed that the belt had disappeared. It seems from this that the South Equatorial Belt may have sunk while layers of bright zone material closed over and obscured it. The cause of the change is not known.

The foregoing describes some of the best known and most interesting phenomena of Jupiter's atmosphere that have been seen by Earth-based observers over the years. The reports of the Jupiter sections of the BAA and the ALPO are packed with descriptions of other phenomena and changes on Jupiter and the reader is referred to these for further information.

Surface markings of the satellites

Many observers using large telescopes in good seeing conditions have seen and sketched dark markings on the surfaces of all four Galilean satellites. Some have even constructed maps. Especially interesting is a series of 245 disc drawings made by H. Camichel, N. Lyot and M. Gentili at the Pic-du-Midi Observatory from September to October in 1941. These observers employed a 15.2-inch (38-cm) refractor under superb seeing conditions. Powers used were ×500–×900. Their findings were as follows:

(1) Io. Paler than Europa and yellowish in colour. Decidedly the larger of the two inner satellites.
(2) Europa. Very brilliant white.
(3) Callisto. Much larger than Io. Pale and of a dull chestnut colour.
(4) Ganymede. The largest satellite. Similar to Io in colour and brightness.

Subsequently these drawings were found to agree quite well with drawings made by W. H. Pickering, Barnard, Antoniadi and Holden. This was especially so in the drawings of Io and Ganymede. A dark spot seen on Europa by Antoniadi in 1927 corresponded with one of the small dark spots found by the Pic-du-Midi observers.

By arranging each satellite drawing in order of its orbital longitude and by comparative study of the locations of the satellites' surface features, it was concluded that all the satellites revolve around their primary with revolution periods about equal to their axial rotation periods and that their axes are nearly perpendicular to the planes of their orbits.

Maps were constructed from these drawings and showed the following features:

(1) Io. Dark polar regions. Many light zones in the equatorial region separated by longitudinal dusky bands.
(2) Europa. Light polar regions. Wide equatorial belt of a dusky appearance. Three dark spots on the equator.
(3) Ganymede. Two white areas near the poles, that near the north pole being the larger. Two dark belts paralleling the equator with at least two dark spots on each of them. The darkest spot lies close to the white north polar area.
(4) Callisto. A small white cap near the south pole, small light patches near the north pole. Dusky equatorial region.

We now turn to the results of spacecraft exploration of Jupiter.

Spacecraft observation of Jupiter

On March 2nd 1972 the Pioneer 10 spacecraft was launched by NASA on a fly-by mission to Jupiter and Pioneer 11 was launched on April 5th 1973 on a mission to Jupiter and Saturn. Pioneer 10 flew past Jupiter on December 3rd 1973 and transmitted close-up pictures and miscellaneous data back to Earth. Pioneer 11 passed Jupiter on December 2nd, 1974 and confirmed the results obtained by Pioneer 10. It continued onwards towards Saturn, the fly-by date being September, 1979.

A new determination of Jupiter's diameter was made by Pioneer 10 and it turned out that Jupiter is somewhat more flattened at the poles than was previously deduced from Earth-based measurements. The polar diameter was found to be 82 967 miles (135 516 km) and the equatorial diameter 88 734 miles (142 796 km).

During November–December 1973 Pioneer 10 transmitted more than 500 pictures to Earth. The earlier pictures did not show much more detail than the best Earth-based pictures but those obtained when the spacecraft was within 48 hours of the closest approach to Jupiter were much better than anything seen before from Earth. This applies not only to better resolution but also to the first ever views of Jupiter from angles impossible from Earth, such as almost directly above the polar regions and phase angles that are never seen from Earth, such as the crescent phase that was imaged as Pioneer 10 receded from Jupiter. The last picture was taken on December 31st 1973.

One of the most important pictures obtained from Pioneer 10 was of a spot in the northern hemisphere similar to the Great Red Spot but much smaller. It closely resembled the Great Red Spot and it seems likely that these and other red spots are meteorological phenomena in the Jovian atmosphere.

The Great Red Spot itself was seen to be an anticyclonic feature and it rotates in an anticlockwise direction. Previous speculations that the Red Spot and the whitish zones are places where there are clouds, anticyclones and ascending masses of gas appear to be confirmed by the Pioneer 10 findings. The dark belts are cyclonic and are sinking masses of gas leading to clouds lower down.

The Pioneer 10 pictures of Jupiter were full of surprises regarding the clouds. Billows and whirls near the edges of zones and belts at middle latitudes showed that these are rapid fluxes in wind direction and speeds as was previously suspected. The Equatorial Zone plume was shown in great detail. In the North Tropical Zone, slanting trends were seen which indicate latitudinal as well as longitudinal motion.

Pioneer 11, the second spacecraft, sent back about 460 pictures of Jupiter and the Galilean satellites between November 18th and December 9th 1974. It came much closer to Jupiter than Pioneer 10 and from a different angle. This allowed the spacecraft to obtain unprecedented views of the near-polar regions.

Pioneer 11 was the first to photograph Jupiter's polar regions and many good pictures of the poles were secured. Views were obtained of the North Polar Region that are unobtainable from Earth. They showed that north from the North Temperate Belt the dark belts and light zones show progressively less organisation and the banded appearance changes to a pattern of circular and oval features further north than 75–80° of latitude.

A picture taken from a distance of 582 000 miles (936 000 km) revealed great detail in the belts and zones which include light and dark cells probably resulting from convection in the South Temperate Zone.

The best view of the Great Red Spot up to that date was obtained at a distance of 320 000 miles (545 000 km). It revealed the Red Spot as containing less fine structure than was visible in similar images at other latitudes. The Great Red Spot appears to be in the most undisturbed zone on Jupiter and this probably accounts for its stability.

In addition to the splendid close-up views of Jupiter afforded by the Pioneer spacecraft, unprecedented views of the surfaces of the four Galilean satellites were also yielded. The pictures obtained of the Galilean satellites afford a much better idea of the albedo and colour characteristics of different parts of their discs than has ever been obtainable from Earth-based telescopic study.

Pioneer 10 obtained images of Europa and Ganymede. The pictures of Ganymede show a dark mare at the south pole and a central mare, both about 480 miles (800 km) in diameter, and a bright north polar region. A less detailed image was obtained of Europa and shows an appearance roughly similar to Ganymede. Bright and dark areas are visible in the image.

Pioneer 11 also obtained images of several of the Galilean satellites showing surface features. It obtained the only good image of Io, a view of the north pole. Pioneer 10 missed this satellite. Io seems to differ considerably from the other Galileans. It has an orange colour and is among the most highly reflective bodies in the Solar System. Dark polar caps were visible. Earth-based observations suggest that the polar regions of Io have a reddish colour. The Pioneer 11 image shows an orange polar region and a whitish equatorial region. Io was found to have an atmosphere which, although it has a density 20 000 times less than Earth's, is about 70 miles (115 km) deep. Io, therefore, is one of the smallest planet-type bodies that possesses an atmosphere. The two pictures of Ganymede obtained by the Pioneer spacecraft reveal little variability of colour over the surface but quite large brightness variations. Several good pictures of Callisto were obtained, revealing small colour and brightness variations over the surface. Density measurements of the Galilean satellites made by the spacecraft indicated that Io and Europa are rocky while Ganymede and Callisto appear to be icy bodies.

The Voyager 2 spacecraft was launched on August 20th 1977 and was intended to be a four-planet fly-by: Jupiter (1979), Saturn (1981), Uranus (1986) and Neptune (1989). Voyager 1, launched 16 days earlier on September 5th 1977 was a two-planet fly-by: Jupiter (March 5th 1979) and Saturn (November 12th 1980).

Voyager 1 came closest to Jupiter on March 5th 1979 and its encounter with the planet ended early in April. Nearly 19 000 pictures were secured as well as many kinds of scientific measurements. Voyager 2 came closest to Jupiter on July 9th 1979. 33 000 pictures of Jupiter and its five largest satellites were taken together by Voyagers 1 and 2. The quality of the pictures taken by the Voyager cameras was much better than expected, the mission yielding photographs having 80 times better resolution than those given by Pioneers 10 and 11.

One period of intensive observation lasted for four days and yielded 3600 detailed images of the cloud belts. Much detail never before seen from Earth was revealed within them. Data on the circulation of Jupiter's atmosphere and detailed geological information on the surfaces of the satellites were also returned to Earth. A search was also made for new satellites. Many data were obtained on the physical make-up of Jupiter and its environment as well as of its atmospheric features and the surface features of the Galilean satellites. The following brief summary of findings is confined almost entirely to atmospheric features and satellite surface features as being the most likely to interest the telescopic observer of Jupiter.

In the first few days of the mission there was discovered the ring surrounding Jupiter, the complexity of structure and motion of Jupiter's atmosphere and the weird surface features of the Galilean satellites.

The Great Red Spot was closely studied. The close-up photography revealed activity that astronomers had suspected for a long time but which had never been clearly visualised from Earth (fig. 9.15). This was that the Red Spot is rotating in an anticlockwise direction, a complete rotation taking about 6 days. Matter near the centre of the Spot shows little movement and this appears to be random while matter at the outer rim of the Spot completes a rotation in 4–6 days.

The Voyager encounters revealed that atmospheric features of many different sizes have apparently uniform velocities, indicating that mass motion or actual movement of matter and not wave motion, i.e., energy movement through essentially non-moving mass, is occurring.

The rapid increase in brightness of atmospheric features and subsequent diffusion of cloud matter was observed. This is probably the result of disturbances setting off convection. Previous studies had indicated that the regions north or south of latitudes 45° north and south respectively are regions of predominantly upward and downward convection. The Voyagers recorded

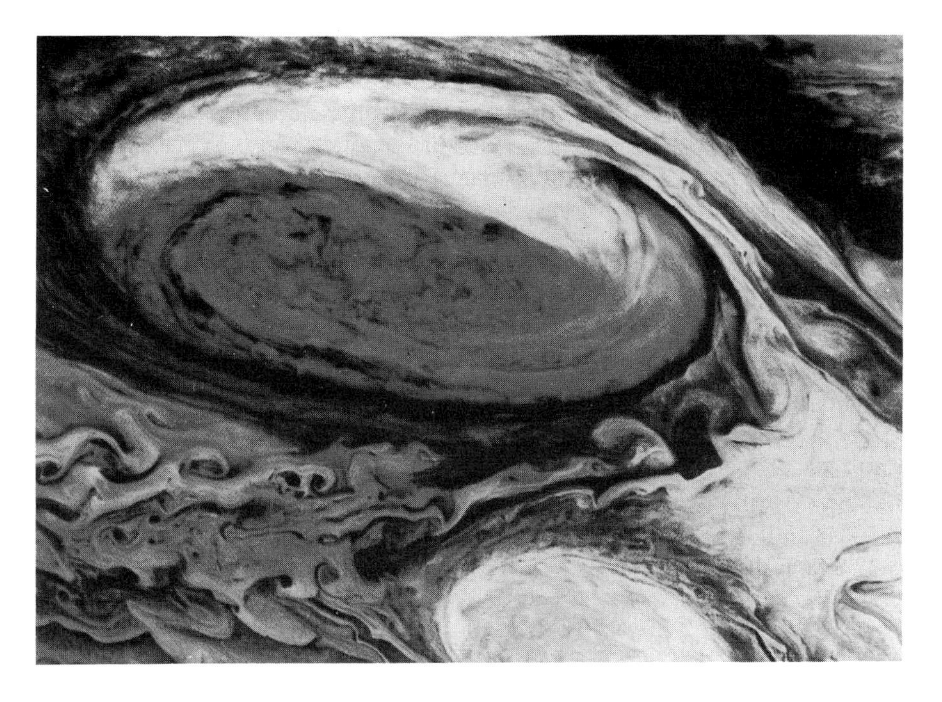

Fig. 9.15 Jupiter's Great Red Spot, July 3rd, 1979 (Voyager 2).

east to west winds as far poleward as 60° north and south latitudes, so that this shows that at least up to 60–75° latitude cannot be dominated by convection activity.

Auroral phenomena were detected in the polar regions and have also been seen in ultraviolet light. The latter were not seen in the Pioneer 10 and 11 close approaches. Lightning bolts in the cloud tops were seen.

Regarding the satellites, nine erupting volcanoes were certainly identified on Io by Voyager 1. Plumes of eruptive matter reached up to more than 190 miles (300 km) above the surface. The distant low-resolution pictures taken by Voyager 1 revealed Europa's surface to be marked by a multitude of criss-crossing linear markings. Higher-resolution pictures taken by Voyager 2 showed these features to be not cracks in the crust as was first thought but purely surface markings. Ganymede was found to be the largest satellite in the Solar System thus denying Titan, Saturn's satellite, that distinction which it had been thought to hold. Two kinds of surface terrain were observed on Ganymede, grooved and cratered. The surface of Callisto was seen to be much cratered. The largest rings appear to have been eroded and virtually no topographic relief is visible in the ghost-like remains of these features. They are distinguishable only by their light colour. Amalthea was found to be elliptical, its dimensions being 170 miles (273.5 km) by 105 miles (168.9 km) by 95 miles (152.9 km).

A dust ring about 20 miles (32 km) thick was discovered surrounding Jupiter. The outer edge is 80 000 miles (129 000 km) from the centre of Jupiter. There is evidence to indicate that ring material may extend as far as 30 000 miles (50 000 km) further inwards and into the upper layers of Jupiter's atmosphere, but the brightest part of the ring is only about 4000 miles (6000 km) wide. Other evidence indicates that ring material may extend as far outwards as the orbit of Amalthea. The brightness of the dust ring is far too low for it to be visible from Earth.

Two more satellites were discovered, Metis and Adrastea, each only about 25 miles (40 km) in diameter, just outside the ring. Another, Thebe, was found between the orbits of Amalthea and Io. Its diameter is about 50 miles (80 km).

Visibility of Jupiter

Jupiter can be well observed every year, its oppositions arriving about a month later each year as previously shown. It can be observed for periods of approximately ten consecutive months between successive conjunctions. Its declination limits are 25° north or south of the Ecliptic; in northern latitudes therefore, summer oppositions tend to be low in the sky. However, inversion layers in the atmosphere frequently nullify the usually detrimental effects of low altitude on planetary images and give superbly steady seeing. Jupiter will be south of the celestial equator during roughly 1992–7.

Particularly good observation is possible at winter oppositions in the northern hemisphere, at which times, Jupiter is high above the horizon and can be visible for as long as 16 hours during a single night. It is therefore possible to observe a complete rotation of the planet in one night – assuming that you have the endurance! Jupiter's visibility sequence is as shown in Table 9.3.

Table 9.3 *Visibility and movements of Jupiter.*

	Days after conjunction with sun
Jupiter in morning sky.	13
Retrograde motion begins.	140
Opposition to sun.	200
Retrograde motion ends.	260
Jupiter disappears from night sky.	386
Conjunction with sun.	399

Jupiter is so bright that it can be seen in the telescope in the morning twilight as little as 13 days after its conjunction with the sun. It gets higher in the sky every night as the distance between the Earth and Jupiter lessens. Retrograde motion begins 60 days before opposition and goes on for another 60 days after opposition. When the retrograde motion halts, Jupiter stays visible in the evening sky for roughly 125 days.

Jupiter is seen as a disc in binoculars and the Galilean satellites are plainly visible. A power of ×50 on a small refractor clearly shows the oblateness of the disc and the two equatorial cloud belts. This low power shows Jupiter's disc as large as the full moon appears to the naked eye. Some find this difficult to believe but it is easily verifiable. Choose an evening when the moon and Jupiter are close together in the line of sight and look at Jupiter through a telescope with one eye and simultaneously view the moon with the other eye (unaided). The two discs will be seen to be very nearly the same size.

Observing Jupiter

Of all the Solar System planets Jupiter is in many ways the most rewarding to observe. It is visible in the night sky every year unlike Mars which can only be observed for a few months every two years, and Mercury and Venus with their limited visibility. It can be observed all night at an opposition and it presents the largest planetary disc in the telescope except for Venus near inferior conjunction. For many months each year Jupiter's disc is over 40 seconds of arc in apparent diameter. Also as mentioned previously, the apparent angular size of its disc does not vary nearly as much as that of Mars in the weeks before and after opposition. In contrast, when Mars attains the favourable diameter of 24 seconds of arc it lies south of the celestial equator and does not attain a height of more than 40°. Such closeness to the horizon does not favour good telescopic observation. The apparent angular size of the disc expands and shrinks considerably on either side of opposition. On the occasions when Mars rises to a height of 50° or more, the apparent angular diameter is only 15 seconds of arc. For northern hemisphere observers, then, Mars is not easy to observe.

True, we do not see surface details on Jupiter but only the top of a cloud layer; this presents an every-changing and fascinating spectacle. The rapid rotation of Jupiter brings new details into view literally almost every few minutes. Jupiter never looks the same from one opposition to the next, the cloud belts and zones exhibiting changes in width, colour and intensity. Other phenomena are quite short-lived and change over a few days or weeks during an opposition. Then there are the four Galilean satellites which some sharp-eyed people say that they can see without optical aid. They are a fascinating sight even in binoculars and usually appear in straight line array on one or both sides of the brilliant disc of Jupiter.

A 2.5- or 3-inch refractor will show the two dark equatorial belts and satellite shadows on the disc of Jupiter. In fact the sharpness and blackness with which the tiny satellite shadows are seen are a good optical test for a 3-inch telescope. Apertures of 6 inches and over are needed to show the thinner cloud belts and associated details. The cloud belts usually have a greyish-brown tint but can show unmistakable colour effects like reddish grey and blue-grey. Shades of red, pink, orange, yellow and green have been noted and even brown, purple and blue. The general appearance of Jupiter and its satellites in a small telescope is shown in fig. 9.16.

The zones are usually a bright ivory white. The dark belts and light zones form circles around the planet in a latitudinal orientation and show fairly rapid local changes of shape due to streaming and turbulence effects. The dark cloud belts seem to stop just short of the east and west limbs of Jupiter owing to the combined effects of the atmosphere and the curvature of the globe that make the ends of the cloud belts seem paler. In the inverting astronomical telescope in the northern hemisphere, Jovian markings appear to drift across the disc from right to left. Perhaps the best-known atmospheric feature is the Great Red Spot, a huge oval body more or less prominent and usually of a faint pink colour. It lies in the South Tropical Zone indenting the south edge of the South Equatorial Belt and touching the north edge of the South Temperate Belt.

A complete Jupiter observing programme will comprise the following studies:

(1) Longitude determination of features by central meridian transit timings.
(2) Latitude determination of features.

Fig. 9.16 General appearance of Jupiter and the Galilean satellites in a small telescope.

(3) Disc drawings, strip and sectional sketches.
(4) Determination of rotational periods of Jovian features from longitudinal drift.
(5) Observations of white ovals.
(6) Observations of the Great Red Spot.
(7) Colour and intensity estimates.
(8) Photographic work.
(9) Study and timing of satellite phenomena.
(10) Radio work.

Determination of the longitudes of Jovian features by central meridian transit timings

The most valuable kind of observations that can be made is the determination of the longitudes of Jovian features by timing their transits across the central meridian. The central meridian is the imaginary straight line joining the poles at right angles to the cloud belts and that bisects the disc of Jupiter into east and

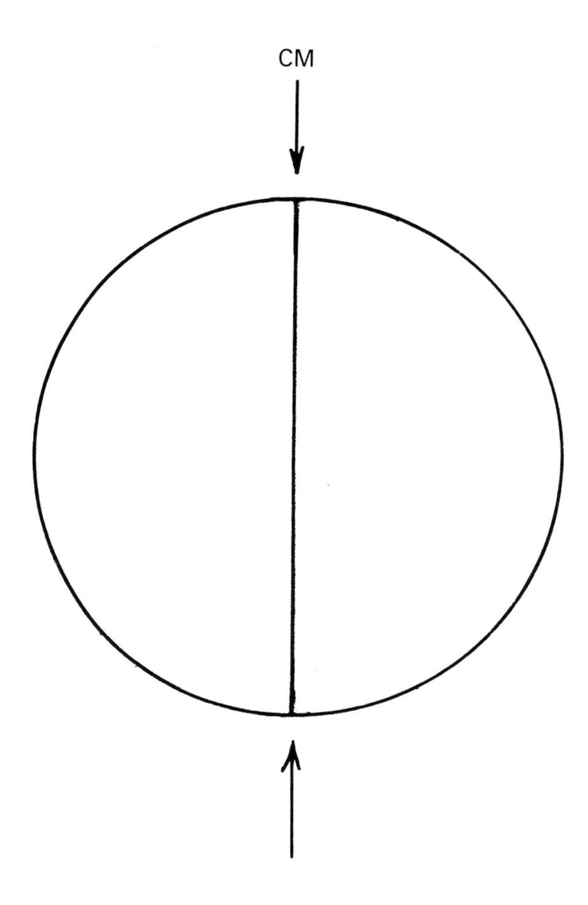

Fig. 9.17 Central meridian (CM) of Jupiter.

west halves (fig. 9.17). Most of what we know about atmospheric phenomena on Jupiter such as the cloud currents has been obtained by this simple and easily learned technique of timing central meridian transits. If the longitude of a given feature is frequently measured over a period of weeks, it is often found that the feature changes its longitude, either drifting east or west or exhibiting irregular changes. It is this kind of observation that has revealed so much about the nature of the currents in the cloudy atmosphere of Jupiter.

The determination of Jovian longitudes is similar to the determination of Martian longitudes but in the case of Jupiter there is no visible solid surface and therefore no permanent surface marking to serve as a marker for a zero meridian of longitude. From observations of Jupiter's clouds we conclude that Jupiter makes one axial rotation in a little less than 10 hours. The only possible way to choose a standard meridian is to select one that rotates about Jupiter's axis in roughly this period. The precise value assigned to this rotation period is not important provided that it is uniform and convenient.

It had been noticed by early observers that the rotation periods of markings on or near the equator of Jupiter differ from the rotation periods of markings in the temperate and polar regions; the equatorial rotation period was ascertained to be generally 5 minutes shorter than the temperate and polar rotation periods. The difference is sufficient to make impracticable the use of a single meridian of longitude for markings in all latitudes. Hence, two arbitrary standard meridians are used with different rotation periods. These are designated System I and System II. All markings within about 10° of the equator are referred to System I and markings between this region and the poles are referred to System II. The rotation time of the System I standard meridian is 9 hours 50 minutes 30.003 seconds and that of System II is 9 hours 55 minutes. 40.632 seconds. This degree of precision may seem odd since the times chosen are arbitrary. The reason for them is that the rotation periods have been calculated from an adopted rotation period of System I of 877.90° of longitude in 24 hours and a rotation of 870.27° of longitude in 24 hours for System II. The value from System II is almost exactly the rotation period of the Great Red Spot between the oppositions of 1890/1.

The *Astronomical Almanac* and the ALPO *Solar System Ephemeris* for the current year provides separate tables giving the longitude of the central meridian at 00 hours 00 minutes UT for both System I and System II for every day that Jupiter is observable. In addition, other tables are provided that give the change in longitude in a given time. Hence, it is a simple matter to calculate the longitude of the central meridian at any time in either System I or System II.

The timing of transits of features across the central meridian of Jupiter gives determinations either directly or indirectly of precise measurements of rotation rates, changes in rotation patterns, the extent in longitude of a feature and the determination of the longitude of a feature to within ±1.2°.

All that is required for timing transits of disc features across the central meridian is a telescope, pencil and notebook and a watch or clock accurate to within a minute of the actual time. Alternatively, a shortwave radio may be used adjusted to WWV time signal frequencies of 5, 10 or 15 megahertz. Other frequencies may be more appropriate outside the United States.

Even a small telescope can be usefully employed but its aperture should be no less than 3.5 inches for a refractor and 4.5 inches for a reflector. Large

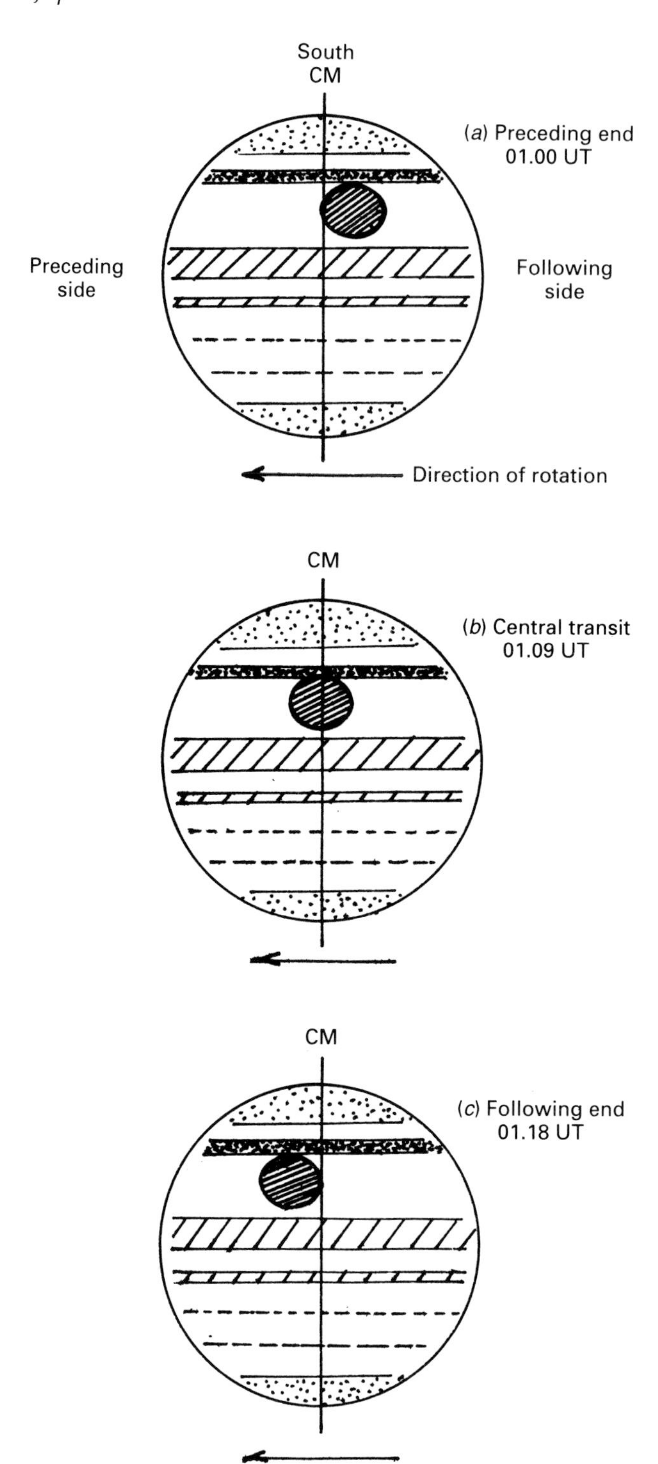

Fig. 9.18 Central meridian transit timing of Jupiter's Great Red Spot: (a) *preceding end (01.00 UT);* (b) *central transit (01.09 UT);* (c) *following end (01.18 UT).*

instruments will, of course, show smaller and/or fainter objects thus increasing the number of central meridian transits that can be timed.

In determining the time of transit of a very small feature (less than 2° of longitude in extent) across the central meridian, only the time of transit of the centre of the feature needs to be recorded. If the feature is extensive in longitude, such as the Great Red Spot, then three timings should be made, one of the leading (preceding) end or edge of the feature, one of the centre and one of the trailing (following) end. See fig. 9.18. The determination of the time of transit of the centre of a feature across the central meridian thus fixes its position in longitude since this will be the same as the longitude of the central meridian at that time.

The rapid rotation of Jupiter causes an easily noticeable displacement of the apparent positions of features near the centre of the disc; the lapse of 3 minutes between looking at a feature and then looking again is enough to make a perceptible difference in its position. Jupiter will have rotated through nearly 2° of longitude during this short interval.

When a feature is observed continuously as it approaches, transits and passes the central meridian, it rarely appears to be on the central meridian for more than 2 minutes. If the mean is taken of the times when an object was first and last determined to be on the central meridian, an acceptably accurate estimate of its time of central meridian transit should be secured.

Practice is necessary in making accurate transit timings; it depends upon the ability of the observer to judge accurately when an object is exactly on the central meridian. Because of inevitable errors of judgement, even experienced observers may estimate central meridian transit times of Jovian features differing by 1 or 2 minutes from the true time of central meridian transit, but not often by more than 3 minutes.

Some observers tend to record transit times a minute or two early while others may be a minute or two late while yet others are about right. The error has been found to be fairly constant or systematic for a given observer and hence it is called the observer's 'personal equation'. It can be allowed for once it is known.

Estimates differing from the true value by as much as 5 minutes may, however, be valuable on occasions in identifying a feature; sometimes a gap may occur in a series of transit timings of a feature because of bad weather. A spot may have been observed for three or four weeks and then bad weather may have intervened for another three weeks. When the weather clears a spot similar in position and movements to the first is seen, suggesting that they may be one and the same object. If during a brief interval of clear weather during the bad three weeks a spot was seen in a similar longitude then even an approximate single estimate of its time of transit across the central meridian will help to establish its identity with the other two spots and that therefore all three are identical.

Is the use of the filar micrometer necessary in making determinations of the longitudes of Jovian features? Estimates of longitudes by the simple visual method described above cannot hope to be comparable in accuracy to what can be done with a filar micrometer and if the only purpose of central meridian timings was the accurate determination of longitude of Jovian features at particular times, then the use of the filar micrometer would be preferred to timing of eye estimates of central meridian transits. However, the main reason for timing transits is more than just longitude determination. It is to detect drifting in longitude of objects over more or less extended times and this has led to the discovery of the complex currents in the atmosphere of Jupiter and the wanderings in longitude of the Great Red Spot. In this kind of work, errors of 1 or 2° in longitude determinations would make no noticeable difference to the results. In the case of very small faint objects the filar micrometer could be useless because the wire web in the device would obscure such features.

A further objection to the use of the filar micrometer is that it can waste valuable observing time. On an evening of good seeing 20 or more features worth recording may transit the central meridian in the space of an hour. Some of these could be important but too faint to permit the use of the micrometer. For most of the others there won't be enough time for adjusting the micrometer for all of them.

There would appear to be no real justification for using more accurate methods of central meridian transit timings than eye estimates since the latter have produced and will certainly continue to produce interesting and valuable results. The filar micrometer reveals its value when measuring very small anomalies in the longitudinal drifting of certain spots. It might also be used to simply *define* the central meridian by setting the wires perpendicular to the Jovian equator and one Jovian radius apart, one wire coinciding with the limb, the other with the central meridian.

Classification and description of Jovian disc features

Every observer of Jupiter should be able to describe accurately the features he sees on the planet. As previously mentioned, the major dark latitudinal bands are called cloud belts or just 'belts' and the bright areas between them are called zones. The nomenclature of these has also been described earlier and is illustrated in fig. 9.3. The appearance and nomenclature of the most commonly occurring smaller features on Jupiter's disc are as follows (see fig. 9.19):

(1) Garlands (wisps). Faint narrow straight or curved dusky streaks extending from the edge of a belt into the adjacent bright zone. They commence as a 'bump' at the edge of the belt. Commonly seen on the south edge of the North Equatorial Belt and extending into the Equatorial Zone and bending in the following direction. The end may bend right over and rejoin the belt.

(2) Festoons (bridges). Thin bands of dusky cloud material connecting two belts across the intervening zone. Elusive and difficult to observe.

(3) Light ovals. As their name suggests, these are light or grey-coloured more or less oval or circular objects. The light ovals may be seen on zones or belts as they are usually the brightest of Jovian features. They are often seen in the South Tropical Zone and in the South Temperate Zone and may intrude on the south edge of the South Temperate Belt. Grey ovals are best seen when on the light zones.

(4) White spots. Well-defined rounded spots, much smaller than ovals, smaller even than the shadows of satellites. Usually found within the North and South Equatorial Belts.

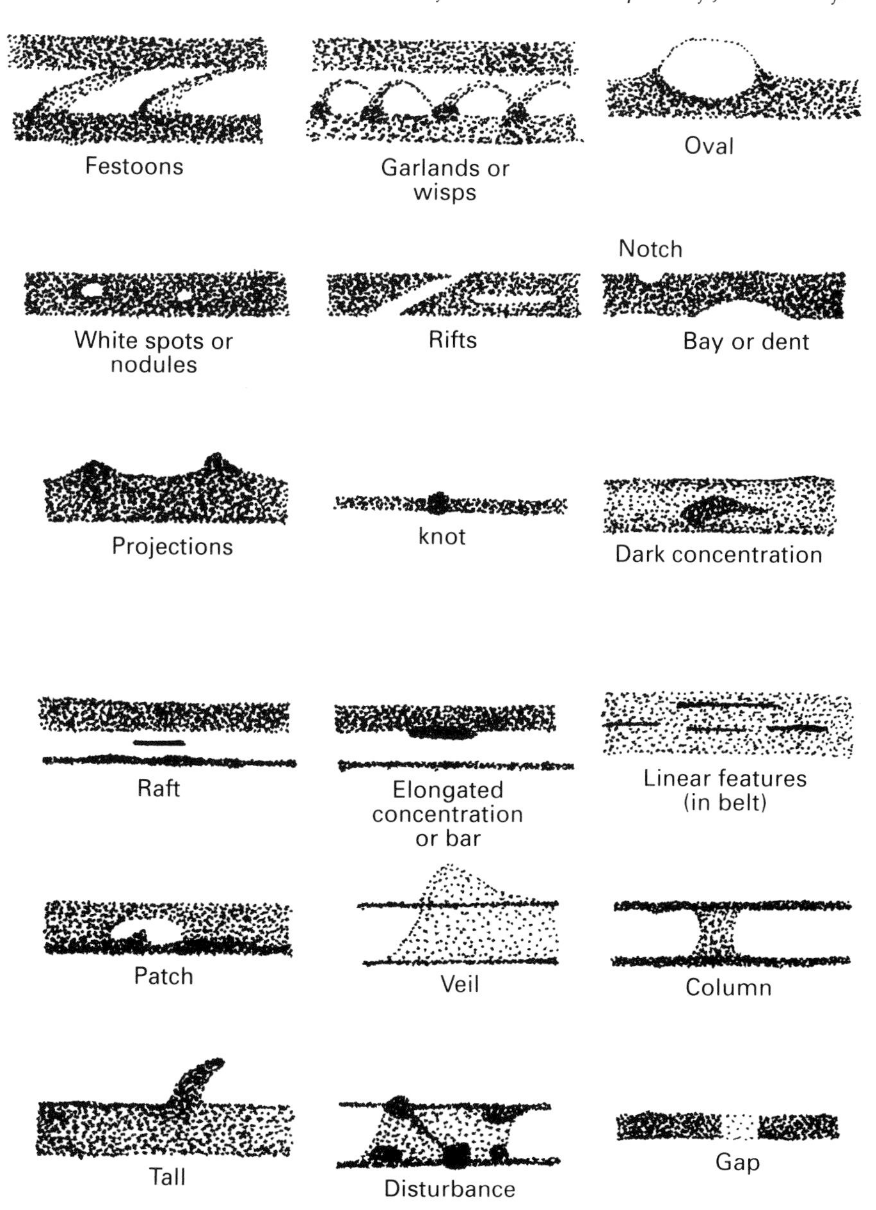

Fig. 9.19 Nomenclature of Jovian disc features.

(5) Rifts. White sharply defined linear features, 1–3° in width lying at an angle of 45–60° on a dark belt extending to both edges so that the two adjacent zones are joined.
(6) Granular spots. Red-coloured glowing spots, smaller than satellite shadows.
(7) Dents. Sharply defined semicircular bays on the straight edge of a dark belt.

(8) Protrusions or indentations. Dark and light irregularities on the edges of cloud belts.
(9) Knots. Knot-like thickenings seen in narrow cloud belts.
(10) Dark concentrations in belts. Ill-defined often extensive darker areas within belts.
(11) 'Rafts'. Horizontal linear markings (HLM). Short straight dark linear features usually seen in the North Tropical Zone just separated from the edge of the North Equatorial Belt. They look like short strokes drawn with a pencil.
(12) Bars. Short, rather thick segments, usually found in belts rather than zones. During disturbances, these may have an orange or reddish colour. Thicker than rafts.
(13) Fine linear markings. Grey or reddish, found in the belts.

A consistent method should be used in recording transits. This can be made easier if suitable abbreviations for the names of features are used. Virtually every Jovian feature can be classed as either a bright marking (W) or a dark marking (D). The transit time of a feature may refer to the preceding end (p), following end (f) or centre (c) of the feature.

The most important part of an observer's description of the central meridian transit of a marking will have been recorded if the description begins either with a W or D plus the appropriate subscript, p, c or f. These can then be followed by abbreviations identifying the nature of the feature. The abbreviations in table 9.4 are those most frequently used in recording transit observations. As an example, the preceding end of a dark area in the South Temperate Belt can be recorded as: Dp (sect.), STB.

If for any reason there is uncertainty about the transit time of a feature this should be indicated by writing est. (estimated) after the recorded time. Include a brief written description of the type of feature and its structure.

From changes in longitude, the rotation period of a feature can be accurately determined if it can be observed for weeks or months. The best method for determining rotation periods is to plot the observations on graph paper with degrees of longitude displayed horizontally and dates displayed vertically. A suitable scale would be to let 0.1 inch represent 5° of longitude and the same length for 2 days. Study of such graphs enables identification of features seen

Table 9.4 *Abbreviations used for Jovian markings.*

W	bright markings	L	large	proj.	projection
D	dark marking	N	north	cond.	condensation
p	preceding or preceding end	S	south	indef.	indefinite
		v	very	sm.	small
f	following or following end	elong.	elongated	sect.	section
conspic.	conspicuous	c	centre	GRS	Great Red Spot
				RSH	Red Spot Hollow

on successive dates, the longitudinal drifts can be demonstrated and rotation periods determined.

If a drift line is linear and has a constant longitude this indicates a constant rotation period which is the same as that of the system used. An irregular or curved drift line indicates a variable period.

Rotation periods can be obtained when longitudinal drift has been established by consulting the conversion tables in *The Planet Jupiter* by B. M. Peek (Appendix IV). Rotation periods may also be calculated from one of the following formulae:

If $D1$ = number of degrees the feature drifts in longitude in 30 days in System I and $D2$ is the same for System II, then the rotation period P of the feature in System I is given by:

$$P = 9\text{h}\ 50\text{m}\ 30.003\ \text{s} + (1.345\text{s})\ (D1)$$

and the rotation period P in System II is given by:

$$P = 9\text{h}\ 55\text{m}\ 40.632\text{s} + (1.369\text{s})\ (D2)$$

$D1$ and $D2$ are positive or negative when the features drift towards increasing or decreasing longitudes respectively.

Determination of latitudes of Jovian features

The latitudes of the belts, zones and other features of Jupiter do not change much even over long time intervals and the changes are never large. Latitude determinations therefore need only to be carried out infrequently. This is fairly straightforward but the telescope must have a good clock drive and RA and declination slow motions are necessary.

The different methods of determination of Jovicentric latitudes are of varying degrees of accuracy and are based on measurements of positions of features on Jupiter made with either an eyepiece graticule or a filar micrometer.

A glass graticule ruled with fine lines is inserted into the focal plane of the telescope eyepiece; positive eyepieces are the obvious choice here because the focal plane lies in front of the optical components. The graticule lines should not be coarse as they may obscure faint cloud belts. The measurements to be made and the procedure used are as follows:

(1) With the image of Jupiter accurately focused rotate the eyepiece so that the lines of the graticule are exactly perpendicular to the cloud belts.
(2) As accurately as possible, measure how many graticule divisions and fractions of a division correspond to the apparent pole to pole diameter of Jupiter.
(3) Note how many graticule divisions are equal to the distance of the belt or other feature from the more distant of the two apparent poles; if the feature is located north of the equator, measure the distance to the south pole. Measuring this instead of the shorter distance to the apparent north pole gives somewhat greater accuracy.

Alternatively, these measurements may be made with a filar micrometer.

Determination of latitudes of Jovian disc features

First method Refer now to fig. 9.20 in which for simplicity it is assumed that Jupiter is spherical and that the Earth is in the plane of Jupiter's equator.

The points P and P′ are the apparent poles of Jupiter (also the actual poles under the above assumptions), O is the centre of Jupiter and the line OG is in the plane of the equator and parallel to PA and P′B. AB is the measured distance in graticule or micrometer divisions between the apparent poles and DC is the measured distance from the south pole P′ of S, the feature on the central meridian (PSP′) of Jupiter whose latitude is to be measured. (It is not recommended that an eyepiece graticule line or micrometer wire be oriented along Jupiter's equator followed by measurement of the distances of the feature from the equator and the equator to pole distance because it is difficult to find the equator exactly if the Earth happens not to be in the same plane. Sometimes a dusky equatorial belt is visible but this does not always coincide with the equator.)

In terms of the eyepiece graticule or micrometric measurements the ratio SE/SO is equal to the sine of the angle SOE, the Jovicentric latitude of S. It is equal to:

$$\frac{2\mathrm{CD} - \mathrm{AB}}{\mathrm{AB}}$$

From a table of sines look up the angle whose sine is equal to this, the required Jovicentric latitude.

This method is approximate only as it ignores the ellipsoidal figure of Jupiter and the inclination of Jupiter's equator to the ecliptic. However, it is better than guesswork or no latitude determination at all.

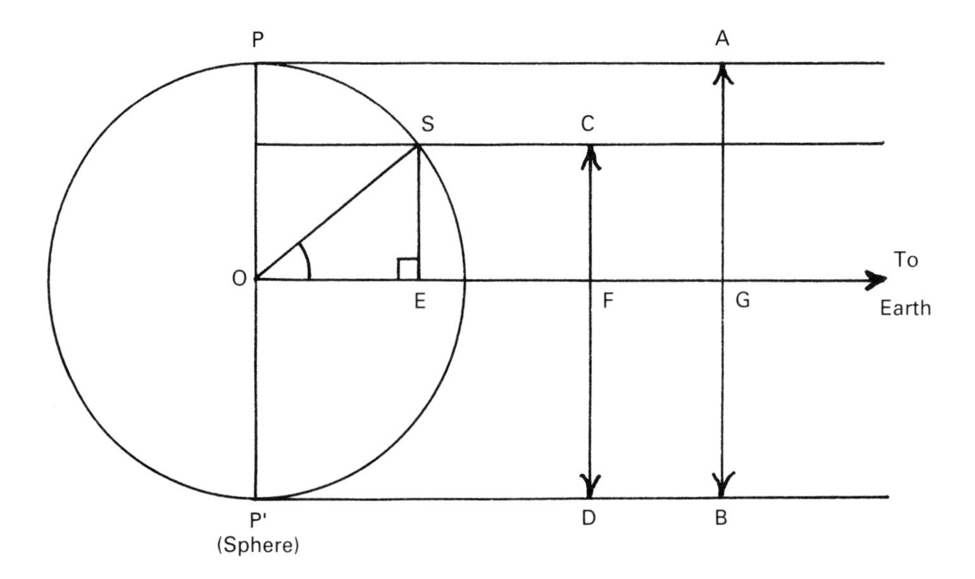

Fig. 9.20 Determination of Jovicentric latitude (first method).

Second method In the previous method the Jovicentric latitude of a feature was easily calculated on the assumption that Jupiter is a sphere because the line OS, the radius of the sphere, is invariant in length whatever the latitude of S. This is not so in the case of an ellipsoid such as Jupiter and the mathematics involved in taking this into consideration may deter some readers. A simpler solution involving scale drawing is therefore offered here instead.

First, draw a fairly large ellipse of the correct oblateness on a sheet of paper $8\frac{1}{2}$ by 11 inches, preferably ruled with faint squares to facilitate orientation of the ellipse. The ellipse can be drawn with a plastic stencil obtainable from drafting and art supplies stores. The oblateness of the ellipses on these stencils is designated in degrees and a 70° ellipse has just about the right oblateness for drawing Jupiter. Expressing oblateness in degrees is based on the fact that an ellipse is generated by parallel projection of a circle onto a plane inclined at an angle to the circle. If the circle is parallel to the plane (zero degrees) its projection will be a circle also. If the circle is perpendicular (edge-on) to the plane, then its projection will be a straight line of length equal to the diameter of the circle. If, now, the circle is gradually tilted from the perpendicular the projection will be an ellipse; at small angles from the perpendicular the ellipse will be very oblate (elongated). As the angle of tilt increases the ellipse projected onto the plane will get less and less oblate and will approximate more and more closely to a circle.

The oblateness of an ellipse is the ratio:

$$\frac{\text{minor axis}}{\text{major axis}}$$

The oblateness of the Jovian ellipsoid is therefore:

$$\frac{\text{polar diameter}}{\text{equatorial diameter}}$$

which is closely equal to 15/16, the degree of oblateness of a 70° ellipse. Expressed as a decimal to two places, 15/16 is 0.94, the sine of 70°, very nearly.

Make the ellipse that you have drawn with a major axis about 15–20 cm in length. This large size permits more accurate drawing and measurements to be made than would be possible in a smaller drawing.

If a stencil this large cannot be found, making a drawing of the largest available 70° ellipse and make an enlarged copy of it on a photocopying machine that has this facility. Having drawn the ellipse carefully mark the points where the major and minor axes intersect its circumference – these will be marked on a stencil – and draw the axes GF and PP′ (fig. 9.21). S′ marks the position of the Jovian feature near or on the central meridian as seen in the telescope. Carefully measure the distance PP′, the polar diameter in millimetres. Then, on the central meridian, plot S′, the feature whose latitude is to be determined so that the distance P′S′ in millimetres is equal to PP′ multiplied by the ratio S′P′/PP′ i.e., by CD/AB, where S′P′ and PP′ are expressed in graticule or micrometer divisions. Imagine now that Jupiter rotates through 90° to the right and that the elliptical outline now represents a section through Jupiter. Draw a straight line through S′ parallel to the equator so that it intersects the circumference of the ellipse at S. PSFP′ is now the central

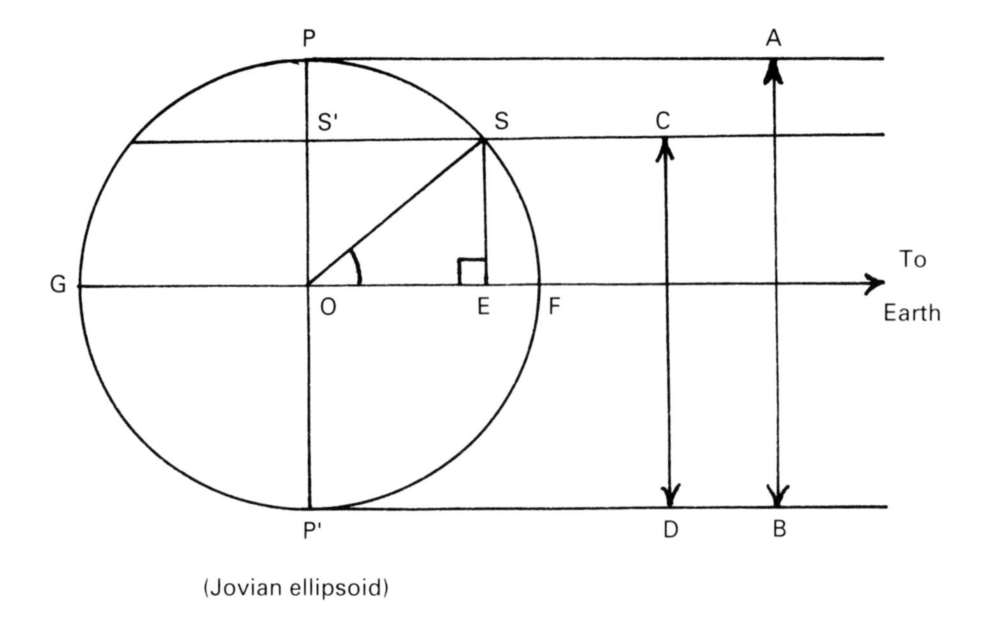

Fig. 9.21 Determination of Jovicentric latitude (second method).

meridian as it would be if Jupiter was viewed from Earth at the right side of the page and S is the feature whose latitude is to be determined.

The line OS is now drawn; the angle FOS is the Jovicentric latitude of S. This may be measured directly with a protractor or calculated from any of the trigonometrical functions of the angle: the sine SE/SO, cosine OE/OS or the tangent SE/OE. Since in this right-angled triangle SOE, the side SO – the hypotenuse – is longest, it can be measured most accurately so therefore the sine or cosine function should preferably be used.

Again, this method is approximate because it ignores the angular inclination of Jupiter's equator to the Earth's orbit and this may lead to errors of up to ±13°.

Third method Rough latitude estimates may be made by measurements from photographs of Jupiter. A good, fairly large photograph of the planet is required from which measurements of the positions of Jovian features are taken with a millimetre ruler from which latitudes are estimated as in the second method. Alternatively, the photograph may be compared with fig. 9.22 and a rough estimate made of the latitude of the feature. It will be evident from fig. 9.22 that in all of the above methods, which depend on measurements of the distance of a feature from one or other of the poles, that the nearer is the feature to the poles of Jupiter, the greater will be the effect on any error of measurement in the subsequent determination of latitude. Further, owing to limb darkening, it is difficult to locate the exact positions of the poles in a photograph. It is not practical to expect measurements of latitudes higher than about 5° north or south to be as accurate and reliable as those made at lower latitudes.

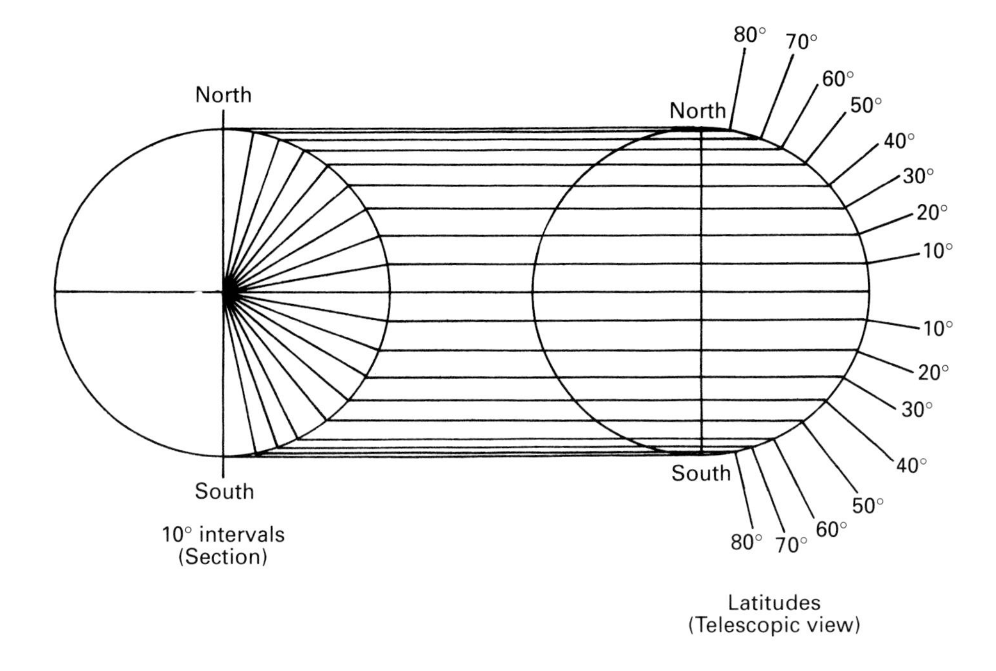

Fig. 9.22 Latitudes on Jupiter.

If both the oblateness of Jupiter and the inclination of its equator to the Earth's orbit are to be allowed for in Jovian latitude determinations then a somewhat complicated calculation involving more advanced trigonometry is required. If you want to go this far see the following books for details:

The Planet Jupiter. Peek, B. M. (revised and with foreword by P. A. Moore), Faber and Faber, London (1981).

Observational Astronomy for Amateurs. Sidgwick, J. B. Enslow Publishers, New Jersey (1982).

Handbook for Planet Observers. Roth, G. Faber and Faber, London (1970).

Disc drawings, strip and sectional sketches

Full disc drawings of Jupiter are part of a complete observing programme, but are quite secondary in importance to transit observations. However, if carefully executed they are valuable as a record of the changing appearance of Jupiter. Many of the major phenomena and disturbances in Jupiter's atmosphere are to be found in the drawings of observers of the past.

The oblateness of the disc of Jupiter must not be ignored in drawings and an outline of suitable size – about 2 inches in equatorial diameter – may be drawn from a 70° ellipse stencil as previously mentioned. The ALPO supplies standard blank forms for making disc drawings and these may be obtained by writing to the Jupiter Recorder. The disc of Jupiter on these forms is of the correct oblateness and printed on a black background which gives a more realistic appearance to the finished drawing. Space is provided for recording the relevant notes and data. Since Jupiter rotates so rapidly on its axis, the observer must work quickly otherwise the drawing will be spoilt by distortion caused by the rapid rotation. Ten–twenty minutes should be aimed at for completion of a drawing.

At most the phase angle of Jupiter is 12° (when at quadrature) which makes on fiftieth of the diameter invisible. A large telescope will reveal a thin dark crescent when occultations of Jupiter's satellites are under observation. In practice, the phase angle can be safely ignored in drawings. More obvious is the stronger limb darkening on the limb facing away from the sun.

Since Jupiter's equator is inclined at an angle of only 3° to the plane of its orbit there will be hardly any visible seasonal changes on Jupiter and alternating views of the polar regions will not be evident.

Before commencing a drawing it is advisable to study the planet for several minutes so that the eye becomes accustomed to the image and light level. Begin the drawing by lightly indicating the two equatorial belts paying careful attention to their relative and absolute widths with respect to the dimensions of the disc and carefully position them in latitude. Then add the other belts, again paying attention to their widths and accurate positioning. Some observers tend to crowd the cloud belts too close to the equator while others have the opposite tendency.

Indicate by shading the relative intensities of the belts and be careful to show the paleness at their ends near the limb of Jupiter. Do not emphasise their edges with heavy pencil lines. You can take your time over this as the rapid rotation of the planet does not have any appreciable effect on the latitudes of the belts. If there are prominent light and/or dark spots on the disc, record the time and then take one or two minutes to draw these spots and position them as accurately as you can. These will provide 'land marks' with which you can now draw and accurately position the finer details on the disc. Look out for 'wisps' and 'festoons' on the south edge of the North Equatorial Belt and carefully draw them.

When the drawing is complete the following data should be added:

(1) The date, starting time and finishing time (UT) that the drawing was made.
(2) The longitude of the central meridian in each system (CM1 and CM2) at the time that the drawing was made.
(3) The telescope used, aperture and magnification and any filters used.
(4) The seeing and transparency conditions.

Interesting disturbances or other features may be represented in larger scale drawings and fine details shown better. Don't forget to indicate the latitude and longitude of the feature.

To save time when beginning an observation, a drawing may be previously prepared with the positions of the principal cloud belts lightly pencilled in and shaded by copying from a previous recent drawing. Modifications or other details may then be added when the telescopic observation is commenced. This practice is justifiable because the latitudes of the belts do not normally show appreciable changes during an apparition of Jupiter.

A strip sketch is made by adding details to an already drawn preceding half

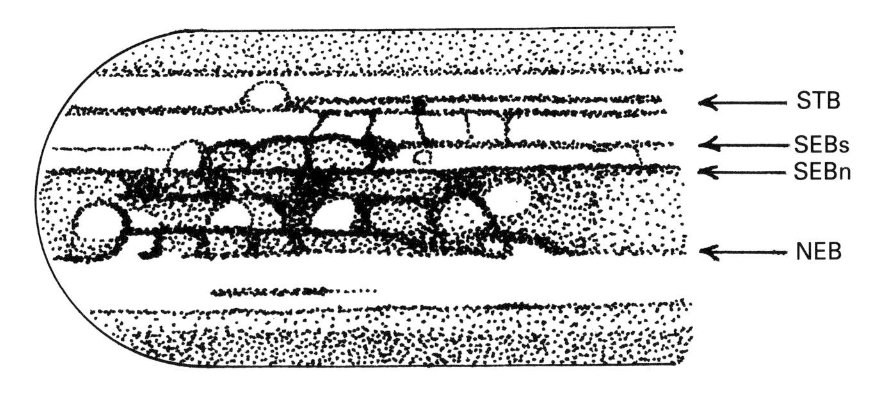

Fig. 9.23 Extended disc (strip) drawing of Jupiter. (Based on a drawing by P.S. McIntosh, 4-inch refractor, November 4th, 1962.)

of the disc as features in successive longitudes are continually brought to the central meridian by Jupiter's rotation and then drawing them at their correct longitude so that the drawing extends as a strip towards the following (right) side (fig. 9.23). Instead of working out longitudes on the spot simply position the details at distances to the right proportional to the times at which they appeared on the central meridian. The actual longitudes can be worked out and indicated later. Strip sketches thus enable a much greater range of longitudes to be shown than is possible in a disc drawing and are especially valuable for recording active belts or zones.

Accurate drawings of active areas on Jupiter made on dates close to one another have potential as a useful method of studying such problems as the nature of Jovian atmospheric phenomena, their peculiar interactions and influences that they have on each other. It is important to record central meridian transit times of such features and adjacent belts and zones should be identified in the drawing. The correct latitude/longitude relationship for strip and sectional sketches of Jovian features is shown in fig. 9.22.

Determination of rotational periods of Jovian features from longitudinal drift

The true rotational period of a Jovian feature can be determined by measuring its longitudinal drift between two dates separated by 20 or 30 days. The method of measuring longitude has already been described. Table 9.5 shows the correction factors for varying amounts of drift during a period of 20 days. All time intervals must necessarily be converted to 30 days so as to correspond with the figures in the table. As an example, supposing that a feature is observed to drift 2° in a 25-day period. Simple proportion will translate the drift to a 30-day period:

$$\frac{2}{25} = \frac{d}{30}$$

therefore $25d = 60$ and $d = 2.4$ degrees in 30 days. The values in the table must

Table 9.5 *Conversion of change of longitude to rotation period*

Longitude Change in 30 days (degrees)	Rotation period System I	Rotation period System II
	9h 50m 30s. 003±	9h 55m 40s. 632±
0.1	0.1345	0.1369
0.2	0.2690	0.2738
0.3	0.4036	0.4107
0.4	0.5381	0.5476
0.5	0.6726	0.6845
0.6	0.8071	0.8214
0.7	0.9417	0.9583
0.8	1.0762	1.0952
0.9	1.2107	1.2321
1.0	1.3452	1.3689
2.0	2.6905	2.7379
3.0	4.0357	4.1068
4.0	5.3810	5.4758
5.0	6.7262	6.8447
6.0	8.0714	8.2137
7.0	9.4167	9.5827
8.0	10.7620	10.9516
9.0	12.1072	12.3206
10.0	13.4525	13.6895
20.0	26.9050	27.3790

(from B. M. Peek (1958) after T. E. R. Phillips.)

be added to the prime value if the feature increases in longitude, i.e., if it drifts westward and must be subtracted from the prime value if the feature decreases in longitude, i.e., if it drifts eastward.

Note that the prime value – meaning no longitudinal drift in either System I or System II during 30 days – is the mean rotational rate for each system. For System I the prime value is therefore 9h 50m 30.003s and for System II it is 9h 55m 40.632s.

If the precise amount of drift is not in the table then values adding up to the value you want are summated. The conversion of 2.4° for example,. is found by adding the values in the table for 2° and 0.4°.

Example of determination of rotational rate from longitudinal drift

Suppose that in a certain year you observed the centre of a Jovian spot to transit across the central meridian on March 12th at 02.10 UT and that this is computed to System II longitude 30.5°. Thirty days after this observation on April 11th the same spot transits at 01.50 UT and this gives a computed longitude of 32.9°. The spot has therefore increased in longitude and has drifted 2.4° westward. The correction factor from the table must therefore be added to the prime value of

9h 55m 40.632s since the spot is situated in System II. From the table it is seen that a drift of 2.4° corresponds to the value 3.2855s (value for 2.0° plus value for 0.4° = 2.7379s + 0.5476s × 3.2855s). Adding this to the System II prime value gives 9h 55m 40.6320s plus 3.2855s = 9h 55m 43.9s (after realistic 'rounding off'). Therefore, in the 30-day period the rotational rate of the feature has been determined to be slightly greater than the mean System II value which shows that the feature decelerated during this time.

Determination of these rotational periods is the single most valuable contribution that can come from amateur studies of Jupiter for it is this type of investigation that reveals the existence of Jovian atmospheric currents and their interactions which are so important to our understanding of the planet's atmosphere.

Observations of the Great Red Spot

Jupiter's Great Red Spot is one of the most interesting of planetary objects. It can attain a size of 30 000 miles (48 270 km) in length and 7000 miles (11 263 km) in breadth. It is usually associated with System II but its motion is irregular and mostly independent of the System II current. Its mean rotational rate is 9h 55m 38s.

Sometimes the Red Spot may be invisible, its position being occupied by the prominent bay, the Red Spot Hollow. When only the Red Spot Hollow can be seen observations should be continued and data on the Hollow recorded.

An observational programme of the Red Spot should include:

(1) longitude determinations;
(2) size determinations;
(3) determinations of longitudinal drift;
(4) colour estimates;
(5) strip sketches of the neighbourhood of the Red Spot;
(6) interactions of the Red Spot with other atmospheric features.

Longitude determinations The method of determining longitudes of Jovian features has already been described. The longitude of the Red Spot is found from noting the time of transit of its centre across the central meridian. In addition the times of transit of the preceding and following ends can be valuable.

Size determinations The size of the Red Spot in longitude is found by recording the times of central meridian transit of the preceding and following ends of the Red Spot. Sometimes, the Red Spot has slightly pointed ends and special care is necessary in timing the central meridian transits of the pointed preceding and following ends accurately.

The longitudinal size of the Red Spot and other Jovian features can be determined first by calculating its longitudinal extent in degrees. As an example, assume that the time elapsed between central meridian transits of the preceding and following ends was 17 minutes. We know that 0.6° of Jovian longitude corresponds to 1 minute of time, therefore the longitudinal extent of the Red Spot is given by $0.6 \times 17 = 10.2°$. It is important to measure the angular size of the Red Spot as it fluctuates rapidly and to a large extent.

It is, of course, possible to convert the angular length of the Red Spot to actual length in miles or kilometres. To do this the latitude of the Red Spot must be determined, since the actual length of a degree of longitude depends on distance from the equator. This would then be followed by a calculation involving spherical trigonometry complicated by Jupiter's oblateness. Whether the calculation of the actual physical length of the Great Red Spot is worth the effort is debatable.

Determination of longitudinal drift The Great Red Spot exhibits considerable drift in longitude and rotation rate. Between the years 1831 and 1955 it drifted somewhat more than 500° in longitude on either side of the central meridian which, as mentioned earlier, is an angular distance equivalent to nearly three complete rotations around the planet. This alone precludes the possibility of the Red spot being directly connected to any feature on a solid surface.

Frequent determination of longitudinal drifting and rotational rate of the Great Red Spot is of crucial importance. The technique described earlier is used as for any other Jovian feature in order to determine accurately the longitude of the centre of the Spot.

The pattern of drift is affected by several factors which include its interactions with the white ovals of the South Tropical Zone, the disturbances at the South Equatorial Belt and other poorly defined influences. Attempts should be made to relate these factors to drift of the Spot, an important part of observational research on this feature.

Colour estimates Of considerable importance in a programme of observation of the Red Spot are colour estimates. Some evidence indicates that the size of the Red Spot may be correlated with its colour and intensity as seen from Earth. The colour of the Spot is variable ranging from orange to salmon pink. Occasionally it appears grey; the grey appearance may be due to the intensity of the Red Spot being insufficient to stimulate the colour receptor cells in the human retina.

Whenever there are major cloud disturbances in the neighbourhood of the Red Spot, its colour changes, if any, should be carefully scrutinised. Changes in colour within the Red Spot should also be noted. Record as accurately as you can the colour that you perceive the Red Spot to be.

Strip sketches of the neighbourhood of the Great Red Spot Accurate renditions of the Great Red Spot are best done as strip sketches which isolate a narrow latitude range rather than as full disc drawings. These are drawn to a larger scale than on disc drawings to show finer detail (fig. 9.24).

The following phenomena should be drawn if they are seen:

Spot shape. The exact shape of the Red Spot should be carefully delineated.

Internal detail. Note if there are any light spots, changes in colour or any central dark areas.

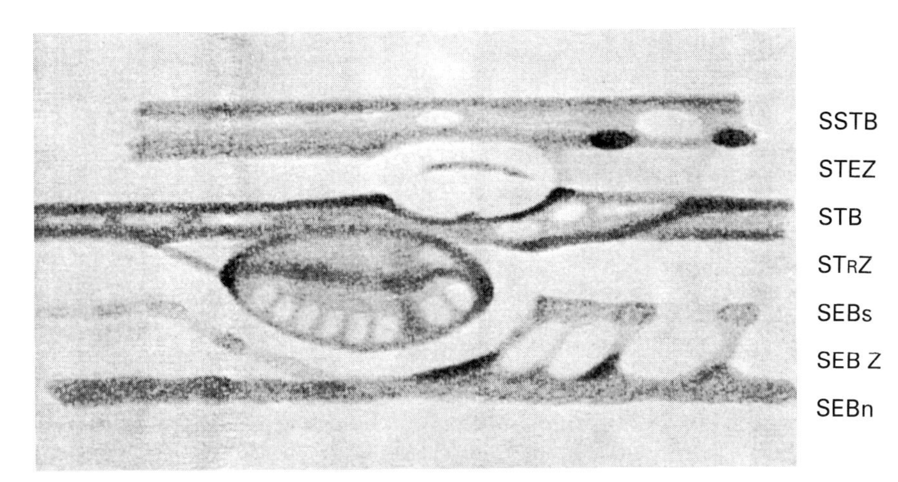

Fig. 9.24 Jupiter's Great Red Spot and surrounding region: January 21st, 1964, 00.29 UT. (E.J. Reese, 16-inch Newtonian reflector belonging to Clyde W. Tombaugh.) (From The Strolling Astronomer **17(9–10)**, *1964.)*

Spot borders. Look for and draw a distinct light or dark border to the Spot.

Interactions. Draw any interactions of the Red Spot with belts or zones in its neighbourhood.

Studies of interactions of the Great Red Spot with other atmospheric features A lookout should be kept for interactions of the Red Spot with other nearby features. In particular, interactions with the bright ovals in the South Temperate Zone and disturbances in the South Equatorial Belt. These features have patterns of drift that are usually in a direction opposite to those of the Red Spot. As a result, the opposite drifts bring features towards each other and this may cause some kind of change or disturbance.

If interactions are observed record the following information in addition to a series of strip sketches made during the interaction:

(a) Date and time that the interaction was seen.
(b) Estimated latitude difference between the feature and the Red Spot.
(c) The rotation periods of the Red Spot and other features during the interaction.
(d) The change in the rotation period of the Red Spot and other features after the interaction, if any.

The drift patterns of both the Red Spot and the feature should be plotted graphically. The period before, during and after the interaction should be included. This gives a clear visual grasp of changes in drift patterns and how they are correlated.

Colour changes and intensity estimates of Jovian features

Considerable variation in the colours of Jovian features becomes evident if the planet is systematically observed. For best results in colour observation a reflecting telescope is preferable to a refractor because of the residual chromatic aberration in the latter; however, both types of telescope are suitable for detecting pronounced colour changes and for determining strong tones.

Colour estimates are probably unreliable when Jupiter is low in the sky owing to atmospheric dispersion. It is not unusual to see a cloud belt of Jupiter fringed with orange-red on one edge and bluish on the other at low altitudes because of this effect. When 45° or higher above the horizon, colour estimates may be accurately and reliably made. The sky should be clear and dark. Note the colours of the zones, the cloud belts and the Great Red Spot and other notable features.

Some evidence points to regularity in the colour changes of the clouds of Jupiter. There seems to be a period of about 11 years on average and this seems correlated with an increase of redness in the belts. There is no satisfactory explanation for this phenomenon except that the period of 11 years approximates the length of the sunspot cycle.

Examples of colours and colour changes that have been recorded are as follows:

The Great Red Spot was orange-red becoming orange-pink in the early 1970s and was a dim pink colour in 1976.

The North Equatorial Belt was a deep brown colour in the early 1970s and was brown-black in 1976.

The Equatorial Zone appeared bright yellow in 1972 and became a dusky brownish colour in 1976.

The South Tropical Zone appeared bright yellow in 1976.

The keeping of records of colour and intensity estimates of the Jovian clouds is somewhat like keeping daily records of the clouds and atmospheric conditions in the Earth's atmosphere; it may not seem to have any immediate scientific value but can be of considerable value to future observers who may discover long-term patterns of change or important correlations between phenomena.

Regarding the belts, a logical plan is to write the names of the belts in a vertical column in order of their intensity. Beside each belt is then written its colour and intensity. A similar list should be made of the zones.

Abbreviations may be used in describing colours. The following have been suggested:

White: W	Ochre: Oc
Orange: O	Red: R
Brown: Br	Blue: Bl
Grey: G	Blue grey: Bl-G
Yellow: Y	Reddish-brown: R-Br

General observing notes

The following aspects of Jovian features should be specially looked for in a systematic observing programme.

Polar regions. Watch for the appearance of white or grey ovals or narrow belts and the sudden appearance of zones in both North and South Polar Regions.

Equatorial Zone. Note changes in the brightness and colour of this region and any bright spots or festoons. The thin Equatorial Band may sometimes put in an appearance. If any prominent features are seen, time their transit across the central meridian.

North Equatorial Belt. Very rapid changes are seen in this belt which is usually dark, wide and with several knots and spots. Several of the festoons seen in the Equatorial Zone spring from the south edge of this belt. Very thin delicate festoons may be seen in the North Tropical Zone and might herald the start of disturbances in the North Tropical Zone and North Equatorial Belt. Short-lived reddish spots develop in the North Equatorial Belt when the belt is active. Of considerable importance is comparison of the rotation periods of these red spots and the Great Red Spot. Often, at times of much activity in the Jovian atmosphere, large telescopes will often reveal a very thin north component of the North Equatorial Belt.

South Equatorial Belt. This belt also exhibits much activity. It usually appears double and is the origin of many disturbances. The darker and wider of the two components is generally the northern (SEBn) and it appears to be connected to the southern (SEBs) by thin delicate festoons and frequently with groups of white spots. Observe carefully the relationship between the Great Red Spot and the SEBs. Note the amount of interaction of the Red Spot as it moves the SEBs to the north. Large and small disturbances originate in the zone between the two South Equatorial Belt components.

Watch for fine festoons starting from the SEBs that cross in a southerly direction into the South Tropical Zone; they almost always indicate that a disturbance is on the way. Try to ascertain if a festoon begins in the SEBs or from inside the zone separating the two South Equatorial Belt components. Within this zone numerous white spots and dusky bars with rotation times differing slightly from that of System II can appear. Transit timings should be made of such features.

North Temperate Belt. This belt can have two different appearances. It may appear as a single belt, fairly wide and dark or it may appear as two very thin dusky belts lying close together. The former appearance is associated with increased activity and the latter with a decrease. The intensity and colour of the two appearances also differ markedly.

South Temperate Belt. This belt is usually fairly broad and is sometimes the darkest feature on Jupiter. It is irregular with curves, interruptions, swellings and knots throughout its length and is occasionally seen to split into two thin dark belts for short distances.

Tropical zones. Watch for festoons originating from the North and South Equatorial Belts extending into the North and South Tropical Zones. Sometimes thin faint belts a few degrees long are seen in these zones. Ill-defined large ovals are frequently seen but are short-lived. They are irregular in shape and changes in size and shape should be noted as well as making determinations of the rotation periods.

Each observational record should include your name, the date and time (UT) that the observation was made and the longitudes of the central meridian in both Systems I and II when the drawing was begun and finished.

Filter observations.

Optical glass filters are useful for increasing the visibility of details that may be almost invisible in white light and for improving contrast of certain features. Colour filters are, of course, also important in estimating colours themselves! Whether or not coloured glass filters are employed, the apodising screen always improves contrast and steadies the image; I never observe Jupiter without using it.

The following filter colours are useful in studying various Jovian features:

Medium blue. Detail in the Great Red Spot, white spots and ovals, the two equatorial belts and temperate belts.

Green. Strongly recommended for all the cloud belts, the Great Red Spot, white ovals and spots, festoons.

Orange. Good for some white spots, some of the belts and for details within zones. Yellow and red filters are not recommended.

Light pollution filters. Some of the presently popular light pollution filters employed by deep sky observers to enhance visibility and contrast of faint galaxies have been found to be useful for planetary observation. The Orion Telescope Center of Santa Cruz, California market 'Skyglow Enhanced Broad Band' light pollution filters that the manufacturers recommend for planetary as well as deep sky work. They claim that these filters are particularly suitable for viewing Jupiter especially if another colour filter is used with them. Without any filter Jupiter's cloud belts may look pale and diffuse lacking detail but with a light pollution filter used with another filter, the manufacturers claim that knots on the belts and festoons are revealed and the limb of Jupiter stands out sharply. These light pollution filters are rather expensive, though.

Photographic work. See chapter 15.

Satellite phenomena

Of the several satellites of Jupiter only the four largest (the Galileans) can be observed with modest telescopes, the other satellites being beyond the reach of all but really large instruments although, as previously mentioned, Himalia has been seen in the larger amateur instruments. The smallest of telescopes, even

Fig. 9.25 Satellite III (Ganymede) shadow in transit across the North Tropical Zone of Jupiter: 8-inch reflector ×225, March 4th, 1992, 02.25 UT. (F.W. Price.)

Fig. 9.26 Jupiter showing the Great Red Spot and the shadow of Satellite I (Io): January 15th, 1968, 08.28 UT, 6-inch reflector, Lunar and Planetary Laboratory. (University of Arizona photograph.)

ordinary binoculars, will easily reveal the Galilean satellites. In fact some keen-sighted people have claimed to have seen them with the unaided eye under good seeing conditions. The possessor of a 2- or 3-inch refractor is well equipped to observe the different configurations of the satellites as they revolve around their primary. They are a continuing source of interest and pleasure. With a 3-inch refractor satellite shadows on Jupiter's disc are easily visible as inky black circular dots (figs. 9.25 and 9.26). The almanacs and year books contain much information about the positions, transits, occultations and eclipses of Jupiter's satellites. The phenomena that the observer should be on the watch for are threefold:

(1) Transits of satellites and satellite shadows across Jupiter.
(2) Eclipses and occultations of the satellites by Jupiter.
(3) Mutual eclipses and occultation of the satellites.

The diameters of and distances from Jupiter of the Galilean satellites are listed in the general section. Table 9.6 lists their albedos and magnitudes.
Owing to its being at the greatest distance from Jupiter, Callisto's behaviour is somewhat different from the other Galileans. Sometimes, when Jupiter's equator is tilted at an angle greater than $\pm 2.2°$ with respect to the Earth, Callisto will be seen to pass above or beneath Jupiter. Also, Callisto will not be eclipsed at solar declinations outside of this range.

Transits The satellites and their shadows pass in front of Jupiter from east (following) to west (preceding). Before opposition the shadow precedes the satellite and falls on the disc of Jupiter before the satellite transits. After opposition the reverse happens; the shadow follows the satellite and will remain on the disc after the transit has ended (fig. 9.27).

The time elapsing between the transit of a satellite and its shadow varies with the satellite's distance from Jupiter and the period of time between the transit and opposition. This could be several hours long and frequently makes identification of shadow transits difficult when Jupiter is nearer conjunction than opposition. As already mentioned Callisto may pass above or below Jupiter and therefore will not transit at all and the same applies to its shadow.

The satellites differ in their appearances when passing across the disc of Jupiter. Europa is quite light and may be hard to see when in transit across one of the light zones. In contrast, Ganymede and Callisto both appear dark when on the disc of Jupiter and may even be mistaken for satellite shadows. Io is intermediate and has a grey tint when in transit (fig. 9.28).

Because of the slight limb darkening of Jupiter, all of the satellites near the beginning of transit will be seen as bright spots against the slightly darker limb. This is easily seen in even a small telescope. As transit progresses the satellites 'disappear' fairly quickly. Io and Europa will be quite 'lost' unless a fairly large aperture is used. They reappear as bright spots at the opposite limb as transit ends. Io may be seen during transit if it lies on one of the zones when it will look like a small dusky spot. If projected against a belt it may not be possible to see it again until the transit ends. When seen against the limb shading of Jupiter it may appear elliptical in shape when seen in telescopes of moderate aperture.

Europa is the only one of the satellites that normally looks bright during an entire transit. It is plainly visible against a belt but even this satellite may be

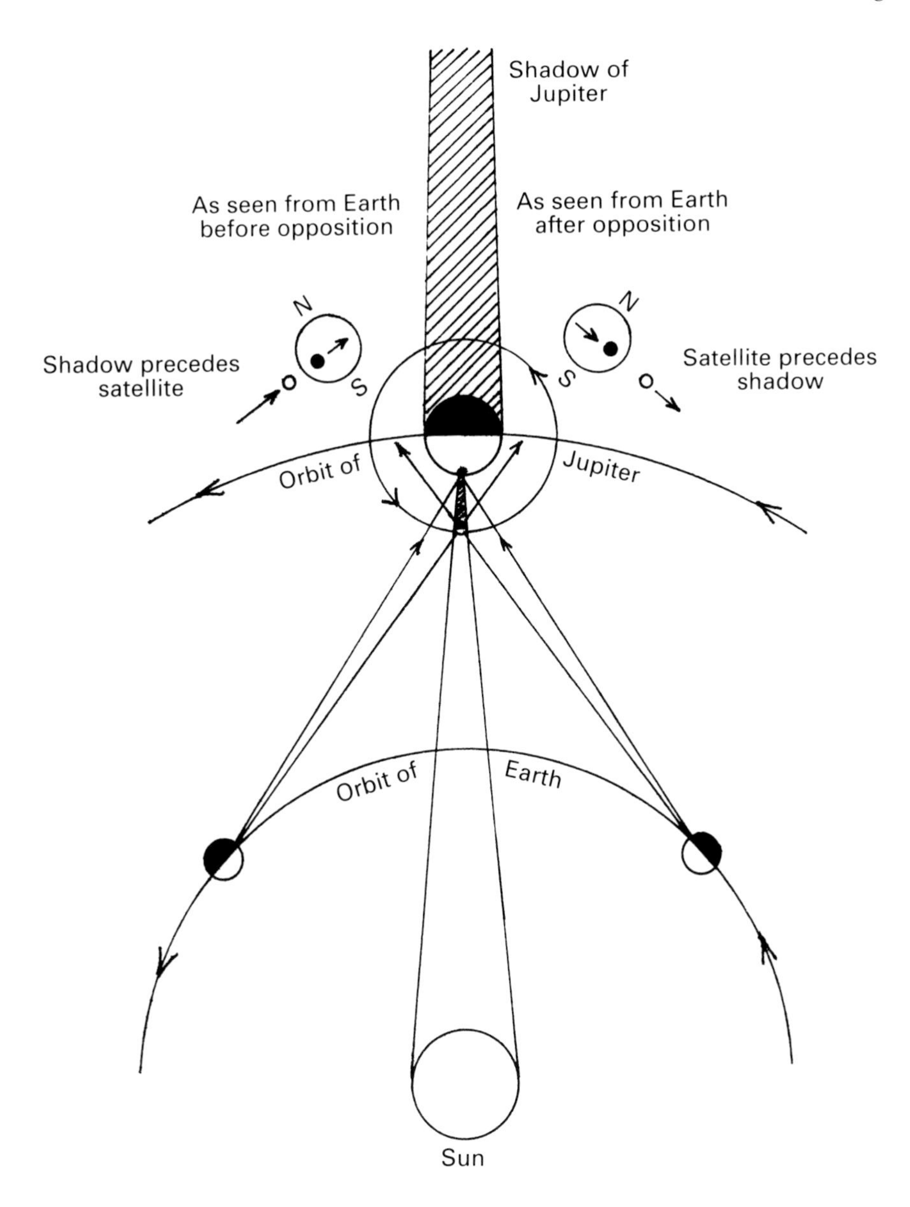

Fig. 9.27 Jovian satellite transit and shadow phenomena.

Table 9.6 *Albedos and magnitudes of Jovian satellites.*

Satellite	Albedo	Magnitude	Satellite	Albedo	Magnitude
I Io	0.57	5.0	III Ganymede	0.34	4.6
II Europa	0.60	5.3	IV Callisto	0.14	5.6
			(Earth's moon)	0.073	−12.7

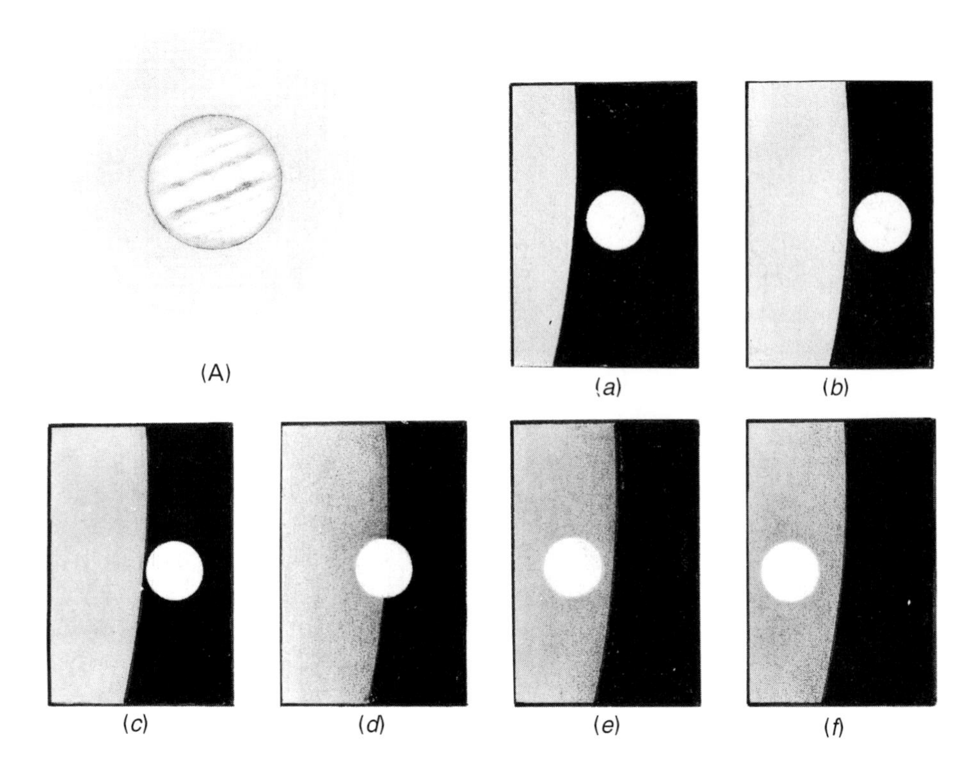

Fig. 9.28 Jupiter – transit of Satellite I (ingress) (A), June 6th, 1970: (a) *22.25 UT;* (b) *22.28 UT;* (c) *22,33 UT;* (d) *22.35 UT;* (e) *22.39 UT;* (f) *22.42 UT.*

difficult to see if on a zone. Ganymede and Callisto always appear as prominent dark spots soon after the beginning of a transit but both appear bright against the limb darkening.

Transit timings are not of much scientific value because the indeterminate character of ingress and egress of the small satellites on the Jovian disc renders accurate timing of transits difficult.

Eclipses and occultations When satellites are on the far side of Jupiter they move from west to east, i.e., from the preceding to the following side. Satellites are said to be occulted by Jupiter when they pass out of sight behind the planet and are eclipsed when immersed in Jupiter's shadow. In the period between Jupiter's conjunction and opposition, satellites are eclipsed before they reach the preceding limb and emerge from occultation at the following limb. Between opposition and conjunction, occultation of a satellite occurs at the preceding limb and emergence from eclipse is seen on the following side of Jupiter (see fig. 9.27 for explanation of this). Ganymede and Callisto can, at times, lie between the limb and the shadow and may thus, for example, emerge from occultation and be eclipsed.

The satellites do not disappear instantly on entering Jupiter's shadow; Io and Europa disappear quickly but the others are slower. The timing of the instant when a satellite becomes visible as it exits from eclipse (emersion) or the last instant of visibility when it enters eclipse (immersion) – the timing of the latter is likely to be more accurate – may be useful in revising ephemerides of the satellites. (The ALPO has been conducting a Galilean satellite eclipse timing programme since 1975. Interested individuals should contact the Office of the Executive Director of the ALPO). The following peculiarities in behaviour of the Galilean satellites should be noted.

Io. Because of Io's closeness to Jupiter entry into and emergence from the same eclipse cannot be observed. This satellite invariably goes from occultation into eclipse or vice versa. The total time elapsing during a combined eclipse and occultation is about $2\frac{1}{4}$ hours.

Europa. Is usually seen entering or exiting from an eclipse, very rarely both. Time elapsing is about $2\frac{3}{4}$ of an hour.

Ganymede. Both entry and exit from an eclipse are observable but not when Jupiter is close to opposition. The average time taken for an eclipse is about $3\frac{1}{4}$ hours but there is considerable variation from this.

Callisto. As with Ganymede, entry into and exit from the same eclipse can be seen except when Jupiter is close to opposition. The duration again shows considerable variation and averages about 4 hours. Sometimes Callisto may not enter Jupiter's shadow at all and sometimes there is a partial grazing eclipse. Every 12 years there are two periods of 3 years each when Callisto does not exhibit eclipses, occultations or transits.

The motions of Io, Europa and Ganymede are correlated; their mean daily motions are related insofar as the sum of the mean daily motion of Io (I) and twice the mean daily motion of Ganymede (III) equals three times the motion of Europa (II). In more exact terms this is expressed as:

$$L_{\mathrm{I}} + 2L_{\mathrm{III}} - 3L_{\mathrm{II}} = 180°$$

where L_{I}, L_{II} and L_{III} are the mean longitudes of the satellites in their orbits which are measured from any fixed arbitrary radius through Jupiter's centre. The motions are thus said to be commensurate. They are precise and permanent and the above relation always holds and follows from the mutual gravitational perturbations between all three satellites. It follows from this relation that all of the three satellites cannot show the same phenomena simultaneously. For example, Europa and Ganymede may both be seen transitting the disc of Jupiter but Io will be occulted. Oddly, Callisto has no effect on this motion relationship between the other three satellites. Its motion is therefore said to be incommensurate with the others. A purely hypothetical configuration of Jupiter's satellites and the appearances as seen from Earth is shown in fig. 9.29.

Mutual eclipses and occultations of the satellites Once in every 5.93 years a terrestrial observer is enabled to look almost exactly along the planes of the satellites' orbits for a few months. This is when the inclination of Jupiter's axis is small and it is near to 8 hours or 20 hours of RA. The satellites will then exhibit mutual eclipse and occultation phenomena. The satellites appear to move to and fro in linear paths on either side of Jupiter and it is obvious therefore that, as seen from Earth, we may expect one satellite to pass in front of

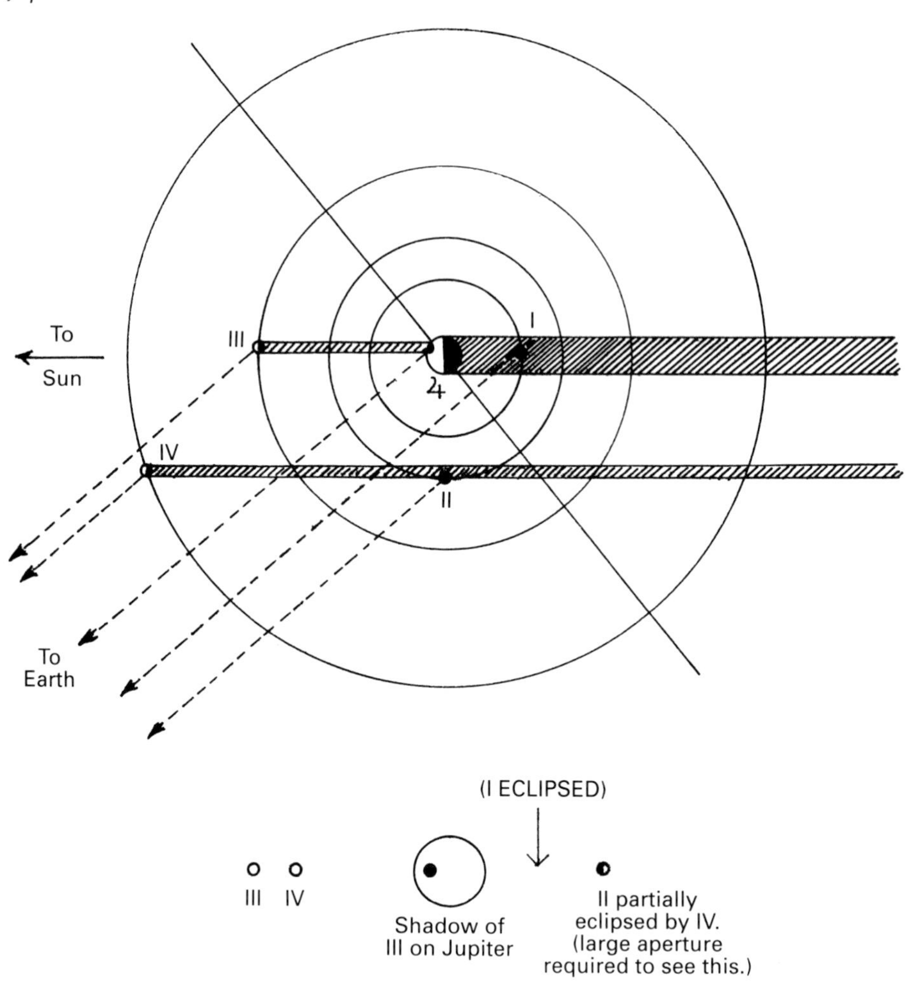

Fig. 9.29 Hypothetical arrangement of Jupiter's Galilean satellites and their appearances as seen from Earth.

another. This is termed a mutual occultation. Such occultations clearly display the differences in albedo of the satellites (fig. 9.30) and this is especially marked if the darker of the two satellites is nearer the observer.

The approach of a mutual occultation is easily perceived, for if two satellites are getting nearer to one another then obviously an occultation is about to occur and when it does the two satellites merge into a single elongated image as seen in the telescope (fig. 9.31). The combined image of the two moons becomes dim when the satellite nearer the Earth overlaps all of, or part of, the other satellite. The amount of dimming is dependent on what fraction of the disc of the occulted moon is hidden and on the reflectivities (albedos) of the two moons.

The combined light loss in a mutual occultation can never be greater than 62% as the satellite nearer to the Earth is still visible during the event. After the moment of maximum dimming of light, the reverse series of events is seen. An

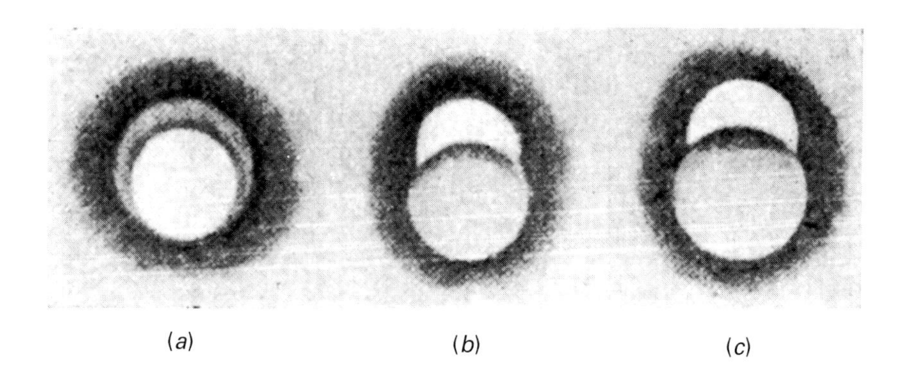

Fig. 9.30 Satellite occultations. (a) *Occultation of satellite IV (Callisto) by satellite II (Europa) January 8th, 1932. (T.E.R. Phillips.)* (b) *Occultation of satellite I (Io) by satellite IV (Callisto) February 18th, 1932. (B.M. Peek.)* (c) *Same as* (b) *Note the duskier tone of IV (Callisto). (T.E.R. Phillips.) (From BAA 29th Jupiter report.)*

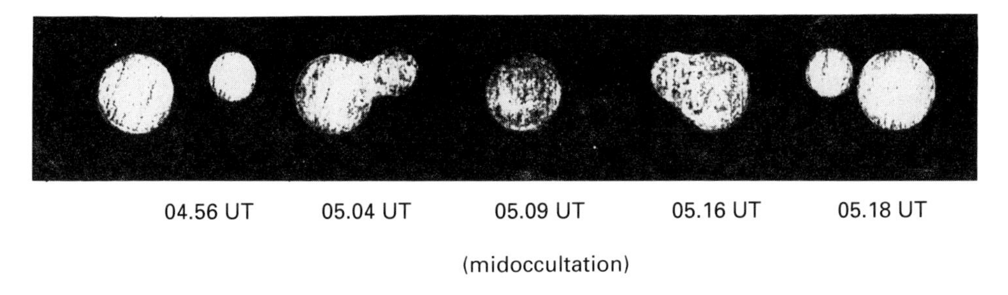

Fig. 9.31 Occultation of satellite III (Ganymede) by satellite II (Europa), November 20th, 1990, 6-inch telescope. (P. Smith.) (From Astronomy **19(6)**, *76, 1990.*

occultation may last for a few minutes or it may last for as long as an hour or more.

Not so easy to anticipate are eclipses of one satellite by another, because the shadows cannot be seen when lying only in space. Also the distances separating the eclipsing and about-to-be-eclipsed satellite may be quite large so that during an eclipse, the two satellites are still seen as separate bodies in the telescope. The satellite being eclipsed becomes dimmer over a time that may be a few minutes or longer than an hour. The degree of dimming of the light of the moon being eclipsed depends on the relative sizes of the two moons and on whether the shadow of the eclipsing body crosses centrally over the eclipsed moon or merely grazes it. Hence, the loss of light can be from virtually zero to complete; in the latter case, the satellite may actually disappear. A mutual eclipse can be so short-lived that an observer may overlook it if not specially looking out for one.

Mutual occultations and eclipses of the satellites are not uncommon when the sun, Earth and Jupiter's equator are coplanar. There occurred a series of mutual satellite phenomena between November 1990 and April 1992 in which no less than 300 mutual events were observed. Worthy of special mention are the following: On January 26th 1991, Io was eclipsed by Europa and straight afterwards was occulted by it – in fact the occultation had started before the eclipse had finished. A rare occultation of Ganymede by Callisto occurred on June 29th 1991, the first of three mutual events during that night. It was a good opportunity to compare Callisto's dark grey colour with Ganymede's much lighter surface.

Two possible but rare phenomena that an observer may have the good luck to see are as follows:

(1) The eclipse of one satellite by the shadow of another while the first-mentioned satellite and both satellite shadows are passing across Jupiter's disc. The appearance will be of the two shadows merging and becoming one while the satellite will turn black. Something like this was seen on the night of June 11th 1991 when Io was eclipsed by Ganymede. Io's light was dimmed by 4% and the shadows of the two satellites were seen transitting Jupiter, at one point actually merging.

(2) One of the outer satellites, either Ganymede or Callisto, not yet immersed in Jupiter's shadow, could be occulted by either Io or Europa while in Jupiter's shadow and therefore invisible. If the occultation is to be total then the outer satellite must be Callisto and the inner Ganymede.

On February 18th 1932, W. H. Steavenson, witnessed the simultaneous eclipse and occultation of Io by Callisto and Jupiter respectively (fig. 9.32).

Observing and recording mutual satellite phenomena. Mutual occultations are always observable even in small telescopes but if your telescope is smaller than a 6-inch you will not be able to see a satellite in eclipse unless the light loss is greater than about 20%. This is so even if you attempt to compare the brightness of the eclipsed satellite with another not being eclipsed.

Of course, the larger the telescope the better the view of either type of phenomena. For example, if on an evening of unusually good seeing, you are observing with a telescope of 12 inches aperture or more, you may be able to

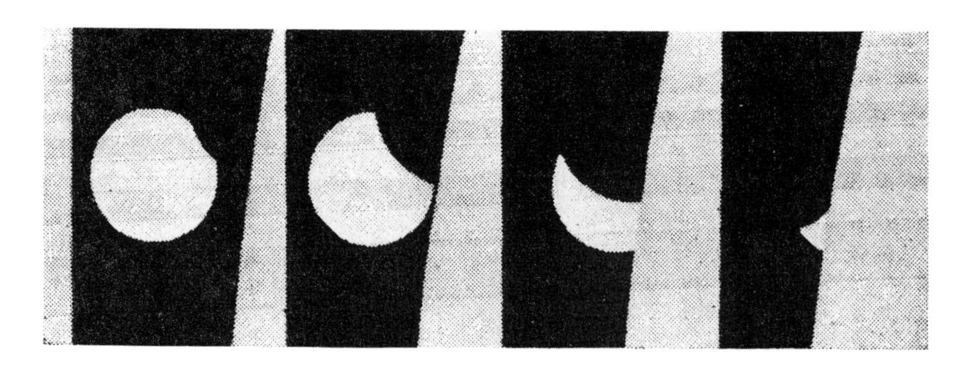

Fig. 9.32 Eclipse of satellite I (Io) by IV (Callisto) during occultation of I by Jupiter. February 18th, 1932. Left: at 9h 34.3m. Others at 1-minute intervals thereafter. (W.H. Steavenson.) (From BAA 29th Jupiter report.)

see the shadow of a satellite moving across the surface of another during an eclipse or a satellite transitting another during an occultation.

Mutual satellite events may be recorded simply by making a series of drawings separated by a few minutes accurately timed to within about one tenth of a minute. Notes about the estimated brightness of the eclipsed or occulted moon's brightness may be included in the observation by comparing the moon's brightness to that of two others. The best situation would be if one of the other satellites is brighter and one fainter than the one being estimated.

Objectivity is added to recording of brightness variations in mutual events by employing photography. Dr John E. Westfall of the ALPO says that he gets good satellite images with a 10-inch telescope using exposures of 4 seconds duration with ISO 200 slide film. Consecutive exposures should be separated by a few minutes. Videotaping can also be used but a fairly large telescope (14 inches of aperture or more) must be used to obtain clear satellite images.

More advanced recording techniques are the use of a CCD camera (see later in this book for details) or a photoelectric photometer. If you use a CCD camera, pictures should be taken with short time intervals between them. Ensure that your equipment is capable of storing several megabytes of data.

When using a photometer V (visual) band readings should be taken at intervals of a second if possible. Rapid alternation between V and B (blue) readings should be used when recording occultations of Io by Europa. Readings made in these two different colours are potentially useful to professional astronomers insofar as they may be enabled to chart volcanic phenomena on Io owing to fresh volcanic matter having a different colour from older matter.

Data that may be usable by professional scientists should be sent to the following:

Light curves, printouts, discs and videotapes.

Smithsonian Centre for Astrophysics,
60 Garden Street,
Cambridge MA 02138

Bureau des Longitudes,
77 Avenue Denfert-Rochereau, 75014–Paris,
France.

Drawings, photographs and copies of VHS or S-VHS videotapes:

Association of Lunar and Planetary Observers,
P.O. Box 16131,
San Francisco, CA 94416.

Satellite surface features Users of telescopes with apertures in excess of 10 inches may see indications of surface markings on the Galilean satellites, but this will not be easy. Smaller telescopes cannot be expected to show anything; after all Ganymede, which shows the largest disc of the Galileans, has an apparent mean opposition angular diameter of only 1.52 seconds of arc. The appearance of Ganymede in a 12-inch telescope has been likened to the view of Mars in a 2-inch instrument.

Several members of the ALPO have observed Ganymede and made drawings that show surface markings (fig. 9.33). Interestingly, the drawings of Camichel, Lyot and Gentili described in the 'History of observation' section show remarkably close general correspondence with the appearances and surface

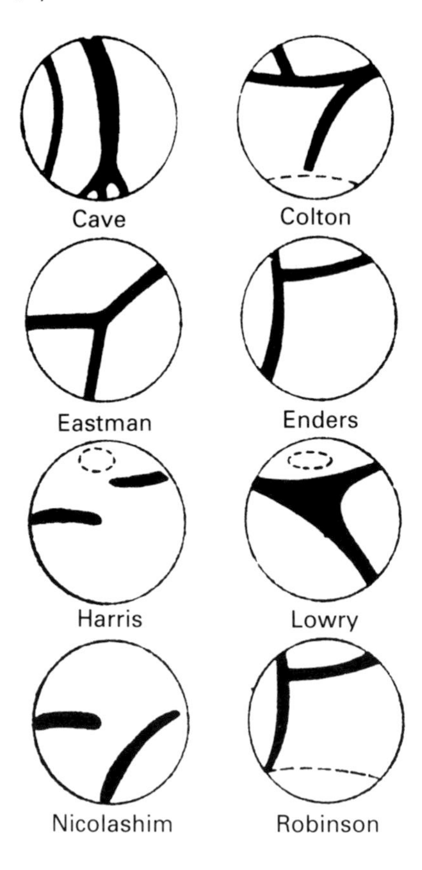

Fig. 9.23 Drawings of surface features on Ganymede by various ALPO observers. (From Handbook for Planet Observers, *Roth, G. D. Faber And Faber, London, 1970.)*

features of the Galilean satellites as shown in the Voyager photographs. However, in the post-Voyager era it would seem that little of scientific value can result from Earth-based amateur studies of the satellite markings.

Radio studies of Jupiter

At times Jupiter is among the most active of celestial objects emitting radiation over a broad frequency spectrum but only those frequencies from 5 to 40 megahertz will be of interest to the amateur observer. Jupiter's radio emissions are probably caused by its atmosphere; there is a strong correlation between them and some System III longitudes and with the longitude of the satellite Io. Quite simple apparatus such as an ordinary communications receiver is all that is required to detect these emissions when Jupiter is active.

The emissions are discontinuous and days and weeks may elapse without anything being detected. Every positive observation is thus more significant and valuable. Whatever it is that causes these emissions can only be ascertained from observations pursued over a long period of time.

Although apparatus for observing at 20 megahertz is simple there are problems that may complicate the work. These are:

(1) Observations at the above frequencies are made in the communications band. Finding a free region is quite difficult. Before setting up an aerial it is best to spend a week or longer observing in a number of frequencies. You can thus find out if that wavelength is reasonably clear of interference.
(2) Thunderstorms: interpretation of weak outbursts from Jupiter can be interfered with. Observational practice is needed in order to differentiate outbursts from Jupiter and those originating in our own atmosphere.
(3) The terrestrial ionosphere: this limits the angle at which radio observations can be made.

Further reading

Books

The Planet Jupiter. Peek, B. M. Macmillan, New York, (1958).

Papers and Articles

How to observe Jupiter and why. Hargreaves, F. J., *JBAA* **60(7)**, 187–9 (1950).

Variations of surface features on Jupiter. Millar, D. W., *JBAA* **54(8/9)**, 162–6 (1944/45).

Periodicity in the activity of Jupiter's atmosphere. Favero, G., Senigalliesi, P., and Zatti, P., *JALPO* **27(11–12)**, 240–5 (1979).

Colour of Jupiter's polar regions. Bayley, D. P., *JBAA* **55(5)**, 116–20 (1944/5).

The colour of Jupiter's polar regions. Smith, C. F. O., *JBAA* **55(1)**, 23–5 (1944/5).

The colour variations of Jupiter's Equatorial Zones. Williams, A. S., *JBAA* **47(2)**, 68–70 (1937).

Possible long term changes in the Equatorial Zone of Jupiter. Tatum, R., *JALPO* **29(7–8)**, 141–3 (1982).

Jupiter through colour filters. Heath, A. W., and Robinson, J. H., *JBAA* **87(5)**, 485–7 (1977).

Disturbances on Jupiter's South Equatorial Region. McIntosh, R. A., *JBAA* **60(8)**, 247–50 (1950).

On the life and continuity of Jupiter's South Tropical Disturbance. Peek, B. M., *JBAA* **73(3)**, 109–12 (1963).

The Great Red Spot on Jupiter. Fox, W. E., *JBAA* **78(1)**, 16–21 (1967).

The Great Red Spot. Schwartzenburg, D., *Astronomy* **8(7)**, 6–13 (1980).

The dance of the Jovian moons. Westfall, J. E., *Astronomy* **19(6)**, 76–9 (1991).

The mutual antics of the Galilean satellites. Westfall, J. E., *JALPO* **34(4)**, 189–90 (1990).

Naked Eye Observations of Jupiter's Moons. *Sky and Telescope* **52(6)**, 482–4 (1976).

10

Saturn

General

Saturn, the most distant of the planets known to pretelescopic astronomers, is the second largest planet in the Solar System and has an equatorial diameter of 74 990 miles (120 660 km). Like Jupiter, Saturn is a 'gas giant' planet. Its axial rotation rate of about 10.25 hours, which is not much longer than Jupiter's results in an equatorial bulge and polar compression like Jupiter's except that it is even more pronounced. The polar diameter is only 67 122 miles (1087 000 km) so that the ratio of the polar to the equatorial diameter is about 10 : 11 compared to 15 : 16 for Jupiter. This is the largest polar compression known of any of the Solar System planets. Comparative sizes of Saturn and the Earth are shown in fig 10.1.

Saturn orbits the sun at a mean distance of 886.9 million miles (1427 million

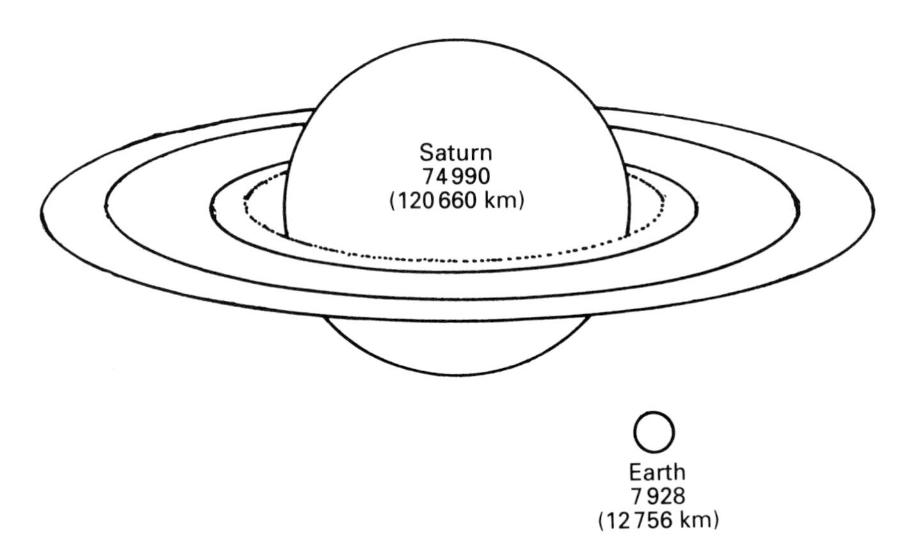

Fig. 10.1 Comparative sizes of the Earth and Saturn (equatorial diameters in miles).

km) and the orbital eccentricity is 0.056. The perihelion distance is 837.2 million miles (1347 million km) and the aphelion distance is 936.6 million miles (1507 million km). The orbital revolution period is 29.46 Earth years (10 759.2 Earth days, the sidereal period) and the synodic period is 378.09 Earth days. Saturn's orbit is inclined at an angle of 2° 29′ 33″ of arc to the ecliptic and the equator is inclined at an angle of 26° 45′ to Saturn's orbit.

Saturn moves about one thirtieth of the way around the sun, i.e., 12°, in an Earth year so that the Earth overtakes it about 12.85 days later each year. Because, as we have seen, Jupiter orbits the sun in 12 years its apparent eastward movement against the background of fixed stars is about 30° per year. It therefore gains about 18° on Saturn annually and gains a complete lap on Saturn every 20 years ($20 \times 18° = 360°$). The two planets therefore appear close together in the line of sight every 20 years and are then said to be in conjunction. A conjunction of Jupiter and Saturn occurred in 1961 and again in 1981. The latter was a triple conjunction such as was seen in 1940–1. If the Earth lies within about 30° of the line connecting Jupiter and Saturn in space and the line passes through the sun, a triple conjunction will be seen from the Earth, a rare event as it occurs only once in every 120 years.

On August 15th 1940 Jupiter and Saturn were in conjunction. The Earth overtook them and they seemed to move westward again against the star background. Jupiter's westward movement was faster and this caused the two planets to be in conjunction again on October 11th. When their eastward motion recommenced, Jupiter passed Saturn for the last time on February 20th 1941.
In the 1981 conjunction the first two conjunctions occurred close together in the early part of the year and the third came many months later.
As with Jupiter, the telescopic view of Saturn is not of a solid surface but the top of a cloud-laden atmosphere. Saturn displays light zones and dark cloud belts parallel to the equator somewhat similar to Jupiter's but they are not as clear or as well defined. This is partly because Saturn is much further from the Earth than Jupiter and partly because the belts and zones are actually less well defined than the cloud belts and zones of Jupiter. The cloud belts exhibit differential rotation so that different latitudes have different apparent rotation periods:

System I (near equator): 10 hours 15 minutes.

System II (higher latitudes): 10 hours 38 minutes.

System III (based on radio emissions from higher latitudes): 10 hours 39.4 minutes.

As with Jupiter, because of Saturn's fluid nature, various atmospheric features have rotation periods of their own and drift in longitude.

Saturn is best known for its spectacular system of rings, a relatively very thin disc-like structure in the plane of Saturn's equator and seemingly consisting of several bright rings that occupy most of the extent of the disc and some narrow dark gaps or divisions, the bright rings and dark divisions all mutually concentric, Saturn itself occupying the centre of the system. The planets Jupiter, Uranus and Neptune are now known to be encircled by rings also so that Saturn is not unique in this respect but its ring system is by far the most well developed and prominent. Earth-based telescopic observers of Saturn recognise three major rings in the Saturnian system. Proceeding outwards from Saturn there is first the least conspicuous of the three rings, the pale Crepe Ring or C ring. The inner border of this ring is about 8740 miles (14 070 km) above Saturn's cloud tops and with a width of 10 850 miles (17 400 km) is the second widest of the three rings. The C ring merges with the B ring at a point about 19 600 miles (31 400 km) above Saturn's cloud tops and there is no major gap separating them. The B ring is the brightest of the three and is 15 800 miles (25 300 km) wide which also makes it the widest of the rings. The B ring is separated from the outermost A ring by a prominent dark gap, the Cassini Division which is 2800 miles (4500 km) wide. The A ring, the second brightest of the three, is 9000 miles (14 400 km) wide. Towards the outer edge of the A ring is a dark gap much thinner than the Cassini Division, known as Encke's Division. Other delicate divisions have been seen in both the A and B rings. More ring divisions are revealed by studying light fluctuations in stellar occultations, e.g., of 28 Sagittarii, by the rings.

Other less prominent rings have been suspected from Earth-based observation such as the D ring which extends across the gap between the inner edge of the C ring and the globe of Saturn and an outer dusky ring beyond the A ring. The existence of these and other Saturnian rings have now been confirmed by spacecraft and these discoveries will be described in the later section dealing with spacecraft exploration of Saturn. Saturn and its rings are shown in fig. 10.2. The rings consist of innumerable solid bodies orbiting Saturn with sizes ranging from microscopic particles to bodies several kilometres in size. These are, of course, totally unresolvable by any Earth-based telescope and this with their enormous numbers gives the visual impression that the rings are continuous very thin solid structures.

As seen from the Earth, Saturn's rings appear to 'open' and 'close' slowly, a complete cycle taking about 29 years (fig. 10–3). This is because of the

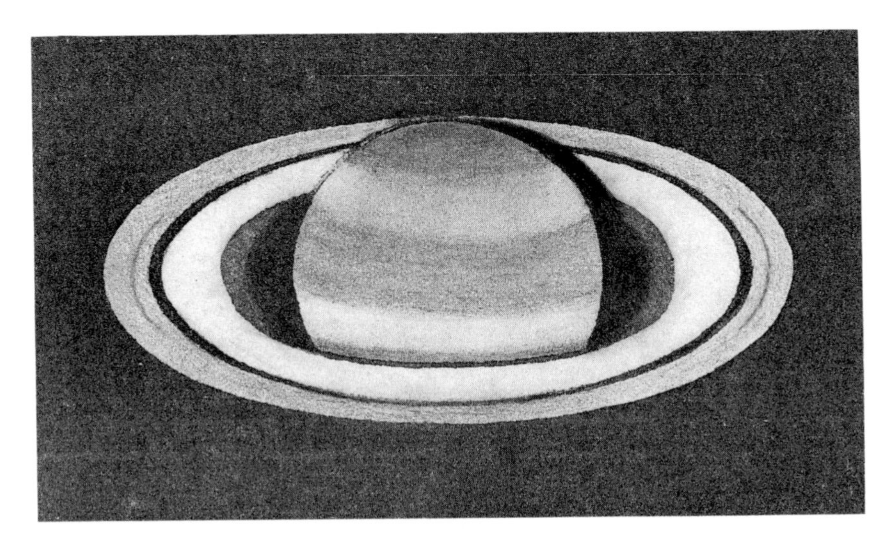

Fig. 10.2 Saturn showing its rings fully 'open' April 20th, 1974, 01.25 UT: 8-inch Newtonian reflector ×225. (F.W. Price.)

Fig. 10.3 Changing aspect of 'phases' of Saturn's rings.

inclination of the rings to the plane of the Earth's orbit so that we sometimes see the northern face of the rings and sometimes the southern face in between times, when the Earth passes through the plane of the rings, they are seen 'edge on' and become almost invisible which demonstrates how very thin they are as compared with their diameter. The time between two edge-on presentations of the rings or between a wide open northern face and a wide open southern face is a little less than 15 years. The rings lie exactly in the plane of Saturn's equator so that at the times when the rings are seen at their widest opening, the Earth and sun are 26° 45′ 'above' or 'below' their plane.

Saturn is accompanied by 17 satellites. Of these, nine are known from Earth-

Table 10.1. *Satellites of Saturn*

Name	Diameter		Mean Distance from Saturn		
	miles	kilometres	miles ($\times 10^3$)	kilometres ($\times 10^3$)	Mag
Atlas	23 × 21 × 17	37 × 34 × 27	85.56	137.67	18.1
Prometheus	92 × 62 × 42	148 × 100 × 68	86.72	139.53	16.5
Pandora	68 × 55 × 38	110 × 88 × 62	88.07	141.70	16.3
Janus	121 × 118 × 96	194 × 190 × 154	94.11	151.42	14.5
Epimethius	86 × 68 × 68	138 × 110 × 110	94.14	151.47	15.5
Mimas I	247	398	115.4	185.60	12.9
Enceladus II	309	498	148.0	238.10	11.8
Tethys III	650	1046	183.2	294.7	10.3
Telesto	19 × 16 × 10	30 × 26 × 16	183.2	294.7	?
Calypso	19 × 10 × 10	30 × 16 × 16	183.2	294.7	?
Dione IV	696	1120	234.6	377.5	10.4
Helene	22 × 21 × 17	36 × 34 × 28	235.0	378.1	?
Rhea V	950	1528	327.7	527.2	9.7
Titan VI	3201	5150	759.2	1221.6	8.3
Hyperion II	224 × 174 × 147	3650 × 280 × 236	921.7	1483.0	14.2
Iapetus VIII	892	1436	2212.6	3560.0	10–12
Phoebe IX	137	220	7824.7	12950.0	16.5

Data from The Atlas of the Solar System. Moore, P. A., Hunt, G., Nicolson, I, and Cattermole, P., Crescent Books, New York (1990).

based observation and the remaining eight were all discovered by the 1980 Voyager project. They are listed with pertinent data in table 10.1.

History of observation

The earliest known recorded observations of the planet Saturn come from Mesopotamia and go as far back as about 650 BC. They are a few primitive naked eye reports made by Babylonian astronomer-priests that were found in the library of Nineveh. Being in cuneiform script on clay tablets they survived the fire that destroyed the library when the Assyrian Empire was overthrown. They refer mostly to Saturn's position in the sky relative to other celestial bodies at the various times when the observations were made. One report about 650 BC states that 'Saturn entered the Moon' which probably refers to an occultation of Saturn by the moon.

In the *Syntaxis* of Ptolemy reference is made to a Babylonian observation of Saturn in an evening of the year 228 BC, most likely in January. Saturn is described as being at a distance corresponding to 2° from a star which is probably Eta Virginis. The *Syntaxis* deals with mathematical astronomy; among other things there is mentioned that the limiting celestial latitudes for Saturn are 3° 2′ north and 2° 59′ south of the Zodiac's central plane. Ptolemy had therefore a fairly accurate estimate of the inclination of Saturn's orbit to the ecliptic.

The celebrated Islamic astronomer commonly known as Albategnius made many astronomical observations between AD 877 and 918 and wrote a book that was translated by C. A. Nallino of Milan under the name *Opus Astronomicum*. In volume 2 of this work is a table indicating Saturn's maximum latitude north and south of the ecliptic as 3° 2′ and 3° 5′ respectively. The mean motion of Saturn in longitude in Roman years is shown as varying from 12° 13′ to 12° 16′ and a complete journey around the Zodiac as taking 29.5 years.
The thirteenth century astronomer Aben-Bagel expressed the view that Saturn shone mainly by light reflected from the sun which is correct. However, he mistakenly stated that Saturn is brightest at opposition when in Gemini–Cancer and faintest in Sagittarius–Capricornus–Aquarius.

In the sixteenth century Copernicus made several positional observations of Saturn and found many discrepancies between his own observations and those of Ptolemy. For example, he found that the mean motion of Saturn between the oppositions of 1514 and 1520 were 75° 39′ but the difference of place from his observations was 68° 1′. The mean motion between the 1520 and 1527

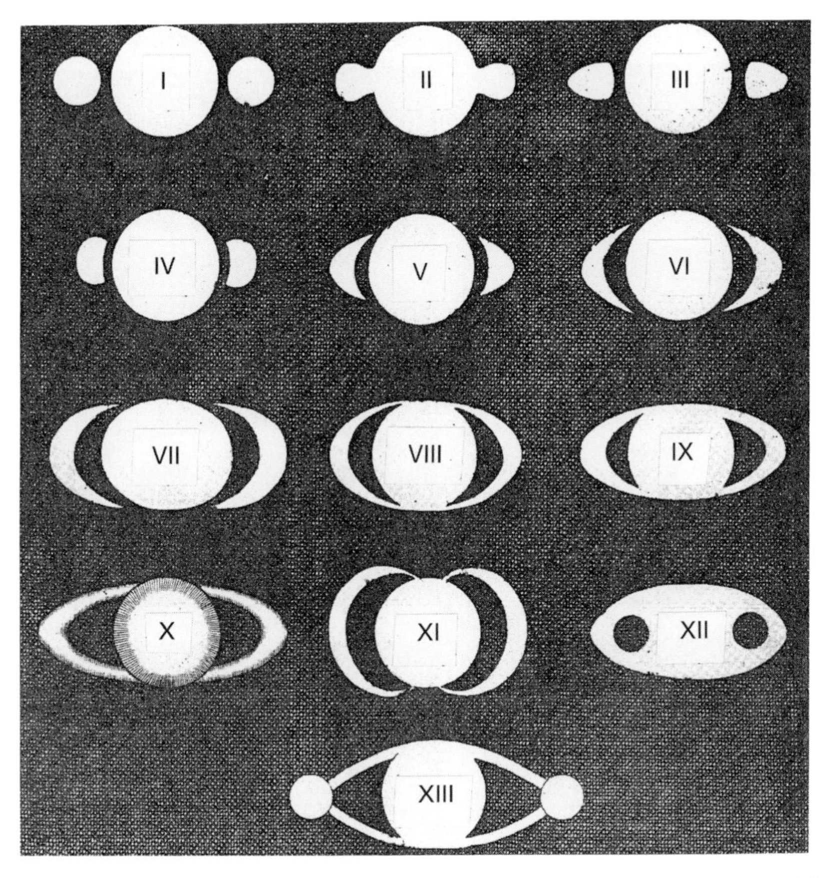

Fig. 10.4 Early drawings of Saturn: I. Galileo (1610); II. Scheiner (1614); III. Riccioli (1641 or 1643); IV–VII. Hevel (theoretical forms); VIII, IX. Riccioli (1648–50); X. Divini (1646–48); XI. Fontana (1636); XII. Biancani (1616), Gassendi (1638–9); XIII. Fontana and others at Rome (1644–5). (From Systema Saturnium, *Huyghens, 1659.)*

oppositions was 88° 29′ whereas the observed difference of place was 86° 42′. Copernicus attempted to reconcile these differences between theory and observation by entering into Ptolemaic mathematical discussions.

The rings and satellites

The first telescopic observation of Saturn was made on a night in July 1610. Using his largest telescope that gave a magnification of ×32 Galileo turned it on Saturn at a time when the planet's rings were slightly open and Saturn was nearing opposition. The imperfect lens revealed the disc of Saturn but could

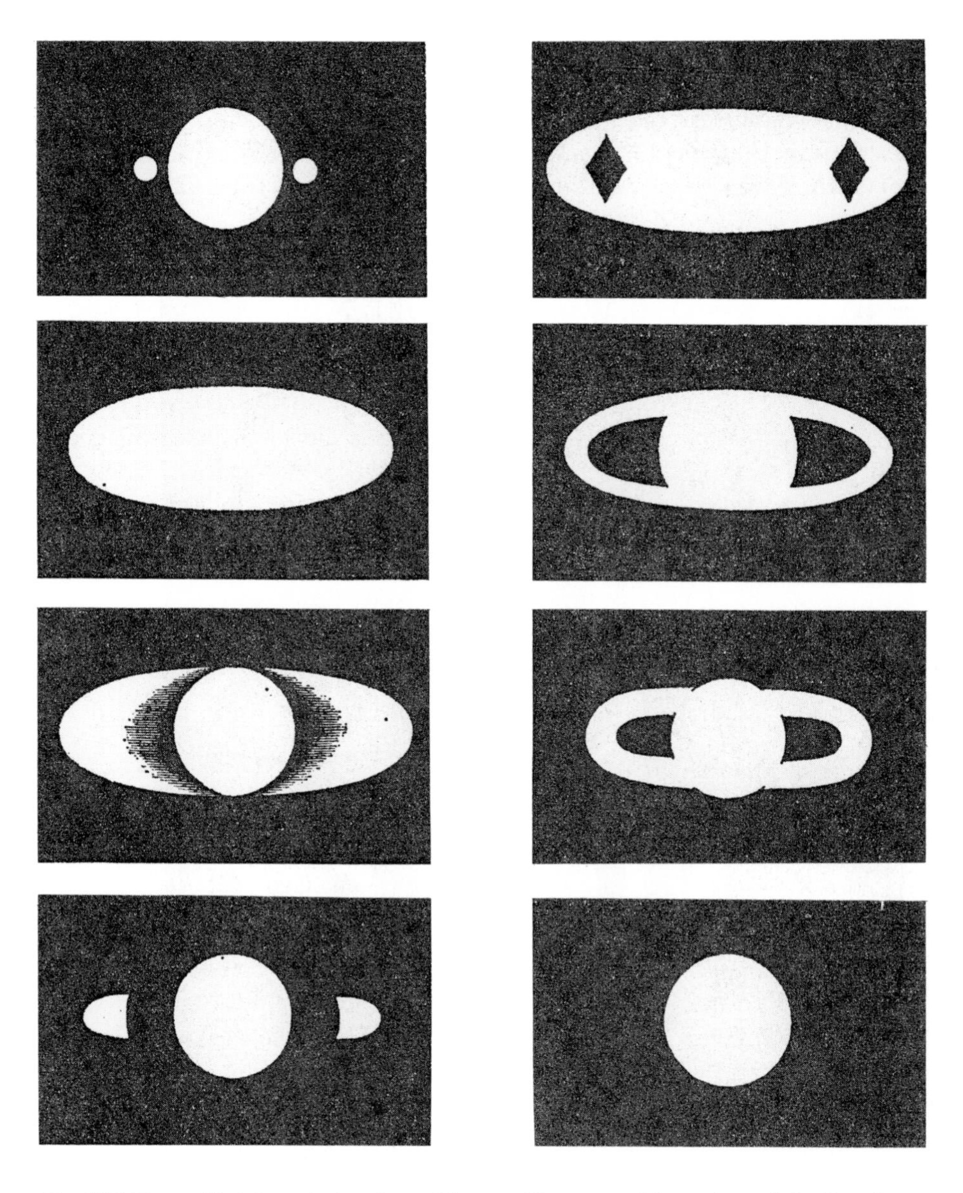

Fig. 10.5 Gassendi's telescopic drawings of Saturn. (From monthly notices of the Royal Astronomical Society **36**, *108–109.)*

only resolve the broad extremities of the nearly edge-on rings. It therefore looked to Galileo as if Saturn had two smaller attendants or satellites (fig. 10.4) but they did not move or vary in brightness. Galileo observed Saturn again two years later and the companions had gone (the rings were now in their edgewise presentation). Subsequently the two companions reappeared (as the rings opened out again) and took on the appearance of two handles attached to Saturn's globe as they opened to their maximum obliquity.

Galileo was soon followed by others in the telescopic study of Saturn; among these were Gassendi, Fontana, Riccioli and Hevelius. (fig. 10.4). According to Riccioli, Francesco Grimaldi was the first to mention, in 1650, Saturn's polar flattening. The Italian physician and telescope maker Eustachio Divini also observed Saturn and made drawings. Some of Gassendi's telescopic drawings of Saturn are shown in fig. 10.5. Many curious theories were put forward to explain the strange telescopic appearance of Saturn by such people as Hevel, Roberval, Fabri Hodierna and even Sir Christopher Wren, the British architect who was also keenly interested in astronomy.

The true explanation of the appearances of Saturn was finally put forward by a young astronomer, Christian Huyghens, when he was only 26 years old. He used a telescope made by himself and his brother that was optically superior to those of earlier observers. This telescope had a focal length of 10.5 feet but had only a small aperture and a power of ×50. Huyghens discovered Saturn's largest satellite Titan with this instrument and announced this in 1656 in a short pamphlet. A few months after this discovery he hit upon the solution of the 'Saturn Problem' and it was the discovery of the satellite that actually helped in revealing the cause of Saturn's curious telescopic appearances. He noted that Titan revolves around Saturn in an orbit in line with the 'handles' (ansae) and takes 16 days to make one revolution. By analogy with other planets and satellites of the Solar System he concluded that Saturn rotates once on its axis in

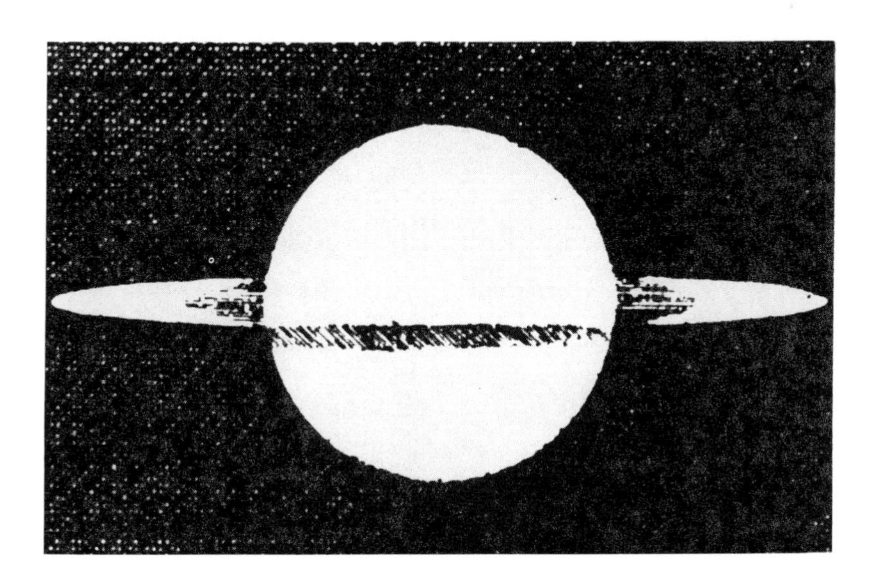

Fig. 10.6 Drawing of Saturn by Huyghens. (From Systema Saturnium, *Huyghens, 1650).*

less than 16 days. Since the ansae as well as the globe of Saturn were always symmetrical it seemed probable that Saturn must be surrounded by another symmetrical structure – a ring. Because the line of the ansae was tilted more than 20° to the plane of the ecliptic this would result in the varying appearances of the ring as seen from Earth – a wide ellipse, a narrow ellipse and sometimes seeming to disappear when presented edgewise, according to the amount of tilt at a given time. A drawing of Saturn by Huyghens is shown in fig. 10.6. The explanation of the 'phases' of Saturn's rings is shown in fig. 10.7. Huyghens also detected the shadow of Saturn's ring on the globe.

That the outer part of the ring of Saturn (ring A) was less bright than the inner part (ring B) was noted shortly after 1660 by two telescope makers, Campani and Divini, according to Antoniadi, who also says that around 1662

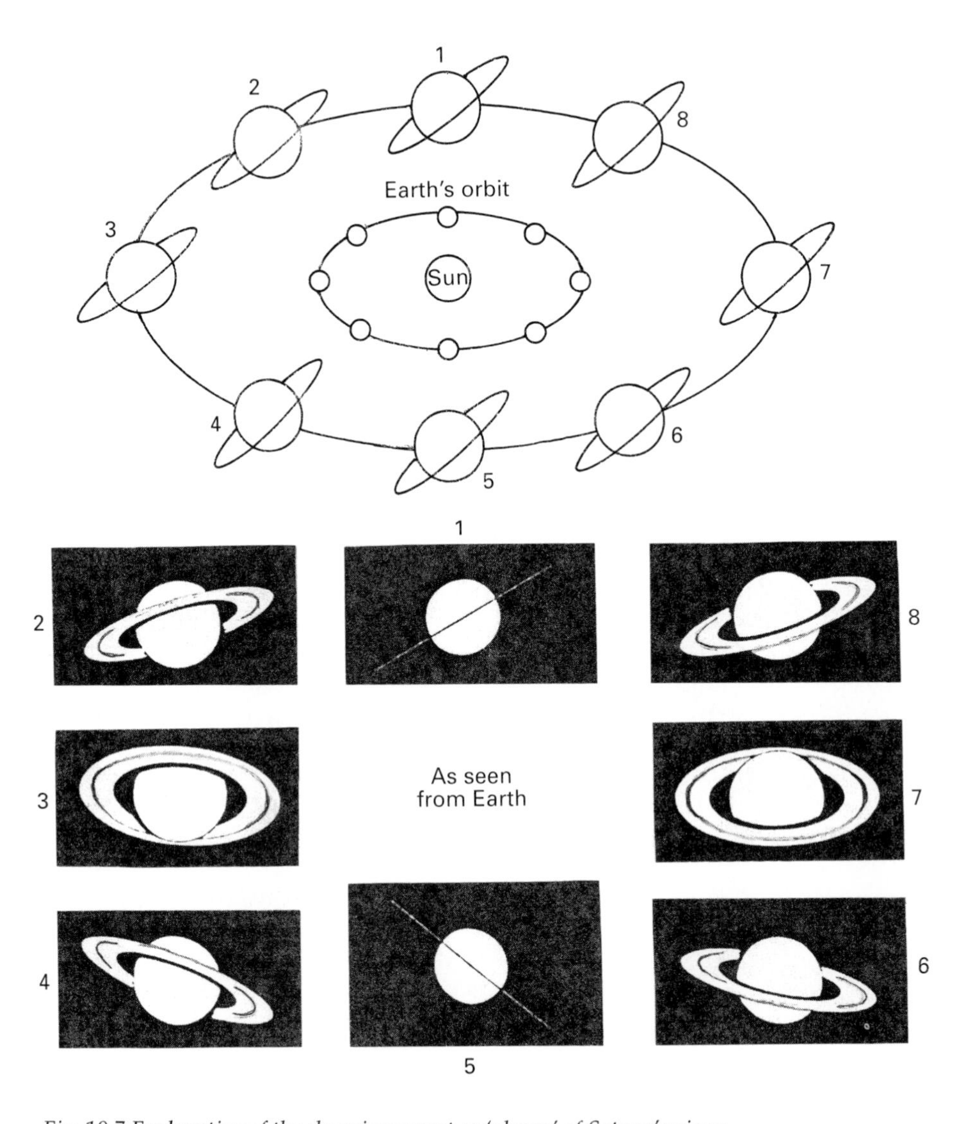

Fig. 10.7 Explanation of the changing aspect or 'phases' of Saturn's rings.

the shadow of the globe on the ring was discovered by Campani and Azout at about the same time.

On June 29th 1666 Robert Hooke made an interesting observation of Saturn using a telescope 60 feet long and he made a drawing of what he saw. The ring is widely open and a black line is shown at the edge of where the ring crosses the globe. Two hundred years later this was interpreted to be an early view of part of the crepe ring. Others appear to have had glimpses of it both before and after Hooke's observation without realising what it was.

In 1671 Cassini discovered a second satellite of Saturn, now known as Iapetus, using one of Campani's telescopes, and in 1672 he discovered the satellite Rhea. Then in 1675 Cassini found the space or division separating the two principal rings of Saturn that now bears the name 'Cassini's Division' (fig. 10.8). His drawing of 1676, considered to be the earliest published drawing of Saturn showing the gap between the rings, indicates that he considered the outer ring (A) to be narrower than the inner ring (B). The drawing also shows for the first time the south equatorial belt on the globe. Despite some discussion many years ago that William Ball may have seen the ring division ten years before Cassini, there is no doubt that Cassini was the true discoverer of the division. The idea that Ball first saw the division is completely wrong and arose from a misinterpretation of a comment made by an anonymous correspondent of Ball's about one of the latter's observational drawings of Saturn.

Two more satellites, Dione and Tethys, were discovered by Cassini in 1684 using aerial telescopes made by Campani. These two satellites are fainter than Rhea and closer to Saturn, closer, in fact, than our own moon is to the Earth. The four satellites discovered by Cassini were named from classical mythology at the suggestion of Sir John Herschel in the early 1800s.

Micrometric measurements of the dimensions of Saturn's rings were made by Bradley in 1719, the final estimates being smaller than the modern accepted values. Among his measurements, those coming closest to the modern values are: 43.7 seconds of arc for the outer ring diameter, 29.2 seconds of arc for the

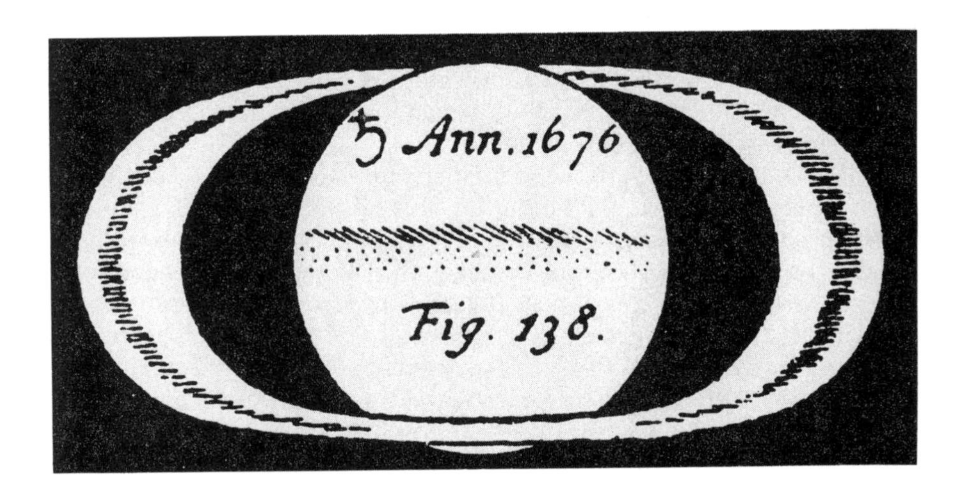

Fig. 10.8 J.D. Cassini's 1676 drawing of Saturn, the first to show the ring division named after him. (From Splendour of the Heavens, *Hutchinson, London, 1923 p.263.)*

inner ring diameter and 19.0 seconds of arc for the diameter of Saturn's globe.

Two more satellites of Saturn, nos 6 and 7, were discovered by Herschel in the late summer and early autumn of 1789 and were named Mimas and Enceladus. Their discovery was facilitated by the superior optical quality and light grasp of Herschel's telescopes and also because Saturn's rings were in the edgewise presentation so that the glare from Saturn and the rings, which would have tended to obscure the satellites, was less than usual. Herschel observed luminous points on the nearly edgewise ring in 1789 and was able to account for many but not all of them as being satellites. He also made estimates of the dimensions of the ring system which are in reasonably good agreement with modern values. His observations of the relative brightnesses of the rings and their colour agree perfectly with modern observations and photographs. He noticed, in 1806, the anomalous concave shape of the shadow of the globe on the rings. Subsequently, many observers have seen curious shapes of the shadow but it appears to be an optical illusion, the seeming narrowness of the shadow on the brightest parts of the rings being caused by an irradiation effect.

'Lucid spots' were also seen on the edgewise ring by W. and G. Bond in 1848, both of whom noted that they were immovable and so could not be satellites.

Peculiar appearances of the ansae were observed just after the edgewise presentation of the ring when it began to open out. Between 1848 and 1849, W. and G. Bond saw the ansae apparently broken up into many fragments with spots like satellites but brighter. They used the 15-inch Merz refractor at Harvard.

Using a much smaller telescope (a 6.3-inch refractor) Dawes, on June 30th 1848, saw the not quite invisible ansae to be of a 'deep coppery tinge' and he spotted a bright point on each arm which he knew could not be satellites. Like the Bonds, he saw, on December 5th and 20th, the easily visible ring apparently broken up into satellite-like points of light.

Between 1819 and 1831 there were many observations of elusive divisions in the rings in addition to Cassini's. Among these were sightings by H. Kater (6.25- and 6.75-inch reflectors) in 1825, J. F. Encke in 1837, who noted a black line on the north face of ring A which he tentatively identified as a division, which was also seen by De Vico in 1838 and by Schabe in 1841, Dawes and Lassell in 1843 (9-inch reflector) and Professor J. Challis at the University of Cambridge using the 11.5-inch Northumberland refractor, the same telescope he used in the search for Neptune in 1846.

Between 1849 and 1861 there were further reports of fine divisions in the rings by Dawes, Lassell, C. W. Tuttle, W. S. Jacob, Secchi and S. Coolidge. However, it now seems certain that the only truly permanent division in the ring system of Saturn is the Cassini Division. Of the others, Encke's is the one most often observed and it is almost constant in position, although it is not always visible. It was confirmed by J. Keeler of the Alleghany Observatory (Pittsburgh, Pennsylvania) in 1888 and is sometimes known as the Keeler Gap. The others do not appear to be true gaps with invariant positions but are most likely ripples on the surface of the rings that appear and disappear from time to time.

Yet another satellite, no. 8, and named Hyperion, was independently and almost simultaneously discovered by W. Bond at the Harvard Observatory and W. Lassell in England, using his 20-foot equatorial telescope, during September of 1848. It was Lassell who chose the name Hyperion. Hyperion was the only satellite discovery made by Bond whereas Lassell had two years previously discovered Neptune's principal satellite and two satellites of Uranus in 1851.

Astronomers of the seventeenth and eighteenth centuries had detected and recorded a faint ring of Saturn just internal to ring B and had seen its dusky projection onto the globe of Saturn. This faint inner ring was almost certainly seen by J. Galle (the astronomer who 'found' Neptune in 1846) in the ansae with the Berlin Observatory 9-inch refractor in 1839. The actual discovery of this third ring of Saturn was made in 1850 by W. and G. Bond in America at Harvard College Observatory with the Merz 15-inch refractor. Dawes independently discovered it with a much smaller telescope. Lassell christened the new-found third ring of Saturn the 'Crepe Ring' It is also known as ring C.

Later, by 1851, Dawes thought he had detected a division in ring C but the astronomers at Harvard do not appear to have seen it and only suspected a division. Dawes thought that he had also seen a division between rings B and C.

On March 26th 1863, J. Carpenter was observing Saturn with the transit telescope at the Royal Observatory, Greenwich. He noticed how unusually bright was the Crepe Ring. This he attributed to the particular angular relationship of the Earth and sun to Saturn's rings at the time which caused very oblique illumination of the Crepe ring resulting in unusually high reflectivity.

Many ideas had been put forward regarding the physical nature of Saturn's rings. The accepted explanation is that put forward by Clerk Maxwell in 1857, namely that they consist of countless numbers of small satellites, all of them circling the planet in their separate orbits. It was J. E. Keeler in 1895 at the Alleghany Observatory who proved by spectroscopic and photographic methods that Clerk Maxwell's particle or meteoric composition hypothesis of Saturn's rings was correct. He also determined the rotation velocities of the ring particles.

Daniel Kirkwood in 1857 predicted that owing to gravitational influences of the satellites of Saturn, perturbations of the orbits of the ring particles should give rise to gaps in the rings whose positions should be mathematically predictable. The orbital revolution periods of the satellites would also be an important factor in producing these gaps. The existence of the Cassini and Encke Divisions is, in fact, in accord with Kirkwood's ideas. The existence of others of these 'Kirkwood's Gaps' can be expected and at least provides observers with an incentive to search for further ring divisions and other phenomena suggestive of regions in the rings where the particles may be sparse.

Some curious appearances of the rings were noted by E. L. Trouvelot between 1873 and 1876. He saw a mottled appearance in rings A and B and peculiar 'dark angular forms' on the inner edge of ring B adjacent to the Cassini Division but the latter are attributable to atmospheric unsteadiness.

In England in the years 1887 and 1888, using an 8.5-inch reflector, T. G. Elger (best known as a lunar observer) saw Saturn's ring A a lavender-grey colour under the best seeing conditions with localised variations in intensity and uniformity of tone. The Encke Division was often seen as was a light line concentric with the ring about halfway between its outer and inner edges, usually a little nearer the outer edge. Elger, like previous observers, saw that the Cassini Division was never really black. In 1887, Elger noticed a characteristic gradation in the apparent brightness of ring B from the outer to the inner edge but this appearance was hardly noticeable the following year. Ring C was seen to have a scalloped or ragged appearance of its inner edge in the preceding ansa. Elger noticed an obvious notch in the shadow of the globe on the rings where it crossed the Cassini Division.

Dr F. Terby of Louvain, who observed with an 8-inch telescope, reported that on March 6th 1889 he observed a curious white spot on the rings adjacent to the globe shadow. It was seen by other observers throughout the world and became known as Terby's White Spot. It has been seen since by other observers and is now regarded as being merely a contrast effect (fig. 10.9).

Using large refractors of between 26 and 36 inches aperture, E. S. Holden, an experienced observer, studied Saturn between 1876 and 1889. In August 1880 he glimpsed a part of the Encke Division about one third of the distance from the outer to the inner edge of ring A. He had a good view of it again in October 1880, two fifths of the width of ring A from its outer edge. He saw ring B as if composed of three rings of different intensities.

During the same period Asaph Hall, using the 26-inch Washington refractor (the same telescope with which he discovered the satellites of Mars in 1877) was never able to see the Encke Division as a gap and saw no division between rings B and C and the shadow of the globe on the rings always looked normal to him.

Using a 6.5-inch Herschelian reflector C. Roberts observed Saturn in April 1896 and saw the Encke Division clearly in both ansae and plainly outside of the centre of ring A. Nothing unusual was seen on either ring A or B. However, in the same month, Antoniadi saw unusual markings on both rings. In ring A he saw very large white spots. A well-defined division and two fainter ones were seen in ring B.

Fig. 10.9 Terby's White Spot on Saturn's rings.

The discrepancies between the observations made by these two observers in the same month and other discrepant observations made close together were explained by E. W. Maunder as possibly being due to the particulate structure of the rings which might well be expected to undergo rapid changes in distribution and reflectivity over short time intervals.

C. Roberts saw the outer edge of the Cassini Division with an odd 'indefinite and indented' appearance and on May 8th 1896 he saw a dark division in ring B as had Antoniadi. Two triangular bright patches were also seen in the Crepe Ring of the preceding ansa. A faint marking was just seen on ring B on May 25th apparently corresponding to one of those seen by Antoniadi.

Further reports of new divisions in the rings occurred during 1897 by L. Brenner, J. M. Schaeberle (Lick Observatory) and P. Fauth (7.5-inch refractor) all of which seem doubtful but it is difficult to decide which were objective sightings and which were illusory.

According to Antoniadi, Flammarion stated in 1895 that he and others had clearly seen the rings of Saturn eccentric with respect to the globe, the eastern space between rings and globe always being wider than the western space. The following year this was verified by micrometric measurements. The effect was especially pronounced in 1900. The matter was investigated by P. Stroobant in 1934 after E. E. Barnard's investigation in 1894 led to a negative opinion.

The wide open rings were studied by Barnard in 1898 with the 40-inch Yerkes refractor. He only saw a dusky shading where the Encke Division is usually seen. The following year the Encke Division was plainly seen on both ansae by Flammarion on July 30th. A series of 'dusky indentations' was seen projecting from the Cassini Division, which appeared grey, over ring A the outer edge of which seemed to shade off gradually into the darkness.

Once again, in 1900, the Encke Division couldn't be found although the indentations from the Cassini Division, still looking grey, on ring A were well seen on several occasions. During 1900, no detail was seen in ring B, the shading gradations that Trouvelot had seen about two years previously apparently had vanished.

The ninth satellite of Saturn, Phoebe, was discovered photographically by Pickering in 1898. Soon after, the discovery of a tenth satellite, Thiemis, was claimed by Pickering but it seems to have disappeared before its existence could be verified. Possibly, one of the asteroids, apparently moving with Saturn, may have been in the telescopic field at the time.

During the edgewise presentation of the rings in 1907, bright 'knots' were seen by R. G. Aitken in the line of the rings on October 19th, apparently the same objects as Barnard's 'condensations' and closely simulating satellites, the inner ones being brighter than the outer. Aitken considered these to be irregularities on the ring surfaces. Their distribution was asymmetrical, the eastern knots appearing closer to the globe, which was in disagreement with the observations of the Bonds and Barnard, respectively in 1848 and 1907. W. W. Campbell saw four symmetrically arranged knots, two on the east side and two on the west, during the week of October 21st–28th. This was confirmed by Lowell at his private observatory at Flagstaff, Arizona. Subsequently Barnard made many observations until the ring reappeared, the results too numerous to detail here. Bond explained the bright 'knots' seen in the edgewise rings in 1848 as being due to the edges of rings A and B being seen through the Cassini

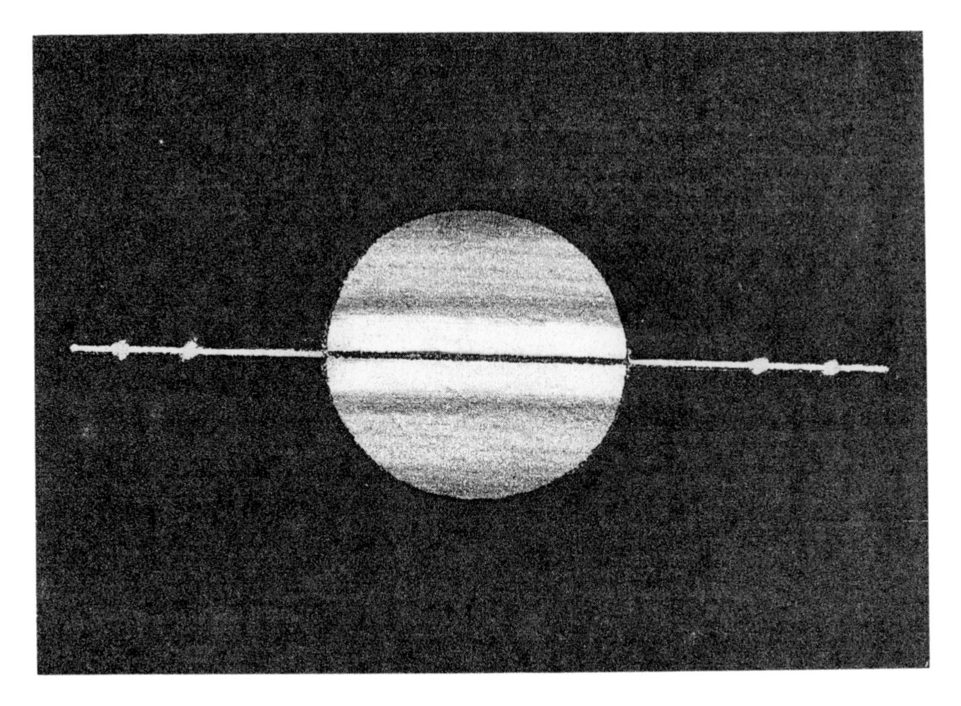

Fig. 10.10 Bright 'knots' in the edgewise rings.

Division but Barnard doubted this. 'Knots' in the edgewise rings are shown in fig. 10.10.

To explain the 'knots' Lowell considered that they are caused by elevated ridges which he called 'tores' (Lat. *torus* = a ridge), on the ring surfaces. He considered that rings B and C were not flat but were tores and assumed that the black core that ran through the middle of the shadow band, observed nowhere else but at Flagstaff, was the flat A ring's shadow bordered by the B and C ring particles above and below its plane. Although he believed this theory of tores to give a better explanation for the 'knots' than any other, there were still observers who disagreed.

During the edgewise presentation of the rings in 1907–8 the existence of a dusky ring outside the others was suggested by a strange appearance surrounding the outer bright ring seen by G. Fournier using an 11-inch refractor on September 5th and 7th in 1907. A little over a year later in October 1908, E. Schaer, who apparently had not heard Fournier's report, independently, announced the discovery of a dusky ring surrounding Saturn's bright rings. Schaer used a 16-inch Cassegrain reflector.

A careful study was made at the Greenwich Observatory in October 1908 with the 28-inch refractor, and on the night of October 10th the outer ring appeared to W. Bowyer to have a dusky appearance at its outer edge. He saw it again on October 15th as also did Eddington. Bowyer again saw traces of an outer dusky ring on October 22nd, 27th and 30th.

Barnard searched unsuccessfully for the outer dusky ring on two nights in January 1909 when Schaer claimed increased visibility of the dusky ring. Using the Yerkes 40-inch refractor, Barnard could see no trace of the ring despite careful scrutiny. Barnard was noted for his discovery of faint objects so that these negative observations even with the enormous light grasp of the Yerkes refractor were puzzling. Antoniadi argued against the existence of a dusky ring external to Saturn's bright rings and most astronomers became sceptical of its existence despite subsequent sightings by other observers with smaller telescopes. Rev T. E. R. Phillips, observing with a 12.25-inch Calver reflector, made an excellent series of observations of Saturn during 1910–11 but saw no trace of the dusky outer ring although he often searched for it.

During the last decade of the nineteenth century, the Saturn Section of the BAA had got off to a good start. Activity then declined owing to the years-long sojourn of Saturn in southern declinations thus making observation difficult in northern latitudes. After about 1910 the Section started active observing again with the return of Saturn to higher declinations. Maximum opening of the rings occurred around August 1912. In reports of the Section's work it was noted that the only ring divisions seen were Cassini's and Encke's. Many other interesting observations were made.

M. Maggini observed Saturn from Arcetri (near Florence) in October 1913 and saw a new subdivision in ring B near to the Cassini Division. On October 19th 1913 the Cassini Division appeared double on the north part of the ring and on October 25th the Encke Division was also seen to be double.

Over the years anomalous appearances of the shadow of Saturn's globe on the rings have often been seen. These were the 'double shadow' and the 'peaked shadow' or 'notched shadow' (Fig. 10.11). In 1914, C. O. Bartrum explained that the double shadow on the ring was not a shadow but merely the limb of the globe of Saturn at opposition which when projected onto the bright ring was apparently dark enough to be mistaken for a shadow in addition to the real one. He accounted for notched and peaked shadows as illusions caused by the blurred appearance of Saturn and its rings even under excellent seeing conditions. The effect could be reproduced by making a model of Saturn and its rings, lighting it and taking photographs with the camera lens out of focus by different amounts. The photographs he obtained showed close resemblances to the peaked shadow on Saturn's rings.

The peaked shadow was seen by G. and V. Fournier in 1913–14. From other observations they concluded that the rings of Saturn were unstable and that the Cassini Division was not entirely empty of particles.

On February 9th 1917 an historic observation was made by M. A. Ainslie and J. Knight. This was of the star BD +21° 1714 (seventh magnitude) as it passed behind ring A. Ainslee's telescope was a 9-inch reflector and Knight's was a 5-inch refractor. This observation proved the translucency of ring A and the emptiness of the Cassini Division, a great triumph for English Saturn observers. Also in 1917 W. H. Steavenson reported that the Encke Division appeared as a clearly defined line and was often seen eccentrically positioned in the A ring by observers using large telescopes.

The translucency of ring B was proved by an observation made on March 14th 1920 by W. Reid, D. G. McIntyre, C. L. O'B. Dutton and H. Reid at Rondebosch, near Cape Town, South Africa. They used a 6-inch refractor to watch the occultation of the star Lalande 20654 (magnitude 7.3) by Saturn and its rings. During the occultation the star was visible through ring B, its brightness showing variations as it passed behind the ring. Thus, the idea put

(a)

(b)

(c)

Fig. 10.11 Anomalous shapes of the shadow of the globe on Saturn's ring: (a) *wrongly curved shadow;* (b) *peaked shadows;* (c) *notched shadows.*

forward in the seventeenth century that Saturn's ring was a single solid structure was finally defeated.

During June–July of 1936, the ansae of Saturn's rings appeared to many observers to be of unequal length. This was observed near the time of passage of the Earth through the ring plane. Photography showed the preceding ansa to be longer than that on the following side. Antoniadi observed with a special reversing viewing system that would nullify optical illusion and saw that Saturn's rings extended to 2.3 disc radii to the west (preceding ansa) but were short by about 20% of their extent on the east (following ansa). Some observers also saw the preceding ansa looking brighter than the following ansa. Also in 1936 bright spots were seen on the ansae and in 1937 the rings were reported to be pale blue in colour. The Crepe Ring in the ansae was seen extending almost half way to the limb of Saturn and the Encke Division was plainly seen in the 1937–8 apparition.

During 1939–46, the War years, American amateurs kept up the work of observing Saturn as the BAA Saturn Section had practically stopped its work. Beginning in 1942 they often saw a rather broad dusky subdivision on ring B near the inner edge. They called it the 'Third Division'. At the same time they sometimes glimpsed a faint shading – the 'Fourth Division' – on the bright outer third of ring B near its inner edge.

Photographs of Saturn taken with the Lick 36-inch refractor in 1939 by H. M. Jeffers and in 1943 by G. H. Herbig showed no trace of the Encke Division although it was easily visible in 1939 in much smaller telescopes and there was no sign of the Crepe Ring in the ansae.

The BAA Saturn Section revived in 1946. In 1946–7 the Crepe Ring appeared unusually intense when seen against the globe while in the Autumn of 1947 it looked unusually thin and faint across the globe.

Focas noted the Encke Division to be variable in 1947; sometimes it was easily seen and at other times with equally good seeing it seemed to have disappeared.

In 1950 W. H. Steavenson observed Saturn with the 25-inch Newall refractor at Cambridge. Although at the time the Earth was near passage through the ring plane, on May 13th he noted a change in the relative brightness of the rings. The inner three fifths of ring A was now the brightest area in the rings. Ring B and the outer part of ring A were about equally bright. The brightness of ring A was confirmed by Professor R. O. Redman, Dr E. H. Linfoot and M. W. Ovenden on May 22nd. It seemed paler on May 28th, however.

During 1952–3 the outer part of ring B was the brightest in the ring system but it dimmed somewhat and then brightened during this period. The Crepe Ring was seen by many but the observation was not easy.

In the BAA Saturn Section Report for 1953–4, M. B. B. Heath recorded that many minor divisions in ring B and one between rings B and C were seen by the American observers D. P. Avigliano and T. A. Cragg with the 24-inch Lowell refractor and 60-inch Mount Wilson reflector respectively. Strangely, in 1954 on a night of practically perfect seeing, G. P. Kuiper saw only the Cassini Division using the 200-inch Mount Palomar reflector but none of these other fine divisions. He considered that the Encke and other fine divisions are only ripples and that 'division' is not strictly the word to apply to them. The only one true division – meaning a real gap – appears to be Cassini's.

During May/July of 1955 Ruggieri observed Saturn's rings using his own 10-inch reflector and the 20-inch refractor of the Merate Observatory. In moments of steady seeing he saw faint radial streaks on both of the ring ansae extending from the Cassini Division to the outer ring edge. Similar appearances had been previously reported by Trouvelot, Antoniadi (fig. 10–12), Rudaux and Maggini. Ruggieri concluded that the streaks were optical illusions as similar streaks can be seen on an illuminated disc with fine concentric streaks at a certain angle of tilt.

A curious phenomenon that is noticed from time to time is the so-called bicoloured aspect of the rings. This is when one of the ansae appears redder than the other. Work by Bartlett around 1954 indicates that it is an illusion caused by the shift of the eye from one ansa to the other. Ring A has a reddish tinge and the complementary blue after-image becomes superimposed on the other ansa when the eye shifts to it.

In 1954, in addition to the Cassini Division – which some observers saw and others didn't! – up to five or six other minor divisions including Encke's were seen by ALPO members. A majority of observers agreed that the outer portion of ring B was the brightest. Most also saw the Crepe Ring and ring D. Seven divisions in the rings were seen in 1955 by at least one observer, C. J. Smith who made a drawing of his observation.

The outer third of ring B was the brightest in 1956 as it was in the two previous years. A drawing by C. J. Smith shows notching in the outer edge of

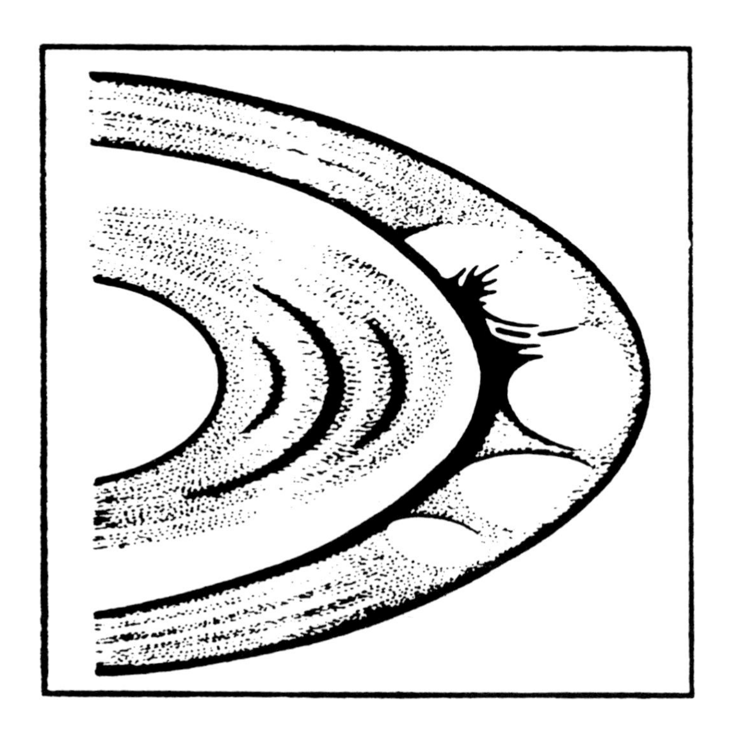

Fig. 10.12 Radial streaks seen by Antoniadi on Saturn's rings, April 18th, 1896. (From Antoniadi, E.M. JBAA **6**, *339, 1896.)*

the Cassini Division, somewhat similar to a drawing by Trouvelot in 1874. Six ring divisions were detected.

Terby's White Spot, which had not been reported for several years, was seen in 1960. It once appeared as a bright notch in the shadow of the globe of Saturn.

Detailed accounts of changes in the visibility and brightness of the rings and the number of ring divisions in each apparition are recorded in the Saturn Section Reports of the BAA and ALPO. To further summarise them all here would make tedious reading so only a brief account of the more interesting ring phenomena will be mentioned from 1960 up to 1988.

The Encke Division, or 'Complex' as it is often called, was seen to be double in 1961 and 1962. It was definitely photographed, apparently for the first time, in 1962 by Herbig, with the 120-inch reflector of the Mount Hamilton Observatory, on June 22nd 1962.

Another faint Terby White Spot was seen adjoining the shadow of the globe on the rings in 1970–1. In this same apparition, the Cassini Division was quite black but the Encke Division was greyish. Most observers agreed that the Crepe Ring had a distinct brownish-grey to reddish-brown colour during 1970–1.

In the 1976–7 apparition the Encke Division was seen best at the ansae and in large telescopes frequently appeared as a vague dark grey feature. The Crepe Ring was well seen and was of a grey to bluish-grey colour. The Terby White Spot was seen as a brilliant white spot on ring B adjacent to the shadow of Saturn on the rings. It was more conspicuous than in 1975–6 and was more frequently recorded. The Terby White Spot seemed to have increased in prominence in the years since, and prior to, 1974–5.

During the 1976–7 and previous apparitions the bicoloured aspect of the rings was confirmed by observation with different coloured Wratten filters in apparent contradiction to the earlier explanation of it in terms of the shift of the eyes from one ansa to the other and the superimposition of after-images of complementary colour onto the actual ring colour. This controversial phenomenon was remarked in observational reports and photographic investigations were called for to establish its objectivity or otherwise.

The Terby White Spot was seen in the 1977–8 apparition and was described as 'brilliant white'.

The rings were well closed up in 1978–9. The Cassini Division was dark greyish but it did not appear as dark as in 1977–8. The Encke Division was not reported although it was shown in at least one observational drawing. A bright white Terby Spot was seen again but it was not as obvious as in 1977–8. Suspected differences in the brightness of the ansae were reported by observers using red and blue Wratten colour filters.

During the edgewise presentation of the rings in 1979–80 two bright spots were seen on the rings about four fifths of the way out from Saturn's globe. These were probably due to light passing through the Cassini Division.

No Terby White Spot or bicoloured aspect of the rings was noted during the 1980–1 or the 1981–2 apparitions when the rings were starting to open again although A. W. Heath saw it in 1981 but this was not confirmed. Interestingly, the Crepe Band, where the Crepe Ring crosses over Saturn's globe, was darker and much more distinctly seen than during 1980–1. The Cassini Division at the ansae was of a very dark greyish-black colour but the Encke Division was not seen by most observers although it was shown in a drawing by Leo Aerts who observed it with a 6-inch refractor under fine seeing conditions on January 20th 1982 and G. Gambato observing from Venice (Italy) also saw it.

The Encke Division was again seen by G. Gambarto at the ansae during the 1982–3 apparition and also by G. Tallone. S. Daniels observing with an 8-inch reflector shows it in a drawing made on April 25th 1983. The Cassini Division was almost black. Curiously, widely different estimates of the brightness of ring A were sent in by different observers and its colour was variously described as greyish, yellowish or green. The Crepe Ring where it crossed the globe of Saturn was seen with difficulty. There were only a few reports of sightings of the Terby White Spot.

The Encke Division was not seen in the 1983–4 apparition and the Cassini Division at the ansae was of a greyish black colour. There were scattered sightings of the Terby White Spot. Although many observers carefully scrutinised the rings with colour filters only W. Haas reported a real difference in the brightness of the two ansae.

In 1984–5 the rings appeared brighter than in 1983 but colours were difficult to detect. The Cassini Division, which was black, was seen at the ansae by all observers, some of whom thought that it looked narrower than usual. Many were able to trace it all around the north face of the rings. The Encke Division was fleetingly glimpsed by some observers. There were infrequent reports of the Terby White Spot. The bicoloured aspect of the rings was not reported during this apparition.

In the 1985–6 apparition, many observers traced the Cassini Division all around the rings. The Division had a black colour. The Encke Division was occasionally seen at the ansae in good seeing conditions. A fairly bright Terby White Spot was seen by many observers. W. Haas was the only observer among several who saw the bicoloured aspect of the rings. There was no difference in intensity of the ansae when he viewed them in integrated light or with a red filter but with a blue filter the west ansa looked brighter than the east.

The Cassini Division appeared a dark greyish black colour in 1986–7 and many observers were able to trace it all around the ring. There were a few sightings of the Encke Division in favourable seeing conditions. Many observers saw a pale yellow-white Terby Spot; this still appears to be a contrast phenomenon and is not likely to be an integral part of the Saturnian ring system. Again, only W. Haas noticed the bicoloured aspect of the rings although several observers had looked for it.

The northern face of Saturn's rings was well placed for observation in the 1987–8 apparition. Only rarely was the Encke Division seen at the ansae even in favourable seeing conditions and the Cassini Division was traceable all the way around and had the dark greyish black colour that it had in 1986–7. Terby's Spot was seen by several observers as pale yellow-white in colour and somewhat bright. Yet again, W. Haas was the only observer to see the bicoloured aspect of the rings.

The globe

Abnormal shape of the globe. A curious phenomenon is the 'square-shouldered' appearance of the globe of Saturn seen by many observers at various times in which a peculiar flattened squarish shape of the planet is

Fig. 10.13 The 'square shouldered' aspect of Saturn. (From The Scientific Papers of Sir William Herschel, *volume 2, p.33, Royal Society and Royal Astronomical Society, London, 1912.)*

apparent (fig. 10.13). Herschel saw this on April 15th 1805 using a power of ×300 on a 10-foot focal length reflector and described it in a paper that he wrote. To him, the polar flattening was much more sudden than it is with Jupiter. After viewing Saturn with different telescopes and magnifications and making careful measurements he was sure that the globe was shaped like a rectangle with rounded corners. This appearance seemed to be due to protuberance of the globe at middle latitudes. Not surprisingly Herschel's observations in both 1805 and 1806 of this 'square shouldered' shape of Saturn stimulated a good deal of controversy among astronomers but his careful measurements of the globe and its peculiar shape plus his considerable reputation as an observer seemed to preclude the possibility of optical illusion. The Astronomer Royal, Sir George Airy, once thought that he saw the square-shouldered shape.

Proctor says that in 1803 Schröter saw that Saturn did not appear to have a perfectly elliptical outline. In the Autumn of 1818 W. Kitchiner observed Saturn with two different achromatic refractors and saw the peculiar square shape of the globe. On this occasion the rings were well closed up and therefore could not have contributed to any optical effect to give rise to the square-shaped appearance. Other observers in the year following saw some odd effects while others saw nothing unusual. Bessel made measurements of Saturn's globe when the rings had nearly disappeared during 1830–3 using the heliometer at Konigsberg. He convinced himself of the elliptical shape of the globe and derived a value of 1/10.2 for the polar compression.

In 1850 it was announced by the Council of the Royal Astronomical Society that as a result of careful micrometric measurements of the globe of Saturn made by Rev. Main at the Greenwich Observatory that there can be no doubt that Saturn's globe is an elliptic spheroid and that there cannot be any reason for supposing the planet to have an anomalous shape.

In late 1880 W. F. Denning observed Saturn with a 10-inch reflector and saw the curious square-shouldered appearance of the globe. He explained the effect as being due to contrast effects of the belts on the globe. The bright belts near the pole produce an obvious 'shouldering out' of the limbs at its ends whereas the dark belts at the pole and near the equator have an opposite effect and seem to compress the globe when their ends approach the limbs. He also saw the square-shouldered appearance on October 15th 1880, caused by large bright areas on both limbs in the South Temperate Region together with the darker shading of the limb northwards and southwards from the bright areas.

Cloud belts and spots. In 1852 Lassell, observing from Malta with his large reflector, described the two main equatorial cloud belts as having a reddish-brown colour becoming suddenly blue-green at the south edge of the more southerly of the belts. The lighter shading of the south pole was seen variegated with tiny stripes and lines.

A bright equatorial spot was seen by Asaph Hall on December 7th 1876 with an apparent angular diameter of 2–3 seconds of arc and it remained visible for several weeks afterwards. It finally extended as a streak on one side. Hall was able to calculate a rotation period for Saturn derived from 19 central meridian transit observations of the spot made by him and five other observers. These observations extended over 61 rotations of Saturn, a period of nearly 4 weeks. Hall's value for Saturn's rotation (assuming that the spot had no motion of its own) was 10 hours 14 minutes and 23.8 seconds, the probable error being 2.5 seconds. No further appearances of spots occurred until the 1890s.

In 1891 A. S. Williams detected some bright spots in Saturn's Equatorial Zone south of the equator. From the motions of four of these spots an approximate rotation period for Saturn of 10 hours 14 minutes and 21.84 seconds was deduced by Marth. W. F. Denning obtained a rotation period somewhat different from this by observing ten bright equatorial spots. The value he derived was 10 hours 14 minutes and 26.6 seconds. These spots were observed by Williams in 1892 and Marth derived a rotation period of 10 hours 13 minutes and 38.4 seconds.

The many dark spots of 1893 were located between 17° and 37° north latitude and Williams noticed that they occurred in pairs, one on each component of the double North Equatorial Belt, one of each pair lying practically due north of the other. Transit observations of the spots gave two groups of rotation periods, one of 10 hours 14 minutes 29.1 seconds and the other of 10 hours 15 minutes 0.7 seconds. Thus, atmospheric matter in the same latitude was rotating half a minute faster in one hemisphere than in the other, an effect found in Jupiter but so great and pronounced a difference was rarely found on the latter.

The bright equatorial spots of 1893 resembled those seen in Jupiter's Equatorial Zone. Williams found that some of these spots, at least, had been continually accelerating over the previous two years at a more rapid rate than noted for spots on Jupiter.

The double North Equatorial Belt was beset with dark spots in 1894. They

Fig. 10.14 Herschel's quintuple belt.

weren't seen as double as often as in previous years and some were seen projecting into the bright Equatorial Zone. Bright Equatorial spots similar to those of 1893 in appearance and location were seen. These were sometimes seen apparently cut into two by the narrow dark Equatorial Band. Interesting differences were again noted in the rotation periods of different groups of dark and bright spots.

Curiously, E. E. Barnard was unable to see the dark and light spots of 1891–4 even with the 36-inch refractor of the Lick Observatory, although he admits to seeing a very small dark spot at the north pole. Everything he could see with the aperture reduced he could see at full aperture – but none of the dark spots. This is difficult to reconcile with the lengthy detailed investigations of these spots by other observers.

In late 1880 when Saturn's southern hemisphere was turned Earthward, W. F. Denning observed Saturn with a 10-inch reflector. He saw white patches on the Equatorial Zone and something similar to the 'quintuple belt' seen by Herschel on November 11th 1793 (fig. 10–14); also several dark belts in the south polar region.

In 1884 and 1885 the English observer N. A. Green observed Saturn with an 18-inch reflector. He found the south pole area a dark grey colour. Nothing unusual was seen on the globe.

Another English observer T. G. Elger used an 8.5-inch Calver reflector to observe Saturn in 1887–8. As compared to appearances in 1887, the pattern of belts across the globe seemed to have changed. The South Temperate Region was seen overlaid by an unusually wide belt of a cinnamon colour. Tiny white spots occasionally mottled the wide bright Equatorial Zone. The grey-blue south polar cap seemed to vary in darkness from night to night and the north half of the Equatorial Zone seemed duller than the south half of the South Equatorial Belt which was the darkest. Its north edge seemed sharper and 'harder' than the south edge.

A series of observations of Saturn were made by E. S. Holden during 1879–89 with large refractors up to 36 inches of aperture (this being the Lick refractor).

In November 1879 he saw nearly the entire southern hemisphere to be of an olive green colour and dark. The remainder of the globe was light. There was a dark patch near the south pole. Rose-coloured tints were seen in the Equatorial Zone. A thin dark band across the Equatorial Zone was seen in 1882.

The Saturn Section of the BAA was started in 1891 and from that time made valuable contributions to Saturn studies. Rev. A. Freeman, Director of the Section from 1893 to 1895, observed Saturn under excellent conditions with a 6.5-inch reflector on March 12th 1892. He saw the South Equatorial Belt as a dark, wide and grey single belt. Both poles were dusky. The dark South Temperate Belt was thin and he saw a 'quintuple belt' in the north region made up of three dark North Temperate Belts with light zones between them. A. S. Williams, also using a 6.5-inch reflector essentially confirmed Freeman's observations. He saw two delicate South Temperate Belts and the two components of the South Equatorial Belt. The south part of the Equatorial Zone was very bright. The narrow dark Equatorial Band seemed to consist of bright and dark spots and patches. When the seeing was excellent, the whole of the north part of the Equatorial Zone was speckled with faint bright and dark mottlings. He also saw a 'quintuple belt' in the northern hemisphere but consisting of a North Equatorial Belt and two broad and prominent North Temperate Belts with light zones between them.

The equatorial region and northern hemisphere were well seen in 1894 and were observed by B. E. Cammell with a 12.5-inch Calver reflector. The south part of the Equatorial Zone was brighter than the north part, the two separated by the mottled Equatorial Belt. Again a 'quintuple belt' was seen north of this, made up of three belts and two bright zones almost exactly as A. S. Williams had seen it in 1892. The north polar cap was dusky and separated from the quintuple belt by a bright zone.

At the oppositions of 1894 and 1895 Barnard made further observations of Saturn with the Lick refractor but still did not see any light or dark spots.

In January 1895 C. Roberts observing with a 6.5-inch reflector had a view of Saturn similar to Cammell's. This time the Equatorial Belt was very dark and narrow without mottling and four belts were seen in the northern hemisphere. These appeared to be a North Equatorial Belt in two components and two North Temperate Belts northward of a wide light zone.

In 1896 and 1897 more dark spots were seen on the North Equatorial Belt by Flammarion and by Antoniadi who was working with him.

Saturn was difficult to observe during 1896–1900 owing to increasing south declination. Observational reports made in 1896 showed little change in Saturn's globe features since the previous year. Antoniadi noted the very dark north polar regions and the double North Equatorial Belt upon which observers had delineated many hazy spots in their observational drawings. Barnard saw the dusky spots on the North Equatorial Belt fairly easily in 1897 but saw only a few in 1898. He saw that the north polar cap was a dark blue colour in May 1897 and surrounded by a bright zone. There was no polar cap visible in 1898 but the polar regions looked dark. Observing with the 40-inch Yerkes refractor on July 7th 1898, Barnard enjoyed an unusually fine view of Saturn. He saw no less than five cloud belts in the northern hemisphere separated by light zones and a dark north polar cap. Like Flammarion he too found the north polar area not too dark in April 1899. Later in 1904 the north polar region was light yellow in colour.

Flammarion's and Antoniadi's observations of 1899 made with the Juvisy 10.25-inch refractor showed the north polar cap to be lighter than in 1895, the double nature of the North Equatorial Belt was distinct but not too much was seen of the Belt's dark spots, neither was the Equatorial Band seen.

The north polar cap was still not too dark in 1900. The double North Equatorial Belt was a prominent feature and the dark spots were again more easily visible. The Equatorial Band had reappeared and the bright Equatorial Zone was unevenly tinted and had a mottled appearance.

Observing in June 1903 with the 40-inch Yerkes refractor Barnard detected the first one of a series of white spots on Saturn; he had never seen such a feature on the planet during the previous ten years. The spot was located in the North Temperate Zone. Without knowing of 'Barnard's Spot' W. F. Denning saw another on July 1st using a 10-inch reflector. Comas Sola saw it on the same evening with a 6-inch refractor at Barcelona.

Several such light spots were subsequently seen all in the same latitude (36° north) but none as conspicuous as the first. K. Graff at Hamburg determined what seemed a rather long rotation period for Barnard's Spot of 10 hours 39.1 minutes but this was later confirmed by others.

Later observations of the other white spots and estimates of rotation periods were beset by identification difficulties. However, the most significant finding was the reliable establishment of the long rotation period of Barnard's Spot. From the observations of these spots A. S. Williams concluded that there was a great equatorial current on Saturn and that its velocity must be enormously greater than Jupiter's equatorial current. It indicates a greater fluidity of Saturn's surface materials than those on Jupiter.

In September 1909 Percival Lowell observed Saturn from the Flagstaff Observatory and saw what looked like faint lace-like markings lying diagonally across the Equatorial Zone. He photographed them on November 4th.

In 1909 Belopolsky photographed Saturn with the 30-inch refractor of the Pulkovo Observatory using two colour screens, an indigo-violet and a yellow-green. The results suggested that the globe of Saturn had an atmosphere but that the rings had none. This was confirmed spectroscopically.

In August and September 1910 T. E. R. Phillips saw that the south polar cap was the darkest feature on the globe. There was a series of light and dark belts on the globe that had a mottled appearance. The bright zone around the polar cap appeared especially bright near each limb.

Williams noticed an interesting difference between Saturn's and Jupiter's markings; white spots on Jupiter appear brighter when they are on the central meridian and become paler or vanish completely because of atmospheric absorption near the limbs. On Saturn, white spots in the South Temperate Zone are seen only with difficulty when on the central meridian but become brighter near the limbs.

From late September 1910 and onwards, considerable changes took place in globe features and by January 1911 the southern hemisphere had an approximately uniform dusky aspect.

The apparition of 1911–12 was observed by T. E. R. Phillips and H. Thomson.

Apart from a central white area in the south polar cap (also seen by Jarry-Desloges) which they both saw independently, there seems to have been hardly any difference in Saturn's appearance from the last apparition.

During the 1913–14 apparition the south polar cap appeared small and dusky and a dusky ring was seen surrounding the polar regions. As many as six belts were seen on the disc, some of them double. The South Equatorial Belt had a vaguely defined south edge and a north edge with irregularities, this edge being connected by wisps to the faint Equatorial Band.

Photographs were taken by R. W. Wood in 1915 using the Mount Wilson 60-inch reflector and filters transmitting infra-red, yellow, violet and ultraviolet. Mist or dust in Saturn's atmosphere was suggested by the results. The Equatorial Zone looked dark in violet light suggesting to Wood either that there is fine dust or a mist extending from the Crepe Ring to the globe or that there is a pale yellow gas, mist or dust in Saturn's atmosphere.

Among observations made of Saturn during 1917 were sightings of white spots and dark condensations in the South Equatorial Belt and the decline of the previously dark and prominent antarctic belt to the south edge of a south temperate shading.

During the 1920s visual observations were made by many observers but principally by Antoniadi with the 33-inch Meudon refractor. He noted changes in the dimensions, colour and overall appearance of the north polar cap during 1924–9. It had a grey colour in 1926 with a clear zone around it and south of this were two brown-coloured belts. One of these was the North Equatorial Belt and it showed dusky spots similar to those seen by A. S. Williams in 1891. Later in the decade Antoniadi saw complex white patches and irregular filaments on the equator and changing appearances of the tint and visibility of the North Equatorial Belt which on one occasion in 1928 was barely visible even in the Meudon telescope.

B. Lyot observed a rose tint on the North Equatorial Belt in June 1926, confirmed by Antoniadi who also noted a similar colour in the Equatorial Zone. Apparently Trouvelot had seen something similar in 1874.

In the summer of 1927 W. H. Wright published results of photography of Saturn in light of different colours. Marked limb darkening was noticed in light of long wavelengths especially near the north pole but the limbs were bright in ultraviolet. Mars and Jupiter both show this effect. Not much difference was noticed in the appearance of belts and zones in light of different colours. The ultraviolet photographs showed a dark wide belt overlying the Equatorial Zone.

Black patches were observed in the North Equatorial Belt by M. S. Butterton using the 9.3-inch refractor of the Dominion Observatory, Wellington, New Zealand. He calculated a rotation period of 10 hours 20 minutes for them.

The most exciting Saturn observation of 1933 was the discovery of a great white spot on August 3rd by W. T. Hay (the well-known comedian, Will Hay) with his 6-inch Cook refractor at Norbury in England. The oval-shaped spot was situated in the Equatorial Zone and was about one fifth of Saturn's diameter in length (fig. 10.15). During the next few days the spot lengthened in the preceding direction and extended over the whole width of the Equatorial Zone. At one time it began to spread into the North Equatorial Belt. The following end remained sharply defined for many weeks. By September 13th

Fig. 10.15 W.T. Hay's white spot of 1933. (From JBAA **44**, *1933 frontispiece).*

the spot had spread so much that it was not a spot but a streak. It remained brighter than the Equatorial Zone. Finally, it looked like nothing more than the ragged following end of a bright part of the Equatorial Zone. Other members of the BAA (B. M. Peek, W. H. Steavenson, T. E. R. Phillips, R. L. Waterfield) and many observers throughout the world made detailed observations of Hay's white spot. W. H. Wright was of the opinion that the spot was eruptive in nature. Photographs of the spot in light of different colours made by him at the Lick Observatory with the 36-inch Crossley reflector showed the spot to be approximately circular in shape in violet and in ultraviolet light but elongated in yellow light. Forty-eight hours after these photographs were taken the spot had lengthened as seen in red, yellow and violet light.

In 1936 Antoniadi observed interesting colour effects on Saturn's disc. On July 2nd the southern hemisphere looked grey but the northern hemisphere was more dusky with some admixture of orange. North of the equator the orange colouring in the Equatorial Zone was especially pronounced. In the reddish-brown North Equatorial Belt there was seen a dark brown spot within which were three dusky spots. On December 19th the southern hemisphere was cream-coloured and the northern purplish-orange. The Equatorial Zone south of the equator was a yellow colour and north of the equator it was purple-orange. The darkest parts of the disc were the polar regions.

The Saturn section of the BAA reported among other things that the North Equatorial Belt was darker, thinner and more well-defined than the South Equatorial Belt which was diffuse and in a higher latitude. A phenomenon seen on Jupiter but very rare on Saturn was observed by L. Andrenko at the Odessa Observatory, USSR. Using a 6.5-inch refractor on September 28th 1936 he saw a small bright white projection on the following limb, seemingly demarcated from the disc by a faint grey border. The sighting was confirmed by three other observers. Next day the projection was again seen but on the preceding limb and during the next four days was still visible. It was seen for the last time on October 1st and it looked much smaller than previously.

The BAA Saturn Section Report for 1939 mentioned that in 1937–8 the cloud belts were very faint and the bright Equatorial Zone had faded. In late 1939 the South Polar Region was dusky, the South Equatorial Belt was faint and broad (maybe double?), a well-defined dark line marking its northern edge.

During 1939–46, the War years, the BAA Saturn Section virtually stopped work and the War was not conducive to observation in Europe but American astronomers continued to observe during this period and Saturn was favourably placed in the heavens. American observers noted that the South Equatorial Belt was fainter than normal and its double nature was difficult to make out in 1942–3 but at other times it was prominent; generally the north component was the larger. W. Haas measured the transit times of two small 'bumps' on the north edge of the north component and deduced a rotation period of 10 hours 17 minutes 44 seconds. A belt at the edge of the south polar shading darkened and became prominent in 1943–4 and then faded. During 1943–6 delicate narrow belts, which at times were rather dark, were seen. One was near the equator and another in the South Temperate Region.

Wisp-like features were seen crossing obliquely over the Equatorial Zone in 1954 and where a festoon terminated on the Equatorial Band a concentration or knot was seen. Transitory light clouds were also seen in the Equatorial Zone. The two components of the North Equatorial Belt were not seen late in this apparition.

Occasional dark spots were seen in the Equatorial Zone in 1955, again where festoons crossed the Equatorial Band. In this year the North Equatorial Belt was the most active of the disc features. Both components were visible. Dark spots were frequently seen in the North Equatorial Belt and in a drawing made by C. J. Smith on June 5th using a 6-inch reflector, a very large dark spot is shown on the south component of the North Equatorial belt.

The rings were wide open in 1956 exposing the northern part of the disc. The Equatorial Zone was decreased in brilliance and there were far less light clouds seen. The Equatorial Band was less conspicuous than in 1955. The North Equatorial Belt continued to be the most prominent feature on the disc, seen double by some and as nearly single by others. Festoons crossing the Equatorial Zone were still seen. It is strange to note that Cruikshank did not see any such festoons during regular observations of Saturn during the 1958 and 1959 apparitions although he was using the Yerkes 40-inch refractor.

A South African member of the BAA Saturn Section, J. H. Botham, was observing Saturn on March 31st 1960 during the early hours. He had never seen any striking markings on the disc of Saturn although he regularly observed the planet, so it was with some amazement that, on this occasion, he saw a large brilliant white oval spot in the North North Temperate Zone. He noted that the spot took about 1 hour and 10 minutes to transit the central meridian. The spot had lengthened into a long strip on the preceding side by April 12th. On May 2nd Botham saw that the spot's outline was diffuse. Botham used 6-inch and 9-inch refractors for these observations.

Botham's spot was independently observed by A. Dolfuss with the Pic-du-Midi 24-inch refractor in April 1960. The last time he observed it was on May

26th, previous to which time it had spread itself out into the North North Temperate Zone, the latter becoming much brighter as a result. Between late August and early September 1960, several bright ovals were seen in the zone. The brightness of the North North Temperate Zone continued into the early part of the 1961 apparition. In a drawing made by J. Robinson on July 25th 1960 using a 16-inch reflector, two projections are shown on the south edge of the North Equatorial Belt with an Equatorial Zone bright oval between them.

Interestingly, Botham's White Spot occurred in much higher latitudes on the globe than earlier ones whose transit times had been measured. Although 20° or more further north than Barnard's and the other northern spots of 1903, its rotation period of 10 hours 39 minutes deduced by Botham was very similar to these others.

After the disappearance of Botham's Spot, another appeared in the North North Temperate Zone about equal in size and brilliance, attaining maximum prominence on August 25th 1960. Its rotation period was 10 hours 38.5 minutes.

Saturn appeared to be fairly quiescent in 1961. Nothing unusual lasting for any length of time was noticed. Some observers saw festoons between the Equatorial Band and the south edge of the North Equatorial Belt.

The ALPO reported that there was a rather limited amount of information available over the next few years and that Saturn had been poorly observed for a few years following 1966. Only a very limited report appeared of activity in the south part of the disc of Saturn during the 1970-1 apparition.

Atmospheric activity on Saturn appeared to increase during the 1976–7 apparition as compared to the previous one. Subtle brightenings, mottling and festoons were seen in the zones and dark spots were sometimes seen in the belts. A white spot was seen in the South South Temperate Zone in November 1976 and several dark spots on the north edge of the South Temperate Belt.

The degree of atmospheric activity during the 1977–8 apparition was about the same as it was in 1976–7. As usual the Equatorial Zone was the brightest zone on the planet. The Equatorial Band was seen only with difficulty.

Somewhat inconspicuous faint mottlings, brightenings and festoons were again seen at times in the more conspicuous zones and dark spots were again reported in the cloud belts.

During 1978–9 Saturn appeared not too different from the two previous apparitions. Again, intermittent brightenings, faint mottlings and faint festoons were detected in the zones. The North Equatorial Belt appeared single and the South Equatorial Belt was the widest of the cloud belts.

The ring system of Saturn was seen in the edgewise presentation during 1979–80. Again, comparatively little activity was seen as in 1977–8 and 1978–9. The South Equatorial Belt was still the widest of the cloud belts and was usually seen as single as well as the North Equatorial Belt and as in many previous apparitions, the Equatorial Zone was the brightest part of the globe. The Equatorial Band was not reported. Festoons, mottlings, inconspicuous brightenings again were seen in the northern zones.

In 1980–1 the northern hemisphere began to be advantageously presented as the rings began to open out again. Not much activity was seen in this hemisphere, only dark spots and elongations were seen on the south edge of the North Equatorial Belt late in the apparition. The mottlings, brightenings and festoons visible in previous apparitions seemed to have died away. The North Equatorial Belt again appeared single and the South Equatorial Belt was inconspicuous and quiescent.

Activity in the two hemispheres of Saturn was still fairly low during 1981–2. The North Equatorial Belt reappeared as two components. Very few belts and zones were seen in the southern hemisphere.

Dark and light spots were observed in the Equatorial Zone and the North Equatorial Belt during 1982–3 and a faint Equatorial Band was suspected. The Equatorial Zone was seen by some observers to be crossed by dark streaks. White ovals were also seen in the Zone. The North Equatorial Belt was conspicuous and split into two components.

The northern hemisphere was again fairly quiescent during 1983–4 as it seemed to have been since 1982. Vague whitish spots were seen in the North Tropical Zone. Short-lived mottlings and elusive dark features were seen in the double North Equatorial Belt. The Equatorial Zone had darkened since 1982–3. Practically nothing of the southern hemisphere was visible owing to the strong Earthward tilt of the northern hemisphere.

The low level of activity in the northern hemisphere observed in 1983 and during the few years prior to that year continued into the 1984–5 apparition. What elusive and poorly-defined detail was visible occurred mainly in the Equatorial Zone, North Equatorial Belt, North Tropical Zone and North Temperate Zone. The Equatorial Zone was somewhat brighter than in 1982–3. Vague white spots and festoons were seen here.

Intermittent activity in the northern hemisphere started up in the 1985–6 apparition. Occasionally discrete detail was visible in the North Tropical Zone, North Equatorial Belt and the Equatorial Zone. Elusive festoons and an ill-defined whitish spot were seen in the North Tropical Zone. Dark and light spots were seen in both components on the North Equatorial Belt and projections were seen on the edges of both the north and south components. Slight darkening of the Equatorial Zone was noted. A whitish oval was seen in the Equatorial Zone and some festoons and occasionally a very narrow Equatorial Band.

About the same level of activity was apparent in the northern hemisphere in 1986–7 as occurred in the previous apparition. Vague hints of whitish spots and festoons were suspected in the Equatorial Zone. The Equatorial Band was not often seen.

A slight increase in atmospheric activity was noticed in 1987–8. The rings were tilted to their maximum extent at this time, fully exposing the northern hemisphere. The Equatorial Zone appeared much the same as in 1986–7 with a few whitish spots and festoons and the Equatorial Band was not often seen. As in several previous apparitions the North Equatorial Belt was seen split into two components. Vague and diffuse features were observed within them as well as several discrete features.

During 1988–9 the northern hemisphere again showed only limited and sporadic activity with about the same frequency of occurrence of local phenomena as in the preceding apparition. The North Equatorial Belt was most often seen as a single belt although it appeared double at times. Many discrete details were seen in the north component. The south component was the darkest belt on the globe as it had been during several previous apparitions. Ill-

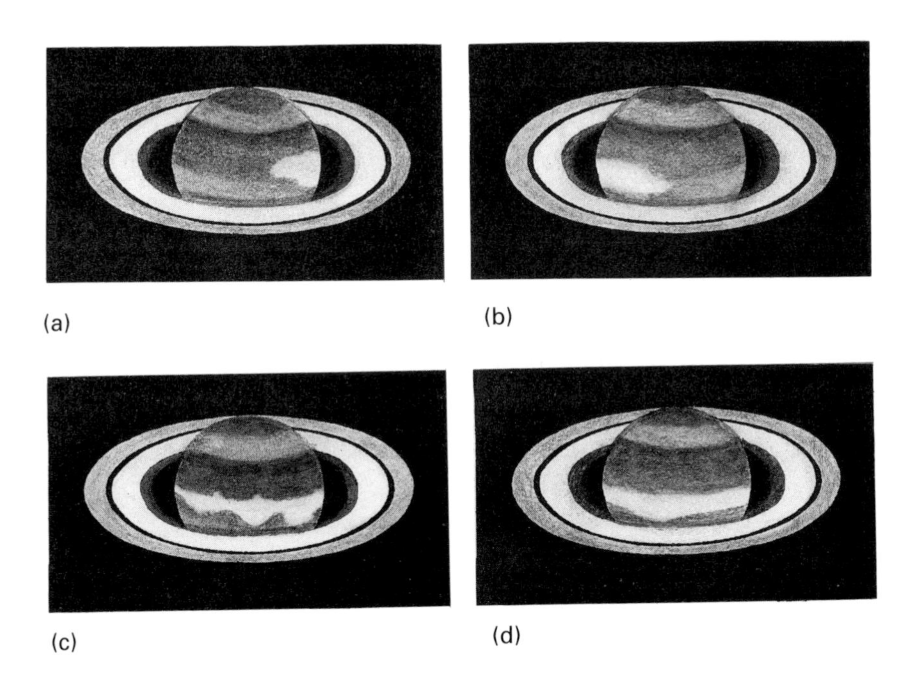

Fig. 10.16 The Great White Spot of 1990, from photographs: (a) *October 8th;* (b) *and* (c) *October 16th;* (d) *October 23rd.*

defined features were seen within it at times. Vague white spots and festoons were again seen in the Equatorial Zone and wispy festoons. The Equatorial Band was rarely seen.

Another 'Great White Spot' appeared on Saturn in 1990 (fig. 10.16). It was first seen on September 24th by A. Montalvo who was observing with a 12.5-inch reflector at Los Angeles. On the same evening the spot was seen by S. Wilber observing from Las Cruces, New Mexico. Wilber reported the discovery before the more cautious Montalvo and so the spot became known as Wilber's Spot. The spot appeared as a bright oval area nearly on Saturn's equator at a latitude of +4° and was easily visible in even small telescopes. It was much larger than the white spot of 1960. The spot remained oval in shape for about three weeks. In early October a 'tail' began to appear extending in an easterly direction.

By October 10th the spot was distorted and began stretching into a bright band around Saturn's equator and some of the bright spot material began to move in a northerly direction. By November the spot material had become so disperse as to give Saturn the appearance of having a bright but featureless Equatorial Zone.

As noted earlier large bright spots appeared on Saturn in 1876, 1903, 1933 and 1960, i.e., at roughly 30-year intervals. The 1990 spot was therefore not totally unexpected. All of these spots were located in the northern hemisphere during the Summer solstice on Saturn. Since Saturn takes 29.5 year to make one orbital revolution the roughly 30-year period between the spots suggests that their appearances may be a seasonal effect. Because Saturn's axis of rotation is tilted at an angle of 26.7° to its orbital plane the planet is subjected to a heating and cooling cycle as is the case with the Earth. Each Saturnian season lasts for almost 7 Earth years, however. In the Earth's northern hemisphere the hottest days occur not at the Summer solstice but several weeks later in August. This time lag is due to the Earth's atmospheric mantle behaving as an insulator which takes time to warm up. It is possible that a similar effect takes place on Saturn. The northern hemisphere of Saturn was 3 years past the Saturnian summer solstice in late 1990. It may be that the accumulation of heat in the lower layers of Saturn's atmosphere precipitates an upward burst of cloud material somewhat analagous to the high thunder clouds that ascend in the Earth's atmosphere on a hot summer day. The analogy may not be a good one, though, because Saturn has an internal heat source sufficient to account for these outbursts. Further, the correlation of this and previous white spots may not be as obvious as it appears; the 1876, 1933 and 1990 spots appeared at the equator and were large but the 1903 and 1960 spots were small and were situated nearer the north pole, so that there may be a 57 year cycle for the repetition of major eruptions in the equatorial region but the reason for it is unknown.

The preceding sections are only a summary of the phenomena observed on Saturn and its ring system over the years but even this should convince the reader that Saturn is a most interesting and challenging planet to study. More detailed information may be found in the Saturn Section Reports of the BAA and ALPO.

The revelations of spacecraft observations of Saturn will now be described.

Spacecraft exploration of Saturn

Saturn was first approached by the spacecraft Pioneer 11 in September 1979 and it came to within 13 000 miles (20 880 km) of the top of the planet's cloud-laden atmosphere. Pioneer 11 spent ten days measuring and photographing Saturn. It discovered two more rings outside of the main ring system, the F and G rings and another satellite, the eleventh, with an estimated diameter of 250 miles (400 km), close to the outer edge of the A ring.

The Voyager 1 spacecraft flew to within 77 000 miles (124 200 km) of Saturn's cloud tops on November 12th 1980. It made detailed observations of Saturn, its ring system and its several moons. It photographed six previously unknown moons of Saturn, some of which had been suspected but not confirmed.

On August 25th 1981, the Voyager 2 spacecraft approached to within 63 000 miles (109 300 km) of the cloud tops of Saturn which was about 14 000 miles (22 500 km) nearer than Voyager 1. The probe studied the ring system and several of Saturn's moons throughout the close encounter that lasted for many days. The more detailed studies of the rings included examination of the curious spoke-like features and the discovery of a narrow ring in the Encke Division. Better pictures were secured of the satellite Hyperion and of Saturn's atmosphere and a large 250 mile (400 km) crater on the moon Tethys. This is the largest crater yet observed in Saturn's system.

The Rings

Perhaps the most interesting and spectacular findings concerned the ring system. Before the Pioneer encounter there were only the three 'classical' rings known to observers – the outer A ring, the B ring inside this and separated from the A ring by the Cassini Division and the inner C or Crepe Ring. The amazing images of the ring system showed that there were literally thousands of rings and each of the known rings was itself composed of hundreds or thousands of rings (fig. 10.17). There were even faint rings discovered within the Cassini Division.

All of this has now made necessary new nomenclature of Saturn's rings and there are now seven rings based on the 'classical' nomenclature (fig. 10.18). The new discoveries are as follows:

(1) Just 2300 miles (3200 km) outside the outer edge of the A ring is the bright but delicate F ring.
(2) Much further out at a distance of 20700 miles (33120 km) beyond the A ring's outer edge is the narrow G ring.
(3) The extremely faint E ring is 55800 miles (145000 km) from Saturn's cloud tops and stretches out beyond the orbit of the moon Enceladus.
(4) Nearest to Saturn is the wide and extremely faint D ring extending outwards from the cloud tops of Saturn and merging with the inner edge of the Crepe Ring. The existence of this had been suspected from Earth-based observation.

Fig. 10.17 Saturn's rings.

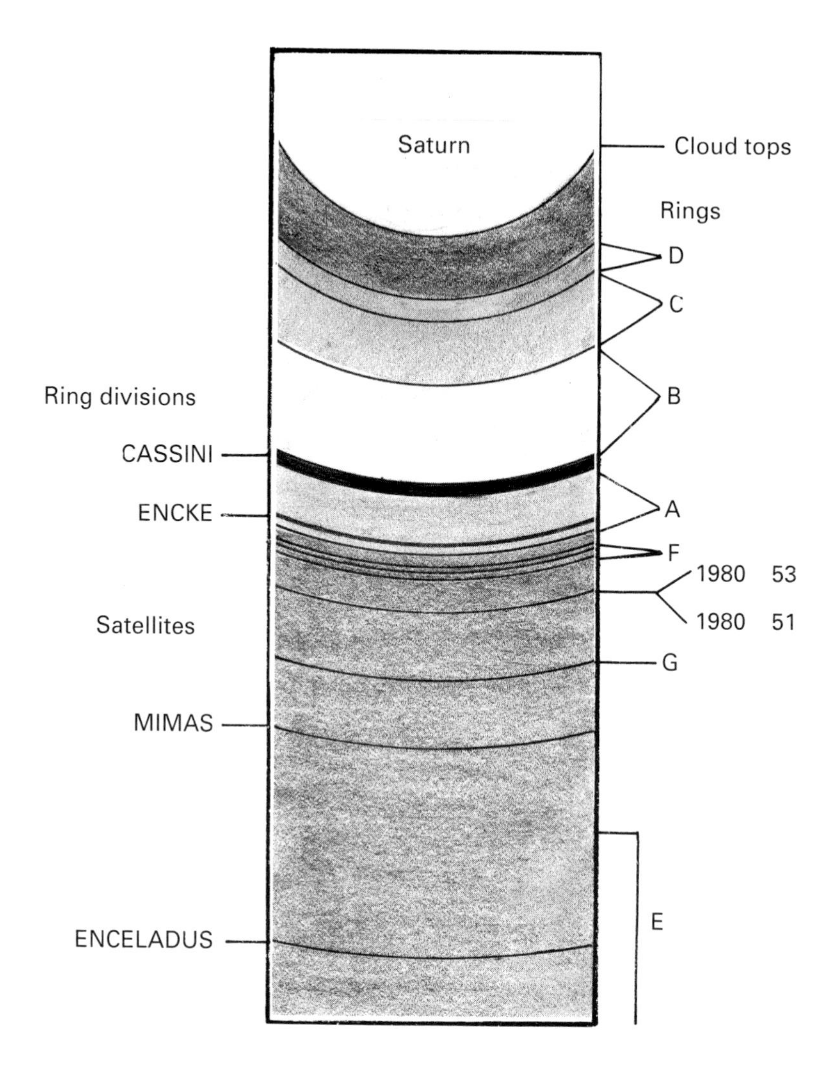

Fig. 10.18 New nomenclature of Saturn's rings.

Earth-based observers of Saturn always see the rings by reflected sunlight. The B ring, which is most reflective, always appears to be the brightest, the A ring is next brightest and the C ring is but slightly reflective and always appears pale. The Voyager spacecraft viewed Saturn's rings from many different angles relative to the sun, both by reflected light and by light transmitted through them and the appearance in both is quite different. The B ring was found to contain literally hundreds of fine bright and dark ringlets without arrangement into any recognisable pattern. Their widths went all the way down to the resolution limit of the cameras aboard Voyager. Computer-enhanced pictures make them look spectacular but most of the ringlets show no more than 10–30% variation from the mean brightness of the B ring so that they are difficult to differentiate.

The view obtained by Voyager from the side of the rings in shadow showed the B ring to be dark and therefore relatively opaque to the sun's light. Only a few bright ringlets shone through. However, the bright crescent of Saturn was seen shining through parts of the B ring by Voyager's cameras. A curious pattern of radial spokes, also seen from Earth, was seen on the B ring in one of the ansae and they rotated in the same direction as the ring. Apparently they overlaid both bright and dark ringlets and were 5–10% darker than the surface of the B ring. The spokes could not be seen from below the plane of the rings but became visible again from Saturn's far side as bright spokes. The spokes were situated at the same distance from Saturn at which ring particles and the magnetic field of Saturn rotate at the same speed. Subsequently it was found that the spokes rotate with the magnetic field and not with the ring particles so that the spokes may be made up of fine particles carrying electrical charges that become lifted outside the ring plane by the magnetic field of Saturn.

The widest gap in Saturn's ring system is the Cassini Division which separates the A and B rings. From the Earth it always looks black or dark grey. The inner edge of the division is in resonance with Saturn's satellite Mimas and the outer edge resonates with Iapetus. The division is about 2486 miles (4000 km) wide. From the sunlit side of the rings, the Cassini Division looks about as dark as the C ring and it has a bluish tinge as has also the C ring. From the dark side of the rings the division looks bright although it looks dark from beyond Saturn so there cannot be a high density of fine particles within it. The radio occultation experiment carried out by Voyager showed that there were 10-metre-sized lumps revolving within it. Much structure was detected within the Cassini Division by Voyager. There are four principal wide bright ringlets separated by narrow divisions. There may be 'beats' within resonances. There is one eccentric ring and maybe as many as 60 narrow dark features within the bright ringlets. There are dark bands at the perimeter of the A and B rings which appear to be devoid of ring particles as they appear dark under all conditions of illumination.

Contrasting with the B ring is the A ring which lacks the numerous fine ringlets caused by gravitational resonances. The Encke Division in this ring is a wide resonance gap. About 497 miles (800 km) beyond the A ring's outer edge revolves the satellite S15, whose presence appears to keep the outer edge of the ring well-defined and sharp. When lit by sunlight from behind, the outer edge of the A ring looks bright which suggests the presence of fine particulate matter.

Pioneer 11 discovered the narrow somewhat eccentric F ring which lies just outside the A ring. It appears bright by back lighting and so consists of fine particles. The ring has a satellite on both of its sides, S13 and S14, which are both quite small and function to keep the F ring within its orbit. Particles wandering either inside or outside the ring collide with one or other of these satellites which 'nudge' them back into orbit. Because these two satellites keep the ring particles on their path they are called 'sheep dogs' because they 'herd' the ring particles and keep them on course. High resolution Voyager photographs revealed that the F ring is not one but three intertwined narrow rings, 'braided' with kinks and twists.

The G ring is narrow and faint and lies less than 621 miles (1000 km) within the orbits of the satellites S10 and S11. It was detected by Voyager when its shadow passed across satellite S11 and by forward light scattering when Voyager receded from Saturn but it was not seen from beneath the rings or from the sun-facing side.

Far outside Saturn's other rings is the E ring which extends to a distance equal to eight Saturn radii from the planet. The ring was detected by Earth-based observation when the Earth passed through the plane of the rings in 1979–80. It is brightest at the orbit of Enceledas, the latter believed to be the source of the particles making up the E ring. The ring was detected by Voyager cameras as the spacecraft receded from Saturn.

Globe features

The globe of Saturn has always appeared paler and less detailed than Jupiter's and this is because the top of its cloud-laden atmosphere is partly hidden beneath a 50 mile (80 km) deep haze that reflects about half of the sun's light that impinges on it. The spacecraft encounters revealed much structure in the cloud tops beneath the haze layer such as red ovals and dark ovals. These had pointed ends and circulation of the anticyclonic type. Like Jupiter, Saturn has long-lasting oval features such as the orange oval found at a latitude of 55° south. This was observed for about 16 weeks. Long 'wave' formations and small eddyings were also revealed. At the lower latitudes there are fewer features but the higher latitudes have a cloud structure similar to that of Jupiter's.

There is a general lack of large-sized structure due to the dissipating effect of the enormously powerful winds on Saturn. The Voyager mission revealed that unlike Jupiter's winds, Saturn's blow in an easterly direction, the same as the direction of Saturn's rotation with a speed of 0.31 miles (500 metres) per second which is 1116.0 miles (1800 km) per hour.

The satellites of Saturn

In the following description the satellites are arranged in order of increasing distance from Saturn.

Mimas, with a diameter of 242 miles (380 km), is an icy satellite and its surface is marked by a ring structure 62 miles (100 km) in diameter with a 5.6 mile (9 km) central mountain peak on the side facing Saturn. The opposite side is covered with craters and the surface is cut through by a series of enormous trenches.

Voyager 1 did not observe Enceladus with sufficient resolution to show with certainty, but it seems to be fairly free of craters and all over the surface are faint linear markings. The surface is the brightest of all the satellites of Saturn. The orbit of Enceladus is located within the E ring's brightest part. Its orbit is eccentric and is in resonance with Dione.

Tethys, diameter 653 miles (1050 km), was not photographed at high resolution. It has a light and a dark side and is covered with craters. Near where the two types of terrain meet is a curious large feature that has been interpreted as a modified ring structure. The density of Tethys suggests that it consists of about 15% rock and 85% ice.

Dione has a diameter of 696 miles (1120 km) and resembles Jupiter's satellite Ganymede in that its dark trailing hemisphere is marked by light wisp-like markings that criss-cross the surface. It also has a very large ring structure and the bright areas are peppered with craters. Judging from its density Dione appears to consist of about one third rock.

Rhea was photographed by Voyager 1 in more detail than any of the other satellites. It has a brownish colour, areas with wisp-like markings and with light and dark terrain. The light and dark areas are absolutely replete with craters but the dark areas do not have as many large craters.

Titan is the largest of Saturn's satellites with a diameter of 3181 miles (5118 km) and can be considered as a planet in its own right. It is slightly smaller than Jupiter's satellite Ganymede and slightly larger than Callisto. Voyager's cameras revealed little about Titan as it is shrouded by a mist and clouds composed of white opaque liquid nitrogen droplets. It is the only Solar System satellite to have an atmosphere which it partly owes to its great distance from the sun; Jupiter's satellites Ganymede and Callisto although of about the same size as Titan, are much nearer the sun and have no atmosphere. The atmosphere of Titan is mostly nitrogen. To Voyager's cameras Titan appeared as an almost blank orange-coloured sphere surrounded by a thin haze layer about 124 miles (200 km) above the visible limb. The solid surface cannot be seen. The north pole was overlaid by a somewhat darker orange-coloured curtain of haze. The southern hemisphere appeared somewhat brighter than the northern, possibly due to the season which would be the beginning of 'Spring' in the southern hemisphere. Titan's density of 1.95 suggests that it is made up of about half silicate rocks and half ice.

Hyperion was photographed by Voyager from a considerable distance and only a tiny disc was revealed. It has a diameter of 193 miles (310 km), has an eccentric orbit and exhibits 'chaotic' rotation.

Iapetus, diameter 895 miles (1440 km), has a light-coloured trailing side and a dark leading side and this has been known for a long time. The contrast is quite extreme. The bright side seems to be made up of ice and reflects one half of the sun's light. The dark side is quite black and reflects only about 3% of the sunlight. Voyager 1 showed its surface to be sparsely cratered. A large ring structure was discovered close to the light and dark side border.

Some of Saturn's satellites such as the satellites designated S10, S11 and S12 are unusual in that they do not occupy separate orbits. S10 was photographed by Voyager in transit across Saturn's disc and it has an irregular shape. The diameter is 124.3 miles (200 km). S10 is somewhat larger than S11 whose dimensions are 43.5 × 83.9 miles (70 × 135 km) and of irregular shape with craters on its surface. The satellites S10 and S11 are peculiar in that they share the same orbit.

In 1990, about nine years after the Voyager 2 photographs of Saturn's rings were obtained, a new satellite orbiting within the Encke Division was discovered and tentatively designated 1981 S13. It was found on 11 of the images taken by Voyager 2, its discoverer Mark Showalter reported. The satellite is about 12.4 miles (20 km) in diameter and is similar to the ring particles in its neighbourhood.

Table 10.2. *Visibility and movements of Saturn*

	Days
Conjunction with the sun	0
Saturn appears in dawn sky	18
Retrograde motion starts	125
Opposition to sun	189
Retrograde motion ends	253
Disappears from night sky	360
Conjunction with sun	378

Visibility of Saturn

As an outer planet Saturn, like Jupiter, can be observed all night when at or near to opposition. The cycle of visibility is shown in Table 10.2. Opposition occurs about two weeks later each year. Saturn's declination limits are 26° on either side of the celestial equator.

Saturn's apparent brightness varies more than Jupiter's because at times the rings reflect most of the light that we see. Hence Saturn appears much brighter to the naked eye when the rings are fully open than when they are edge-on.

Observing Saturn

General

Of all the Solar System planets Saturn is perhaps the most intriguing because of its ring system, which although not unique in the strictest sense as we now know of rings around other planets, is unique in the sense of being the only planetary rings prominently visible with Earth-based telescopes. Any telescope used with as low a power as ×30 will reveal the rings. Even the casual observer on seeing the ringed planet in a small telescope for the first time is enthralled at the sight. The rings have an apparent width equal to Jupiter's apparent angular diameter.

The rings are easily visible in a 3-inch refractor. The Cassini Division is visible in such a telescope when the rings are fairly widely open.

The shadow of the globe of Saturn on the rings is quite easily seen in amateur telescopes. Sometimes this is the most easily observable feature of Saturn next to the rings themselves. It is hardly visible at opposition; 1–4 months on either side of opposition is the best time to see the shadow which is strikingly apparent at these times and gives the spectacle of Saturn and its rings a three-dimensional quality.

In the telescope Saturn shows a disc with only about one sixth of the area of Jupiter's disc when viewed with the same magnification. If the telescope used has an aperture of less than 3 inches there is little likelihood of seeing any features on Saturn's disc apart from the shadow of the rings.

For systematic work on Saturn and the recording of potentially valuable observations an aperture of about 4 to 5 inches must be regarded as the minimum. Small apertures can sometimes be more useful than large apertures when comparing visibility of projections on the north edge of the South Equatorial Belt. However, any telescope with an aperture of less than about 5 inches while giving beautiful views of the ring system do not have sufficient resolution to reveal significant detail.

With increasing aperture Saturn's bright equatorial region, the dusky equatorial bands and the dark polar regions become visible. Fig. 10.19 is a fine drawing by Ernst E. Both of Saturn showing these features as seen with an 8-inch refractor and fig. 10.20 is a Hubble Space telescope photograph of Saturn. Saturn has cloud belts similar to Jupiter's but they are more quiescent and show less contrast against the light zones. The nomenclature of Saturn's cloud belts and bright zones is shown in fig. 10.21. In telescopes less than 4 inches in aperture, the chances of seeing more than one or two cloud belts is slight. On rare occasions a light or dark spot may suddenly appear on Saturn's cloud tops.

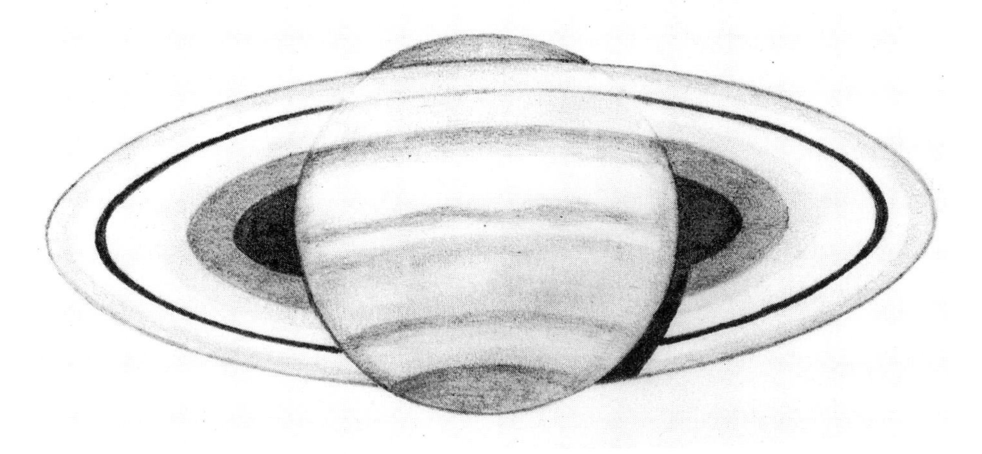

Fig. 10.19 Saturn, September 18th, 1962, 0.30 UT (E.E. Both, Kellogg Observatory, Buffalo Museum of Science.

Fig. 10.20 Saturn photographed by the Hubble Space Telescope's Wide Field/Planetary Camera.

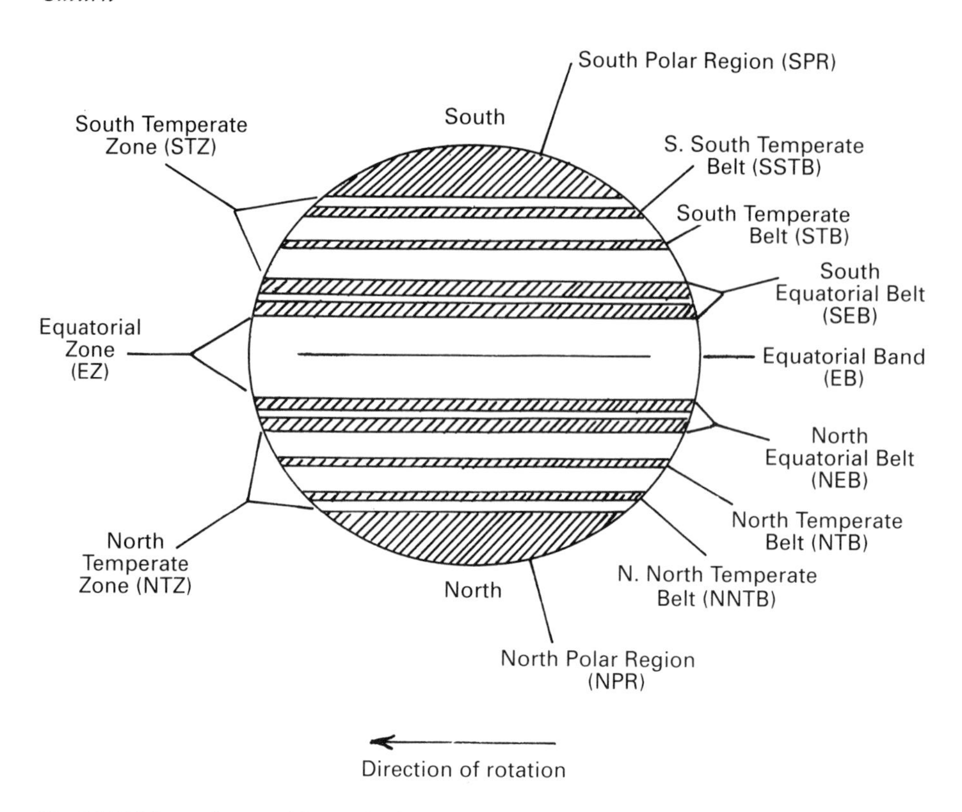

Fig. 10.21 Nomenclature of Saturn's cloud belts and zones. (Parts of the temperate zones adjacent to the North and South Equatorial Belts are sometimes called the North and South Tropical Zones respectively as in the case of Jupiter.)

It is possible to observe again a particular region on Saturn after the planet has made five, seven, twelve and fourteen rotations on its axis. The longitude of interest will be at the same position on the disc to within about three hours of the time when it was first observed.

A Saturn observing program should include the following observations. These should be done by every observer no matter how large or small the telescope used.

(1) General survey of the planet and ring system taking note of:
 (i) The shape of the globe shadow on the rings and the ring shadow on the globe.
 (ii) Changes in the usual shape or position of the belts on the globe.
 (iii) The position of any light or dark surface feature or ring feature that appears to rotate with the planet.
 (iv) Changes in visibility or outline of the Crepe Ring (ring C).
 (v) The location of any dusky markings or minor divisions – such as Encke's – in the ring ansae.
 (vi) Sightings, if any, of a faint ring beyond the principal bright rings.
 (vii) Any star that looks as if it will be occulted by the globe or rings.

(2) Drawings of the entire planet whenever the seeing is good enough.

(3) The relative intensities of the three main rings, the global belts and zones and shadows should be estimated and recorded.
(4) Note and describe any colour effects.
(5) Brightness estimates of the satellites.

Drawing Saturn

Observations of the globe and rings of Saturn are best recorded on a ready-made 'blank' drawing of the globe and rings with the Cassini Division already lightly indicated. At least one author has justifiably remarked that Saturn and its rings is an exasperating object to draw. Not only do we have to portray a series of ellipses – the globe of the planet and the rings – but also the slowly and continually changing aspect of the rings which is noticeable after the lapse of even a few weeks of observation.

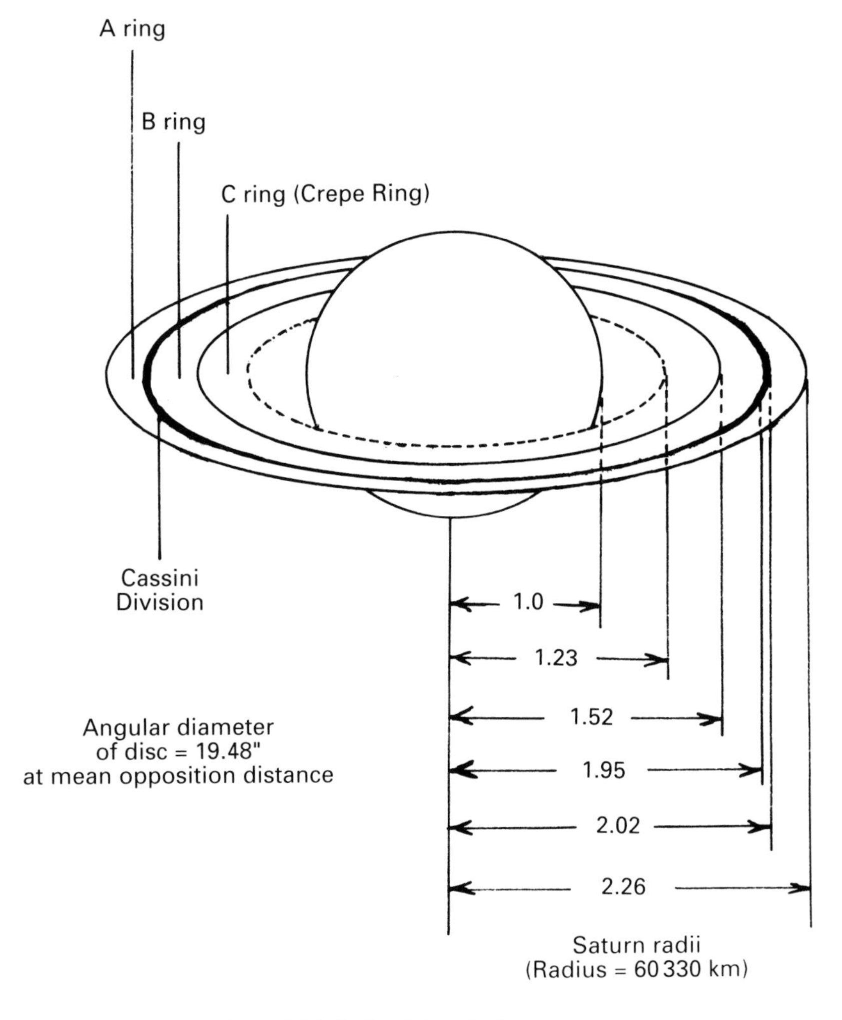

Fig. 10.22 Saturn – rings visible in Earth-based telescopes.

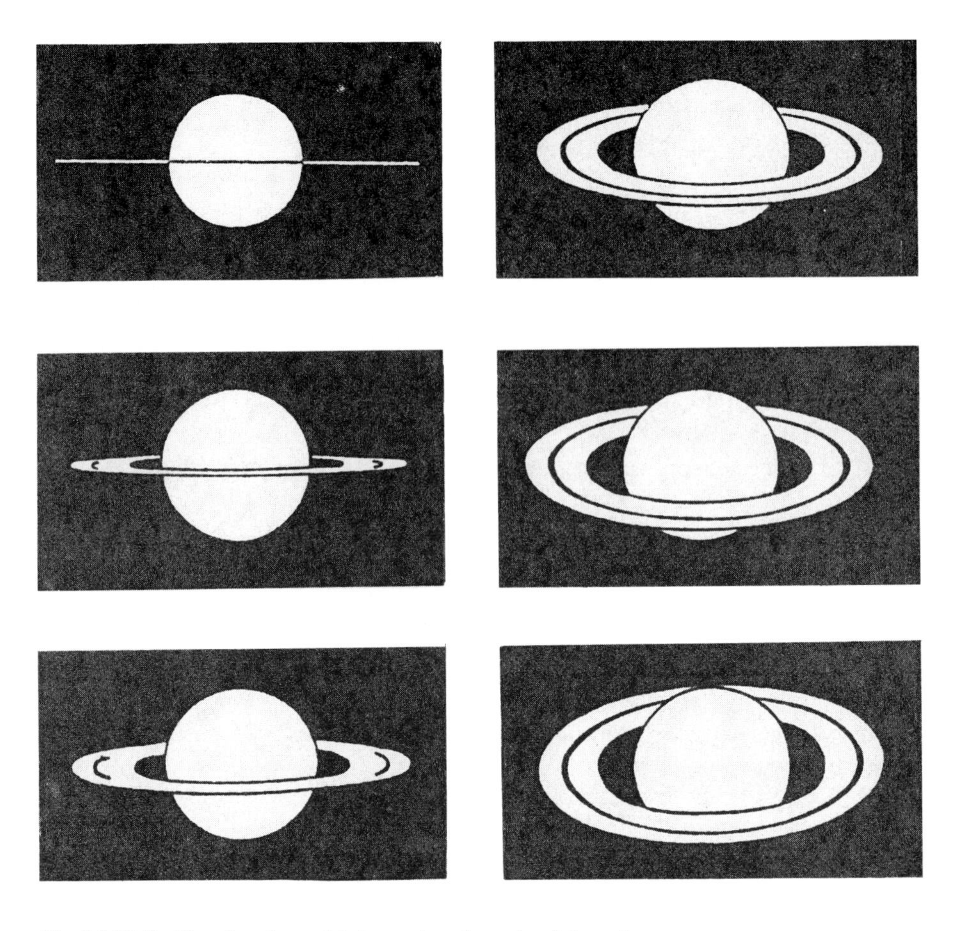

Fig. 10.23 Outline drawings of Saturn at various ring 'phases'.

By far the easiest way to draw Saturn and its rings is to use ellipse templates. For the globe an ellipse somewhat more oblate than Jupiter will be required but there is no need to be over precise. The real difficulty is posed by the rings. Not only do we need a set of templates for ellipses ranging from say 5° (rings nearly closed) to a little over 20° (rings wide open) but they have to be in four different sizes to mark the boundaries of the principal ring divisions – the outer and inner boundaries of ring A and the inner boundaries of rings B and C. Further, their sizes have to be in the correct proportions both with respect to themselves and to the globe of Saturn. These relative proportions are shown in fig. 10.22 in which the radius of Saturn is unity and the semimajor axes of the ring ellipses are expressed in terms of the Saturnian radius. Some difficulty may be experienced in getting a set of ellipse templates with just the right proportions but this need not be overwhelmingly so as each template usually has several sizes of ellipse for each degree of oblateness. You should aim to get the proportions fairly accurate but again there is no point in being overprecise. Alternatively you may prefer to make photocopies of the set of Saturn outline drawings in fig. 10.23 any one of which should be a sufficiently close approximation for practical purposes to the actual aspect of Saturn at any particular time. Individual outlines can be enlarged to whatever size you want if your photocopier has that facility.

Observations of the globe

As previously mentioned we don't see the surface of Saturn in the telescope but the top of its cloud-laden atmosphere. Because of pronounced different rotation rates at different latitudes, the cloud layers exhibit dark belts and light zones in a pattern essentially the same as Jupiter's. What little information we have indicates that rotation velocity on Saturn is more dependent on latitude than is the case with Jupiter. The equator rotates almost one hour faster than the belts and zones in higher latitudes (see table 10.3).

Saturn does not have as much detail and contrast to show as Jupiter and this makes reliable determinations of rotation rates more difficult to obtain than with Jupiter. One reason for the apparent lack of detail as mentioned earlier is the much greater distance separating us from Saturn so that Saturn presents a decidedly smaller disc than Jupiter. Other reasons are that there is much less thermal convective action than on Jupiter so that there is less activity and colour change and Saturn has a bright haze layer that reduces contrast.

The high latitudes on Saturn do not show much activity but in the equatorial and tropical latitudes large white spots and ovals are frequently seen. Then there are sometimes minor outbreaks of activity that are worth watching for and when they do occur are worth following and recording while they last.

Cloud belts and zones Saturn's cloud belts and zones are not as sharply defined as those of Jupiter. Moderate apertures will reveal the Equatorial Belts and perhaps a belt to their north or south.

The disc of Saturn is traversed across its middle by the bright almost white Equatorial Zone which is the most prominent of the zones and is visible in almost any telescope. Occasionally it can become abnormally dark as it did in 1964. Next in prominence are the North and South Tropical Zones which are usually difficult to spot in most amateur telescopes. The Temperate Zones are a little darker and the cloud belts, of course, are darker still. The zones are difficult to see because of the lack of contrast between them and the cloud belts have rather ill-defined edges that seem to merge imperceptibly into the bright

Table 10.3. *Differential rotation rates on Saturn*

Latitude (degrees)	Feature	Rotation time
±(0–10)	Equatorial zone	10hr 12 min to 10hr 16min
±(10–20)	Equatorial belts	10hr 15min to 10hr 20min
±(35–40)	Tropical zones	10hr 30min to 10hr 38min
±57	–	11hr 00min to 11hr 15min

Data from G. Roth, 1966

zones on either side. The two Equatorial Belts often have fluted borders on the sides adjacent to the Equatorial Zone. Thin belts can frequently be detected in higher latitudes but these belts are usually featureless. The polar regions are normally dark but sometimes will become bright. In 1963 the South Polar Region brightened and almost equalled the Equatorial Zone in brightness. Most activity on Saturn takes place in the equatorial region.

The intensity of the dark cloud belts can change unpredictably and so it is important to note and record the darkness of the belts. Sometimes a belt may appear where you have seen none before. This should be carefully delineated in a drawing to show its approximate latitude.

Look for thin festoons, which are sometimes visible in amateur telescopes, mostly bridging the North and South Equatorial Belts, but they are not as common or as prominent as those on Jupiter. They are frequently seen springing from the North and South Equatorial Belts in the direction of higher latitudes. Large festoons are seen quite frequently and coincidentally at times of white spot activity. These should be observed and recorded as carefully as possible and their central meridian transits timed.

The determination of the longitudes of features on Saturn's disc is done in the same way as for Jovian features. The exact timing of transits of features across the central meridian is even more important because of the comparative scarcity of long-lived prominent features on Saturn. As with Jupiter, there are two rotation zones employed for reducing central meridian transits to longitude:

System I (Equatorial Zone and Equatorial Belts).

Sidereal rotation period = 10 hours 14 minutes 1.08 seconds.

Sidereal rotation rate = 844 degrees per day.

System II (remainder of planet).

Sidereal rotation period = 10 hours 38 minutes 25.4 seconds.

Sidereal rotation rate = 812 degrees per day.

The ALPO publishes a table annually giving the central meridian longitudes of Systems I and II daily at 00.00 UT. Included are corrections for phase, light time and the Earth's Saturnicentric longitude.

Central meridian timings should be made for spots of any kind, dark or light; the edges of the Equatorial Belts are especially productive of spots.

Equatorial white spots and ovals White spots do not appear often and it is their rarity that makes searching for them a worthwhile activity. Their discovery enables you to see for yourself eruptions of thermal origin, the study of which may one day result in discovery of the mechanism and nature of the processes going on inside the 'gas giant' planets of the Solar System. They also make possible determinations in the changes of rotation of the equatorial current.

Useful observations of equatorial white spots can be undertaken with even a small telescope but regular observation of the equatorial and tropical regions is necessary to ensure a reasonable chance of discovering them. They are usually indistinct but have just enough prominence to reveal themselves against the bright Equatorial Zone. If you see a spot or oval, proceed as follows:

(1) Record the date and time (UT) when you first saw the spot.
(2) To avoid unconscious 'wishful seeing', contact another observer who may be able to verify your observation.
(3) Make a drawing showing the position of the spot on the disc of Saturn.
(4) Attempt a central meridian transit timing of the spot to the nearest minute. Do this for the preceding and following ends as well as for the centre of the spot. An approximate estimate of the dimensions of the spot will thus be given, also data that are usable in determining the rotation period.
(5) Assuming that the spot will transit the central meridian again after about 10 hours 14 minutes, estimate when you will be able to see another central meridian transit. When this time arrives, observe and time the transit and again record your data as obtained in steps (1)–(4). Determine whether there appears to be any drift from the expected transit time.
(6) Copies of your observations should be forwarded to:

IAU Telegrams
Smithsonian Observatory,
60, Garden Street,
Cambridge MA 02138.

Lunar and Planetary Laboratory.
University of Arizona,
Tucson AZ 85721.

or the Saturn Sections of the ALPO or BAA.

Ring shadow on the globe Note carefully the extent and boundaries of the shadow of the rings on the globe of Saturn. That part of the ring system which is between us and Saturn casts a shadow on the globe which may be seen on either the equatorial or polar side of the rings. When on the equatorial side it looks rather like one of the dark belts on the planet. Be careful not to be deceived into thinking that it is a belt; it has often been reported as such by unwary observers.

Colour on the globe Generally, the globe of Saturn has a dull yellow colour but occasionally it may show some colour changes. Filters should be regularly used in observing the globe for colour. The Wratten 25 (red) and 44A (blue) are recommended. If you make frequent records of intensity estimates of the belts and zones with the red and blue filters and also without them, different relative intensities may be noticed and this indicates different tints. The zones and belts may also appear to have different widths depending on the colour of light in which you observe them.

Record any unusual colour effects and endeavour to scrutinise the globe for colour phenomena every time that you observe Saturn. By so doing you will increase your chances of spotting white ovals. Details of colour changes observed should be sent for evaluation to the BAA or ALPO Saturn recorders.

Latitude determinations The Saturnicentric latitude of any globe feature and of the belts and zones may be determined in the same way as for similar features on Jupiter. Moderately large telescopes are desirable for this work. The Saturn Section of the BAA recommends especially the regular latitude determination of the following: (latitude measurements of any unusual features such as dark or light spots are also needed).

Central part of the Equatorial Zone.

The north and south edges of the North and South Equatorial Belts and North and South Temperate Belts.

Edges of the north and south polar bands.

Edges of the North and South Polar Regions.

Although rather indefinite and difficult to locate exactly, an attempt should be made to measure the latitude of the middle of the C ring against the globe.

Attempts to measure even careful drawings do not give as reliably accurate results as measurement of the telescopic image with an eyepiece micrometer or reticle. However, improved accuracy is possible if the measurements and drawings of several observers are used so that mean values and 'smoothing out' of errors may be obtained. This underlines the value of the pooling of results from the Saturn Sections of the BAA and ALPO.

J. B. Sidgwick gives the following procedure for latitude determination: Let h = angular distance of the feature north (+) or south (−) of the centre of the disc. Observation gives the quantities a and b, the distances of the feature from the two ends of the central meridian so that

$$h = \frac{(a - b)}{2}$$

r = polar semidiameter (from the *Astronomical Almanac*),

B = Saturnicentric latitude of the Earth (from the *Astronomical Almanac*)

B' is such that $\tan B' = 1.12 \tan B$

B'' is such that $\sin (B'' - B') = \frac{h}{r}$

The Saturnicentric latitude of the feature is given by the expression

$$\tan b = \frac{\tan B''}{1.12}$$

and the Saturnigraphic latitude by

$$\tan b' = 1.25 \tan b$$

The Rings

Ring components As mentioned previously, there are three main rings around Saturn that are visible in even small telescopes, an outer A ring, the B ring which is the middle and brightest of the rings and an inner C or Crepe Ring, which is transparent and dim.

The A and B rings are separated by the dark gap known as the Cassini Division, the widest and most prominent of the ring divisions. When the rings are photographed in ultraviolet light the Cassini Division disappears which shows that the division contains very fine particles that are dispersed so thinly that it is necessary to use violet light in order to see them.

Studies of the ring components may usefully involve the following:

(1) Estimations of the relative brightnesses of the three main ring components and also of the globe belts, zones and polar regions at the same time. This is valuable work as both ring components and global belts and zones are believed to exhibit variation.

You should also make studies to determine if variability in ring brightness is periodic. If the ring components do vary in brightness then a determination should be made of whether all ring components change at the same time and proportionally to other ring components. Should this prove to be the case it might point to the origin of the brightness changes being caused by periodic gravitational phenomena caused by the effects of the orbiting satellites.

Brightnesses can be expressed on a numerical scale in which 0 is brilliant white and 10 is black shadow. A good way to record ring brightness is to make a drawing of the globe and rings (this need not be perfectly accurate or correctly proportioned) and simply write the brightness of the various parts upon it. Patrick Moore recommends making two estimates for each division. Start with the darkest feature and progress to the brightest then begin again but this time starting with the brightest feature.

In practice it is difficult to make intensity estimates with the degree of accuracy implied by this numerical system. It would seem better to use rings A and B (the B ring is always the brightest part of the Saturnian System) as standards with which to make brightness estimates of the belts, zones and polar regions of the globe. To record that a zone was slightly brighter than the A ring conveys more than a number on the 0–10 scale.

The brightness of the rings is strongly correlated with Saturn's phase and should be correlated with their tilt. Their greatest magnitude is at opposition when the phase angle = 0°.

(2) Make an estimate of the colour of each ring component as compared with the others. Observations with filters and photographs show that ring C has a brown or reddish brown tint, ring B always appears brilliantly white and ring A exhibits bluish tints.
(3) Note the degree of visibility of the Crepe Ring so that variations in this can be noted over extended time periods. Definite variations occur in this ring; at times it is easily seen and at others it is invisible. A profitable investigation would be for you to determine if the visibility of the Crepe Ring and the brightening of other ring components are correlated.
(4) Make a rough estimate of the width of the Crepe Ring in relation to the width of the B ring.
(5) Note whether or not a gap separates the Crepe Ring from the B ring.
(6) Look out for intensity variations ('ripples') in the rings. It is possible that the forces that produce these are of the same kind that are responsible for gaps in the ring system but are not as strong. Note for how long a ripple is visible. Draw its position relative to the Cassini Division or others if any are visible. If 'spokes' are seen time their rotation.
(7) Each time that you observe the ring system look out for Terby spots. These are large bright elliptical areas that occasionally are seen in ring B and are easy to distinguish from the duller ring upon which they are seen. They usually last only a short time. Carefully watch out for any that appear and search for them even if none are thought to be present. The brightness of Terby spots and the probability of their discovery is increased by using a Wratten no. 12 yellow filter.

Accurate measurements of the rotation rate of the ring components are needed and information helping to do this is provided by observing any white spot for a period of a week and timing when the spot is on the central meridian of the ring as viewed from the Earth.

Ring divisions Table 10.4 is a list of the main divisions in Saturn's rings, their dimensions and mean opposition angular dimensions.

Table 10.4 shows the positions and dimensions of some of the suspected ring divisions. It is possible that there may be others in addition to these. Take some time to get acquainted with the positions and possible visibility of these divisions. Apart from the Cassini Division the only other really well established ring division, so far as Earth-based observation is concerned, is the Encke Division in the A ring. Even this is difficult to see at times. Naturally, the ring divisions are easiest to see when the rings are opened out to their fullest extent.

Make careful search for divisions other than Cassini's and Encke's and record their positions accurately on drawings. An intriguing property of the ring divisions is that they may be easily visible at some times but at others they cannot be found even with large telescopes under similar seeing conditions. Accurate observation of the visibility and positions of ring divisions are urgently needed. This is especially so since the Voyager 1 spacecraft revealed much particulate material within the rings gaps. The 'population density' of the particles in the gaps may vary and this would manifest itself as brightening or darkening in these areas.

The Cassini Division is visible at the ansae until a little before the occurrence of the edge-on presentation of the rings. It seems not to be so prone to variation in aspect or visibility as are other ring divisions.

The following features of Saturn's ring divisions should be observed and recorded over an extended time period:

(1) Make an accurate drawing of all visible ring gaps starting with the Cassini Division.
(2) Compare the intensity of any gaps to the almost black Cassini Division. Determine whether the divisions are equally dark. If differences are seen try to correlate these with the positions of satellites and/or ring brightenings. Also note if the ring divisions are perfectly uniform in width or whether some parts seem broader or narrower than others.

Table 10.4. *Divisions in Saturn's rings.*

Division	Distance from globe		Width
	seconds of arc	km	seconds of arc
Encke	19.0	129500	0.35
Centre of the Cassini Division	17.2	118700	0.53
IV	15.6	107000	0.18
III	13.5	93500	0.65
V	12.7	87500	0.20

(3) If a gap is seen to vary in visibility more than once, refer to your observational records to see if there is any indication of a periodic change in visibility.

(4) Sometimes a division may suddenly become much more prominent. If you see this study other parts of the ring, especially if the bright B ring is the one affected, and note if other previously difficult-to-see gaps have become more prominent at the same time. Such observations are important.

(5) If minor divisions are visible, see if they are equally prominent in both ansae. If not, observe whether a gap visible in one ansa is visible in the other. Quite often there will be a difference in the two that is difficult to explain.

The edge-on presentation of the rings The edgewise presentations of the rings (fig. 10.24) are justifiably regarded as rare astronomical events and are therefore worth observing carefully. This occurs about every 14 or 16 years and

Fig. 10.24 Saturn – rings edge-on: (a) *March 3rd, 1980, 23.10 UT, 115 mm OG, 3186;* (b) *May 29th, 1980, 21.30–22.30 UT, 115 mm OG, ×786. (R.M. Baum, Chester.)*

warrants special observational studies. Through all of the sidereal revolution period of Saturn of 29.5 years, the coincidence of the plane of the Earth's orbit and the rings actually occurs at intervals of 13.75 years and 15.75 years. The inequality of these periods is due to the elliptical form of Saturn's orbit. In the 13.75 year period, Saturn's southern hemisphere and the south side of the rings are turned Earthwards and Saturn passes through perihelion during this period. In the 15.75 period, the northern hemisphere and the north face of the rings are turned Earthwards and Saturn passes through aphelion during this period.

You should note the visibility of the rings during edgewise presentation. Record when the entire edge-on ring system was last visible and the precise day on which they reappear. Note also how far from the globe can the rings be seen and the aperture of the telescope necessary to see them.

Intensity clumpings (bright 'knots', fig. 10.10) are features that can be readily seen in the edgewise ring presentation and are visible although other parts of the rings may not be. Take care not to mistake satellites in the ring plane for these. The visibility of faint satellites is increased when the rings are in the edgewise presentation.

The bicoloured aspect of the rings This phenomenon is manifested as an apparent brightening of one ring ansa as compared with the other. It can be detected with some filters but not with others. Although the effect has long been thought to be due to attenuation of certain wavelengths of light by the Earth's atmosphere, studies of the phenomenon made during 1974 and 1976 show that at least part of the effect may be due to actual physical changes in the rings themselves.

It is best to study the bicoloured aspect when Saturn is near the meridian so that the rings are parallel to the horizon. First ascertain if any apparent brightening of one ansa over the other is due to atmospheric refraction of light. (If Saturn is near to the horizon a brightness due to atmospheric refraction is often noticeable with a red filter but not with a blue one.) If not, observe the ansae through a set of filters and record changes of brightness seen with the whole set of filters.

The best filters to employ in studying the bicoloured aspect are red (25), orange (21), yellow (12), green (58), blue (80A) and violet (47). It is best to start with the violet filter and to use the others in sequence until all have been used. Always keep to the same sequence in successive observations.

Record the following data:

(1) Your name, address and type of telescope used.
(2) Date and time (UT).
(3) The altitude in degrees of Saturn above the horizon and whether it is the east or west horizon.
(4) The results obtained with the set of filters.

In recording the results of the filter observations it is best to make a table of the filters beginning with the violet and to note beside each which one of the ansae appeared brightest when viewed with that filter – if such was in fact seen.

The best way to distinguish a genuine brightening of a ring ansa from the bicoloured aspect caused by atmospheric refraction is to observe Saturn on a particular evening, first when it is low in the east, then when it is at its highest in the sky and then when it is low in the west. Compare the results of these three observations. If the brightening looks the same in all three then it is probably genuine. If a brightening is seen when Saturn is low in the sky which disappears when it is high, then the effect is most likely attributable to the atmosphere.

If a genuine brightening caused by physical changes in the rings is observed, you should watch the rings on consecutive nights until the effect disappears. It is worthwhile making an attempt to photograph the phenomenon either with colour film or with black and white film used with colour filters. Your observations of the bicoloured aspect of the rings should be sent to:

The Saturn Section Recorder,
Association of Lunar and Planetary Observers,
P.O. Box 16131,
San Francisco, CA 94116 USA.

The satellites

Saturn is attended by at least 17 satellites. The largest, Titan, is of the eighth magnitude and is easily visible in any telescope. It makes one revolution around Saturn in about 16 days. At its east and west elongations Titan's apparent distance from Saturn is above five ring diameters.

It is useful to make estimates of the magnitudes of the satellites even if only relative to each other (fig. 10.25). Telescopes larger than about 2.5 inches of aperture should show Rhea which is of the tenth magnitude. It is less than two ring diameters from Saturn. Iapetus appears five times brighter at western elongation (magnitude 10.1) than it does at eastern elongation (magnitude 11.9). This is because one side of the satellite is as reflective as snow whereas the other side is like dark rock. At its brightest Iapetus is about twelve ring diameters westerly from Saturn. You will need several nights of observation in order to be quite sure that you have in fact seen Iapetus because it is difficult to tell it from a star. The satellites Tethys and Dione may be glimpsed in a 6-inch telescope. The others need large apertures and photography to detect them.

Forthcoming oppositions of Saturn

Year	Date	Magnitude
1993	August 19th	0.5
1994	September 1st	0.7
1995	September 14th	0.8
1996	September 26th	0.7
1997	October 10th	0.4
1998	October 23rd	0.2
1999	November 6th	0.0
2000	November 19th	−0.1

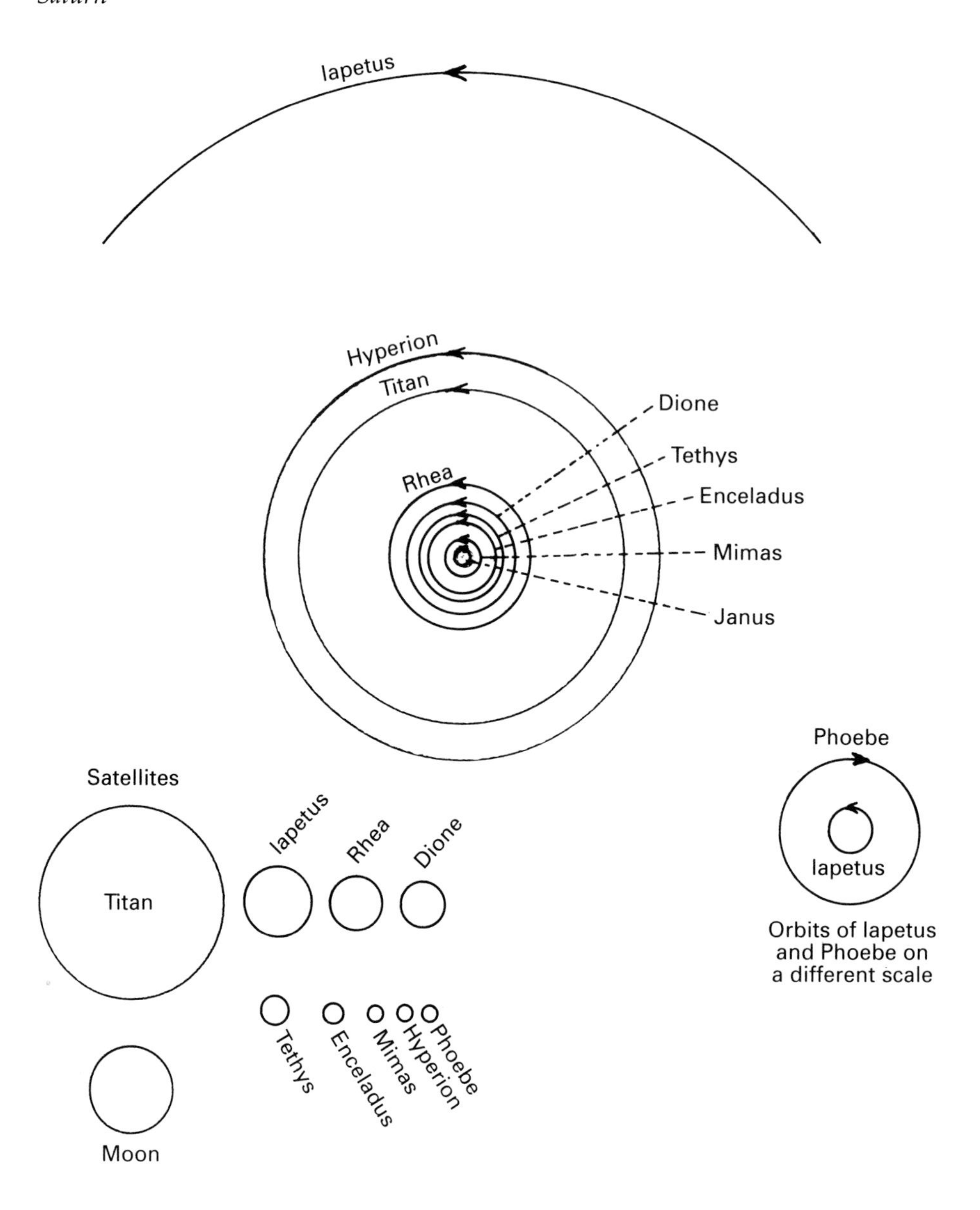

Fig. 10.25 Orbits and comparative sizes of Saturn's principal satellites and the Earth's moon.

Further reading

Books

The Planet Saturn. A. F. O'D. Alexander, Faber and Faber, London, (1962).

Visual observations of the planet Saturn: theory and methods: *The Saturn Handbook,* (5th revised edition). Benton, J. L. Jr, Review Publishing Company, Savannah, Georgia (1988).

Papers and articles

Amateur observations of Saturn. Budine, P. W., *Strolling Astronomer* **15(5–6)**, 80–2 (1961).

A beginner's guide to visual observations of Saturn. Benton, J. S. Jr, *JALPO* **30(5–6)**, 89–96 (1984).

Nomenclature of Saturn's belts and zones (Southern Hemisphere). Lavega, A. S., *JALPO* **27(7–8)**, 151–4 (1978).

Observing the gas giants. Schwarzenberg, D., *Astronomy* **9(3)**, 39–42 (1981).

On the reversed curvature of the shadow on Saturn's rings. Jenks, A., *Sidereal Messenger* **9**, 255 (1890).

Schröter and the rings of Saturn. Ashbrook, J., *Sky and Telescope* **36(4)**, 230–1 (1968).

The white spot on Saturn's rings. Johnson, H. M., *JBAA* **51**, 309 (1941).

A new ring around Saturn? Cragg, T. A., *The Strolling Astronomer* **8**, 22 (1954).

The composition of Saturn's rings. Anonymous, *Sky and Telescope* **39(1)**, 14 (1970).

Saturn, Lord of the Rings. Burnham, R., *Astronomy* **19(8)**, 72–5 (1991).

A closer look at Saturn's rings. Berry, R., *Astronomy* **10(2)**, 72–9 (1982).

On the observations of the reported dusky ring outside the bright rings of the planet Saturn. Baum, R. M., *JBAA* **64**, 192 (1954).

The 'braided' F ring of Saturn. Dermott, S. F., *Nature* 290, 454 (1981).

The periodic variation of spokes in Saturn's rings. Porco C. C., *Eos* **63**, 156 (1982).

The Saturnian moons. Burnham, R., *Astronomy* **9(12)**, 6–24 (1981).

New evidence for the variability of Titan. Noland, M. *et al.*, *Astrophysical Journal* **194**, L157 (1974).

The origin and development of the Dollfus white spot on Saturn. Sitler, J., *JALPO* **16(11–12)**, 251–2 (1963).

New white spot on Saturn. Anonymous, *Sky and Telescope* **39(1)**, 56 (1970).

The Great White Spot on Saturn. Anonymous, *JALPO* **33(4)**, 191 (1990).

The new Saturn system. Morrison, D., *Mercury* **10**, 162 (1981).

Voyager 1 at Saturn. Kerr, R. A., *Science* **210**, 1107 (1980).

Voyager 1 at Saturn. Berry, R., *Astronomy* **9(1)**, 6–22 (1981).

Voyager 1 at Saturn, riddles of the rings. Gore, R., *National Geographic* July 3–31 (1981).

Voyager: Science at Saturn. Berry, R., *Astronomy* **9(2)**, 6–22 (1981).

More science from Saturn. Berry, R., *Astronomy* **9(3)**, 16–22 (1981).

Voyager 2 encounter with the Saturnian system. Smith, B. A., *Science* **15**, 499 (1982).

Voyager 2 at Saturn. Berry, R. and Burnham, R., *Astronomy* **9(1)**, 6–30 (1981).

Voyager discovers spokes in Saturn's rings. Anonymous, *New Scientist* **88**, 276 (1980).

Iapetus: Saturn's mysterious moon. Maran, S. P., *Natural History* **93**, 92 (1984).

The discovery of Janus, Saturn's tenth satellite. Dollfus, A., *Sky and Telescope* **34(3)**, 136–7 (1967).

11

Uranus

General

Although visible to the naked eye as a sixth magnitude star, Uranus appears to have been unknown to the ancients. The planets previously described are all brighter than most naked eye stars and move rapidly among them and so attracted the attention of the peoples of old. Uranus, on the other hand, being faint and moving very slowly among the fixed stars did not attract attention. In the dark unpolluted skies of the ancient world it would be quite lost amongst the myriad stars that peppered the clear night skies of long ago. Uranus was, in fact, the first planet to be discovered, by William Herschel in 1781, who at first thought that it was a comet. He made the discovery with a home-made 6.2-inch reflector.

Uranus, like Jupiter and Saturn, is classed as a 'gas giant'. It is an enormous rapidly rotating world with a low density and a thick atmosphere. Its equatorial diameter is 31 763 miles (51 111 km) and the polar diameter is 30 811 miles (49 575 km). Uranus is therefore noticeably ellipsoidal, the ratio of the polar to equatorial diameter being 0.97. Compared to the Earth Uranus is about four times larger (fig. 11.1).

Four of the Solar System planets have their rotational axes tilted from the perpendicular to their orbital planes by between 20° and 30°. The axis of Uranus is tilted 98° to the plane of the orbit and this makes it unique among the planets. Strictly speaking the axial rotation of Uranus is therefore retrograde but is not generally so regarded. Since the axis lies nearly in the plane of the orbit it follows that twice in a single orbital revolution the sun will lie almost exactly in the plane of the equator and at times in between, one pole or the other will point within 8° directly to the sun. The pattern of insolation changes on Uranus is therefore unusual to say the least, although at this distance from the sun the temperature changes don't amount to much. Also, as seen from Earth, we have alternating equatorial and polar views of Uranus (fig. 11.2). The north pole was presented Earthwards in 1985 and in 2007 Uranus will be seen from Earth in its equatorial presentation.

Atmospheric features of Uranus rotate once in 16.0–16.9 hours but the magnetic field does a complete rotation in 17.24 hours. The orbital eccentricity is 0.047 so that the distance of Uranus from the sun varies from 1867 million miles

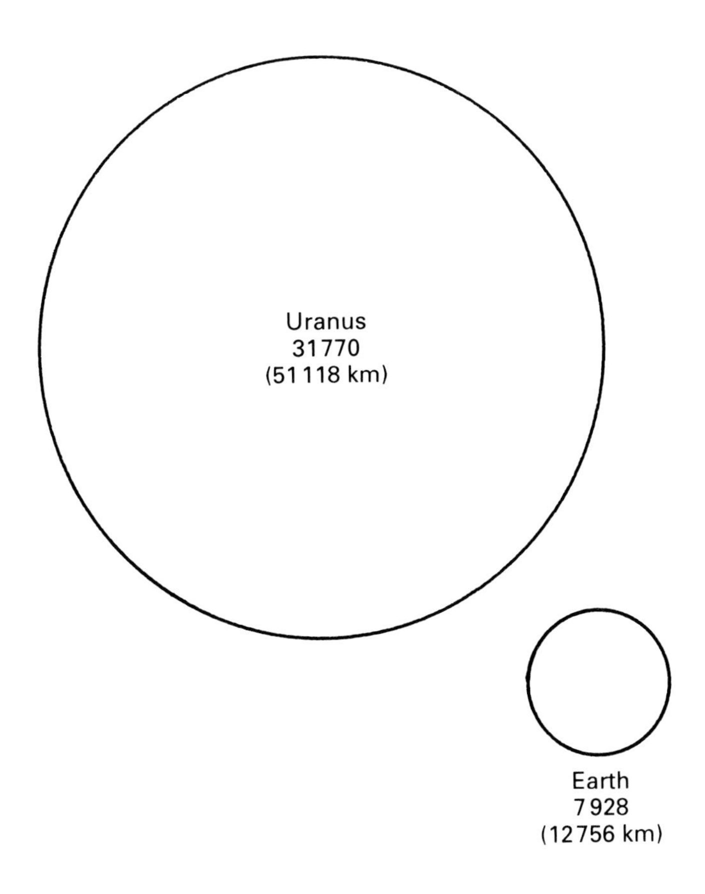

Fig. 11.1 Comparative sizes of the Earth and Uranus (equatorial diameters in miles).

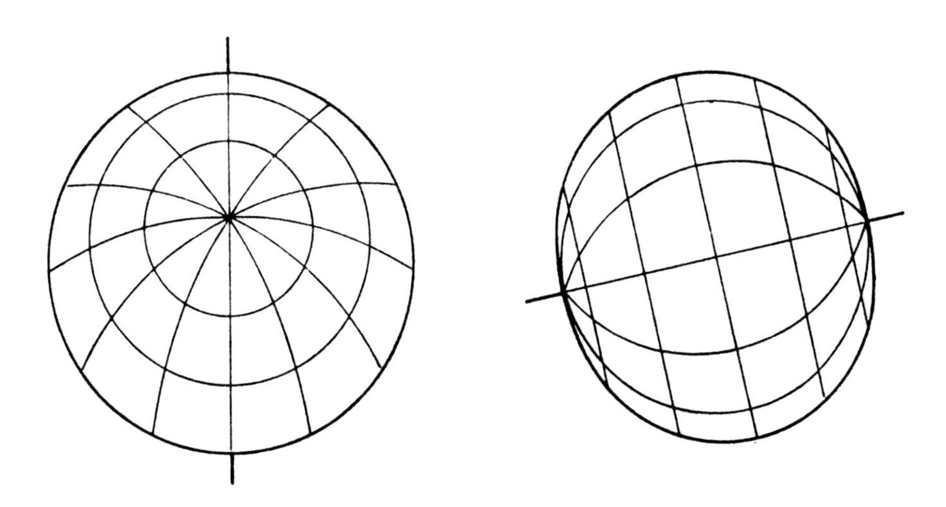

Fig. 11.2 Polar and equatorial presentations of Uranus. (From The Planet Uranus. *Alexander, A.F.O'D., Faber and Faber, London, 1965.)*

(3004 million km) at aphelion to 1700 million miles (2735 million km) at perihelion.

The orbit of Uranus is inclined at an angle of 46 minutes of arc to the plane of the ecliptic. The mean orbital velocity is 4.2 miles (6.758 km) per second and one orbital revolution takes 84.01 Earth years, the sidereal period. The mean synodic period is 369.7 Earth days. The last aphelion passage was in 1925 and the next will be in 2009. The last perihelion occurred in 1966 and the next will be in 2050. Uranus was 1607 million miles (2586 million km) from the Earth when at its closest point in 1966.

That Uranus exhibits brightness fluctuations has long been known and is due to several concurrent periodic changes – those caused by physical phenomena and those due to geometric factors. An example of a geometric factor is the continually changing configuration relative to each other of Uranus, Earth and the sun; the planet will appear brighter when at opposition than at conjunction and this amounted to a 0.22 magnitude difference in 1992. It will appear brighter at perihelion than at aphelion, a 0.2 magnitude difference according to Meeus.

As previously mentioned the unusual axial inclination means that we have gradually and continually changing views of the planet – polar views alternating with equatorial views and all intermediate orientations. Because of the polar flattening the disc of Uranus looks slightly larger in polar presentation than when in equatorial presentation. This can be expected to give an increase of brightness of 0.03 magnitude. Uranus appears brightest when one of the poles is aimed straight at the sun (although there appears to be a darkish polar cap) and dimmest when the equator is directly in line with the sun when the axis is at right angles to the line joining Uranus to the sun.

The geometric causes of brightness variations may be summarised:

(1) The changing Uranus–sun distance has a period of 84.01 years, the planet's sidereal period, and causes an amplitude variation of 0.41 magnitude.
(2) The changing Uranus–Earth distance with a period of 365.24 days, the Earth's sidereal orbital period, causes an amplitude variation of 0.22 magnitude.
(3) The changing axial rotation position of Uranus with a period of 42.01 years (half the sidereal orbital period) causes an amplitude variation of 0.03 magnitude (based on geometry only).

Periodic changes in the albedo of the atmosphere of Uranus appear to be one physical cause of changes in the magnitude of the planet. This was first detected by observers at the Potsdam Astrophysical Observatory in Germany. Similar effects are known to occur on Jupiter and Saturn but are more difficult to substantiate in the case of Uranus.

The continuing motions of the atmosphere of Uranus undergo rhythmic pulsations, as do those of Jupiter, with a repeat period of 8.4 years. This causes brightness variations with an amplitude of 0.3 magnitude.

Light changes attributable to the axial rotation of Uranus were first detected by Campbell at the Lick Observatory in 1917. Light variations having a period of 10 hours 49 minutes and amplitude of 0.1 magnitude were once erroneously attributed to an axial rotation period of the same duration. The axial rotation period of Uranus is now known to be 16.0–16.9 hours.

Small differences in the surface shading of Uranus are detectable during one axial rotation and these areas do not reflect light as strongly as other areas. In 1933 a German observer, W. Becker, claimed to have found an approximately 8-year period of brightness fluctuation with an amplitude of 0.3 magnitude. G. Kuiper and D. L. Harris denied this in 1961. However, observers in the BAA Remote Planets Section studied the brightness of Uranus from 1952 to 1955 and found it to be significantly brighter than the accepted 'official' magnitude of 5.8. The magnitude of Uranus was subsequently revised to 5.5 and later studies in 1969 gave support to this. More recently, photoelectric photometry of Uranus was reported in the *JALPO* (vol **35(4)**, page 155, 1991), a value of 5.40 ±0.14 being reported.

Variations of the brightness of the sun itself may be a cause of Uranian brightness variations. Activity on the sun is also suspected of strongly influencing these events because during sunspot maxima the light fluctuations are more pronounced. Light changes due to axial rotation are not observed when the poles of Uranus point to the sun.

Before the Voyager 2 survey and prior to 1948, four satellites of Uranus were known. In order of increasing distance from the primary they are Ariel, Umbriel (both discovered by Lassel in 1851), Titania and Oberon (discovered by Herschel in 1787). In 1948 a fifth small satellite was discovered by G. Kuiper and named Miranda. It is closer to Uranus than Ariel. The orbits of four of these five satellites around Uranus lie almost exactly in the plane of its equator, and are shown in fig. 11.3. Miranda's orbital inclination is 4.2°. Because of the alternate polar and

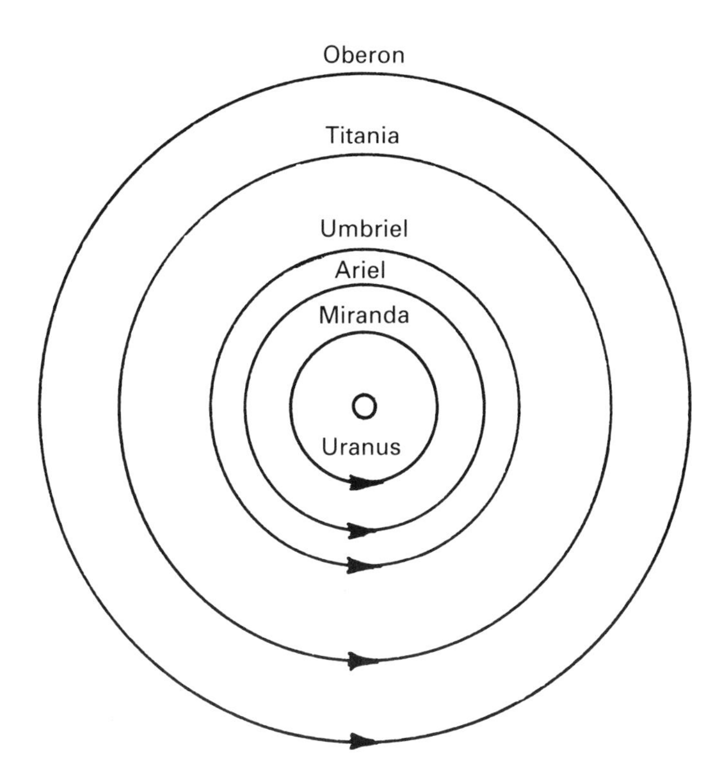

Fig. 11.3 Orbits of the five major satellites of Uranus.

Table 11.1. *The Satellites of Uranus.*

Name	Magntiude	Diameter (km)	Man distance from Uranus (km)	Orbital period
Cordelia	24.1	26	49770	8h 02m
Ophelia	23.8	30	53790	9h 02m
Bianca	23.0	42	59170	10h 26m
Cressida	22.2	62	61780	11h 08m
Desdemona	22.5	54	62680	11h 22m
Juliet	21.5	84	64350	11h 50m
Portia	21.0	108	66,090	12h 19m
Rosalind	22.5	54	69940	13h 24m
Belinda	22.1	66	75260	14h 58m
Puck	20.2	160 × 154	86010	18h 17m
Miranda	16.3	480	129390	1d 9h 55m
Ariel	14.16	1158	19120	2d 12h 29m
Umbriel	14.81	1172	266300	4d 03h 28m
Titania	13.73	1580	435910	9d 16h 56m
Oberon	13.94	1524	583520	13d 11h 07m

(Data from *Astronomical Almanac*, 1992, pp. F2–F3.)

Table 11.2 *The rings of Uranus.*

(See also fig. 11.4)

Ring	Semimajor axis (km)	Eccentricity	Inclination (degrees)	Width (km)
6	41870	0.0014	0.066	*ca.* 2 (1–3)
5	42270	0.0018	0.050	2–3
4	42600	0.0012	0.022	2
Alpha	44750	0.0007	0.017	7–11
Beta	45700	0.0005	0.006	8–11
Eta	47210	almost 0	almost 0	2
Gamma	47660	almost 0	0.006	1–3
Delta	48330	0.0005	0.012	7–9
Epsilon	51180	0.0079	almost 0	22–93

The semimajor axis and eccentricity data are from the *Astronomical Almanac*, 1992, p. F63.
The figures in the last column (width) are from G. Hunt and P. A. Moore with the exception of the last figure, which is from Lane, A. L. *et al.* 'Photometry from Voyager 2: Initial Results from the Uranian Atmosphere, Satellites and Rings'. *Science*, **233**, 65–70 (1986).

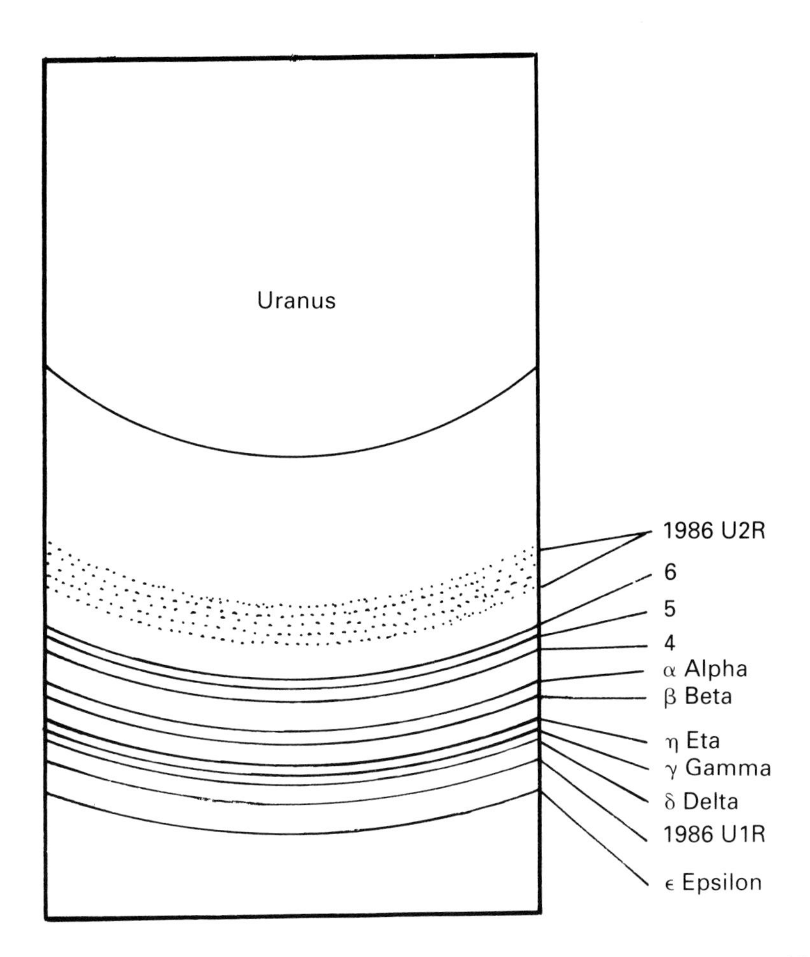

Fig. 11.4 The rings of Uranus. Rings 1986 U1R and 1986 U2R were discovered by Voyager 2 in 1986.

equatorial presentation of Uranus towards the Earth, the apparent paths of the satellites around their primary seem to change shape gradually with time. When Uranus is in polar presentation the orbits of the satellites appear to be circular as in 1946 and 1985. When in equatorial presentation the satellites appear to follow an almost straight line path on either side of Uranus as in 1966 and will do again in 2007. In between these two extremes the satellite paths appear as ellipses of varying eccentricity.

Ten more satellites even closer to Uranus were added to the list as a result of the Voyager 2 mission. The Uranian satellites are listed in table 11.1 with their diameters and distances from Uranus.

In addition to the satellites Uranus is encircled by several rings (fig. 11.14). They are narrow, thin and dark and are not all circular. The ring furthest out from Uranus is asymmetric; the small satellites Ophelia and Cordelia act as 'shepherds' to it. The main rings are surrounded by many hundreds of ill-defined almost transparent dust bands. The nomenclature of the rings with their widths and distances from Uranus is shown in table 11.2

The discovery of Uranus

The planet Uranus was discovered by Friedrich William Herschel who is generally regarded as being the greatest of all amateur astronomers. Born in Hanover in 1738 he emigrated to England in late 1757. He settled at Bath where he remained for 16 years earning a livelihood as a composer and teacher of music, an orchestral player and choirmaster. He was also organist of the Octagon Chapel in Bath. He became a naturalised Englishman in 1793.

Astronomy was Herschel's hobby. He could not afford to purchase a telescope so he taught himself the art of grinding and polishing telescope mirrors and became adept at constructing telescopes. It was with one of his home-made telescopes that Herschel discovered the planet Uranus.

Prior to 1779, Herschel had used a 4.5-inch reflector to make a survey of all stars from the first to fourth magnitudes. Starting in the late summer of 1779 he then commenced a survey of stars down to the eighth magnitude using a 6.2-inch reflector. On the evening of March 13th, 1781 he noticed what he described as a curious nebulous star or perhaps a comet near Zeta Tauri (fig. 11.5). The following three nights were not clear enough for observation but on March 17th he saw that the object had moved. He considered it to be a comet.

Herschel noticed that the object showed a definite disc which enlarged in proportion to increases in magnification whereas the images of stars did not. Although the disc had no tail or other appendages, Herschel continued to think that it was a comet because many comets had been discovered but never a planet.

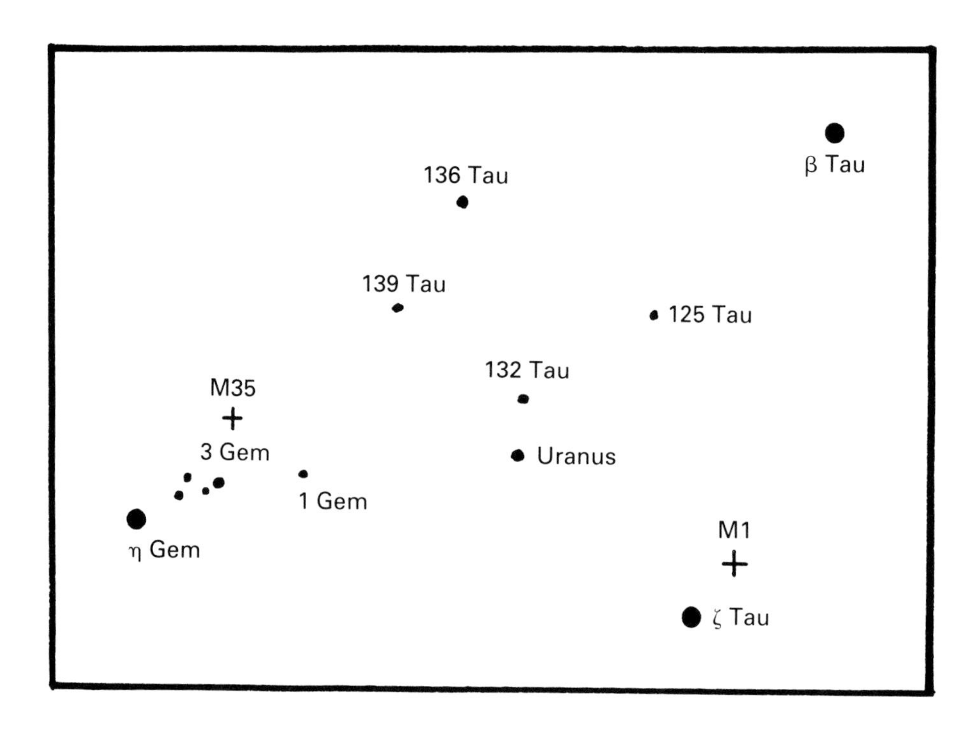

Fig. 11.5 The star field where Herschel found Uranus on March 13th, 1781.

In a published paper he gives 12 micrometric measurements of the diameter of the 'comet' and these showed that the object had steadily increased in diameter and so he concluded that it was approaching the Earth. However, this simply illustrates how difficult it is to measure planetary diameters accurately as was shown by H. Struve one century afterwards in a paper on Saturn.

During the ensuing weeks Herschel continued to observe the 'comet' as often as he could, delineating its path on a star chart and calculating its speed. Further observations of the path of the object among the stars and its uniform motion showed that it could not be a comet. It must be a planet.

When the King, George III, was advised of Herchel's discovery, he made a grant of £200 annually to Herschel who could now devote himself to astronomy full time without having to earn his living as a musician. To show his gratitude, Herschel wished to name the new-found planet after his royal patron – the Georgian Star or Georgium Sidus. He considered this to be more suitable than choosing a name from Graeco-Roman mythology as it indicated the period and country of its discovery and was more modern. Others considered that it should be called Herschel after its discoverer. Finally, tradition prevailed and Herschel's planet was named Uranus after the ancient Greek deity.

Soon after the discovery of Uranus, in August of 1782, Herschel moved to Datchet and finally settled at Slough. The house he occupied there is still standing and may be seen on the road from Slough to Windsor.

Herschel was never knighted by George III for his discovery but at last in 1816 recognition came when the knighthood was conferred on him by the Prince Regent.

Prediscovery sightings of Uranus

Before the 1781 discovery of Uranus, several observers had seen the planet without realising what it was. No less than 22 records of the position of Uranus had been made between 1690 and 1781, the year of its discovery. The first Astronomer Royal, Flamsteed, saw Uranus six times in 1690, 1712 and 1715 without realising its planetary nature. Other good observers also 'missed' it because of their rather poor quality telescopes or failure to notice the movement of the planet. One such was Tobias Mayer who saw and catalogued Uranus as a star in 1756. Mayer would have discovered Uranus 25 years before Herschel if he had observed the star for several consecutive nights and noticed its motion.

Another of these early sightings was made by Le Monnier in 1769 and, astonishingly, was written on a paper bag that had contained hair powder! He had recorded the position of Uranus, thinking it to be a star, on four consecutive nights and two more times within nine days. These six 'stars' were actually the one moving planet. Unfortunately, Le Monnier did not compare his observations; if he had done so, Uranus would have been discovered 12 years earlier in 1769.

History of observation

On January 12th 1787, using a telescope of improved light grasp, Herschel found two satellites of Uranus. Herschel had modified his telescope of 240-inch focal length by removing the secondary mirror and tilting the main mirror slightly. He was thus enabled to view the image directly by looking with the eyepiece down at the primary mirror. This was only the second night of his search for satellites. Of the stars he had charted in the vicinity of Uranus on the previous night, two had moved. To convince himself that these were really satellites he tried to watch one of them actually moving.

On February 7th Herschel watched one of these objects continuously for 9 hours. He saw that it moved with the planet and was able to follow the satellite along part of its orbit (amounting to about 10° in the 9 hour period). He watched the other 'star' on the next night for more than 3 hours and noted that it moved more quickly than the other satellite and so was probably in an orbit closer to Uranus than the other satellite. He calculated that the outer satellite orbited Uranus in 13.5 days and the inner in 8.71 days. They were named respectively Oberon, the first to be discovered and Titania. The reality of these satellites was well established by Herschel's observations. Subsequently he made a careful search for others. It was difficult to differentiate a genuine satellite from a faint star that chanced to be near Uranus. Finally he announced that there were four other satellites bringing the total to six. The evidence for their reality was insufficient however, and their existence has been disproved.

While observing the satellites and their movements in February and March of 1787, Herschel saw that the disc of Uranus was definitely oblate and he also saw two rings around the planet perpendicular to each other but did not consider them to really exist. After another year he saw quite unmistakeably a single ring that appeared to give Uranus an equatorial bulge. Suspecting an instrumental optical artefact, he rotated the telescope tube through a right angle but the ring remained in the same orientation. He saw the ring on several occasions between 1787 and 1893 but was never completely convinced of its real existence.

The ring would be most visible between when it was tilted at an angle of 45° to the observer and when it was 'edge on'. Herschel viewed it when it was in a favourable orientation and tilted at an angle to the Earth between 27° and 49° using telescopes of between 6.2 and 18.5 inches of aperture. The optical defect called astigmatism, which would be significant in an off-axis optical system as was used by Herschel, could have produced a linear flare on a bright object simulating a ring but Herschel's telescopes were of such long focal length and were used with high powers and Uranus was so dim that this effect was not likely. Nothing further seems to have been done and astronomers forgot about the ring until March 10th 1977 when a real ring was discovered – see later.

For about 50 years no telescopes larger than Herschel's were turned upon Uranus and no further progress was made on the matter of the reality or otherwise of the satellites. Finally in 1851, Lassell discovered two new faint satellites within the orbits of the other two brighter ones using a 24-inch reflector of his own construction; there was no trace of any of Herschel's additional satellites. The satellites were also looked for by Otto Struve in 1847 using the 15-inch refractor at the Pulkowa Observatory. He believed that he had caught glimpses of them before they were definitely found by Lassell but was unable to follow them with sufficient continuity to permit their orbital revolution times to be measured. Lassell found their revolution times to be 2.5 days and 4 days respectively. The satellites are named Ariel (near to Uranus) and Umbriel respectively. They are justifiably regarded as among the most difficult objects to see in the Solar System, of those that are visually observable at all.

During the early 1840s J. H. Mädler pioneered a series of diameter measures of Uranus with the Dorpat refractor soon after he became director of the observatory. The intention was to determine the degree of polar flattening of the planet. In all, a total of 330 diameter measures were made on 18 nights during three oppositions, from which he obtained values of 1/10.85 ± 0.9 (1842), 1/9.92 (1843) and 1/9.45 (1845).

During 1852–3, 1862 and 1864–5, Lassell observed Uranus from Malta with his 24-inch reflector and made several measurements of the planet's diameter. Although the disc looked sharply defined to him, he could not make out any definite surface features.

Some indications of the visibility of Uranus were given by Rev. T. W. Webb in 1863. He said that with the naked eye it appeared as a tiny sharp but steady point. A power of ×144 on a 3.7-inch refractor revealed a disc and with a 5.5-inch refractor the disc was visible with only ×55 and quite definite at ×170.

The first time that fairly definite disc features were seen was in observations made between 1870 and 1872 by W. Buffham with a 9-inch reflector. At one time the disc appeared non-uniform and on another occasion two round bright spots were seen. The ellipticity of the disc was obvious. He subsequently saw a light streak in the position previously occupied by the two spots and saw it again on other later occasions. In December 1871 Buffham had an impression of a bright region traversing the disc from north to south. Captain W. N. Noble, who observed Uranus between 1857 and 1875, always saw the disc of Uranus as having a pale blue colour except on March 9th 1875 when he saw it looking whiter than he had ever seen it before and even with a hint of yellow. He used a 4.2-inch refractor.

S. Newcomb observing Uranus in 1874 with the 26-inch Washington refractor found the planet to be a sea-green colour. Variations in tint were never seen. Observations of the colour of Uranus were made in 1877–83 by A. Safarik using a 4.5-inch refractor and a 6.5-inch reflector. He noted that the colour varied from dull blue-grey to a very bright grey-blue.

In a letter written in 1869 A. Secchi describes his discovery of a prominent absorption band, apparently near the red end of the spectrum of Uranus. W. Huggins found other lines in 1871 and a total of six were described by him. Dark and bright bands in the spectrum were studied in 1889 by A. Taylor, A. Fowler and Sir Norman Lockyer. Lockyer considered that the spectrum of Uranus was due to emission and that the dark bands were due to a deficiency of radiation in different parts of the spectrum. Later, in 1895, H. K. Vogel who had carefully studied previous work by Keeler and photographed the spectrum, declared that Lockyer's assertion that the Uranus spectrum is due to emission was quite without foundation.

In 1883 Safarik pointed out that the current position of Uranus was favourable for a fresh determination of its shape and urged astronomers with large telescopes and good micrometers to make measurements. He himself made eight observations between March 1877 and April 1883 and the disc of Uranus looked decidedly elliptical each time. Following Safarik's suggestion Schiaparelli investigated the shape of Uranus using the 8-inch Merz refractor at the Royal Brera

Observatory in Milan. Although he found the ellipticity of the disc to be ovious it was difficult to measure accurately. He made many micrometric measurements and obtained a value for the ellipticity of 1/10.98 ± 0.93 which was further refined to allow for errors he gave as 1/10.94 ± 0.67.

Schiaparelli also mentions how in steady seeing conditions he verified the presence of spots and colour variations on the disc of Uranus. These were confirmed by his assistant. Although admitting the excellence of the 8-inch Merz objective he said that it was useless trying to delineate surface features or to try to determine a rotation period from them. In 1883 C. A. Young used the Halsted Observatory (at Princeton) 23-inch refractor to study the disc markings and ellipticity of Uranus. Under good conditions he detected very faint belts similar to those of Jupiter and Saturn. His colleague McNeill suspected the existence of two polar belts in addition to two equatorial belts. Often it seemed that the belts were inclined at quite a large angle to the line of the satellites, a puzzling observation.

E. E. Barnard made diameter measures of Uranus in 1894 and 1895 at a time when the ellipticity would be expected to be slight. Using the 36-inch Lick refractor measures of the polar and equatorial diameters plainly showed the disc to be elliptical and indicated the polar flattening to be even greater than Saturn's. He measured the diameter on nine occasions during 1894 and from the results deduced an ellipticity of 0.05. This is less than Saturn's (0.105) but may be the projected ellipticity.

The existence of two equatorial belts, a brilliant zone between them and dusky polar regions was observed by Paul and Prosper Henry who observed with a 15.2-inch refractor. They confirmed the angular tilt of the belts to the line of the satellites. The tilt was measured and found to be 40°.

During 1893–5, the amateur astronomer A. Henderson of Liverpool observed Uranus with his 10.5-inch reflector and he described the colour of the disc as 'pale cobalt blue' and that the central part was brightest. He often suspected a bright zone stretching from the south-east to the north-west of the disc. On May 8th 1893 when the definition was perfect he drew a wide bright zone between the darker east and west sides of the disc.

Between 1894 and 1898 E. C. Pickering determined the stellar magnitude of Uranus using the meridian photometer of the Harvard College Observatory. The results he obtained for the five determinations ranged from 5.48 to 5.56 giving a mean opposition magnitude of 5.51. Extensive photometric work was also carried out by Muller.

The relative brightnesses of the Uranian satellites were studied by E. E. Barnard using the Lick Observatory refractor. He found the two inner satellites (Ariel and Umbriel) difficult. In 1894 Titania and Oberon seemed about equally bright but in 1895 he found Ariel always 0.5 magnitude brighter than Umbriel. In the same year Titania was twice as often brighter than Oberon. However, he noticed that of the two, the lower in the field of view always looked brighter but were equally bright if horizontally placed in the field. It therefore seemed that it was his eyes that made the lower of the two equally bright points look brighter by 0.5 magnitude and that his initial finding of the two being equally bright was correct.

In 1894–5 Schaeberle found Ariel to be 0.5 magnitude brighter than Umbriel and Oberon 0.5 magnitude brighter than Titania. Aitken in 1900–1 found the reverse in the case of Titania and Oberon, the two magnitudes being respectively 13 and 13.5 and those of Ariel and Umbriel were respectively 14.5–15 and 15–16. C. Wirtz observed Uranus with the Strassburg 20-inch refractor in 1905 when the planet's declination was 24° south. The faintness of Titania and Oberon made them difficult to observe so it is not surprising that he could find neither Ariel nor Umbriel. He found Titania to be brighter than Oberon by from 0.2–0.6 magnitude, average 0.5 magnitude.

Improved photography of the spectrum of Uranus by V. M. Slipher in 1902–9 with the Lowell 24-inch refractor led to the discovery of the planet's axial rotation period in 1911 by Lowell and Slipher. A rotation period of 10 hours and 50 minutes was determined.

In the Autumn of 1913 the French planetary observer G. Fournier enjoyed good views of the faint disc of Uranus. He observed with a 20-inch refractor. The disc had a bluish colour. At first glance light zones and darker belts were clearly seen and a broad bright zone at the equator with dusky belts at its edges. (A similar appearance was seen by W. H. Steavenson in 1915 with a 10-inch telescope. See fig. 11.6). A light zone stretched all the way to the south pole but the north region to the pole was shaded. In November he clearly saw the polar flattening. In the light equatorial zone a mottled appearance was seen. Both the north and south parts of the disc were vaguely shaded. The belts looked tilted as they had on October 18th but were now more so.

The light variations of Uranus were utilised to arrive at an axial rotation period for Uranus by L. Campbell who made a series of photometric observations of the planet. The light of Uranus was found to vary by about 0.15 magnitude in a period of 10 hours 49 minutes. Pickering noted that this gave independent confirmation of Lowell and Slipher's value of 10 hours 50 minutes which they obtained by a spectroscopic method. The conclusion was that the variations in brightness are due to the unequal brightness of different parts of Uranus. R. L. Waterfield suggested that these brightness variations may give some insight into the physical condition of Uranus which at that time was still not known. Not much could really be learned from the difficult-to-observe disc features which

Fig. 11.6 Uranus showing a broad white zone between two dusky belts, September, 1915, 10-inch aperture (W.H. Steavenson). (From Splendour of the Heavens, *Hutchinson, London, 1923.)*

are only seen in broad outline. Waterfield proposed that the appearance of a large white spot and its gradual elongation into a belt as was seen on Jupiter would explain the decrease in range and sharpness of the maxima of the periodic brightness increase that had been observed in Uranus in a series of photometric studies that he commenced in August of 1915. This would indicate that Uranus had a fluid surface and the magnitude of the light variations were indicative of physical changes on a much larger scale than on Jupiter.

E. E. Markwick, a Director of the Variable Star Section of the BAA, made 84 observations of the apparent brightness of Uranus between 1916 and 1921 using ×8 prismatic binoculars and estimating the magnitude of Uranus by comparing it with the brightness of nearby stars of known magnitude from 5.28 to 6.22. In 1916 he found little variation and nothing like the frequent maxima observed by Waterfield in 1915. He doubted that the elongation of a white spot could account for the variations. It seemed that a white spot would not be large enough relative to the disc of Uranus to cause such brightness variations. However, Waterfield could not think of any other explanation. He stated that Uranus decreased slightly in brightness during 1920–2.

A photographic investigation of the colour of Uranus and Neptune was commenced in 1917 by G. A. Tikhoff and the results were published in 1922. The method used was to study the intensity of photographic images in light of different colours. The purpose of this was to find out if the atmospheres of Uranus and Neptune scatter light like that of the Earth which gives our skies a blue colour. It was already known that the planets' own atmospheres absorb some of the radiations of longer wavelength. The results enabled him to conclude that the light on Uranus and Neptune is diffused by a mechanism similar to that for diffusion in our own atmosphere and that the bluish colour is almost wholly due to the absorption bands in the red end of the spectrum.

M. B. B. Heath observed Uranus in 1921, October 24th, with a 8.5-inch reflector. The greenish disc appeared to him as slightly elliptical and there was a wide somewhat brighter band vertically across the middle.

The Earth was nearly in the equatorial plane of Uranus in 1924 and Uranus was observed by Antoniadi using the 33-inch Meudon refractor. The planet appeared to him to have a yellowish colour but with some greenish-blue in it. He saw greyish polar caps and two faint dusky belts, one on each side of the light equatorial zone and they looked exactly parallel to the line of the satellites. He concluded that the belts seen at an angle to the satellite line by previous observers was illusory and caused by the difficult seeing conditions. The planet looked decidedly flattened to Antoniadi, more so then Jupiter but less than Saturn. He sketched the equatorial and pole-on views of Uranus that were presented at approximately 21-year intervals (fig. 11.7).

The rotation period was again measured spectroscopically by J. H. Moore and D. H. Menzel during 1927–30. At this time the equator of Uranus made a much smaller angle with the sight line than it did in 1911 when Lowell and Slipher made their spectroscopic measurements of the axial rotation period and there was a more favourable opportunity. They calculated that the rotation period is between 10 hours 40 minutes and 11 hours which was in close agreement with the 1911 values but Moore and Menzel considered the agreement to be fortuitous; the values obtained may have been as much as half an hour in error owing to the difficulties of dealing with such a small planetary image.

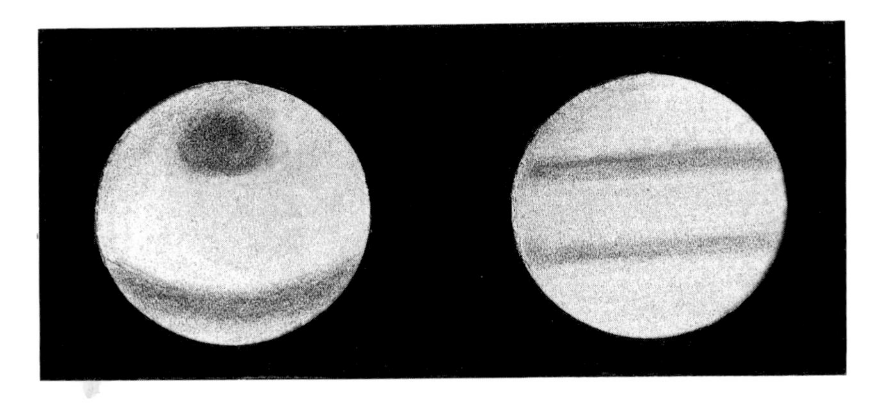

Fig. 11.7 Observational drawing of Uranus by Antoniadi showing polar and equatorial views (33-inch Meudon refractor). (Redrawn from L'Astronomie **50**, *256, 268, 1936.)*

In 1932 Wildt suspected that some at least of the dark absorption bands in the spectrum of Uranus may be due to methane. However, no absorption bands like those of Uranus had been demonstrated in methane in the laboratory either in the visible or ultraviolet part of the spectrum. For many years Slipher was continually improving photography of the spectrum of Uranus and wrestling with the problem of the origin of the dark absorption bands. He and A. Adel finally succeeded in 1934. They studied the spectrum of a source of continuous radiation passed through a 45-metre path of methane at a pressure of 40 atmospheres and found that 14 of the absorption bands, including the strong ones, in the spectrum of Uranus coincided with the methane spectrum thus proving its presence in the atmosphere of Uranus.

In 1933, W. Becker studied records of light changes of the planets from Mars to Neptune made over many years. He found the following patterns for Uranus:

(1) A 42-year cycle caused by polar flattening and possibly also by albedo variations in different parts of the planet. This gave a sinusoidal light curve with an amplitude of 0.29 magnitude.
(2) A residual variation having an apparently decreasing period of 8.4 years generating a sinusoidal curve with an amplitude of 0.31 magnitude.
(3) Sudden variations of amplitude about 0.15 magnitude that frequently occur after stable periods lasting for many weeks.

The light change period of 8.4 years was confirmed by J. Ashbrook who made a series of observations of Uranus with an opera glass between 1936 and 1947.

Most of the measurements of the diameter of the disc of Uranus had been made with filar micrometers. In spite of precautions the scatter of results obtained by an individual observer and the considerable disagreement between the mean values obtained by different observers showed that the filar micrometer is not suited to this type of work.

Greatly improved precision in planetary disc diameter measures was achieved by H. Camichel using an apparatus called the disc meter described and illustrated in a paper in which he published the results of disc measures of Uranus made in 1944–7. Essentially, the planetary disc seen in the telescope is simultaneously compared with an artificial disc whose size can be varied. When the two discs appear identical then the artificial disc has the same size as the planetary disc. If the seeing conditions are excellent then the angular diameter of the planetary disc can be calculated with a probable error of less than 1 in 100. Using the disc meter Camichel measured the disc of Uranus on three nights in December 1944 and two nights in September 1947 using the Pic-du-Midi 24-inch refractor. The scatter between determinations was much less than by filar micrometric methods. Camichel determined a disc diameter of Uranus at unit distance of 64.91 seconds of arc in 1944 and 65.12 seconds of arc in 1947. This is the equatorial diameter as Uranus was then in its polar orientation toward the Earth.

Around 1947 G. P. Kuiper measured the disc size of Uranus by the double-image micrometric method which brings the two images of the disc meter into contact. With the 82-inch reflector of the McDonald observatory he obtained a value of 65.4 seconds of arc for the angular diameter of Uranus at unit distance, a value in close agreement with Camichel's.

A fifth satellite of Uranus named Miranda was photographically discovered by G. P. Kuiper on February 16th 1948 with the 82-inch McDonald reflector. Its revolution period around Uranus was found to be very nearly 33 hours 56 minutes. The orbit is roughly circular and coplanar with the other satellite orbits. The distance of Miranda from Uranus is about 80 400 miles (123 000 km). Miranda is of the 17th magnitude.

On February 23rd 1949, Armellini observed two small white spots on Uranus near its equator and these were seen again during the following two days. They were too small to be measured. Armellini observed with the 16-inch refractor of the Monte Mario Observatory, Rome. In 1954, A. P. Lenham remarked that the rotational brightness fluctuation cycle of Uranus with a period of 10 hours 49 minutes and a magnitude amplitude of 0.1 'seems to have disappeared in recent years'.

The magnitudes and colours of Titania and Oberon were measured photoelectrically, apparently around 1950, by D. L. Harris with the 82-inch McDonald reflector and he found Titania to be somewhat brighter than Oberon. The mean opposition magnitudes were determined to be 14.01 for Titania and 14.2 for Oberon. The reflectivity results indicate that their surfaces are not quite neutral reflectors of the sun's radiations and that the reflectivity is lower in blue–ultraviolet wavelengths than it is in red wavelengths.

In early 1969 Uranus appeared about as oblate as Jupiter. Using a 12.5-inch reflector, R. G. Hodgson found Titania and Oberon to be easy objects. In 1974, the orientation of Uranus had changed. It no longer looked oblate and the satellites were difficult to find. Also, in April 1969, Hodgson detected dusky poles on a night of unusually good seeing but he never certainly saw equatorial belts with his 12.5-inch reflector.

Between 1972 and 1982 the Lowell Observatory programmes revealed a long-term brightness fluctuation with an amplitude of 0.1 magnitude. This was what was expected owing to variations in the planet's apparent projected area from the near equatorial to the polar presentations.

On March 10th 1977 Uranus occulted the 9th magnitude star SAO 158687 in Libra. Unexpectedly, this event furnished evidence that Uranus was surrounded by rings. The event was watched at many different observatories including the Kuiper Airborne Observatory (KAO) and monitored photoelectrically. About 40 minutes before the star was due to be occulted by Uranus, observers on the KAO from Cornell University recorded a disappearance of the star of 7 seconds duration. During the following nine minutes there were four more disappearances, each of about one second's duration. Thirty minutes later the star passed behind Uranus and was hidden for 25 minutes. After the star reappeared there was another short series of disappearances with periods similar to the first series. The existence of rings of obscuring matter encircling Uranus was thus indicated and has since been amply confirmed by space probe photography and imaged from Earth (Palomar Observatory in 1978 and Siding Spring in 1984).

Starting in 1954 but mostly during 1969–86, several hundred visual estimates of the brightness of Uranus were made by members of the Asteroids and Remote Planets Section of the BAA, using small telescopes (less than 4 inches) and powers lower than about ×50. Estimates were made by comparison with nearby stars. No regular periodic variation was detected and within observational error, no certain large changes more than 0.2 magnitude were observed during this period. No brightness fluctuations were detected that correlated with the sunspot cycle having an amplitude that was detectable by visual observation. The disc of Uranus was variously perceived to be grey, bluish-grey, white, very pale green, greenish-white, greenish-blue. Limb darkening was distinctly seen by D. L. Graham.

D. Gray and J. H. Robinson made disc drawings at the equatorial presentation, both using 10-inch reflectors, that show indications of belts, light equatorial regions and darker polar regions. Gray's drawing was made on April 7th 1969 and Robinson's on May 13th 1974. The near polar aspect was drawn by F. C. Butler (8.7-inch reflector) in 1979 on May 21st and shows one band and a dusky half of the disc but these details were said to be 'uncertain'. The polar aspect drawn by Gray using a 16.3-inch Dall–Kirkham on June 16th 1985 shows a dusky centre to the disc and the polar aspect drawn by A. J. Hollis on September 8th 1986 with a 12-inch reflector shows three dusky patches, one in the centre of the disc.

Based on failure to detect detail in photographs of Uranus made during the 1960s and 1970s, B. A. Smith and H. J. Reitsema concluded that the disc of Uranus is 'probably featureless in visible light'. Whatever is the truth, the features are certainly very difficult to detect.

During the 1980s, few observational reports of Uranus were published, possibly a result of its closeness to the southern horizon for northern hemisphere observers as it came to opposition in those years when Uranus was difficult to find in the teeming star fields of the southern Milky Way. Some observational drawings were made by the Spanish observer Christopher Tobal during 1986–8 using a 4-inch refractor and an 8.7-inch Newtonian reflector. Some of the drawings depict nearly linear markings traversing the disc a little off centre. They cannot be related to latitude since they did not correspond to the orientation of Uranus in 1986–8. Other drawings made a few days later showed dark curved linear features that might be related to latitude. Although these were made with the larger telescope, it is difficult to accept that the cloud pattern of Uranus could have changed in such a short time. Although Tobal reliably drew what he saw one cannot help wondering whether the disc details are instrumental or subjective in origin.

Spacecraft exploration of Uranus

The Voyager 2 spacecraft was launched on August 20th 1977 and encountered Uranus on January 24th 1986. It returned an enormous amount of data relating to the structure and composition of the atmosphere of Uranus, the magnetosphere, the temperature and pressure of the atmosphere, disc features, the rings and the satellites. The following brief summary will deal only with disc features, the rings and satellites, as probably being most relevant to Earth-based observers who will find comparisons and correlations of Earth-based telescopic observations and Voyager 2 imagery most interesting.

Globe features. The Voyager 2 spacecraft approached the south pole of Uranus. Not much was revealed in the way of atmospheric features because the clouds of Uranus lie deep within its atmosphere, deeper than those of Jupiter and Saturn and beneath a fog of hydrocarbons. This correlates well with the elusiveness of disc features in telescopic observation. Some atmospheric features were made visible by employing colour filters and very high image contrast. Violet filters showed the south pole to be darker than the equator and middle latitudes; the north pole was in darkness at the time. Orange filters reveal a somewhat lighter ring encircling the pole at higher latitudes. Strong limb brightening was seen in images obtained with a methane band filter that passes light wavelengths absorbed by methane.

The well-known greenish colour of Uranus is due to a small amount of methane in its atmosphere which consists mostly of hydrogen and helium. Methane absorbs red light strongly so that blue-green wavelengths are reflected from the planet.

The Voyager imagery revealed that clouds at different latitudes move at different speeds around the planet but this does not indicate the actual axial rotation rate of Uranus as a whole. Those closer to the equator at 27° latitude take 16.9 hours for a complete rotation while at 40° latitude the clouds make a complete rotation in 16 hours.

A cloud 2500 miles (4000 km) in length seen near 35° south latitude indicates convection that lifts aerosols into the stratosphere of Uranus as in the case of Jupiter and Saturn, which is then 'pulled out' by wind shear. Voyager 2 'observed' auroral phenomena at Uranus. The ultraviolet spectrometer detected auroral radiation at the dark north pole and found that it rotated with the planet and 'electroglow' emissions from the sunlit hemisphere.

A measurement of the axial rotation period of Uranus has never been reliably or accurately determined. A few days before the close approach the first certain indication of rotation was given by radio noise that exhibited regular intensity variations. A rotation rate of 17.29 hours ± 0.10 hour was derived from data accumulated during 15 days (*Science*, July 4th 1986, page 85).

Satellites. The sunlit surfaces of the five larger satellites of Uranus were photographed by the Voyager cameras and as was the newly discovered satellite Puck. The smallest of the five, Miranda, is 300 miles (480 km) in diameter. The satellite Puck with a mean diameter of 86 miles (154 km) was imaged at the last moment so to speak, with a single frame that was to have been used for Miranda, from a distance of about 300 000 miles (500 000 km). The resulting picture showed it to be spheroidal and with a mottled appearance. The resolution wasn't sufficient to show if this was due to craters. A single large crater was seen on the terminator.

Miranda was seen to have several large circular dark regions ('circi maximi', 'ovoids' or 'coronae') on its surface and valleys bordered by very high cliffs. Upon its surface Miranda shows all the strange terrains that have been seen on many of the strange Solar System worlds. The really unique features are the 'circi maximi'. Three of these can be seen on the Voyager pictures. Two are circular and the other, trapezoidal in shape, contains the so-called 'chevron' feature. The tendency is for them to be dark on the inside and they are composed of ridges, grooves or cliffs parallel to the bordering rim. Perhaps these three features in the Voyager pictures are three different stages in the development of a single kind of feature that appears to grow by a process proceeding outwards from the centre by fault formation.

Ariel has the brightest surface of the moons of Uranus; it also has a large reflective range. The darkest parts are 25% reflective and the brightest are 45% reflective. There are ray craters on its surface, scarps, faults, valleys cutting across each other and possibly flow features. There is a population of smooth-edged craters. During its history the surface of Ariel seems to have been moulded by many different geological processes. In addition ice flows or water have been involved in moulding its valley floors.

The darkest of the satellites, Umbriel, has a dull grey colour and is covered with overlapping ring formations up to 124 miles (200 km) in diameter. Its almost invariant reflectivity is 18%. For a body with an ice-covered surface, as Umbriel is, this is very low. Near the limb of Umbriel is an unusual lighter ring about 93 miles (150 km) across with a reflectivity of 25% and of a dark grey colour. It may be the floor of a crater.

Titania, with an albedo of 27%, has a neutral grey colour and its surface is variegated with ray craters, large circular depressions and graben-like trenches. The latter are part of a hemisphere-wide system of faults that appear to have been caused by tensional faulting. A large circular depression was seen near the terminator but on the whole, the surface is devoid of large craters. What craters there are are small and don't show much size variation.

Oberon is rather dark and its reflectivity is about 24%. It is covered with much darker patches. A single mountain at least 12 miles (20 km) high was seen on the otherwise smooth circular limb (*Science* July 4th 1986, page 57). Many large craters have dark areas on their floors, possibly indicating outflow of dark molten matter after the craters were formed. A single large scarp can be seen.

Rings. The rings of Uranus are very tenuous, possessing little mass and, as described earlier, were discovered only in 1977 by observation of a stellar occultation by Uranus. During its close encounter the photopolarimeter abroad Voyager 2 recorded occultations of two stars by the rings. The results gave fine structural data of the major rings and evidence pointing to the existence of hundreds of thin almost invisible minor rings encircling Uranus. Previously, nine Uranian rings were known; the Voyager cameras detected a very faint tenth ring called 1986 U1R between the epsilon and delta rings and revealed the gossamer-like sheen of fine dust within the ring plane. The ten major rings were resolved by a factor of 100–1000 times better by the photopolarimeter occultation study than by the best spacecraft camera images. Analysis of data provided by radio occultation experiments at two different wavelengths indicated that the rings differ from one another in their characteristics. The outermost and widest of the rings, which is eccentric, is the epsilon ring and is about 26.7 miles (43 km) across. Actually, its width varies from 14.0 miles (22.6 km) to 46.6 miles (75.0 km) according to *Science* (July 4th 1986) page 81, but is given as 22–93 km on page 67. On both sides of it were found shepherd satellites that were tentatively named 1986 U7 and 1986 U8; they are now known as Cordelia and Ophelia respectively.

The clearest camera images showed the epsilon ring to have a double structure which agrees with the occultation experiment. There is also fine 'chaotic' structure within the ring. According to Hunt and Moore, radio occultation data indicate that the majority of the ring particles are about 1 metre in diameter. The epsilon ring contains more material than all of the others combined. The next ring inward is the delta ring which was shown from one occultation experiment to have three principal components but one only from another occultation. Four components were found in the alpha ring, three wide and one narrow and bright. The beta ring has two wide components and one narrow. The gamma ring was shown to be very narrow and only 0.62–1.86 miles (1–3 km) across with edges 0.062 mile (0.10 km) wide. The gamma ring was invisible in transmitted light but bright by reflected light. Contrasting with this was the eta ring which was wide and diffuse.

Ten very slender ring arcs (partial rings) were also revealed in about 5% of the total data. They measure no more than about 0.6–10 miles (1–16 km) in width. One of these ring arcs, the largest, had already been detected by Earth-based stellar occultation observations. It was imaged by the Voyager 2 cameras many days prior to the close approach and is usually described as 'newly-discovered' by Voyager 2, hence its designation 1986 U1R,

Visibility of Uranus

Uranus comes to opposition about 4 days later each year. Its declination limits are about 24° north and south. Its mean opposition magnitude is about 5.7 (5.6 according to recent ALPO photometry). It is easily seen with binoculars and a 3-inch refractor will show its small featureless greenish disc which is nearly 4 seconds of arc in diameter.

Observing Uranus

Uranus was not well placed for observation for northern hemisphere observers during the 1980s owing to its low altitudes at oppositions during that period but things should be better from 1990 onwards. At the time of writing (1991) Uranus is starting to move northwards and this will continue for the next 40 years. The opposition positions of Uranus are moving into the northern autumn evening sky, a time of year that is favourable for observation.

Disc studies

Uranus offers little opportunity for the ordinary amateur. The only features ever seen on its dim green disc are one or two faint belts; apertures of at least

10 inches are required and superb seeing. Many amateur drawings of Uranus made with small telescopes show many spots and belts that simply cannot be reality. It is as well that these drawings remain unpublished. Some authorities debate whether Uranus has disc features or dusky polar regions that can be seen from Earth. However, this is not to say that telescopic observation is futile, only that large enough telescopes are a must in this type of work, probably of a size not commonly found in amateur hands. Good quality drawings and photographs made with such instruments under excellent seeing conditions are sorely needed if only to help resolve this debate about the reality of disc features seen and also the question of the rotation period. Any sightings of definite surface features should be telephoned immediately to the Recorder of the Remote Planets Section of the BAA or ALPO. It cannot be emphasised too strongly that believable disc drawings of Uranus require large telescopes under unusually good seeing conditions. Apertures of at least 16 inches are needed if disc drawings are to be reliable. Surface markings have been definitely seen with smaller apertures but what the observers draw are probably subjective interpretations of detail glimpsed at the limit of resolution of the telescope used.

A valuable study would be to monitor the disc of Uranus to see if there is clear-cut reliable evidence of the reappearance of a pattern of dark belts and light zones as Uranus steadily comes to an equatorial presentation toward the sun in the years following 2000. Such a pattern is observed on Jupiter and Saturn and a similar one was seen on Neptune by the Voyager 2 spacecraft.

Disc oblateness (polar flattening) is extremely difficult to measure accurately and obviously is best done when the Earth lies in the equatorial plane of Uranus as it was in 1966 (the next time will be in 2007). Micrometry is unreliable and results of observations over the years show much variation in estimates. An alternative approach might be to study data from several top quality occultation observations. Oblateness studies are important because of the relation of oblateness to axial rotation period which is disputed.

Apart from these areas and magnitude measurements that have been done with 8-inch telescopes, other studies require large apertures in excess of 12 inches and frequently costly equipment in addition, such as micrometers, photometers, Schmidt cameras and plate measuring apparatus. Also, observers in these fields need to be experienced and disciplined.

Magnitude studies

As already noted the brightness of Uranus varies in a random manner with some regular components by several tenths of a magnitude. Fortunately, for the rest of the 1990s, observers wishing to make magnitude studies of Uranus will find several comparison stars near to it in the sky. Observers wishing to study magnitude variations of Uranus by the visual method will have to select comparison stars anew at each opposition since the position of Uranus on the ecliptic changes by about 4.3° annually. However, visual magnitude estimates of Uranus are not accurate or of scientific value, notwithstanding what is written in many books, because the magnitude fluctuations are too small to be reliably discerned. Sensitive photometers are required which can reliably measure fluctuations as little as 0.01 magnitude. Pacific Photometric Instruments, OPTEC and DOAA Enterprises manufacture such instruments or they may be home made (see the photoelectric photometer described by R. Bryant in *The Minor Planets Bulletin*, **4**,

29–31, (1977). Valuable magnitude studies can only be done by observers with photometric equipment (see chapter 16). The monitoring of the brightness of Uranus can be done in white light or with precision filters at selected wavelengths, e.g., U, B, V, R (see chapter 16). There are two kinds of investigation that can be carried out:

Rotational short-term variations. Up to about 1976 it was thought that Uranus had an axial rotation period of 10 hours 49 minutes. However, Hayes and Belton's observations indicate a period of 24 ± 3 hours. L. Trafton finds a value of 23 ± 5/2 hours (See *Icarus*, **32**, 383–401 (1977). Precision photometric studies may assist in resolving this problem. Voyager 2 imagery gave 16.0–16.9 hours for atmospheric features.

Long-term variations. Of the long-term light variations of Uranus, some may be due to brightness changes of the sun as previously mentioned, since Uranus is a relatively quiescent planet and is always virtually at the 'full' phase as seen from Earth. Presumably all of the planets and satellites would be similarly affected. A study extending over long periods of time of these long-term fluctuations may therefore be of considerable value in investigating long-term variations in the Solar Constant. This is of great importance not just for photometry of the Solar System but also in connection with long term changes in the terrestrial climate.

Satellite studies

Observationally, the satellites of Uranus have been much neglected. They are difficult to observe and really dark skies are needed but when the seeing is good Titania and Oberon should be visible in an 8-inch telescope. A 10-inch may be needed if the sky is not really dark. Apertures of 12–14 inches should reveal Ariel but Umbriel won't be seen with less than 16 inches. Miranda can only be seen with large observatory telescopes if at all.

In his little book *Hours With a Three-Inch Telescope* (2nd edition, Longmans, Green & Co., London, 1887) Captain William Noble describes how a Mr I. W. Ward, a man with unusually keen eyesight, actually glimpsed the two outer satellites of Uranus with only 4.3 inches of aperture on several occasions during 1876! The reality of this extraordinary visual achievement was completely verified by comparison of Mr Ward's positional drawings of the satellites with their computed positions. With this exception Captain Noble stated that probably no human eye had ever seen the two outer Uranian satellites with less than 7 inches of aperture.

It would be worth while investigating which of the two outer satellites appears to be brighter. Different values at different times may be due to the changing positional relationship of Uranus and its satellites. The satellites may be compared with one another for studies of relative brightness or with nearby comparison stars. The glare of the disc of Uranus can be a nuisance – more than you might think – in making accurate estimates of satellite brightness. Try to observe the satellites when they appear at about equal distances from the planet. Studies of the Uranian satellites are needed in which their relative brightnesses are compared and their absolute brightnesses compared to stars of known magnitude within the varying fields through which Uranus will pass during the next few decades.

Occultation observations of the satellites are also valuable. Eclipses are interesting to observe in the rare years when both Earth and sun lie in the equatorial plane of Uranus. Satellite shadow transits would also be interesting were it not for the small size (0.1 second of arc) of the shadows on the disc of Uranus which makes it very unlikely that they would be seen.

Stellar occultation studies

Observers experienced in occultation work can make valuable contributions even with small telescopes in studies of occultations of stars by Uranus and its rings; the apparent number, position and width of the rings may thereby be investigated. Photometric measurements made with large telescopes may also be valuable in determining the opacity of the rings.

If you wish to pursue this line of work, keep a look out for occultation predictions in *Sky and Telescope*, *Astronomy*, the ALPO *Solar System Ephemeris* and the *IAU Circulars*. You may also need a tape recorder and a stopwatch registering to 0.1 second. A short-wave radio receiver can be used to provide an accurate time source.

Experience is necessary in occultation work. It is good to practice timing lunar stellar occultations. Your telescope's aperture should be sufficient to reveal stars one full magnitude fainter than the occulted star.

Positional studies

Only highly competent and well-equipped amateur observers and professional astronomers who can carry out long-focus astrometry can do this type of work really well. Precise position determination is the only kind worth doing and requires the use of plate measuring equipment.

Further reading

Books

The Planet Uranus. Alexander, A. F. O'D., Faber and Faber, London (1965).

Uranus and the Outer Planets. Smith, B. A. and Reitsema, H. J., (ed. G. Hunt). Cambridge University Press, Cambridge (1982).

Uranus and Neptune: The Distant Giants. Burgess, E., Columbia University Press, New York (1988).

Atlas of Uranus. Hunt, G. and Moore, P. A. Cambridge University Press, Cambridge (1989).

Uranus: The Planet, Rings and Satellites. Miner, E. D., Prentice Hall, New York (1990).

Papers and articles

Herschel and the rings of Uranus. Capen, C. F., *Astronomy* **7(1)**, 42–5 (1979).

The moons of Uranus, Neptune and Pluto. Brown, R. H., and Cruikshank, D. P., *Scientific American* **253(1)**, 28 (1985).

Discovering the rings of Uranus. Elliott, J. M., Dunham E. and Millis, R. L., *Sky and Telescope* **53(6)**, 412–16 (1977).

Uranus: On the eve of ecounter. Crosswell, K., *Astronomy* **13(9)**, 6–22 (1985).

Voyager: discovery at Uranus. Berry, R., *Astronomy* **14(5)**, 6–22 (1986).

Uranus: The voyage continues. Berry, R., *Astronomy* **14(4)**, 7–22 (1986).

Science. Voyager 2 issue (July 4th 1986), pages 39–109.

Big, blue: The twin worlds of Uranus and Neptune. Dowling, T., *Astronomy* **18(10)**, 42–53 (1990).

The work of the ALPO Remote Planets Section. Hodgson, R. G., *JALPO* **27(5–6)**, 85–8 (1978).

The remote planets: 1987–90 report. Schmude, R. W., *JALPO* **35(3)**, 153–6 (1991).

A quinquennial observational report of the ALPO Remote Planets Section for the years 1985–1989. Hodgson, R. G., *JALPO* **35(3)**, 97–9 (1991).

Uranus, Neptune and Pluto: Contributions that ALPO members can make. Schmude, R. W., *JALPO* **35(2)**, 67–8 (1991).

The 1991 apparition of Uranus. Schmude, R. W., *JALPO* **36(1)**, 20–2 (1992).

Multicolour photoelectric photometry of Uranus. Appleby, J. F. and Irvine, W. M., *Astronomical Journal* **76**, 616 (1971).

12

Neptune

General

Like Uranus its close relative, Neptune is a gas giant planet, a huge rapidly rotating largely fluid world with a density more than twice that of water, greater than that of Uranus, and with a considerable atmosphere. The equatorial diameter is 31 403 miles (50 538 km) and the polar diameter is 30 589 miles (49229 km) so that the ratio of the polar to the equatorial diameter is 0.98. Neptune is about four times bigger than the Earth (fig. 21.1) and is slightly smaller than Uranus. The rotational axis is inclined 29.6° to the plane of its orbit so that it does not share the remarkable axial tilt of Uranus. The axial rotation period is 16 hours 3 minutes (16 hours 7 minutes for the magnetic field). Neptune's distance from the sun varies from an aphelion distance of 2819.2 million miles (4537.0 million km) to a perihelion distance of 2771.4 million miles (4460.2 million km).

The mean orbital velocity is 3.37 miles (5.43 km) per second and the orbital eccentricity is 0.009. The orbit is inclined 1° 46′ to the plane of the ecliptic. Neptune's orbital (sidereal) revolution period is 164.8 Earth years and its synodic period is 367.5 Earth days.

Neptune is encircled by eight satellites and a system of rings. Two of the satellites, Triton and Nereid, were discovered by Earth-based observation, the remainder by the Voyager 2 spacecraft. The orbits of Triton and Nereid around Neptune are shown in fig 12.2. There are three rings and some incomplete rings or 'ring arcs'. The inner ring is diffuse but the outer two are more clearly defined. Details of the satellites and rings are given in tables 12.1 and 12.2

The discovery of Neptune

The prediscovery observations of Uranus mentioned in the previous chapter were reliable within their limitation and in 1840 Bessel vindicated their accuracy. These observations, extending as they did over very many years, should have been most useful as they gave widely separated positions on the path of Uranus and so should have made possible accurate calculation of its orbit so that its future positions could be accurately predicted. A. Bouvard attempted to calcu-

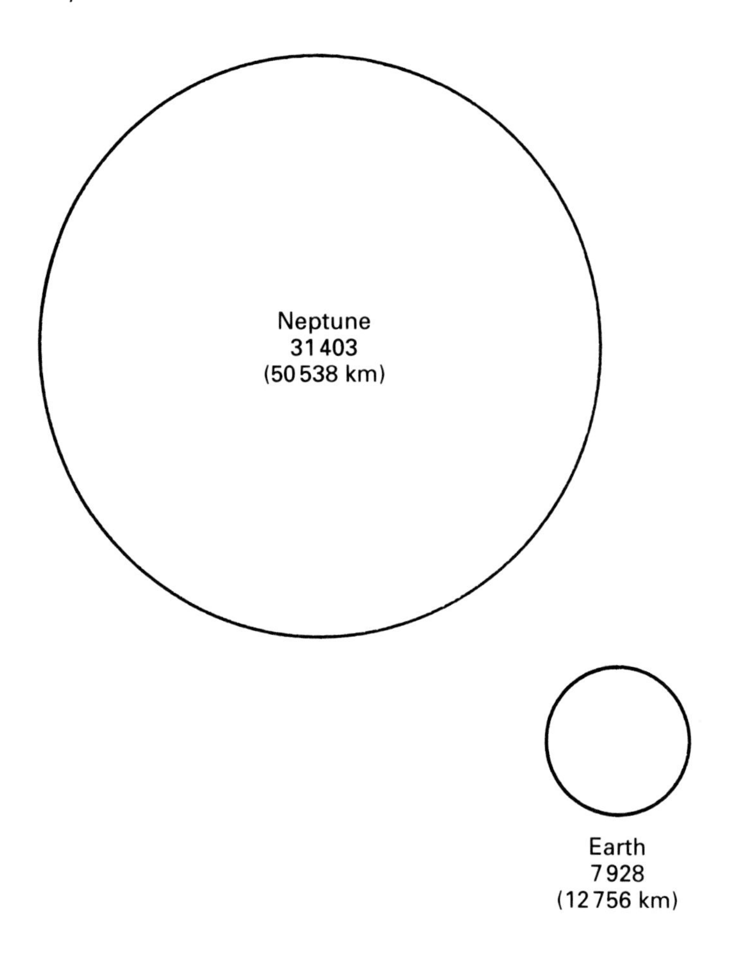

Fig. 12.1 Comparative sizes of the Earth and Neptune (equatorial distance in miles).

Table 12.1. *The Satellites of Neptune.*

Satellite	Diameter		Mean distance from Neptune		Orbital eccentricity	Sidereal period (days)
	miles	km	miles	km		
Triton	1690	2700	219 300	354 760	0.000	5.877
Nereid	210	340	3.45×10^6	5.51×10^6	0.751	360.136
1989 N1	260	420	73 100	117 600	low	1.122
1989 N2	124	200	45 700	73 600	low	0.555
1989 N3	87	140	32 600	52 500	low	0.335
1989 N4	99	160	38 500	62 000	low	0.429
1989 N5	56	90	31 100	50 000	low	0.311
1989 N6	31	50	30 000	48 200	low	0.294

Many of these data are revised values from the *Astronomical Almanac* for 1992, pages F2–F3.

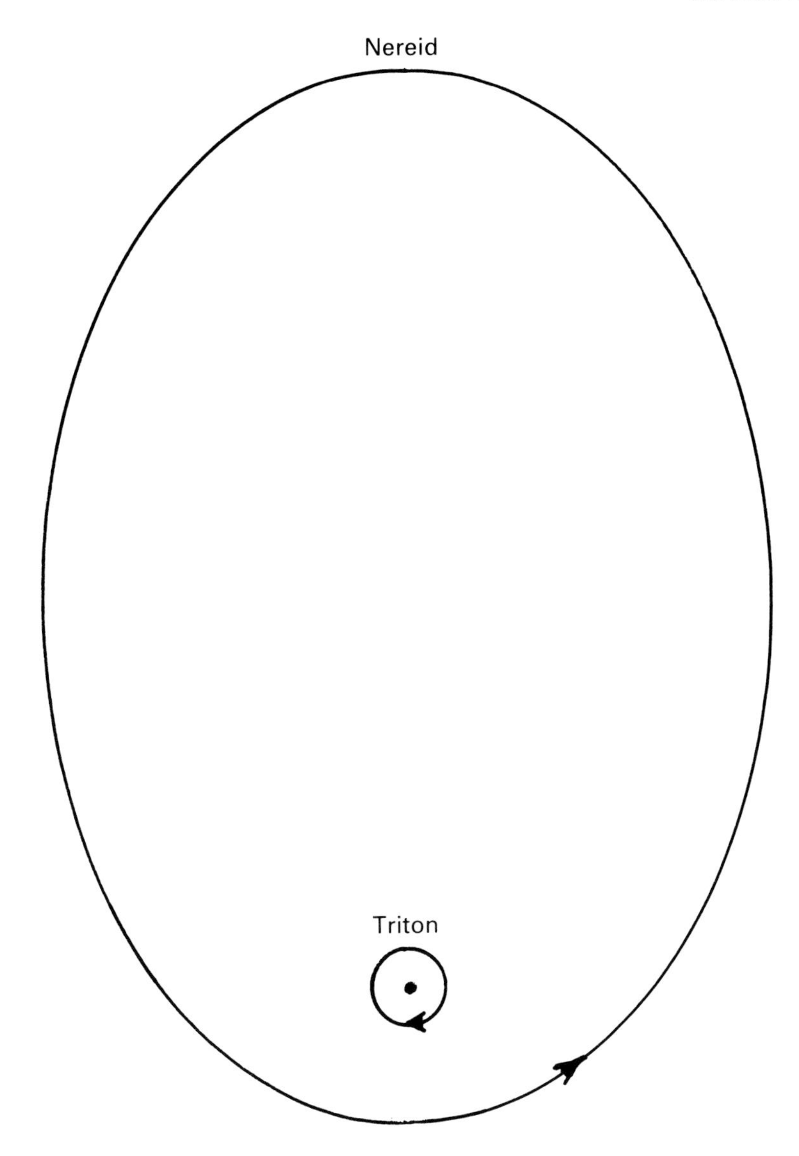

Fig. 12.2 The orbits of Neptune's satellites Triton and Nereid.

Table 12.2 *Rings of Neptune.*

Ring	Radius		Width	
	miles	km	miles	km
1989 N1R	39 080	62 900	31	50
1989 N2R	33 060	53 200	68	110
Plateau (1989 N5R)	35 730	57 500	2486.0	4000
Fuzzy (1989 N3R)	26 040	41 900	1060	1700

Data from *Science*, December 15th 1989, page 1419.

late the orbit of Uranus using the early observational data (1690–1771) and later positional observations made between 1781 and 1820. He made appropriate allowance for the gravitational attractions of Jupiter and Saturn on Uranus but could not compute an orbit that would fit the early and later observations even approximately. He was therefore compelled to base his Uranus tables, published in 1821, on the latter observations only. His own suggestion was that the irreconcilability of the early and later observations were due either to inaccuracy of the early ones or to an unknown disturbing effect on the motion of Uranus.

Positional observations of Uranus after 1820 showed that the planet was consistently falling further and further behind its predicted place in the sky although it was near its calculated place in 1829 and 1830. By 1834, the cause of this was being attributed to the gravitational attraction of a planet beyond the orbit of Uranus. George Biddell Airy, the new Astronomer Royal, wrote to Bouvard in 1837 that the errors of longitude of Uranus were 'increasing with fearful rapidity' (*The Discovery of Neptune*, Grosser, M., 1962, page 93) but he doubted that these were due to the effect of a remote unseen planet and that it would be extremely difficult to locate such a planet even if it existed; the required data concerning the irregularities in the motions of Uranus for the calculations necessary to determine the unknown planet's position would take centuries to acquire. Airy doubted the accuracy of the calculations of the orbit of Uranus and, along with other astronomers, even wondered if the Law of Gravitation was valid at such enormous distances from the sun.

By 1845, Uranus was a full 2 minutes of arc out of place which though practically undetectable by the naked eye was enough to grievously worry astronomers and mathematicians.

On June 26th 1841 John Couch Adams, a mathematics student at the University of Cambridge, resolved to investigate the irregularities in the motions of Uranus and if possible to determine if they were due to an unknown planet. He embarked on this task after taking his degree in January 1843, and arrived at an approximate solution before the year ended. Through Professor Challis, the Director of the Cambridge Observatory, he made application to the Astronomer Royal for complete details of the errors in the position of Uranus. These were quickly provided. Adams, who was only 26 years old in 1845, had practically solved the problem by September of that year. From its influence on the motions of Uranus, he had calculated the probable mass, orbit and position in the sky of the trans-Uranian planet. Adams presented Professor Challis with a copy of his calculations and results and asked to talk about them in person to the Astronomer Royal. Challis wrote a letter to introduce Adams but Adams never made a firm appointment to see the Astronomer Royal. This was a mistake; Airy was busy with routine duties of many sorts and although Adams called at the Royal Observatory, Greenwich, on three occasions during September and October, he was unable to see the Astronomer Royal; he was away the first two times and although in on the third occasion, appeared not to have received the message from Adams. Adams left a brief account of his results and departed upset at Airy's seeming refusal to see him.

Airy was not too impressed with Adam's results as he did not know how they had been calculated. Two weeks later he wrote to Adams stating his interest and asked a question that was crucial to Airy – did the solution arrived at by Adams provide an explanation for the error in the radius vector of Uranus? Adams, now discouraged, did not deign to reply for a whole year to what he thought was a trivial question with an obvious answer. Meanwhile, not completely satisfied with his first calculations, Adams did them all over again and though often interrupted sent the improved but essentially similar results to Airy in September 1846.

Although both Professor Challis and the Astronomer Royal had been presented with the solution to the most puzzling astronomical problem of their day, they did not follow it up. Challis afterwards said that telescopic searching for a planet based on the results of mathematical deductions would involve much hard work with scant chance of success. Airy doubted the accuracy and care of mathematicians and was overly concerned about the risk of error. He had also been inhibited from acting on Adam's results by the latter's failure to answer his question which he considered very important; he did not understand why Adams was silent on this and thought that maybe his question had somehow spoilt the solution to a problem that he had always thought was impossible to solve.

The problem was independently solved by another young mathematician, a Frenchman, Urbain J. J. Le Verrier, who was 35 years old in 1846. He believed equally with Adams that Newton's Law of Gravitation was valid everywhere in the Universe and that there was an unknown planet beyond the orbit of Uranus. Le Verrier presented his solution to the French Academy and it was published in June 1846. When it was received in England, Airy read it and was greatly pleased since Le Verrier had positioned the trans-Uranian planet within 1° of the place predicted by Adams. Airy no longer doubted yet put his crucial – Littmann calls it 'inane' – question again to Le Verrier but unhappily did not mention Adam's solution. A full reply was soon received from Le Verrier and he asked for a search to be made for the unknown planet, offering to calculate and send Airy details of where in the sky the planet would be found. Airy initiated the search although he did not reply to Le Verrier. He urged Challis to use the 11.75-inch Northumberland refractor at Cambridge as he considered the Greenwich telescopes to be inadequate. Unfortunately, no reliable star chart was to be had so that Challis had himself to make a chart of all faint stars down to the 11th magnitude in the rather large area of the sky assigned by Airy. Challis plotted the positions of over 3000 stars and was so busied with this work that he had not time to compare observations. As a result he missed being the first to find Neptune as he had actually plotted its position on July 30th and August 12th when his telescope was pointed at the place Adams had indicated. He simply did not realise that the two 'stars' that he had seen were really Neptune in two different positions. Credit for the discovery of Neptune might still have gone to England had not W. Lassell, a highly skilled amateur observer, been incapacitated with a sprained ankle (according to Grosser, Littmann and Burgess) early in September.

On September 23rd 1846, which was exactly one year after Adams first called on the Astronomer Royal, the assistant astronomer at the Berlin Observatory, Dr J. G. Galle, received a letter from Le Verrier showing where the new planet should be seen and suggesting that telescopic search be made for it. The suggestion was made later in the observatory itself. The young astronomer D'Arrest stated that the area of the sky to be searched may be on a new and as yet unpublished star chart. This was correct and the search for the disturbing planet started on that same evening. Galle observed with the 9-inch refractor, calling out the stars while D'Arrest verified them on the chart. Galle called out the position of an 8 magnitude star that D'Arrest declared was not on the map. This proved to be Neptune. It was located less than 1° from where Le Verrier had predicted, which was about 1° north of a point one third of the distance from the star mu Capricorni to iota Aquarii (fig. 12.3). Its motion was confirmed on the following night and its disc was measured.

Le Verrier's achievement caused great joy in France but this was dampened when Sir John Herschel announced the independent solution put forward by Adams. An unseemly controversy followed. Unfair attacks were made, especially upon the Astronomer Royal. National pride was whipped up in both France and England. However, Adams made no exaggerated claim for recognition of his part in the discovery of Neptune. Both he and Le Verrier became the best of friends and subsequently equal honour for the mathematical triumph was accorded to both of them.

Fortunately, at the time of its discovery (1781), Uranus had recently moved into that half of its orbit which is nearer to Neptune. (It was in heliocentric conjunction in 1821 according to Meeus and will be again on April 21st 1993. There will be geometric conjunctions in 1993 on January 26th and on September 17th and 28th). Uranus passed by Neptune in 1822, making its closest approach. This happens once in every 172 years. Therefore, Bouvard's calculations of the orbit of Uranus was made at a time (1781–1820) when the orbital speed of Uranus was greater than usual and was being speeded up by Neptune's gravitational pull. Then in 1822 Neptune would be pulling against Uranus and would cause it to

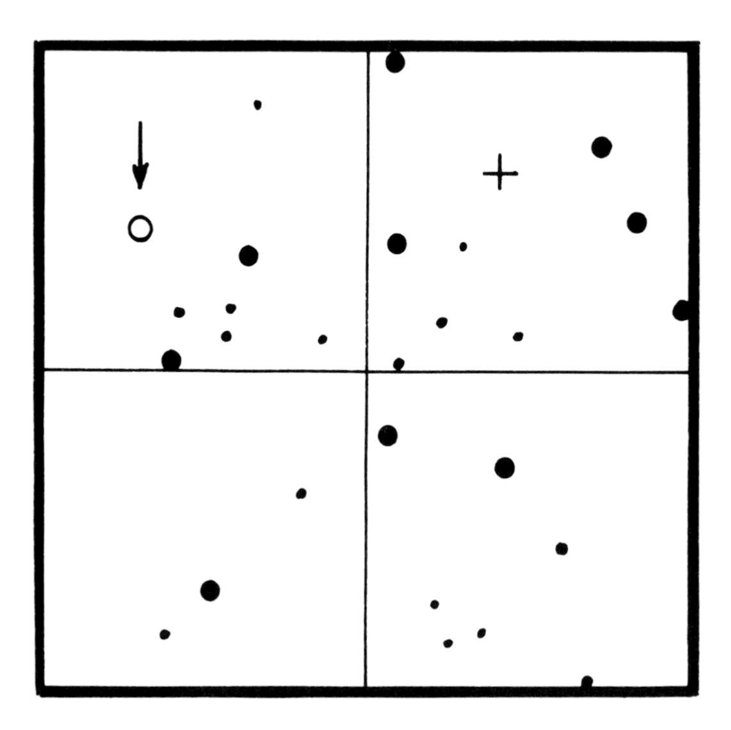

Fig. 12.3 Part of Galle's star map showing where Neptune was found (arrow). Le Verrier's calculated position of Neptune is shown by the cross.

decelerate. If Uranus had been on the other side of its orbit in 1781 the effect of Neptune on the motion of Uranus would have been so slight as to have gone unnoticed until the late nineteenth century. It is possible that Neptune might then have been found telescopically before its position was mathematically deduced. In any case, it might have been as much as half a century after its actual discovery before it was found.

Prediscovery sightings of Neptune

It was Galileo who, in 1612, made the first ever recorded observation of Neptune when he was observing the planet Jupiter and its satellites with one of his telescopes. It so happened that Neptune stood quite close to Jupiter in the line of sight at the time because Neptune was occulted by Jupiter in 1613 as was calculated by S. C. Albers in 1979. In 1980 C. T. Kowal and S. Drake studied Galileo's observations; on one of Galileo's sketches made on December 28th 1612 they found marked a 'star' in the position expected for Neptune, not actually in the correct place, which would have been beyond the edge of the drawing, but Galileo indicated the true position with a broken line. In 1613 on January 28th, Galileo marked the position of the star SAO 119234 (magnitude 7.2) which was near to Jupiter. He marked it 'a', another object marked 'b' must have been Neptune because Galileo noticed that the object 'b' which he had seen on the previous night appeared then to have been further from 'a'. Galileo had therefore witnessed Neptune's motion. D. Hughes in 1980 doubted that Neptune had really been detected by Galileo. There is no doubt, however, that Galileo did detect Neptune's motion but he did not realise its significance, for the object 'b' is almost certainly Neptune.

The question of whether Neptune had been seen by others in the telescopic era since Galileo's time was investigated by S. C. Walker of the US Naval Observatory. He traced the path of Neptune among the stars back in time to ascertain if that part of the sky where it was located at a given date was under telescopic scrutiny in the process of star map making. Not until he went as far back as 1795 did he find evidence that this had happened. On May 8th and 10th of that year, Lalande in Paris had been observing that part of the sky where Neptune was situated. Walker deduced that one of the stars seen by Lalande was near to the place where Neptune would have been. When the same part of the sky was studied telescopically by Hubbard at Washington, the 'star' was missing. However, this did not quite constitute conclusive proof that the 'star' Lalande saw was actually Neptune. Lalande was not sure of the accuracy of the position he had recorded for the star so he indicated it with a colon. The record of a missing star may therefore have been the result of positional error. Fortunately, Lalande's original records were still in safe storage at the Paris Observatory. A study of these revealed that Lalande had not, after all, made a mark of uncertainty in his records of the observations of May 8th and 10th 1795. He had actually observed Neptune on 8th and 10th of May. The object had moved slightly during this interval so the observations were discrepant and Lalande thought that one of them must be erroneous. He was quite unaware that in that discrepancy he had a major discovery before his eyes. (The 'star' was also shown by K. L. Harding in an 1822 star atlas). He had rejected the first observation and copied the other one as unreliable because of the discrepancy and without further examination.

For many years Sears Cook Walker and Professor Benjamin Pierce investigated the motions of Neptune caused by the other planets.

History of observation

William Lassell discovered Neptune's main satellite on October 10th 1846 and observed it often during the following months. He concluded that it makes one orbital revolution around Neptune in 5 days and 21 hours and its greatest apparent distance from its primary is 17 seconds of arc. He said that the preceding half of the satellite was brighter than the following half but this has not been substantiated. Harris mentions that the difference amounts to about 0.25 magnitude. At its mean opposition distance it has a magnitude of about 13.5. The name Triton for the satellite was mentioned by Fouche in 1905 and proposed for it by C. Flammarion. Triton's orbital motion around Neptune is retrograde.

Lassell thought that he had spotted a second satellite on one occasion in 1852 but he was probably mistaken. Sir J. Herschel stated in 1869 that two satellites of Neptune discovered by Lassell had also been seen by American and European observers but P. A. Moore was not able to verify this after careful literature search. There was an unconfirmed report of a second satellite by J. Schaeberle who observed with the Lick refractor on September 24th 1892.

Lassell also thought that he had detected a ring around Neptune during observations with his 24-inch reflector. Although he had discovered Neptune's satellite Triton on October 10th 1846, soon after the discovery of Neptune, his first observations were actually on October 3rd. It was then that he suspected that Neptune was surrounded by a ring. During the ensuing weeks he made further observations that strengthened this belief.

In November 1846, Lassell gave a report before the Royal Astronomical Society. One of those present at the meeting was J. R. Hind who himself had made frequent observations of Neptune with a 7-inch refractor and on December 11th 1846 also reported on the existence of the Neptunian ring or at least an oblong appearance of the disc.

Professor J. Challis observed the ring on January 12th 1847 with the Cambridge Observatory's 11.75-inch Northumberland refractor. He communicated his results to Lassell.

Something odd about the telescopic appearance of Neptune was remarked upon by W. C. Bond, M. F. Maury and F. di Vico but the reports are too vague to be considered reliable as sightings of a ring.

W. R. Dawes doubted the existence of a ring but was convinced when he viewed Neptune with Lassell's reflector. In 1847 the Royal Astronomical Society considered that the ring really existed.

Further observations were made by J. Nasmyth and Lassell in 1848 which gave them both an even stronger conviction of the ring's existence. However, Lassell failed to see it later that year, neither did he see it in 1849; observing conditions then were poor, however. He really began to doubt that the ring existed by 1850. Observing Neptune with the same big reflector on November 4th 1852 in the clearer skies of Malta, Lassell had again an impression of the ring but considered that it might be illusory although he wrote that he had a 'very fine' view of Neptune. Then, on December 15th, he saw that the ring seemed to change its orientation when he rotated the tube of his telescope and so concluded that the ring was after all an optical artefact. Thereafter, Lassell made no further allusions to the ring after making a report on its illusory nature and Hind, Challis and Dawes made no further observations and nor did anyone else. Lassell had made a mistake and he was quick in admitting it. The illusory ring seems to have been produced as a result of difficulties in preventing flexure of the 24-inch mirror when the telescope was moved. The apparent confirmation of the Neptunian ring by Challis and Hind must have been because they went to their telescopes with a preconceived notion of what to look for after Lassell had informed them of the presence of the ring. This is not an imputation of dishonesty on their part any more than seeing canals on Mars, although they don't exist, is dishonest. It is an unfortunate but common unconscious fault due to the way our minds interact with – and misinterpret – what our eyes see in a telescopic planetary image especially if we go to the telescope expecting to see a particular phenomenon.

Although Dawes and Nasmyth also saw the ring with Lassell's reflector they never saw it with any other telescope, which is significant. Whatever it was that Lassell saw, it certainly bears no relation to the ring discovered by the Voyager 2 spacecraft.

In 1899 on October 10th, equatorial belts were claimed to have been seen by T. J. J. See with the 26-inch Washington refractor. He made a drawing showing faint dusky bands and a dusky area at the same latitude as the Great Dark Spot discovered by the Voyager 2 spacecraft in August 1989.

In 1894 E. E. Barnard could detect no polar flattening of the globe of Neptune with the 36-inch Lick Observatory refractor although he detected the polar flattening of Uranus. Neither could he see features on the disc of Neptune even with the Yerkes and Lick telescopes.

Although a 'gas giant' planet, Neptune is so remote that only a tiny somewhat ill-defined disc can be seen in the telescope. This made it difficult for early observers using filar micrometers to measure the apparent angular diameter of Neptune. Most early measurements gave the size as 2–3 seconds of arc which gives an actual diameter of about 30 000 miles (50 000 km). Encke measured the angular size as 3.2 seconds of arc and Main found it to be 4.36 seconds of arc.

Until quite recent times it was generally thought that Neptune was slightly larger than Uranus but it is now known that it is somewhat smaller but more massive. In 1969 G. E. Taylor gave the equatorial diameter of Neptune as 31 650 miles (50 940 km) and 31 020 miles (49 920 km) as the polar diameter. Since 1952 G. E. Taylor has pioneered the stellar occultation method of measuring the diameter of small planetary discs. If the distance and velocity of the occulting planet is known then the duration of a stellar occultation enables the diameter of the planet to be calculated. Many attempts have been made utilising occultations of asteroids and satellites of the planets.

The equatorial diameters of Neptune and Uranus were respectively given as 31 403 miles (50 538 km) and 37 763 miles (51 118 km) with a maximum uncertainty of 12.4 miles (20 km) by the General Assembly of the IAU in 1988 by M. E. Davies and coworkers. The polar diameter of Neptune was given as 30 820 miles (49 600 km) with an uncertainty of about 18.6 miles (30 km). These values are in good agreement with Taylor's 1969 measures.

Attempts to measure the axial rotation period were made almost impossible by the absence of really definite disc features. Brightness fluctuations of Neptune were detected by M. Hall in 1883. He saw Neptune as a bluish disc and noted magnitude changes from 7.6 to 8.3. In 1884 Hall attempted to measure the axial rotation period of Neptune from these light variations and gave a value of 7 hours and 55 minutes, which is much shorter than the modern accepted value. The spectroscopic method was used to measure Doppler shifts by J. H. Moore and D. H. Menzel in 1928 who gave a value of 15.8 hour ± 1 hour. This was a good estimate considering the difficulties; the actual value is 17.8 hours. S. H. Hayes and M. J. S. Belton studied the tilts of spectral lines and arrived at a value of 22 ± 4 hours for the axial rotation period but this value was subsequently found to be erroneous, limb darkening possibly being a factor. An assumed period of from 15 to 20 hours was adopted. Light curve studies enabled D. P. Cruikshank in 1978 to propose two possible values for the axial rotation period: 0.7572 day or 0.8160 day with an uncertainty of 0.002 day.

A search for a second Neptunian satellite was taken up unsuccessfully by W. H. M. Christie who used the Mount Wilson 60-inch reflector. In 1949, G. P. Kuiper detected a second satellite photographically using the 82-inch reflector of the McDonald Observatory and this was confirmed by subsequent photography. The satellite, named Nereid, has a magnitude of 19.5. Its diameter was estimated by Davies in 1988 as 428.8 miles (690 km) with an uncertainty of 223.7 miles (360 km). It has a highly eccentric orbit and makes one revolution around Neptune in 359.9 Earth days. In 1987 M. W. Schaefer and B. E. Schaefer using CCD cameras and a 0.9-metre reflector at Cerro Tololo in Chile made photometric observations of Nereid and found its brightness to vary by over 1.5 magnitudes indicating an irregular shape or albedo differences on its surface. The rotation period is apparently between 8 and 24 hours.

In 1954 G. P. Kuiper measured the diameter of Neptune's satellite Triton as being 2361.7 miles (3800 km). Subsequent estimates by other observers gave differing results, some as much as 3729.0 miles (6000 km) and some as little as 1553.8 miles (2500 km). The presently accepted value is 2175.3 miles (3500 km) with an uncertainty of 310.8 miles (500 km). The presence of an atmosphere around Triton was well established by 1988.

Also in 1954 there appeared a report of a sighting of surface detail on Neptune. Observing with the 30-inch Thaw refractor of the Alleghany Observatory, University of Pittsburgh, on May 30th 1954, O. C. Ranck sketched the planet and showed a bright equatorial region flanked by two dusky bands and a shaded limb. The observation was confirmed by the Observatory Director, N. E. Wagman and also by C. McClellan. Other observations of surface features by Cragg, Epstein and Ranck in 1953 may be found on pages 103–4 of the *Strolling Astronomer* of July 1953. A drawing made by Cragg using the Mount Wilson 60-inch reflector on April 17th 1953 shows a bright equatorial band and dusky poles (fig. 12.4).

Fig 12.5 shows an observational drawing of Neptune made in 1950 by Charles C. Capen using a 12.5-inch reflector. In 1974 Capen published an observational drawing of Neptune in the form of a slide, by the Hansen Planetarium, showing a dark spot that may have been the Great Dark Spot discovered by the Voyager 2 spacecraft in August, 1989.

Filters and CCD equipment (see chapter 15) were used by B. A. Smith,

Fig. 12.4 Neptune as seen by Thomas Cragg in April, 1953: 60-inch reflector (Mount Wilson). (From Sky and Telescope **77(5)**, *486, 1989.)*

Fig. 12.5 Neptune as seen by Charles C. Capen in April, 1950: 12.5-inch reflector. (From Sky and Telescope **77(5)**, *486, 1989.*

H. J. Reitsema and S. M. Larson who reported recording cloud features in 1979. The first time that a cloud feature on Neptune was used to determine its axial rotation period was by B. J. Terrile and B. A. Smith in 1983 prior to the Voyager spacecraft results. Visibility of cloud features was made possible by using a CCD camera and two coronagraphs with a broad band infrared filter centred on 8900 angstroms in conjunction with the 2.5-metre reflector of the Las Campana Observatory, Chile. The images that they secured of Neptune revealed cloud patterns resembling those obtained in 1979. A number of haze patches were seen in the middle latitudes of both hemispheres. For the axial rotation period a value of 17 hours 50 minutes was derived with an uncertainty of 5 minutes.

The changing energy output of the sun appears to influence the brightness of Neptune. In 1977, G. W. Lockwood discovered that Neptune brightened by from 0.005 to 0.025 magnitude per year between 1972 and 1976 and suggested a connection to the declining half of the solar cycle at this time. Lockwood and D. T. Thompson reported in 1986 that during the next solar cycle the brightness and albedo of Neptune changed periodically and had an amplitude of 4%. It was negatively correlated with the sun's output of ultraviolet radiation.

Prior to the Voyager 2 mission many astronomers had studied stellar occultations by Neptune but these gave no conclusive evidence of rings around Neptune. Definite indications of the existence of a ring around Neptune were obtained in 1968 by E. Guinan and J. S. Shaw of Villanova University, who went to New Zealand to observe the occultation of a star by Neptune. The intention was to learn something about the planet's atmosphere. The occultation was of the magnitude 7.8 star BD–174388 and occured on April 7th 1968. The results were not studied for years afterwards. Apparently, Guinan thought that the data were insufficient for his purpose. It was not until 1981 when he made a close examination of the results that he noticed an inexplicable fading of the light of the star that lasted for 150 seconds after Neptune was well past the star. None of Neptune's satellites could account for it. Guinan therefore concluded that there was a ring of obscuring matter surrounding Neptune. He believed the ring to be small and nearer to Neptune than the rings of Saturn and Uranus. Calculation showed the ring's inner edge to be 2039 miles (3281 km) above the planet's cloud tops.

Between 1981 and 1985 observations of six occultations (May 24th 1981, June 15th 1983, September 12th 1983, July 22nd 1984, June 7th 1985 and June 25th 1985) showed good evidence of the presence of obscuring ring material surrounding Neptune but oddly, the light of the occulted stars dimmed on one side of Neptune but not on both. This indicated the presence of incomplete rings or 'ring arcs', which had not been met with before. Some planetary rings, especially the epsilon ring of Uranus, vary greatly in density and width at different parts of the circumference but are still complete.

Spacecraft exploration of Neptune

The Voyager 2 spacecraft encountered Neptune at close quarters on August 25th 1989. Thousands of photographs were taken. They showed that Neptune was a much more interesting world than Uranus and was surprisingly active despite its location in the intense cold of the outer Solar System. Overall, Neptune

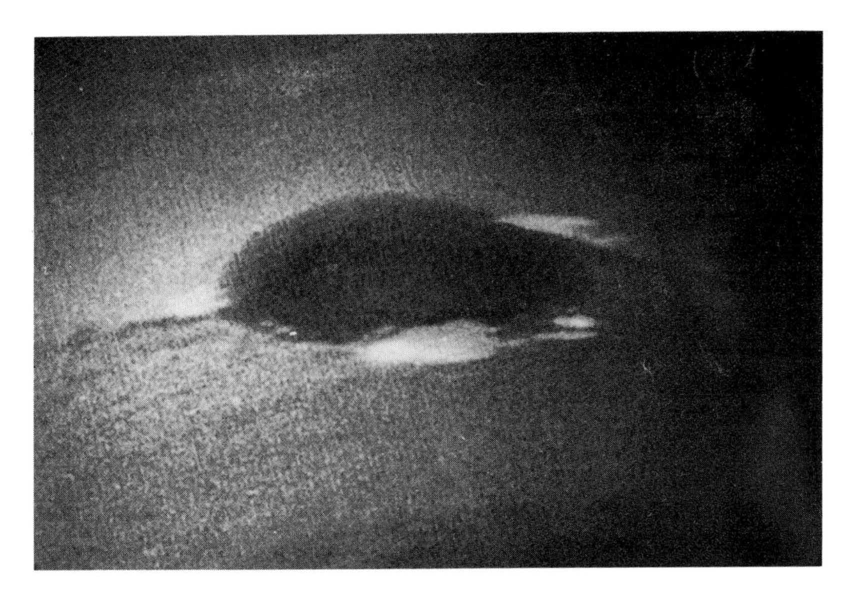

Fig. 12.6 Neptune's Great Dark Spot (colour-enhanced Voyager image).

appears to have dark temperate belts, light tropical belts and a bright polar hood in its southern hemisphere.

Globe features. Westward blowing winds of at least 700 miles (1126.3 km) per hour were recorded. Large vortices are immersed in this wind system, the biggest of which is the 'Great Dark Spot', an enormous Earth-sized whirlwind exhibiting counter-clockwise rotation with a period of 10 days (fig. 12.6). It elongates and contracts over an 8-day period. It has a 38° extent in longitude and 15° in latitude. (For comparison – Jupiter's Great Red Spot extends over 30° in longitude and 20° in latitude.) This corresponds to 7500 miles (12 000 km) in longitude and 5000 miles (8000 km) in latitude. Relative to Neptune it is about the same size as Jupiter's Great Red Spot. Also like Jupiter's Red Spot, the Great Dark Spot is located at 21° south latitude and is a long-lived feature; at least it has persisted for 9 months that we know of and maybe for much longer. At least two Earth-based observations of Neptune, by T. J. J. See (in 1899) and C. F. Capen (in 1974), seem to show a feature resembling it as previously mentioned.

Over the span of seven 18-hour rotations of Neptune the Great Dark Spot exhibits marked changes in shape and extent. The Great Dark Spot shows dramatic differences in shape when photographs taken simultaneously in different wavelengths of light are compared.

Within the Great Dark Spot there are none of the stratus clouds of methane that envelope most of Neptune. The spot appears to be a gigantic cavity that enables us to see deeper into the atmosphere. About 30 miles (50 km) above this huge cavity, clouds like our terrestrial cirrus appear and disappear.

Near the Great Dark Spot and to the south is a bright spot designated S1. Almost certainly this is the spot observed from Earth for several years by Heidi Hammel of the Jet Propulsion Laboratory. It too is quite changeable in its aspect but always stays just to the south of the Great Dark Spot. S1 and other bright clouds near the Great Dark Spot are high altitude clouds. Also to the south of the Great Dark Spot is another bright spot at latitude 41° south and named S2. Because of its rapid 16-hour circuit of Neptune it has been nicknamed 'Scooter'. It is the largest of a group of small white spots.

Another smaller, possibly even darker spot, Dark Spot Two (DS2) is located to the south of the Great Dark Spot and above it rises a plume of air rich in methane that condenses and forms white clouds. This appears as a white spot in the centre of DS2.

At latitude 76° south a system of clouds – the South Polar Feature – encircles the pole which though a permanent feature undergoes considerable variations in form and size.

Rings. The Voyager 2 spacecraft showed that like Uranus Neptune is circled by a system of rings. There is a sheet of dust in the ring plane. The ring closest to Neptune is the so-called 'fuzzy ring' because it is very diffuse. It is about 4000 miles (2500 km) in breadth. Outside of this ring are the two brightest rings designated 1989 N1R and 1989 N2R. They are very thin and the ring particles are kept in strictly confined paths by two small 'shepherd' satellites. The outermost ring is puzzling because within the ring are three condensations of bright ring material – ring arcs – containing much more material than the remainder of the ring. High-resolution images reveal bright points of light within the arcs each maybe about 6 miles (10 km) or smaller in diameter. The other rings did not appear to have any associated moonlets. Why the material in the arcs does not spread out evenly in the ring is a mystery. None of the usual explanations like corotation resonances or shepherding satellites can account for the arcs in ring 1989 N1R. Neither can moonlets within the arcs be the explanation because the arcs looked bright when seen looking back at Neptune from a high sun angle, not dark as would be expected if there was a moonlet within the arcs casting a shadow.

A thin broad disc of fine dust stretches from halfway between the two outer rings and inward toward Neptune. The edge of the dust sheet, known as the 'Plateau', is somewhat brighter than the rest. The other rings also contain much dust and are bright when viewed from high sun angles. However, the Plateau ring looks brightest in light scattered towards the sun so it must be composed of large particles and little dust. Neptune's rings are shown in fig. 12.7 and some data are listed in table 12.2

Satellites. Voyager 2 discovered several more satellites bringing the total to eight. Table 12.1 lists all of Neptune's satellites with data. Nereid is now third in order of size. The satellite 1989 N1 is 260 miles (420 km) in diameter whereas Nereid is 210 miles (340 km) in diameter. Satellite 1989 N1 would have been discovered by Earth-based telescopic observation had it been farther from Neptune. Both it and 1989 N2 have cratered surfaces and are not truly spherical. They are dark, reflecting only 6% of the incident sunlight.

Voyager imagery showed Triton's surface to be covered with very complex land structures and albedo markings. Much of the Equatorial Zone consists of low-relief polygonal ridges and valleys, the so-called 'cantaloupe' terrain. It has an enormous south polar cap which is almost a perfect white. At the time of the Voyager encounter it was summer in Triton's southern hemisphere and in spite of the great cold the polar cap seemed to be evaporating and disintegrating.

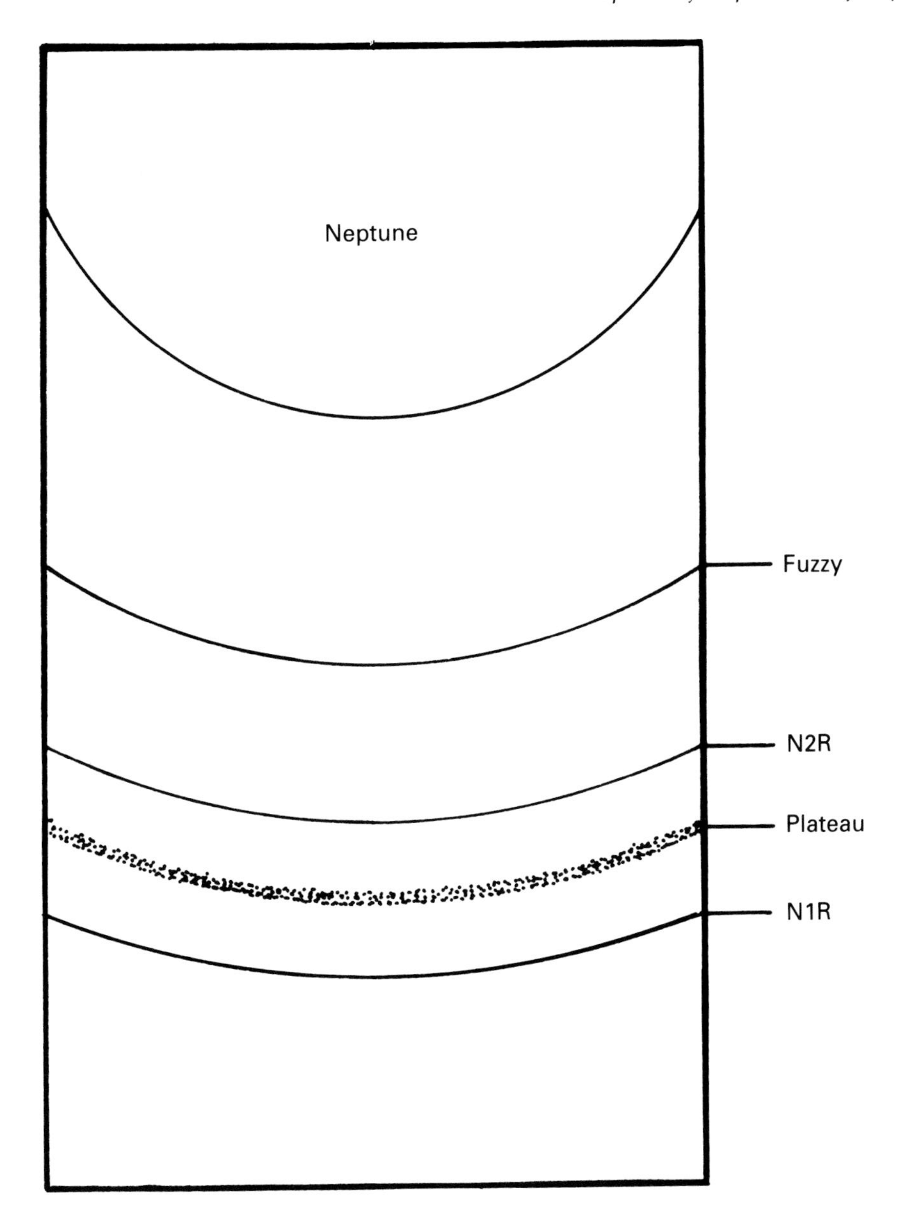

Fig. 12.7 The rings of Neptune.

Triton is the most massive satellite in the Solar System. It is unusual in having retrograde motion like an asteroid but it is much too large to be a captured asteroid. The orbit may be unstable and it is possible that a catastrophe at some remote epoch in the Solar System's history may have been responsible for the retrograde motion, Nereid's highly elongated orbit and maybe even for Pluto's if we consider Pluto to be a one-time satellite of Neptune. It is unclear as to whether Triton is a normal satellite with abnormal motion or a formerly independent body captured by Neptune. As commented by P. A. Moore, it certainly is a 'very remarkable object'.

Visibility of Neptune

Neptune comes to opposition about 2 days later each year. The declination limits are about 25° north and 25° south. Its maximum magnitude is 7.7 and so it is invisible to the naked eye but 7 × 50 binoculars will show it.

Observing Neptune

The apparent angular diameter of the disc varies between 2.0 and 2.2 seconds of arc. There is some disagreement as to what is the minimum aperture needed to show the disc. It was said to have been seen by some with a 3-inch refractor whilst its detection by others with a 6-inch refractor was dubious. There is difficulty in deciding whether a disc is discernible or not. Seeing conditions and the observer's experience are important here. Under most circumstances the disc is probably invisible with apertures of about 8 inches or less. Powers of ×200 or more are needed to show the disc. Under good seeing conditions it definitely has a non-stellar appearance with apertures over about 9 inches used with sufficient magnification (about ×500). A magnification of ×800 makes Neptune look the same size as the full moon to the naked eye but it is much dimmer, of course.

A good way to convince yourself of the non-stellar appearance of Neptune is first to study the carefully focused images of faint nearby stars to accustom the eye to them and then to turn the telescope to Neptune. Use a power of about ×100. Neptune won't 'jump out' at you from a star field as does Uranus but once you have seen it you should have no difficulty in recognising it again on future nights.

Not surprisingly, the disc will look featureless to you. In fact very few, if any, features have ever been certainly recorded on Neptune's disc by Earth-based telescopic observation.

Finding Neptune's satellite Triton (magnitude 13.5) will be about as difficult as finding Titania, one of the satellites of Uranus, but glare from Neptune will be less than that of Uranus. Triton's greatest apparent angular distance from Neptune is 17 seconds of arc. Photographs of Neptune made by F. J. Mellilo on June 15th and 21st 1985 using an 8-inch Celestron Schmidt–Cassegrain telescope show Triton; quite an achievement for an 8-inch telescope! (Westfall says that he couldn't see Triton with his 10-inch telescope from the city in California where he lives but it was visible in a 14-inch in the Sierra – personal communication).

The satellite Nereid has a magnitude of +19 and is so small that attempting to find it with anything less than a 100-inch telescope is doomed to failure. The magnitude of +19 is the theoretical limiting magnitude for such a telescope.

Magnitude studies

You can make magnitude estimates of Neptune by comparing it with stars of known magnitude near to it but as with Uranus, valuable work in this field can only be done with photoelectric photometric equipment (see chapter 16). This can be done with white light (no filters) or with precision colour filters, which enables the colour of Neptune to be determined at selected wavelengths, e.g., U, B, V, R and I (see chapter 16). Studies with these filters can provide information about Neptune's atmosphere and if conducted over several years may show subtle changes going on in the atmosphere in addition to the solar radiation that Neptune receives. The colour of Neptune can be determined by photoelectric or photographic photometry or by CCD imaging (see chapter 15). Of the above filters the I (near infrared) may be especially useful as it includes the absorption band of methane.

The ALPO *Solar System Ephemeris* gives maps of the paths of Uranus and Neptune in the sky with respect to photometric comparison stars.

Stellar occultation studies

Experienced occultation observers using even small telescopes can make useful contributions in stellar occultation studies of Neptune and its rings. The apparent number, position and width of the rings may thus be investigated. Photometric measurements made with large telescopes can be of value in studying the opacity of the rings. Occultation predictions will be found in *Sky and Telescope, Astronomy* and in the *IAU Circulars*.

You will need a tape recorder and a stopwatch registering to 0.1 second and a short wave radio receiver is useful for giving accurate time signals. Before embarking on occultation work practice is necessary. Try timing lunar occultations of stars. The telescope you use should reveal stars a full magnitude fainter than the occulted star. For visual timings the comparison star should not be more than one magnitude fainter than Neptune.

Further reading

Books

The Discovery of Neptune. Grosser, M., Harvard University Press (1962).

The Planets: Some Myths and Realities. Baum, R. M., David and Charles, Newton Abbot, England (1973).

The Planet Neptune. Moore, P. A., J. Wiley and Sons, New York (1988).

Planets Beyond: Discovering the Outer Solar System. Littmann, M., Wiley Science Editions, New York (1988).

Uranus and Neptune: The Distant Giants. Burgess, E., Columbia University Press, New York (1988).

Far Encounter: The Neptune System. Burgess, E., Columbia University Press, New York (1991).

Papers and articles

Two 'special' issues of *Science* detail the results obtained from the Voyager 2 mission:

The Neptunian System *Science* **246(4936)**, 1417–501, (December 15th 1989).

Neptune's Satellite Triton *Science* **250(4979)**, 410–43, (October 19th 1990).

The history of the discovery of Neptune. Adams, J. C., *Memoirs of the RAS*, **16** (1847).

The discoveries of Neptune and Pluto. Jackson, J., *Observatory*, **19**, 364 (1955).

The great unexplained residual in the orbit of Neptune. Rawlins, D., *The Astronomical Journal* **75(7)**, 856–7 (1970).

Galileo's sighting of Neptune. Drake, S. and Kowal, C. T., *Scientific American* **243(6)**, 74–81 (1980).

The moons of Uranus, Neptune and Pluto. Brown, R. H., and Cruikshank, D. P., *Scientific American*, **253(1)**, 28 (1985).

Long-term brightness variations of Neptune and the solar cycle modulation of its albedo. Lockwood, G. W. and Thompson, D. T., *Science*, **234**, 1543 (1986).

Arcs around Neptune? Murray, C. D., *Nature*, **324**, 209 (1986).

Neptune revealed. Berry, R., *Astronomy*, **17(12)**, 22–34 (1989).

Neptune. Kinoshita, J., *Scientific American* **261(5)**, 82–91 (1989).

Big, blue: The twin worlds of Uranus and Neptune. Dowling, T., *Astronomy*, **18(10)**, 42–53 (1990).

Neptune through the eyepiece. O'Meara, S. J., *Sky and Telescope*, **77**, 486–7 (1989).

Prediscovery images of Neptune's Great Dark Spot. Graham, F. G., *JALPO*, **35(2)**, 69 (1991).

The Work of the ALPO Remote Planets Section. Hodgson, R. G., *JALPO*, **27(5–6)**, 85–8 (1978).

A quinquennial observational report of the ALPO Remote Planets Section for the Years 1985–1989. Hodgson, R. G., *JALPO*, **35(3)**, 97–9 (1991).

At the time of writing an additional Neptune Report is due to appear in the *JALPO*, **36(2)**, probably in May 1992.

Uranus and Neptune: Contributions that ALPO members can make. Schmude, R. W., *JALPO*, **35(2)**, 67–8 (1991).

13

Pluto

General

Pluto, the outermost planet of the Solar System, has an estimated diameter of 1431 miles (2302 km) and a mass equal to 0.0025 Earth masses. It has an apparent visual magnitude of 13.7. Its axial rotation period is 6 days 9 hours and 17 minutes. The rotational axis is tilted at an angle of 118° to the plane of its orbit. Pluto orbits the sun at an average distance of 3674.48 million miles (5913.52 million km). Its orbital eccentricity of 0.25 is so high that Pluto will be closer to the sun than Neptune between 1979 and 1999. The orbit is inclined at an angle of 17.1° to the plane of the ecliptic. Pluto makes one complete orbital revolution around the sun in 248.54 Earth years and moves with an average speed of 2.95 miles (4.7 km) per second.

Pluto has one satellite, Charon that revolves around its primary at a distance of 12 204 miles (19 636 km). It makes one revolution around Pluto in the same time that Pluto rotates once on its axis. The minimum diameter given for Charon is 739 miles (1190 km). It has an apparent visual magnitude of 16.8. The plane of the orbit of Charon around the primary is tilted at a large angle (98.8° to the plane of Pluto's orbit. This means that twice in every orbital revolution of Pluto, Charon's orbital plane will be presented edgewise to the Earth and Pluto and Charon will show mutual transit and occultation phenomena. The first time this happened since Pluto's discovery occurred in 1985 and it won't happen again until the year 2109.

The search for a trans-Neptunian planet

In 1834 Peter A. Hansen suggested that the increasing residuals in the longitude of Uranus in its orbit could not be explained by the gravitational effects of one planet, i.e., by Neptune which had not yet been discovered. Residuals are the discrepancies that remain between the predicted and actual positions of a planet after the disturbing effects of other bodies have been allowed for. This belief that there may be a planet beyond Neptune came a full 12 years before the discovery of Neptune itself. Hansen's insight stimulated Percival Lowell 70 years later to make theoretical studies that led to his predicting the existence of a planet beyond the orbit of Neptune.

About 30 years after Neptune's discovery in 1846, many astronomers became interested in the idea of planets beyond Neptune. One of them, D. P. Todd in 1877, studied the orbital perturbations of Uranus and his analysis indicated the existence of a planet beyond Neptune at a distance equal to 52 times the Earth's distance from the sun and with a diameter of 50 000 miles (80 000 km). Because astronomical photography was not very efficient in his day, Todd searched for the planet visually with the 26-inch Washington refractor. Between November 3rd 1877 and March 5th 1878 he scrutinised a 40°-long strip of sky 2° wide along the invariable plane of the ecliptic. Although several objects were seen that may have been the searched-for planet they showed no movement and so were rejected as not being planetary. The result of the investigation was negative.

During the 30 years from 1900 outwards, Percival Lowell and William H. Pickering were especially prominent among searchers for a trans-Neptunian planet. Lowell, who dubbed the trans-Neptunian planet 'Planet X', predicted its existence in a memoir that he published in 1915 although he had searched for the planet since 1905. Lowell and his helpers embarked upon a series of calculations lasting nearly a year in which an attempt was made to predict the position of Planet X. The method depended on measuring perturbations in the motions of other planets caused by the unknown planet's gravitational pull; such was the way in which Neptune's position was deduced by its effect on the orbital motion of Uranus. Since its discovery in 1846, Neptune had not yet made one complete circuit of its orbit so that its orbit was not known well enough to ascertain whether it was being perturbed. Although it was much more distant from Pluto, Lowell had to turn to Uranus which would be less influenced by an unseen remote planet. The results of the studies of the perturbing effect on the orbital motion of Uranus indicated that the Planet X was a giant with a mass 6.7 times that of Earth's and about half the mass of Uranus or Neptune. Two possible positions opposite each other in the sky, one of them in the constellation Gemini, were given for Planet X.

During the 1920s Pickering published several papers in *Popular Astronomy* on the subject of what he called Planet O which resembled Planet X except that Pickering had predicted it to be a fainter object. He also proposed a Planet P in 1911 situated at a distance of 123 times the Earth's distance from the sun and making a complete orbital revolution in 1400 years. Subsequently he made several revisions of the characteristics of the Planet P. A search was conducted for Planet O at the Harvard Observatory in January 1928 but it was unsuccessful. The American astronomer T. J. J. See even predicted the existence of *three trans-Neptunian planets!*

Lowell searched for his Planet X during 1905–7 with a Brashear 5-inch photographic objective. He made 3-hour exposure photographs along a strip of sky extending several degrees on either side of the invariable plane of the ecliptic using the Brashear equatorially mounted and clock driven. The developed plates were examined with a hand lens for a star image showing a small shift that would betray its planetary nature. Unfortunately, Pluto was not in the part of the sky scrutinised by Lowell during the search period. Also, since Pluto's orbit is highly inclined to the general ecliptic plane, Lowell was doomed to failure from the start in this particular search.

Lowell came into possession of a 40-inch reflector in 1909. He and C. O. Lampland used this in a search for Planet X beginning in 1911. Lowell had calculated that the hypothetical planet would be in the constellation Libra. Study of the photographic plates was greatly helped by the blink microscope comparator obtained from the firm of Carl Zeiss in Germany. With this instrument, two identical photographic plates of the same star field taken at different times can be examined as if superimposed one after the other in rapid succession. Any star image that has moved in the interval between when the plates were photographed will show up as a point moving alternately one way and then the other as the plates come under rapid alternate scrutiny. The non-stellar nature of the object will thus be betrayed.

The blink comparator that Lowell and Lampland used was the same one employed by Clyde Tombaugh in the successful search for Pluto during 1929–43. The search by Lowell and Lampland was abandoned after a year but was resumed from April, 1914 to July, 1916, this time with a 9-inch aperture wide-field camera that Lowell managed to borrow from Swarthmore College's Sproul Observatory. Duplicate plates were taken at 2-week intervals by E. A. Edwards and T. B. Gill and the search was supervised by Lampland. In 1915 the search was shifted to part of the constellation Taurus where Lowell had predicted Planet X would be found as a result of revised calculations. Pluto was actually in this region that is very rich in stellar images but its images were overlooked. (Subsequently Pluto's images were found on the relevant plates after its discovery when its position was more accurately known owing to improved computations).

Lowell was keenly disappointed with the failure to find Planet X as he had expended considerable effort and time during many years to theoretical calculations and search for the trans-Neptunian planet. Lowell died in 1916 and during the next 13 years nothing else was done in the search for Planet X.

The discovery of Pluto

The search was once more taken up when Clyde Tombaugh was invited by the Flagstaff Observatory to use the recently acquired 13-inch telescope in the hunt for Planet X. Because of uncertainties in the predictions of where Planet X would be located, Tombaugh decided that the search must take in the entire Zodiac. He commenced the photographic search with the 13-inch telescope on April 6th 1929 guiding on the star Delta Cancri and taking duplicate plates after an interval of about a week. This would give a pair for blink comparison to enable a moving body to be detected. Several months later Tombaugh began the mammoth and extremely tedious task of 'blinking' the pairs of duplicate plates. On February 18th 1930, the duplicate plates he had taken on January 23rd and 29th were examined by him in the blink comparator. These plates covered part of the constellation Gemini. After he had covered about one quarter of the pair of plates, Tombaugh spotted a 15th magnitude point showing movement in the alternating views. The two images were separated by a distance of 3.5 millimetres. It was the looked-for planet. The parallactic shift indicated that it was well outside Neptune's orbit. The star field in which Tombaugh found the planet was one forty-thousandth part of the entire area of the sky that he had examined in the blink comparator. Thus was discovered the planet known today as Pluto.

However, Pluto is *not* Lowell's Planet X because Lowell expected Planet X to be 6.7 times as massive as the Earth whereas Pluto's mass is only 1/400 of the Earth's mass and is nowhere near sufficiently massive to cause perturbations in the motions of other planets. After analysing Lowell's calculations the Yale mathematician Ernst Brown declared that although Pluto was found close to the place predicted for Planet X by Lowell, this was merely an astonishing coincidence and that Pluto's discovery was an accident.

History of observation

As soon as Pluto was discovered it was realised that it was not the giant Planet X predicted by Lowell. Pluto was too small and faint; it showed no perceptible disc even in the largest telescopes. Calculations that were based on the greatest diameter that Pluto could have without it showing a perceptible disc at the Earth's distance and an estimate of its mass showed that the planet's density must exceed that of lead, a result which was extremely difficult to accept. In order to reconcile this with observation it was suggested at one time that Pluto had a 'shiny' specular surface and the image we see in large telescopes is not Pluto itself but only the image of the sun reflected in the surface, Pluto's outline being much larger but invisible from Earth.

Astronomers calculated from Pluto's assumed reflectivity of 4% that it must be smaller than the Earth as it appeared to be too dark for a planet that most likely was snow-covered. The estimates of Pluto's diameter were further scaled down to about half the size of the Earth after careful visual study of Pluto with the 200-inch Palomar telescope during 1950. This enabled the reflectivity estimate to be increased to 13%. Some observers noticed that Pluto's brightness varied by about 10% and that this was cyclical and indicated an axial rotation period of 6.390 ± 0.003 Earth days.

Pluto's orbit was also unexpectedly close to the sun and was more elliptical than that of any known planet (fig. 13.1). Its orbit brought it closer to the sun than Neptune at times and it was inclined at an angle of 17° to the plane of the ecliptic, much more than any other planet. In 1955 the axial rotation period was more precisely determined as 6.38673 Earth days from studies with sensitive light detection apparatus. It also appeared that Pluto was gradually darkening and that this had been happening since the 1950s. A more recent (1989) value for the axial rotation period is 6.387245 –0.000021 Earth days.

During the 1960s and 1970s improved astronomical equipment and techniques permitted more refined studies of what became known as 'The problem of Pluto'. Computer techniques enabled specialists in celestial mechanics to ascertain that Pluto's orbit resonates with Neptune's; each time Neptune makes three orbital revolutions, Pluto makes two. The strong ellipticity and tilt of Pluto's orbit and its resonance with Neptune makes it seem likely that Pluto is a planetismal, i.e., a large primordial fragment remaining by chance from the fragments that accreted to form Neptune when the Solar System was in its early stages of formation.

In 1976 D. Cruikshank, C. Pilcher and D. Morrison of the University of

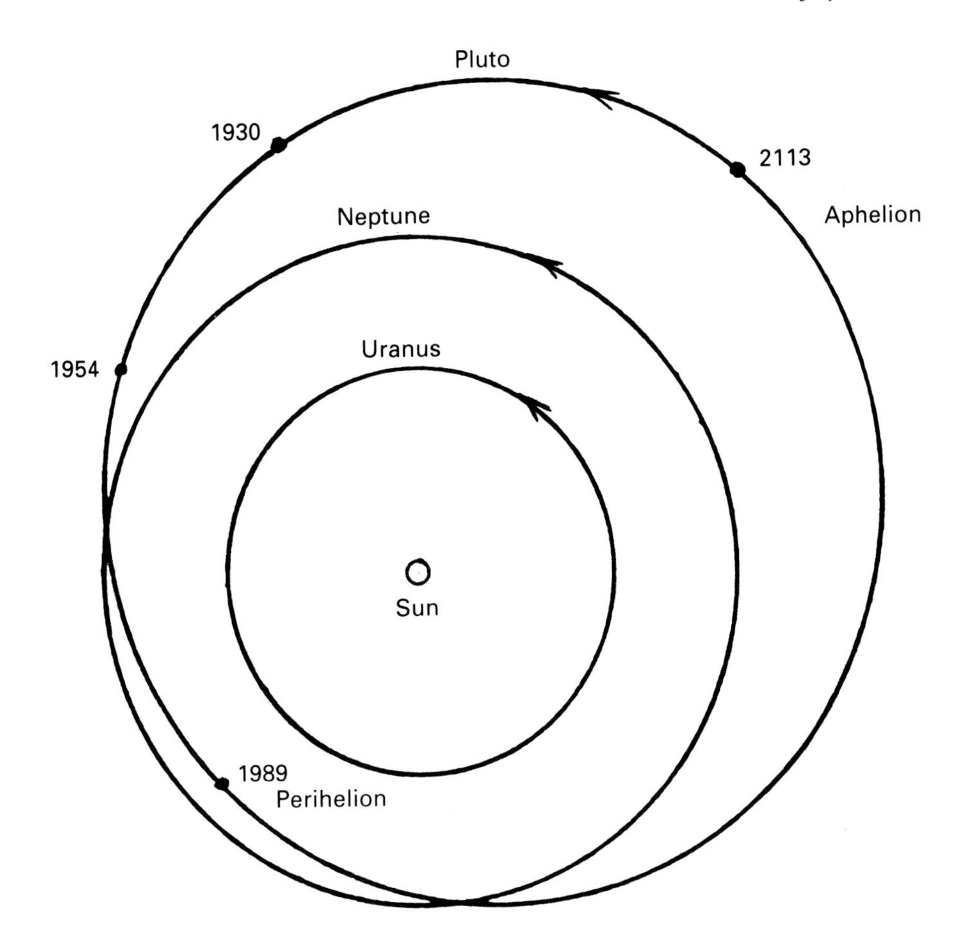

Fig. 13.1 The orbit of Pluto.

Hawaii discovered that Pluto was covered with frozen methane. This is reflective so that Pluto is faint because it is small, not because it is non-reflective. Pluto's mass was estimated by them to be 0.004 that of the Earth. Therefore, Pluto could not have been the object causing the perturbations of Neptune's orbit – but these perturbations still remain.

Stellar occultations were then used as an approach to determining Pluto's diameter. At intervals of a few decades, Pluto occults a star and from the duration of the eclipse of the star determined at different observations a value for Pluto's diameter can be calculated. On April 28th 1965 there was a 'near miss'; Pluto did not quite occult the star but had the diameter of Pluto been greater than 4223 miles (6800 km), it would have done so. Thus an upper limit was now placed on Pluto's diameter. On June 9th, 1988 the brief occultation of a 12th magnitude star was watched by astronomers in the southern hemisphere. There was no instantaneous cut-off of the star's light but it dimmed gradually. The reverse was observed when the star emerged from behind Pluto. This was taken to indicate an atmosphere around Pluto which was calculated to extend for over 870 miles (1400 km) outwards from Pluto.

For many years after the discovery of Pluto the US Naval Observatory worked with a view to producing improved orbital data for Pluto. Plates were taken with the 61-inch astrometric reflector at their station at Flagstaff. J. W. Christy was measuring Pluto's precise location on plates secured during three nights in April and May of 1978 when on June 22nd he saw a 'bulge' on the tiny images of Pluto. Nearby star images were perfect so that it seemed unlikely that the bumps on the Pluto images were imperfections in the plates. Neither were there any background stars nearby in line with the images of Pluto that would simulate a bump as was shown by a study of the prints made with the 48-inch Schmidt Camera of the Mount Palomar Observatory. The bumps always appeared either on the north or south points of the images of Pluto. Christy found similar bumps on Pluto on plates taken in 1965 and 1970.

The only explanation for these bumps was that they were images of a satellite in an orbit very close to Pluto. The bump or extension on Pluto's images were apparently 2 magnitudes fainter than Pluto itself and this indicates that the satellite of Pluto has a diameter of about two fifths that of Pluto. On the assumption that they have a similar composition, density and reflectivity, this means that the satellite is the largest in the Solar System as compared to its primary. The satellite was named Charon. (It is amusing to recall that in 1950 Kuiper and Humason made an unsuccessful visual and photographic search for a satellite of Pluto. Kuiper concluded that there could not be a satellite brighter than magnitude 19 between 0.3 and 2.0 seconds of arc from Pluto!)

The existence of Charon was not formally accepted by other astronomers because its image on photographic plates could not be separated from Pluto's. Although very unlikely, B. Marsden suggested that the bump might be an enormous mountain on Pluto. However, in 1985, Pluto and Charon began exhibiting mutual eclipse phenomena (fig. 13.2) which convinced skeptics of the separate existence of Charon.

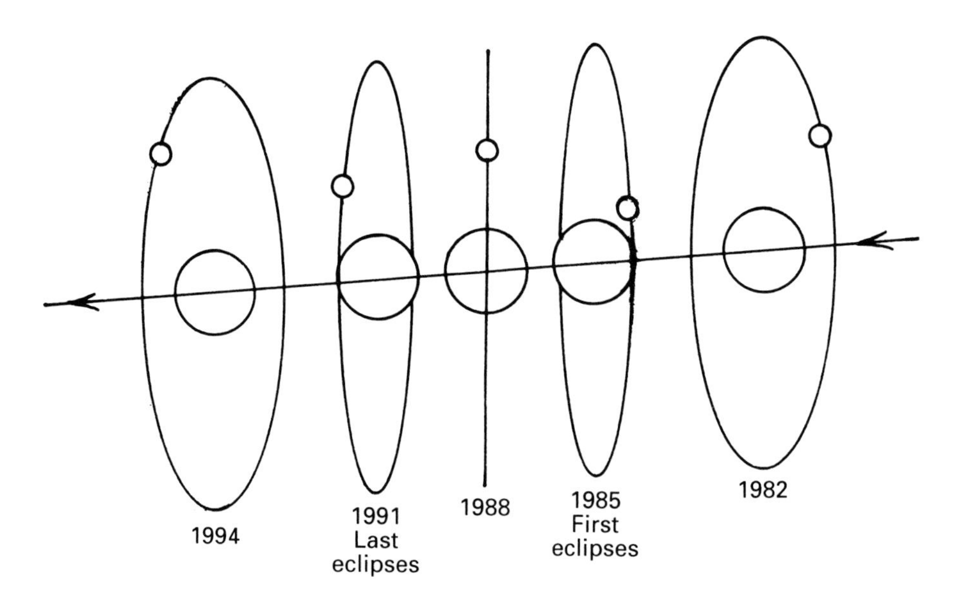

Fig. 13.2 Mutual eclipse phenomena of Pluto and Charon.

Charon is rotationally locked with Pluto and makes one revolution around Pluto in 6.387245 Earth days. Its distance from Pluto is estimated as about 12 206 miles (19 640 km). The discovery of the satellite permitted further breakthroughs in knowledge of Pluto. Knowing Charon's orbital revolution period and distance from Pluto enabled the sum of the masses of the two bodies to be calculated. The value obtained for Pluto's mass came out at about 1/400 of the Earth's.

On April 6th 1980 there occurred another near occultation of a star by Pluto but it was Charon that actually eclipsed the star. Timing of this occultation which lasted 50 seconds enabled a minimum diameter of 745 miles (1200 km) for Charon to be computed. The observation was made by A. Walker from South Africa.

Another phase in the study of Pluto will be started by the fly-by of a spacecraft sometime during the early twenty-first century which will then enter orbit. It has a large antenna that will face Earthwards and then numerous images and other data transmitted to Earth should answer many of the questions still remaining about Pluto.

It may be questioned whether Pluto is a true planet. It certainly is not like any of the other Solar System planets. Pluto may best be considered as a rather interesting asteroid and is so considered by B. Marsden. It could be the biggest member of a zone asteroids beyond Neptune's orbit but we don't know very much about them.

At least two of the other asteroids resemble Pluto. The asteroid 2060 Chiron (not to be confused with Charon) is one such which moves in an orbit round the sun between Saturn and Uranus. At perihelion it approaches the sun more closely than it does Uranus. Also, there is another, 944 Hidalgo, discovered in 1920, which has a very eccentric orbit around the sun. At perihelion it comes nearer to the sun than Jupiter and at aphelion it flies beyond Saturn's orbit. Now that a lot more is known about Pluto and especially because of its very small mass, we must admit that if there is a large planet beyond Neptune it has not yet been discovered. Tombaugh continued the search for trans-Neptunian planets for 13 years after the discovery of Pluto with negative results. Whether or not there are still deviations of Neptune from its calculated path is debatable owing to uncertainties of the amount, sign and even existence of residuals. However, one still cannot help wondering if the ninth planet of the solar system still awaits discovery.

Visibility of Pluto

Pluto is too faint to be visible to the unaided eye, attaining a maximum brightness of magnitude 13.7 at its closest approach to Earth. 1990 was an especially good year to search for Pluto (incidentally, the sixtieth anniversary of its discovery); it reached perihelion on September 5th 1989, an event that takes place once every 248 Earth years. By May of 1990 the Earth was positioned in its orbit exactly between Pluto and the sun so that the Earth was as close to Pluto as it could get. This is the opposition position and the exact time that it occurred was on May 7th at 05.00 UT. However, many claimed that 1989 was the best year to observe Pluto because when opposition occurred on May 4th that year,

Pluto was just four months away from its perihelion. Hence, Pluto would appear to have been closer to Earth in 1989! At the 1989 opposition Pluto was closer to the sun than it was in the Spring of 1990 but in 1990, opposition fell three days later than in 1989. The Earth was just three days nearer to its yearly aphelion position on July 4th. Therefore although Pluto drew further away than in 1989, the fact that the Earth was also further from the sun more than cancelled the difference which, however, is quite small and equal to a distance of only four Earth diameters, less than the distance at opposition in 1989.

Observing Pluto

Finding and identifying Pluto is an observing challenge. It is visible in telescopes of 8 inches of aperture or more and may be visible in a 6-inch telescope under dark skies. It appears as a faint star-like point of light and can be photographed with a 200-mm lens.

The *Observer's Handbooks* published anually by the Royal Astronomical Society of Canada and others, give charts showing the position of Pluto against the background of fixed stars around the date of opposition each year. These assume much greater importance for finding and identifying Pluto than the other much brighter planets owing to the faintness of Pluto and its close resemblance in the telescopic field to a faint star. In addition periodicals like *Astronomy* publish detailed star charts showing the day to day positions of Pluto relative to faint stars thus making it fairly easy to pin-point the tiny image of Pluto. Detailed instructions for locating Pluto are given by fixing on a nearby prominent star and then by a stepwise 'star hopping' sequence the observer is enabled to find Pluto. However, the finding of a faint point of light in the expected place is not actual proof of having seen and identified Pluto. Its movement relative to the other fixed stars must be detected. To do this make an accurate drawing of the stars in the vicinity of the object that you suspect of being Pluto and mark the position of Pluto accurately. Observe the same star field again on the next night. Don't wait longer than this. If the suspected object has moved then it was Pluto. Needless to say, a dark moonless night with steady seeing is very desirable when you go Pluto hunting.

As noted earlier, twice during one orbital revolution period of Pluto, Charon's orbit will be presented edgewise to the Earth and Pluto and Charon will exhibit mutual transit and occultation phenomena giving rise to brightness variations that may be monitored photoelectrically (see chapter 16). These transit/occultation cycles last for 5 or 6 years. The most recent cycle started in 1985 and was the first to be observed since Pluto's discovery. There will not be another until the year 2109 by which time I imagine that none of my readers, however young they are now as I write this in 1992, will be here to observe it!

Other observational work that can be done is visual or photoelectric photometry but this requires large apertures e.g., 17.5 inches and over. Pluto's light varies rotationally by about 30%. Timing the admittedly rare stellar occultations is another useful kind of observational work.

There is virtually nothing else that the amateur astronomer can do with Pluto but even if you have only succeeded in locating Pluto and have convinced yourself that you have really seen it, then that in itself is an accomplishment.

Further reading

Books

Out of the Darkness. Tombaugh, C. and Moore, P. A. Lutterworth Press, London (1980).

Planets X and Pluto. Hoyt, W. G. University of Arizona Press, Tucson, Arizona (1980).

Papers and articles.

Photographic search for a planet beyond the orbit of Neptune. Roberts, I., *Monthly Notices of the RAS* **52**, 50 (1892).

The discoveries of Neptune and Pluto. Jackson J., *The Observatory* **75**, 126–7 (1955).

Reminiscences of the discovery of Pluto. Tombaugh, C., *Sky and Telescope* **19(3)**, 264–70 (1960).

Search for a trans-Neptunian planet. Foss, A. P. O. and Shaw-Taylor, J. S., *Nature* **230**, 266 (1972).

The case Against Planet X. Goldrich, P. and Ward, R., *Publications of the Astronomical Society of the Pacific* **84**, 737 (1972).

The Work of the ALPO Remote Planets Section. Hodgson, R. G., *JALPO* **27(5–6)**, 85–8 (1978).

Some thoughts on Planet X. Moore, P. A., *JBAA* **91**, 483 (1981).

Uranus, Neptune and Pluto: Contributions that ALPO members can make. Schmude, R. W., *JALPO* **35(2)**, 67–8 (1991).

A quinquennial observational report of the ALPO Remote Planets Section for the years 1985–1989. Hodgson, R. G., *JALPO* **35(3)**, 97–9 (1991).

The remote planets: 1987–1990 Report. Schmude, R. W., *JALPO* **35(4)**, 153–6 (1991).

Mysterious Pluto. Berry R., *Astronomy* **8(7)**, 15–22 (1980).

Pluto – Enigma on the edge of the Solar System. Croswell, K., *Astronomy* **14(7)**, 7–22 (1986).

The pursuit of Pluto. Croswell, K., *Invention and Technology* **5(3)**, 50–7 (1990).

In pursuit of Pluto. Talcott, R., *Astronomy* **20(5)**, 73–6 (1992).

14

Constructing maps and planispheres

General

Provided that you have sufficient observational drawings of a planet that comprise one complete axial rotation, you can combine them all into a map or planisphere. Mars is especially suitable for this because of its well-marked surface features. Maps of Mercury have been constructed by several amateurs. Jupiter is also well suited because of its prominent atmospheric features, some of which are fairly permanent.

With some, Mars and Mercury for example, the surface features may take months or even years to gather whereas with Jupiter a complete axial rotation of 10 hours may be charted in a single night when the planet is at or near to opposition.

When making a chart of a planetary surface we are faced with the same problem as with making terrestrial maps – the representation of a curved spherical surface on a flat sheet of paper. The more of the planetary or terrestrial surface that is charted the greater the difficulty of rendering a distortion-free picture – or at least of minimising distortion. Various *projections* have therefore been devised to achieve the representation of the surface of a planetary or the terrestrial globe on a flat sheet as accurately as possible, all of which have advantages and disadvantages,

The horizontal orthographic projection

The horizontal orthographic projection is the one usually employed in making planetary maps. With this the planetary surface features are plotted onto a flat plane as if they were projected by parallel rays. In fact this is the view we get of a globe at infinite distance.

Begin by drawing a circle of almost any radius that you wish. This can be, say, 5 cm. Now divide the circumference into 10° intervals of latitude parallel to

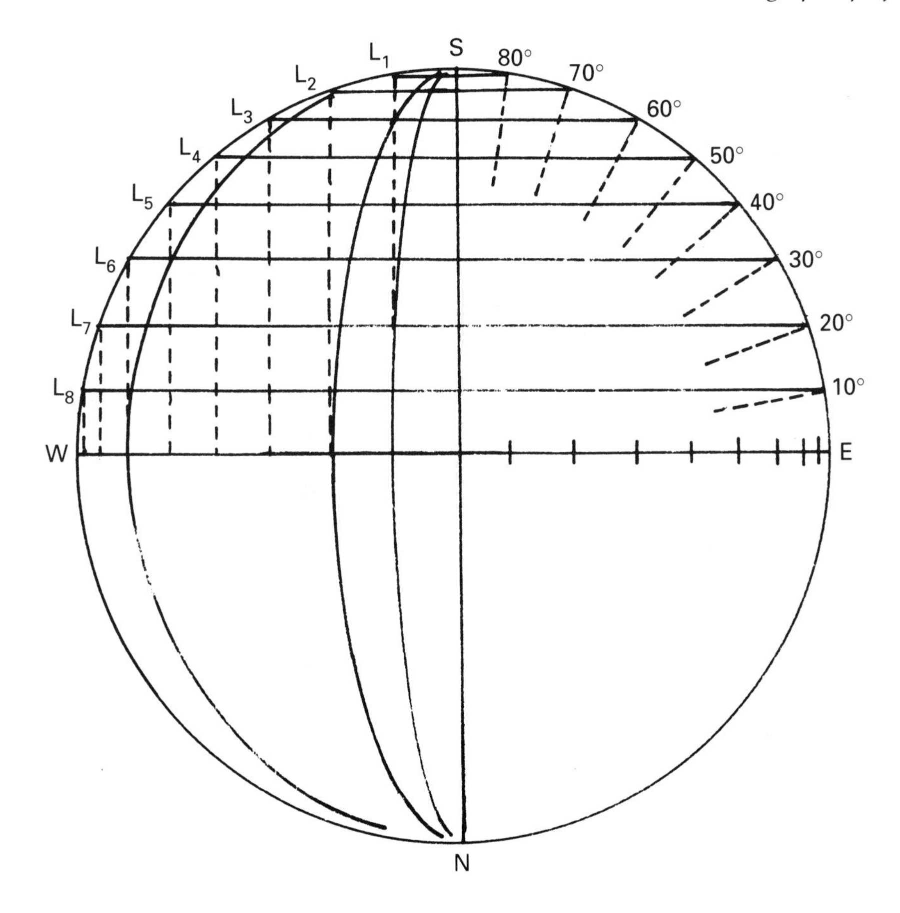

Fig. 14.1 The horizontal orthographic projection.

the equator and mark the north, south, east and west points (fig. 14.1). At the points on the circumference where the parallels or latitude intersect it (L_1, L_2 etc) drop perpendicular lines to the equator. The parallels of longitude may now be constructed and these will be 10° apart. The longitude lines on the planetary sphere are all great circles passing through the poles but will appear as ellipses of varying eccentricities owing to the parallel projection. Their major axes are all of identical length and equal to the planet's polar diameter. The minor axes all differ as shown in the diagram. The ellipses can be drawn from ellipse templates as the major ellipse axes are invariant and equal to the diameter of the circle drawn or ellipse compasses can be used.

It is best not to carry the longitude lines further than latitude 80° north or south otherwise they will crowd together as they approach the poles and the result will be a confused mass of close converging lines. Note also that the vertical distances between the circles of latitude get smaller the further they are from the equator and the longitude meridians are closer together near the limbs of the planet. Hence, areas defined by latitude and longitude are not going to be all the same size on different parts of the map.

Cylindrical projections

Cylindrical projections may also be used which show the entire surface of the planet whereas the previously described map shows one hemisphere only.
The parallel cylindrical projection (fig. 14.2).

In this the intervals between the latitude lines are projected horizontally onto the cylinder. The equator appears horizontal and is its true length. A point of the equator at a given longitude marks the centre of the map and is chosen as the zero point of a system of rectangular coordinates upon which the map is plotted. Latitude lines are horizontal and are all of equal length, the same as that of the equator. Longitude lines are perpendicular to the equator, parallel to one another and separated by equal intervals. The poles are included and the intervals between the latitude lines get smaller as the poles are approached. The equation representing the vertical distance (D) of any latitude line from the equator in this projection is given by:

$$D = R \sin l$$

where l is the latitude in degrees and R is the radius of the sphere.

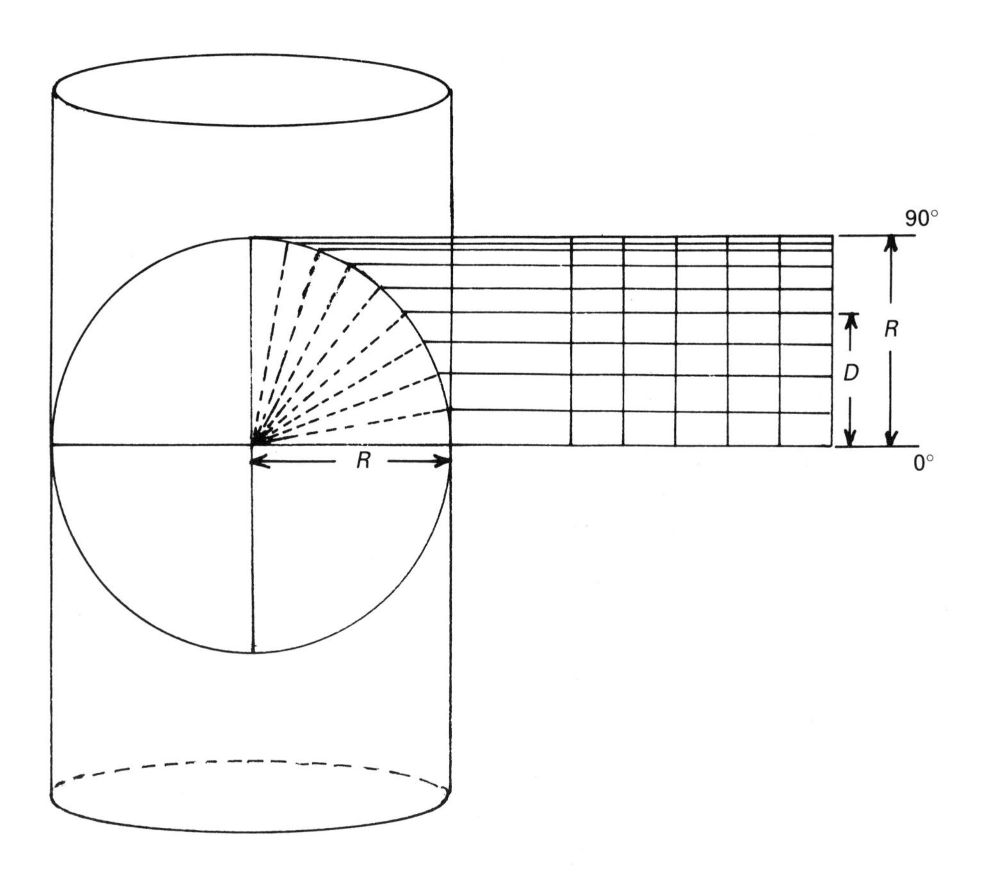

Fig. 14.2 The parallel cylindrical projection. (D = R sin l *where* l *is the latitude in degrees.)*

The Mercator projection (fig. 14.3).

This is another type of cylindrical projection in which the parallels of latitude are projected onto the cylinder by lines radiating from the centre of the sphere so that the lines of latitude become more widely spaced as the poles are approached. This projection therefore greatly exaggerates the size of features in the higher latitudes. It is not practical to carry the northern and southern limits of the map beyond more than about 65° or 70° of latitude so that the poles are not included (fig. 14.3). The maps of Mars shown in figs. 7.21 and 7.25 are represented on this projection.

When constructing, say, a Mercator projection map of Jupiter the central

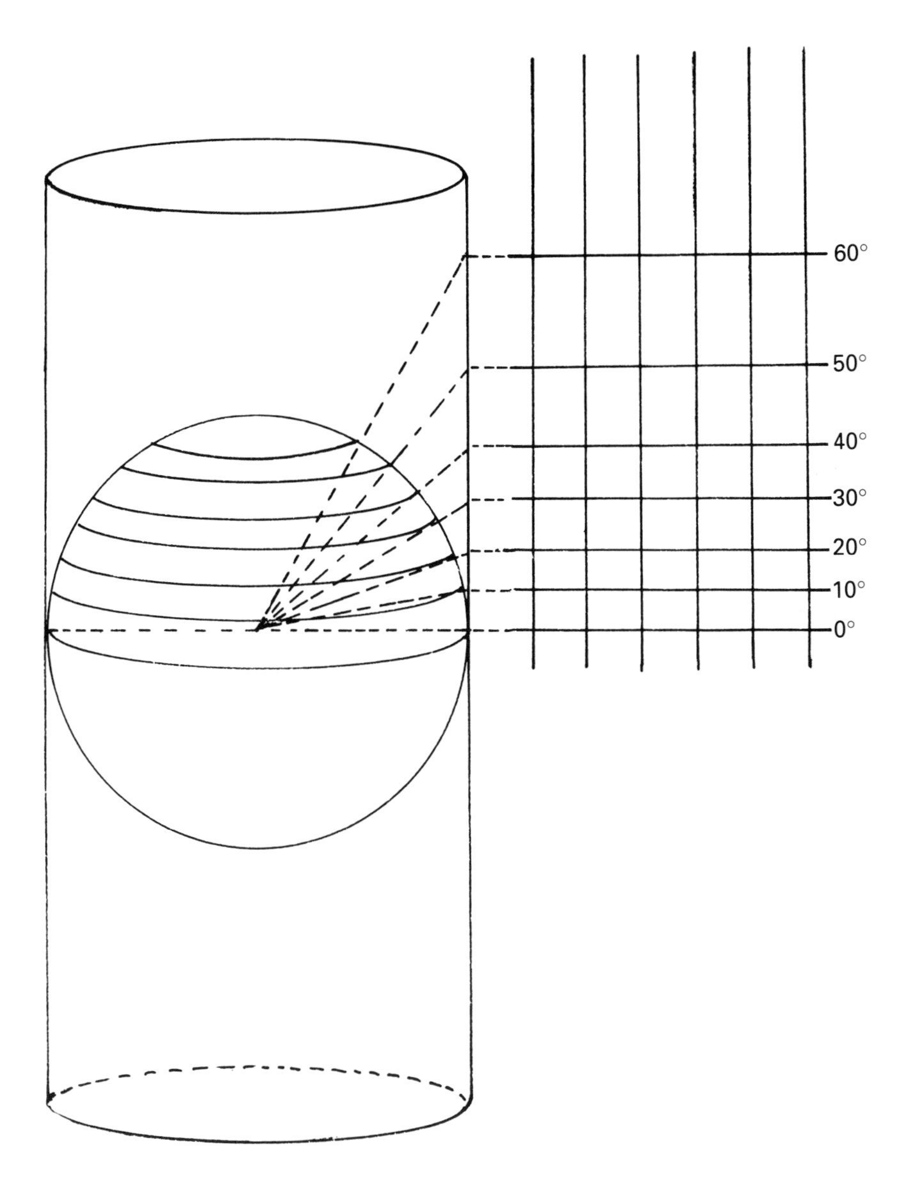

Fig. 14.3 The Mercator projection.

meridian of the map is made to correspond to the arbitrary central meridian of Jupiter. Next, the cloud belts whose latitudes and widths have been determined are added. The Great Red Spot, festoons, ovals and other individual cloud features whose longitudes have been determined are then plotted. To minimise errors resulting from the slightly different rotation periods of Systems I and II, it is best to construct a Mercator map of disc features of Jupiter from disc drawings in which latitude and longitude determinations have been made during as short a time interval as possible. Ideally this could be done during a single night at or near to the time of opposition when a complete axial rotation of Jupiter can be observed.

In constructing a Mercator projection map of Mars compiled from your own disc drawings you can use any of the reliable published maps of Mars to help in precise positioning of the surface features you have recorded in your disc

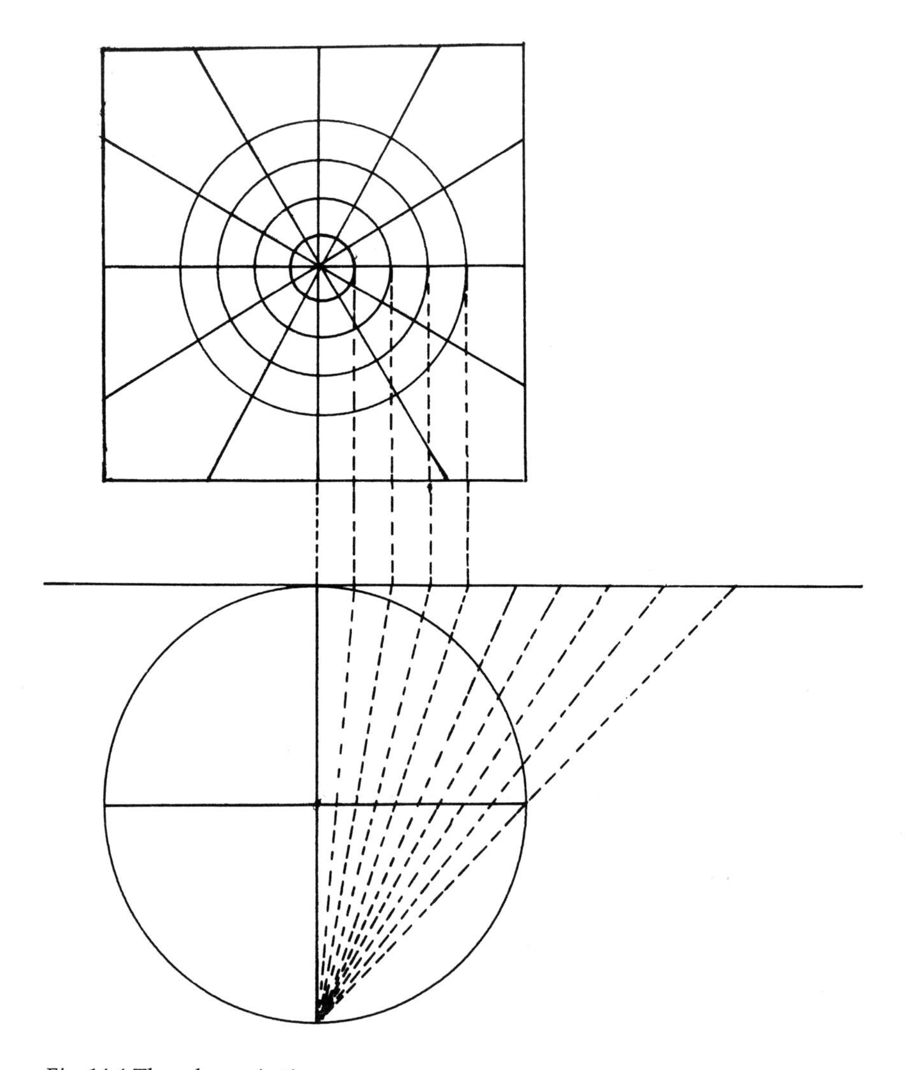

Fig. 14.4 The polar projection.

drawings onto the Mercator grid but remember to be honest – no copying onto your map of additional detail from the published map!

The polar projection

A projection for representing the polar regions of a planet is shown in fig. 14.4. A plane touching the sphere at one or other pole, and this perpendicular to the planet's axis, has circular latitude lines projected onto it from a set of radiating straight lines having their origins at the opposite pole. The longitude lines are projected as straight lines radiating from the pole on the projection plane.

Further reading

Books

Handbook for Planet Observers. Roth, G. D. Faber and Faber, London (1970). Chapter 2, pages 64–71.

15

Planetary photography and videography

General

In addition to visual planetary observation and recording of observations by drawing, the amateur astronomer may also wish to record observations by photography. Until fairly recently, planetary photography involved the use of conventional photographic film exclusively as the recording medium but with the advent of video photography (videography) and the availability of charge coupled devices (CCD), astronomical photography in general has undergone a revolution. The principles of each of the types of planetary photography will be described here.

The planetary photographer's camera

The best kind of camera to use for planetary photography is any one of the many makes of 35 millimetre single lens reflex (SLR) type, the majority of which will be found suitable for this kind of work. Whatever make of SLR you choose, it should have the following features:

(1) Removable lens.
(2) It should have a 'bulb' or 'time' exposure.
(3) The focusing screen should be as clear as possible to facilitate critical focusing.
(4) There must be a suitable socket or other provision for attaching a cable release.
(5) The shutter should operate quietly and smoothly with minimal vibration of the mirror.

Small and moderate-sized high-quality telescopes reveal an astonishing amount of detail on planetary discs but when attempts are made to capture the detail on photographic film the result is usually disappointing. Something always seems to be lost as compared with what can be seen by direct visual observation.

There are two basic ways of photographing through the telescope, each of which has certain advantages and disadvantages:

(1) *Prime focus photography* in which the eyepiece of the telescope and the camera lens are both removed and the primary image produced by the telescope objective (lens or mirror) is focused on the film (fig. 15.1). The telescope is thus being used as a big telephoto lens.
(2) *Projection photography* in which the enlarged image projected by the telescope eyepiece or Barlow lens is received by the film (figs. 15.2, 15.3). When eyepiece projection is used, the camera lens may be left in place if desired (fig. 15.4); both the telescope and the camera are focused on infinity in this method. Special adapters are commercially available for attaching the camera to the telescope after the camera lens has been removed. Different adapters will be required depending on whether you remove the telescope eyepiece as in prime focus photography or leave it in place as in eyepiece projection photography. Camera adapters and many other photographic and observing accessories are obtainable from the following supplier:

Orion Telescope Center,
2450 17th Avenue,
P. O. Box 1158,
Santa Cruz,
California 95061

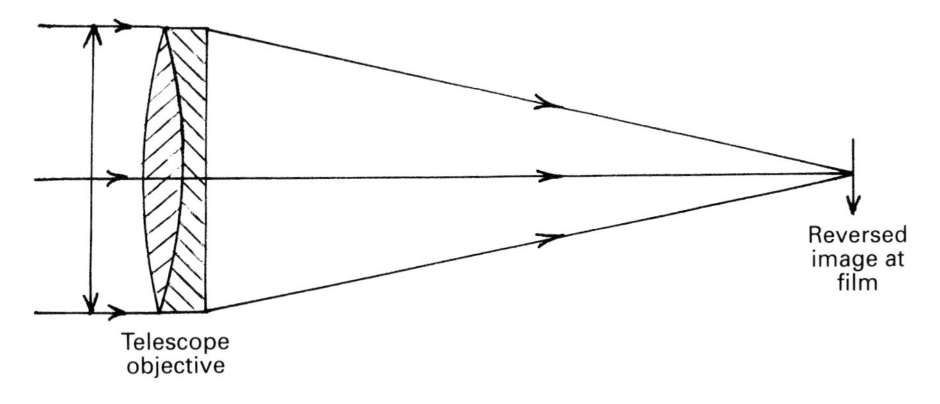

Fig. 15.1 Prime focus photography.

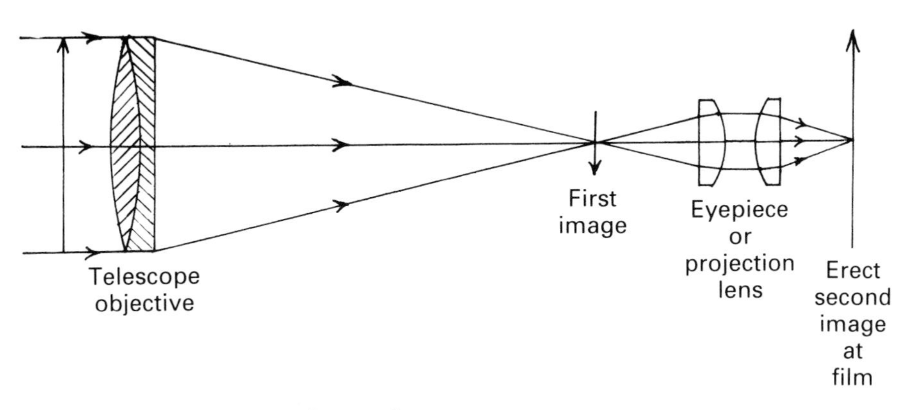

Fig. 15.2 Eyepiece projection photography.

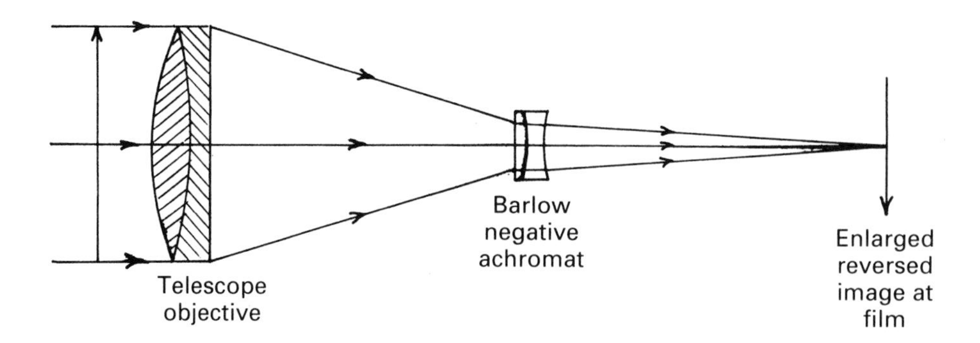

Fig. 15.3 Barlow lens projection photography.

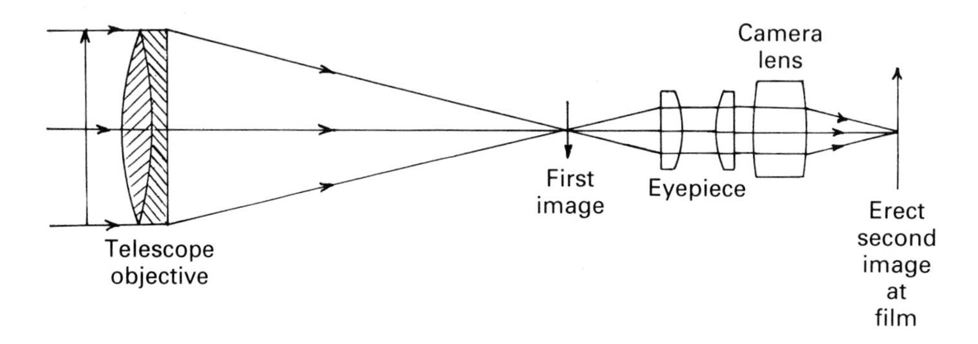

Fig. 15.4 Eyepiece projection with camera lens in place.

The bright but small image at the telescope's prime focus is sharp and requires relatively short photographic exposure times which minimises blurring of the resultant image by atmospheric turbulence. Exposures can be so short that guiding of the telescope may not be necessary. The very small planetary image needs a fair degree of enlargement, however, for the details to be seen in the print and so, depending on the film, it may be a bit 'grainy'.

The larger magnified image produced by eyepiece or Barlow projection may still be quite small and is much dimmer than the prime focus image and so requires longer exposure times that may be from 0.25 second to 15 seconds. The risk of image blurring by atmospheric turbulence is therefore much greater and guiding of the telescope is necessary. However, the larger scale of the photographic image is such that only a relatively small degree of enlargement may be necessary when printing. Only high quality eyepieces and Barlow lenses should be used in projection photography.

There are five essentials for achieving good high resolution photography of the planets:

(1) *Excellent telescope optics.* The telescope must be of the finest optical quality. A second-rate refractor objective or reflector mirror will diffuse light that spoils fine detail and image contrast is reduced. Dirt on the lens or mirror has a similar bad effect.

(2) *Perfect focusing.* This is especially important in projection photography. Most 35-millimetre SLR cameras have ground glass focusing screens that do not transmit much light. These are fine for well-lit subjects but focusing of dim lunar or planetary detail is difficult. Some of the more advanced 35-millimetre SLR cameras have interchangeable focusing screens. If you have this type of camera, change the ground glass screen for one that is transparent. This will facilitate focusing of planetary images and will be much easier on your eyes.

(3) *Accurately guided exposures.* If you are photographing with a 12.5-inch telescope, which theoretically can resolve detail of less than 0.5 second of arc, unguided exposures must be shorter than about 1/30 of a second. A motor drive is necessary for longer exposures. An excellent drive permits exposures of up to 20 seconds on Jupiter that enables recording of detail down to 1 second of arc. Such accurate driving needs a carefully balanced telescope. Exposures of this length must be free from vibration such as is caused by the camera shutter.

(4) *Correct film and F-ratio combinations.* The film used should be capable of resolving all the fine detail in the telescopic image. The resolution of the film is customarily expressed as the number of lines per millimetre that can be seen on it as just separated. The film should have a resolution about three times that of the telescope. This is because film manufacturers' resolution test charts are usually of high contrast such as alternate black and white parallel lines. Features on planetary surfaces, however, are much less contrasty.

The following formula may be used to calculate the number of lines per millimetre that a given telescope can resolve:

$$\text{Lines per millimetre resolved } (R) = \frac{1450}{f}$$

Where f is the effective focal ratio of the telescope. Notice that the resolution is a function not of the telescope's aperture but of the effective focal ratio.

The effective focal ratio of a telescope is increased by using a Barlow lens inserted where the eyepiece would normally be placed. They are available in designs that increase the effective focal length of the telescope from twice to four times. Thus, an F/15 refractor is converted to F/60 by a ×4 Barlow lens. Some suitable films for different F-ratios are shown in Table 15.1.

Table 15.1 *Suitable films for different F-ratios.*

Focal ratio of telescope	R	
F/20	72	Kodak High Contrast copy film.
F/40	36	Kodak Panatomic-X
F/80	18	Plus-X, Tri-X
F/160	9	Any film

The planetary image will be much larger at the longer focal ratios but dimmer, especially at F/50 and over, and so longer exposure times or fast films will be needed. The same applies to the larger images produced by eyepiece projection photography. Control over the size of the image is possible in eyepiece projection; the image is made larger by increasing the distance of the emulsion from the eyepiece by attaching an extension tube between the telescope eyepiece and camera and then refocusing.

(5) *Excellent seeing conditions.* If poor seeing prevents detail at the limit of the telescope's resolving power from being seen by the eye, then it cannot be photographed. Perfectly steady intervals must last long enough for detail to be recorded on the film. The entire image must also be quite still or the picture will be blurred. Capturing detail permitted by these moments of perfect seeing on the film can only be realised by the rather hit and miss procedure of taking several exposures in the hope that at least one of them will have been made during an interval of really steady seeing.

Choice of film

The requirements for an ideal astronomical film are as follows:

(1) *Speed.* Film speed is expressed by the ASA number; the higher this is, the faster is the film and the shorter the exposure time needed for a given subject.
(2) *Grain size.* Fine-grain films are desirable so that fine detail can be successfully recorded.
(3) *Contrast.* Contrasts of features on planetary surfaces are often quite low so that a high contrast film is preferred.
(4) *Colour sensitivity.* Colour sensitivity is important even in black and white films. Planets exhibit almost the entire range of colour so that panchromatic films should be used for planetary photography. They are able to record all colours equally well except at the extreme spectral range.

No film exists with all of these characteristics simultaneously fulfilled; for example, fine grain films are not the fastest. However, not all of these requirements have to be met. The planets are relatively bright objects and so the film speed is not as critical as when you want to photograph dim nebulae or faint stars.

Generally speaking, exposure times of 10 seconds or less are adequate with even slower films (low ASA ratings) so that fast film is not essential. On the other hand, the capturing of fine detail such as is found in Jupiter's cloud belts demands high resolution which means that fine-grained films are essential.

Many black and white films meeting these requirements are available but there is a rather poor selection of suitable colour films. The most easily obtainable black and white film is Kodak's Tri-X, which has fairly good resolution, fine grain and is panchromatic. It is a good film with which to start and experiment.

Characteristics of some films

Kodak High Contrast Copy Film is fast, has very good resolution, very fine almost undetectable grain and high contrast. It can be 'push processed' to give 3.5 times its normal speed using Ethol UFG which retains increased effective speed and contrast.

Kodak SO-410 (now 2415), originally produced for photomicrography is even faster, has very good resolution, very fine grain and is panchromatic with extra sensitivity in the red wavelengths.

Colour films are not recommended for high-resolution work. However, it sometimes happens that the only visible contrast on a planet is due to minor differences in colour rather than to actual contrast so that colour films may be the only type that will record this. Very good colour films are Kodak Ektachrome 200 and 400 (ASA 200 and 400). These are better than High Speed Ektachrome 160. Tests show that the 400 has grain equal to the 200 and better resolution. It can also be 'push processed' to ASA 800 by Kodak with but little loss in the end result. Kodak High Speed Ektachrome is fairly good but gives rather fuzzy images if overenlarged. Kodachrome 25 and 64 are much finer grained but do not respond well to many of the wavelengths essential for successful planetary photography.

Black and white film processing

The usual black and white films such as Plus-X and Tri-X include instructions for development in the packaging.

High-contrast negatives with very fine grain and speed of ASA 160 are obtained when the 2415 film is processed as follows:

(1) Develop in D-19 for 4.5 minutes at 68° F.
(2) Place film in the stop bath 0.5–1.0 minute.
(3) Transfer film to rapid fixing solution for 3–5 minutes.
(4) Wash film 10–15 minutes and soak in Photo-Flo for 1 minute before hanging to dry.

Lower contrast results are obtained but finer grain with an ASA rating of 80 if 2415 film is developed in HC 110 dilution D for 8 minutes at 68° F.
2415 film can be 'pushed' to a rating of ASA 400 in full strength Ethol UFG for 8 minutes. High contrast, high resolution and fine grain are all retained.

Photography of Individual Planets

Apart from recording their phases, Mercury and Venus have nothing much to excite the amateur planetary photographer. Uranus, Neptune and Pluto are too dim and remote to even bother with. The most rewarding planets to photograph are Mars, Jupiter and Saturn.

Mars

Although usually observed in dark skies, Mars can be photographed in twilight in green, red and infrared light. Cloud features are 'brought out' by ultraviolet

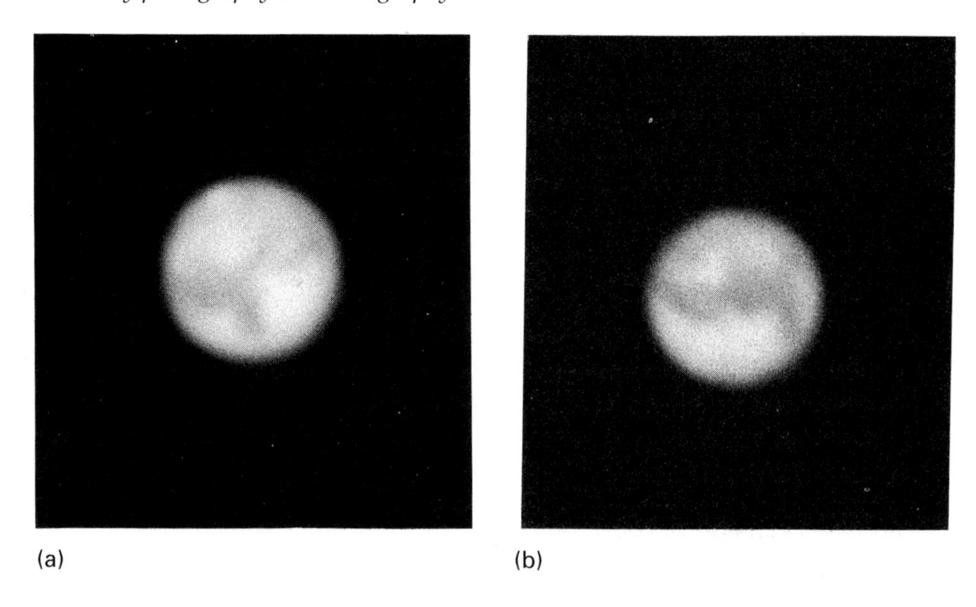

Fig. 15.5 Mars, 1988: (a) *September 26th, 04.30 UT;* (b) *September 30th, 03.15 UT; 14.25-inch Newtonian reflector. (Photographs by Eugene Witkowski, Buffalo, NY.)*

or blue light and limb haze is emphasised in ultraviolet. Two white light photographs of Mars taken with a 14.25-inch Newtonian reflector are shown in fig. 15.5.

Green light reproduces the visual impression of Mars. There is fairly good contrast between the dusky areas and the 'deserts' and white markings are well shown. Red and especially infrared pictures yield extreme contrasts and exposure times must be accurate if the bright and dark features are both to be well rendered. Only the most prominent clouds are recorded at these wavelengths.

Jupiter

Jupiter may be photographed near sunrise and sunset but not during the day. Its multicoloured disc features make it a good subject for colour photography used in conjunction with filters. By this means different features can be selectively enhanced. Red and infrared photographs show well the festoons at the North and South Equatorial Belt edges against the bright Equatorial Zone background. Part of the equatorial bulge of Jupiter is lost in red light so that the full extent of its limb is not revealed. Contrasts on Jupiter show least in light of long wavelengths; the Great Red Spot hardly stands out from the zones. The full extent of the limb of Jupiter is brought out in blue light and the Great Red Spot is prominent. Good contrast is shown between the cloud belts and bright zones. Panchromatic films and green light do not show as much contrast but the festoons are better seen. In ultraviolet light pictures, a pronounced brightening of Jupiter's sunward limb has been recorded. See fig. 15.6 for a photograph of Jupiter showing satellite I in transit; this was also taken with a 14.25-inch Newtonian telescope.

Fig. 15.6 Jupiter, January 16th, 1989, 02.00 UT: shadow of satellite I (Io) in transit; 14.25-inch Newtonian reflector. (Photograph by Eugene Witkowski, Buffalo, NY.)

Saturn

Compared with Jupiter, Saturn is much less colourful and has less variety of disc features. Nevertheless, colour photography in light of different wavelengths shows bluish polar regions, reddish equatorial regions and brownish cloud belts. Infrared or ultraviolet light give the best overall disc feature contrast. The south polar region is very dark in infrared photographs. The equatorial region is light in infrared but dark in ultraviolet. Saturn's A and B rings photograph well in light of all colours. The C ring shows up best in blue light. An interesting phenomenon that shows up well in photographs is the phase-related brightness changes in the rings. They brighten considerably relative to the globe when the planet is near opposition.

Colour filters are effectively used to improve contrast and to emphasise certain features. Unfortunately, increased exposure times are necessary – two or three times or even more than that required without their use – to compensate for the light that they absorb. See manufacturer's instructions.

Exposure times

Only estimates and ranges of exposure times have been previously given because it is impossible to give them exactly for the planet you wish to photograph. So much depends on the brightness of the planet, colour filters

Table 15.2 *Approximate photographic exposure times (seconds) for some planets.*

Film speed	Effective focal ratios of imaging system			
(ASA)	F/15	F/25	F/45	F/100
Venus				
64		1/60–1/125	1/30	1/8
125		1/125–1/250	1/50	1/15
250		1/250–1/500	1/125	1/30
500		1/500–1/1000	1/250	1/60
1000			1/500	1/125
Jupiter				
64	1/4	1/2	2	8
125	1/8	1/4	1	4
250	1/15	1/8	1/2	2
500	1/30	1/15	1/4	1
1000	1/60	1/30	1/8	1/2
Mars				
64	1/15	1/8	1/2	2
125	1/30	1/15	1/4	1
250	1/60	1/30	1/8	1/2
500	1/125	1/60	1/15	1/4
1000	1/250	1/125	1/30	1/8
Saturn				
64	1/2	1	4	16
125	1/4	1/2	2	8
250	1/8	1/4	1	4
500	1/15	1/8	1/2	2
1000	1/30	1/15	1/4	1

used, if any, the film speed and the aperture and focal ratio of your telescope. Only actual trials will tell. It is best to take several different exposures and to keep careful records of the results obtained for future reference (see table 15.2). As well as taking a shot for the exposure time suggested in table 15.2, try taking others at one and two stops faster and slower on either side. One of these shots is bound to be good.

Some specific examples may help to give you an idea of what exposure times to use: with High Speed Ektachrome having a daylight speed rating of 160, the colours of Mars are captured with a 0.5 second exposure at F/50. With the same film, a 1 second exposure of Jupiter shows much colour detail in the belts and zones. If the air is perfectly steady try Kodachrome 25 or Kodachrome 64 – both of these are slide films. Their fine grain favours resolution of planetary details and shorter exposures with smaller images should be tried with these.

You will notice some discrepancies between these exposure times and those for the same planets and film speeds given in table 15.2. This is partly due to differing opinion among experts on the subject of what is the right exposure for a given planet and also to the fact that the exposures recommended in table 15.2 are based on the *average* brightness of the planets. The actual brightnesses show more or less variation. The best thing to do is to use these recommendations as guidelines and to experiment until you get the results that you want. Always take several frames of the same planet during any one photographic session. There is always the chance that one of these will have been shot during a moment of perfect seeing and which will therefore show more detail than the others taken during only average seeing conditions. If you want to get really good photographs you can't afford to be stingy with film!

In connection with exposure an important point to remember is the influence of the focal ratio on exposure time. Although an F/8 8-inch mirror has a bigger aperture than an F/5 6-inch mirror the latter will be the 'faster' system for photography because it gathers more light than the 'slower' F/8 system. This is because the light gathering power of an optical system is directly proportional to the square of the radius of the objective and is inversely proportional to the square of the focal ratio, i.e.,

$$R \text{ is proportional to } \frac{r^2}{f^2}$$

where r is the radius of the mirror or object glass of the telescope, f is the focal ratio of the same and R is the relative light gathering capacity of the telescope. Therefore, for the F/8 8-inch reflector $R = 0.25$ and for the F/5 6-inch reflector, $R = 0.36$ so that the 6-inch telescope has 1.44 times the light gathering power of the 8-inch telescope.

However, the longer focal length is preferable to the shorter for resolving fine detail so that there has to be a compromise between the speed of the optical system and its resolving power. For planetary photography resolving power is perhaps the greater consideration because the planets are relatively bright objects to photograph. Overall, it is best to have as large a lens or mirror as you can afford because this increases both the light collecting capacity and the resolving power of the telescope.

Video and CCD photography (videography) of the planets

The development of video cameras and camcorders (video cameras used in conjunction with a videotape recorder/player), their use in conjunction with charge coupled devices (CCDs) and the recording of planetary images on tape rather than on photographic film has virtually revolutionised astrophotography in general. Some of the drawbacks of film as the image recording medium are avoided in these new techniques as will be shown later.

Celestial bodies that can be efficiently recorded by a video camera are restricted to those that are fairly bright such as the nearer planets. A CCD photograph of Jupiter taken with a 14.25-inch Newtonian reflector is shown in fig. 15.7. The high sensitivity of a video camera is expressed in *lux*; the lower the lux number the greater is the camera's sensitivity. For example, the Panasonic PV-320 camera has a sensitivity of 7 lux but many cameras now use CCDs and this greatly increases sensitivity to lux values of 3,2 or even 1.

Both tube and CCD cameras focus their images on a sensitised matrix

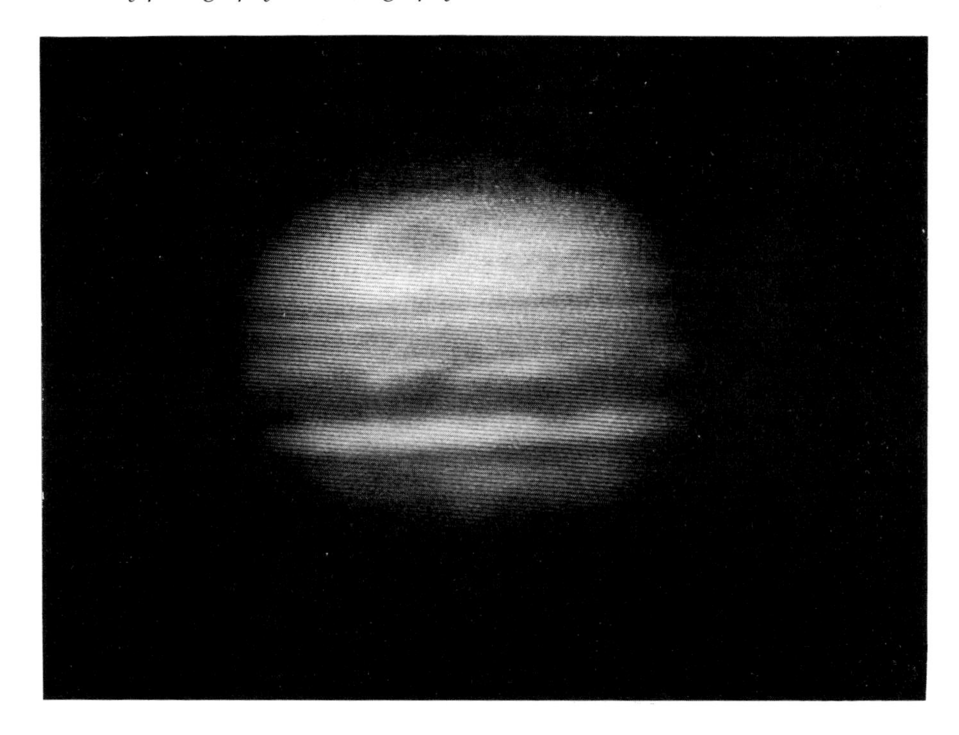

Fig. 15.7 Jupiter, April 6th, 1992, 02.38 UT (seeing: poor): showing the Great Red Spot; 14.25-inch Newtonian reflector, CCD video image. (Photograph by Eugene Witkowski, Buffalo, NY.)

consisting of fine lines. Tube type cameras have always been used for television and present day makes go by the trade names Vidicon, Nuvicon and Satiscon. The tubes work by directing an image through a precision glass plate and focusing it on the target which is a screen sensitive to light. The electric charge on the target is altered by the light and is recorded as a current. At the other end of the tube is an electron gun which emits electron beams that sweep over the target surface one line at a time to record the entire image. The tube design of camcorders gives very good resolution, sensitivity and colour definition compared to the alternatives. However, tube type cameras have now been virtually replaced by those employing the solid state sensors – charge coupled devices (CCD) or metal oxide semiconductors (MOS).

The CCD is a silicon microchip that converts light energy into digital data, i.e., electronic signals that may be recorded, modified and shown on a television screen. On its crystalline surface are hundreds of horizontal lines of submicroscopic photoelectric cells that sense light. These tiny photocells are the picture elements or *pixels*. The lines are electrically joined together in series. A CCD takes a picture every 1/30 second. When this happens the impinging light creates a small charge of electricity in each of the photocells. At the end of the 1/30 second the cells transfer their charges in a way resembling tiny buckets. At 'image-readout', the first 'bucket' ejects its electric charge off the chip. At the same time, the one next to it shifts its charge to the second and so on to the end.

This is repeated until the charge of the end photocell on a scan line has left the chip. All of the lines of the CCD chip are similarly scanned and the whole image leaves it. This initiates a signal that can be recorded and then translated into a picture on the television screen.

A major advantage of using CCD image recording as compared to ordinary photography is that CCD cameras require much shorter exposure times. This is because they are far more efficient as light energy converters than a photographic emulsion. Whereas in an ultrafast photographic film such as ISO 3200 only 2–3% of the impinging light photons interact with the emulsion's light sensitive atoms to give exposed film grains, no less than 40–80% of impinging light photons are converted by a CCD into a stored electrical signal. In terms of film sensitivity a CCD would be rated at ISO 20 000–100 000! Therefore, while a photographic film requires exposures of 1/2–10 seconds to record high-resolution images of planets, a CCD imaging device needs only 1/100 second to record a bright enough image and this is faster even than a CCD videocamera and fast enough to 'fix' those precious short moments of fine seeing. Furthermore, there is no reciprocity failure with CCDs, i.e., they do not lose their very high speed over the duration of the exposure. A photographic film over a 5–10 minute exposure on a dim object may lose a large fraction of its original speed, which is the plague of deep sky astrophotographers. Yet again, CCDs are much less susceptible to saturation that comes from overexposure. Detail will still be visible in excessively bright parts of an image. The fact that CCD images can be stored in ordinary computer floppy or hard discs that are erasable and reusable repeatedly means that the fuss and mess of processing and film hypering are avoided.

The reorder used in conjunction with a video camera can be built in or can be a separate VCR unit. When built in the camera-plus-recorder is called a camcorder. These are battery-operated and quite easily portable. With a separate camera and VCR, line current is often needed, especially with VCRs. However, you can connect a 'video enhancer' between the two and this will sharpen the image.

Structurally, a CCD camera is quite simple. The chip is firmly mounted in a hermetically sealed case and is protected by a precision glass window of optical quality. This keeps out dust and moisture; the latter could condense or even freeze on the imaging surface of the CCD if it is used for long exposures because then it has to be cooled to reduce 'dark current' that otherwise would obliterate the signals from faint objects.

The CCD camera cannot be merely connected to a television monitor to show an image. 'Image control electronics' are needed as a 'go between'. This is also known as a frame digitiser, the function of which is to convert the data from the CCD and to change it to a format compatible with the video display. In addition, it generates the signals needed to control the CCD for exposure, focusing modes, image enhancements and so forth. The controller varies from one to another make or model of CCD imager. The Lynxx-PC (Spectrasource Instruments) and ST-4 (Santa Barbara Instruments) rely on a 'host' computer necessarily supplied by purchasers of these models to perform the bulk of the video processing and control of the camera. The controlling electronics of the Lynxx-PC are mounted on a board accompanying the unit and you plug this in to your personal computer (IBM or compatible). In the same way, the ST-4 electronic controls are in a unit that has to be connected to either a PC or Macintosh via a standard serial post as for a printer or modem. The higher grade Lynxx models need no separate computer because the image controller has the ability to show and manipulate images on its own. If images are to be permanently preserved and manipulated it will still be necessary to connect the unit to a computer that you must provide.

Using a CCD camera

To use a CCD camera the astronomical object that you wish to record must first be located. Take a high-power eyepiece to centre the object, then attach the camera after removing the eyepiece. Focusing cannot be done directly as you cannot look through a CCD camera. A set of images must be taken, gradually refining the focus with each one. An exposure must be 'downloaded' once it is made, i.e., the image data supplied by the camera have to be sent to a computer for storage or enhancement. This might take as long as 15–20 seconds with the ST-4 that uses a 'serial' computer interface or as short as 3 seconds for cameras using a 'parallel' computer interface. The Lynxx with its board plugged inside the computer is even faster.

The time taken for downloading can slow up the process of focusing. To speed things up you can take advantage of the fact that the controllers of all CCD cameras are able to 'tell' the camera to take half – or quarter – frame exposures. Less of the image is seen but it is sufficient to show whether or not the image is focused properly. When using this half – or quarter – frame mode the download time is cut to one half or one quarter of that for the full frame mode.

When exact focus is achieved replace the camera with the eyepiece but don't alter the focus control. Slide the eyepiece backwards and forwards to refocus and use a marker of some kind to indicate the position of the best focus. Every CCD camera outfit is supplied with software that enables you to alter the brightness and contrast of an image. This ability to manipulate an electronic image is easily its biggest advantage. You can do this by calling up the so-called 'histogram mode' on the computer monitor screen. This is simply a graphical representation of how many pixels have a certain intensity level. A study of the graph will help you to extend or reduce the tone value range. An image may be extended or 'stretched' by selecting those pixels that lie inside a narrow intensity range, for example from the 100 to 150 level. You make the faintest pixels appear darker than they were initially and the brightest pixels much brighter, a process analagous to intensifying contrast film. An application of this is the enhancement of planetary features of low contrast. A good way to cut out sky fog and noise is to move the range of values down or up by a fixed amount; those pixels less than a particular value can be ignored and shown as black, for example.

In this technique it is best to have as many brightness steps as possible in the initial image. As an example 4096 levels of grey can be given with a camera having '12-bit' resolution. An '8-bit' resolution camera yields 256 levels at most. The former higher number of brightness levels may seem excessive at first sight; you can only display 256 levels anyway on the majority of computer monitors. The explanation is that the value of having 4096 levels of grey is realised when you choose a narrow range from them and then 'stretch' the contrast. If you are attempting to enhance a brightness range which is divided up into a finer set of steps your final enhanced image will probably have a smoother grading of tones. On the other hand excessive 'stretching' of an image generated by a camera having 8-bit resolution will result in an image with quite a digital appearance because of exaggerated stepping of the brightness levels.

If you decide to use a camcorder your first step in entering the field of planetary videography is to select a suitable one. Deep sky astrovideography requires great sensitivity for recording dim objects but for planetary work less sensitive models should be suitable. Sensitivity alone is not the most important attribute anyway; different camcorders react differently to dim light and show varying amounts of deterioration of the image as light becomes dimmer. Camcorders using CCD chips to sense light have greater initial sensitivity but have more noise as light becomes dimmer.

When purchasing a CCD camera describe to the sales assistant the type of performance that you require and ask to be allowed to try the camera in low light levels by testing it in a dark part of the store. Although many different makes of CCD cameras are said by the manufacturers to have similar light sensitivity, some models outperform others in dim light conditions.

As a start to planetary videography, the afocal method is worth experimenting with. First, focus the telescope on a suitable planetary object. Then attach the camcorder to a strong tripod and point the lens of the camera straight into the telescope eyepiece. This will give acceptable results. The polar caps and dusky markings of Mars, Jupiter and Saturn's rings and the satellites of these planets show well. The changing phases of Venus are easily recorded and Mercury, Uranus and Neptune are revealed as small bright discs.

Whatever type of camcorder you possess you record images on a video tape by first setting up the telescope and the camcorder loaded with tape and with lens cap removed on its tripod. Point the camera lens straight into the eyepiece of the telescope. Now set the zoom lens of the camera to its shortest focal length to give a wide field that includes everything that you can see in the eyepiece. Move the camera while looking through the viewfinder to get the image onto the screen. This manoeuvre takes a little practice to do efficiently. Centre the planet's image on the screen and adjust the zoom lens 'out'. You will see the circular eyepiece field image seemingly enlarge outwards to beyond the field of view of the camera. Also, you will notice that at long focal lengths, image centring is more exacting than at short focal lengths. However, you will have to recentre the image every so often because the fixed camera mount is not following the apparent motion of the planet due to the Earth's rotation. To obviate this an eyepiece of long focal length should be used which results in a wide-angle beam of light issuing from the telescope which cuts down on the frequency of readjustment of the camera's position. A 12-millimetre orthoscopic eyepiece is perhaps best as it permits good uninterrupted viewing time with good image magnification.

The degree of magnification that you use affects the appearance of the image on the tape. Too small an image will produce a saturated appearance, almost without colour and lacking detail. Too large an image will be fuzzy also lacking detail with the colours mixed up. The optimal amount of 'zooming' to use is determined by the object being viewed and the telescope's characteristics as well as your own personal taste. The best thing to do is to experiment until you can produce tapes that please you. The preferred way to assure properly exposed tapes is to look through the camera's viewfinder and set the focal length of the zoom lens to one that gives you good detail in the image. The best way to optimise exposure time is to preview your tape on the same television receiver as you will be using for viewing the tape. If you are using, say, an 8-inch aperture telescope with a camcorder of 7 lux sensitivity, the image size is not as important as getting the correct exposure. It is a good plan to change focal length at intervals while you are recording on tape so that you can see the range of results that are possible.

When you come to view your tapes on the television screen you will need to adjust the colour, contrast, hue and fine tuning controls to get the best possible rendering of image detail. These settings will be quite different from the ones that you will use for ordinary television viewing of brilliantly coloured pictures in well-lit rooms. For a good colourful rendering of a low contrast planetary image, you will obviously need a different adjustment. This will differ again from the rendering in high contrast of a shot of the moon's surface.

The secret of recording good images of the moon or planets is to match the telescopic image to the type of light detector correctly. For very many years planetary photographers have realised that long focal ratios and films of medium speed gave the optimum match between image and detector. In the same way, recording the moon and planets on videotape is best done using long focal ratios and the most sensitive video camera that is available. However, it is different when we come to CCDs. The very small pixels of the CCD make great image enlargement of a planet unnecessary. The image can therefore remain small and bright thus shortening exposure times. With CCD cameras, the best results are obtained by employing short focal ratios and short exposure times. CCDs therefore benefit from the short-lived moments of perfect seeing just as in ordinary planetary photography and visual observing.

In order to catch all detail in the telescopic image the detector, whether film, video tape or CCD, must be capable of reproducing the finest image detail. There is a rough rule that states that when two measurements are made per resolution element, nearly all image information is recorded and further enlargement of the pixels in CCD cameras like the Lynxx Electrim and ST-4, which are rectangular each measuring 0.00054 × 0.0006 inch, will yield negligible further detail. At a focal ratio of F/40 a perfect telescope will have an Airy disc whose brightest part is 0.0010 of an inch in diameter. This is sufficient to cover 2 CCD pixels.

An 8-inch telescope at F/40 gives an image of Jupiter that covers about two thirds of the diameter of the TL-211 CCD chip used in the above three cameras. Jupiter's image size is therefore just about optimum. Exposure times can be calculated using a formula slightly different from that used by planetary photographers.

$$\underset{\text{(seconds)}}{\text{exposure time}} = \frac{P}{P_{max}} \times \frac{F^2}{SB}$$

where B is the object brightness (Jupiter = 30 Mars = 60 Saturn = 10), P_{max} is the maximum pixel value in the image, P is the wanted pixel value for the object, F is the effective focal ratio, and S is the speed of the CCD. For the Lynxx camera:

$$P_{max} = 4095\,,\; P = 3200,\; S = 80$$

It is best to use eyepiece projection. The best match between the Tl-211 CCD chip in the three CCD cameras mentioned earlier and the optical system of the telescope used is at F/40. The bright centre of the Airy disc covers two pixels under these conditions. However, this is true for perfectly steady seeing conditions. When the seeing is tremulous it will be better to use shorter F ratios from F/30 to F/20 according to how unsteady is the seeing. Shorter F values require shorter exposures and so help to reduce the effects of unsteady seeing.

Yet another advantage of using the CCD camera technique over ordinary photography is that each image you take is seen moments later on the monitor screen. If the image is not focused properly or is blurred by bad seeing you can take another picture. This is more economical and gives faster results than taking several identical exposures of the planet on photographic film in the hope that one frame will have recorded much finer detail in a moment of perfect seeing. It has also made obsolete the oft repeated assertion that the human eye at the telescope eyepiece can instantaneously see fine detail in moments of perfect seeing that no photographic film can record because of atmospheric turbulence even in short exposures. In fact CCD imaging is more like visual observing than taking photographs.

The Electrim Corporation has introduced a high-resolution version of their EDC-1000 solid state imaging device. The new version, the EDC-1000 HR, permits users to produce digital images with up to 754 × 488 pixels. It needs an IBM-XT, AT or compatible but needs no frame grabber board or other third party hardware or software. The frame transfer CCD chip used in the EDC-1000 HR is configured into 244 lines and 754 elements per line and 8 bits (256 gray levels) per pixel. Among other things it features computer-controlled exposure times. Many amateurs have obtained digital CCD images that actually *exceed* in detail the best visual images obtainable with the same telescopes. The *surface* of Venus can be imaged at wavelengths that are within the range of these devices and of obvious importance in Jupiter and Saturn studies is that these cameras encompass the methane absorption band at near to 8900 angstroms.

However, there are some disadvantages with CCD imagers. First, only black and white pictures are yielded. Colour images can be produced only by taking three identical black and white pictures with red, green and blue filters and then electronically combining them on a colour television screen. Second, inexpensive imagers such as the Lynxx and ST-4 employ silicon chips with only a relatively small number of pixels – 192 × 165. This is contained in an area measuring only 2.6 millimetres square which is less than 1/100 of the area of a 35-millimetre film frame (36 × 24 millimetres). Because of the excessive smallness of these chips their field of view is very narrow; at the primary focus of an 8-inch F/10 Schmidt-Cassegrain telescope an area only 4 minutes of arc square can be imaged. Locating the object and centring it in such a small frame is by no means easy. There is no need for eyepiece projection when you realise that an 8-inch SCT with a 2000-millimetre focal length needs the addition of only a ×2 or ×3 Barlow lens to give planet images that are frame filling.

Fig. 15.8 Charge coupled device (CCD) for use with the LyNXX series of planetary and lunar imaging systems. Format: 192 × 165 pixels, pixel size 16 × 13.75 micrometres, active area 0.104 × 0.104 inches (0.264 × 0.264 centimetres), efficiency QE > 50% at 650 nanometres, charge transfer efficiency 99.998%. (Courtesy of Spectra Source Instruments.)

Suppliers of CCD Cameras

The addresses of the three CCD camera suppliers mentioned earlier are as follows:

Electrim Corporation. PO Box 2074, Princeton, N. J., USA 08543

Santa Barbara Instrument Group. 1482, East Valley Road (Suite 601), Santa Barbara, California, USA 93108

Spectra Source. PO Box 1054, Agoura Hills, California, USA 91376-1045.

Two Spectra Source CCDs are shown in figs. 15.8 and 15.9.

Video-assisted drawing (VAD) of the planets

Video assisted drawing (VAD) of planetary disc details is a combination of videography and recording of visual observations by drawing. It can be used by both beginning and experienced planetary observers to enhance their skill in

Fig. 15.9 Charge coupled device (CCD) for use with the LyNXX series of planetary and lunar imaging systems. Format: 1024 × 1024 pixels, pixel size 12× 12 micrometres, active area 0.48 × 0.48 inches (1.2 × 1.2 centimetres), efficiency QE > 50% or 50 nanometres, charge transfer efficiency 99.998%. (Courtesy of Spectra Source Instruments.)

observing. The beginning amateur planetary observer can become more accurate and efficient in planetary drawing by first making a brief videotape while observing. When the drawing made at the telescope is finished, the observer can then watch the taped images immediately afterwards and use them while the details are still fresh in the mind and to check the positional accuracy and outlines of planetary markings and to correct them if necessary. Previously this has been done by photovisual observation, i.e., the use of photographs as an aid to planetary drawing. Major features may be accurately placed and then fine details added that were seen in the telescope but not recorded on the photograph. However, there is, of course, a much longer delay between making the exposure and the final photograph.

Videotaping should be done using the same colour filters, if any, with which the telescopic observations and drawings were made. A variety of filters may be used with Mars, for example, but the Wratten 47 (violet) has been found to be too dense for use with the commonly available cameras with sensitivity of 7 lux.

Training more observers to use the VAD technique could lead to much more serious and valuable planetary observation.

Further reading

Books

Astrophotography for the Amateur. Covington, M., Cambridge University Press, Cambridge (1985).

Introduction to Observing and Photographing the Solar System. Dobbins, T. A., Parker, D. C. and Capen, C. F., Willmann-Bell Inc., Richmond, Virginia (1988).

Manual of Advanced Celestial Photography. Wallis, B. D. and Provin, R. W., Cambridge University Press, Cambridge (1988), pages 151–8.

Papers and articles

Planetary photography. Pope, T., *The Strolling Astronomer* **22(7–8)**, 124–9 (1970).

Hints on planetary photography for amateurs – I. Minton, R. B., *Sky and Telescope* **40(1)**, 56–9 (1970).

Recent advances in planetary photography. Capen, C. F., *JALPO* **27(34)**, 47–51 (1978).

A short discourse on planetary photography. Sanford, J. *JALPO* **28(7–8)**, 129–34 (1980).

Planetary projection photography: a simple device featuring a vibrationless shutter and seeing monitor. Parker, D. C., *The Astrograph*, **II(6)**, 83–6 (1980).

The quest for the universal developer. Parker, D. C., and Capen, C. F., *JALPO* **28(9–10)**, 179, 182–4 (1980).

A new twist in planetary photography – unsharp masking. Beish, J. D. and Parker, D. C. *JALPO* **29(7–8)**, 145–7 (1982).

Photographing the giant planets. Burnham, R., *Astronomy* **11(5)**, 35–8 (1983).

A simple technique for obtaining violet light photos of Mars. Parker, D. C., Beish, J. D. and Capen, C. F., *JALPO* **31(9–10)**, 181–3 (1986).

Tools for the astrophotographer. Talcott, R., *Astronomy* **16(1)**, 78–81 (1988).

High resolution lunar and planetary photography. Dilsizian, R., *Astronomy* **16(1)**, 70–5 (1988).

Hypered film for planetary photography. Reynolds, M. and Parker, D. C., *Sky and Telescope* **75(6)**, 668–9 (1988).

A primer for video astronomy. MacFarlane, A. W., *Sky and Telescope* **79(2)**, 226–31 (1990).

A camcorder assist for planetary observers. Troiani, D. M. and Joyce, D. P., *Sky and Telescope* **80(4)**, 409–10 (1990).

Planetary video images and video assisted drawings. Troiani, D. M. and Joyce, D. P., *JALPO* **35(1)**, 1–3 (1991).

Planetary imaging with a small CCD camera. Parker, D. and Berry, R., *JALPO* **36(1)**, 1–8 (1992).

Going digital with color. Berry, R., *Astonomy* **20(7)**, 80–5 (1992).

16

Photoelectric photometry of the minor planets, planets and their satellites

General

The development of photoelectric light measuring instruments (photometers) which have electronic light detectors and current amplifiers has made it possible for the weak current generated by the light of planets, their satellites and the asteroids focused by the telescope objective on the detector to be amplified sufficiently to cause deflection of a galvanometer needle or to give a digital readout as in more recent equipment.

Photoelectric photometry dates from about the latter part of the nineteenth century. The first ever photoelectric measurement of a star's light was made in 1892 with a 24-inch telescope belonging to an amateur. The observation was made by a professional astronomer, G. M. Minchin and two amateurs. Subsequent to this first amateur venture into photoelectric photometry, professional astronomers took it over completely during the next 60 years and it became widely available to them during the 1930s. Photoelectric measuring instruments are now available at prices affordable by most amateurs who are thus enabled to contribute much valuable data in this area of research.

Although astronomical photometry is done by professional astronomers using large telescopes, there are several advantages in amateur pursuit of this type of investigation:

(1) There are more amateurs than professionals available to do this kind of work.
(2) Amateurs don't have to 'book' time on large observatory telescopes which are available for short periods only. There is greater freedom and flexibility from owning their own telescopes.

(3) Amateurs are able to spend more time on prolonged studies of selected objects.
(4) Smaller telescopes are better for study of bright objects.

The photoelectric photometer and its components

The essential structural components of a photoelectric photometer are as follows (fig. 16.1):

(1) *The aperture diaphragm.* This is placed at the telescope's focal plane. It is a thin opaque diaphragm pierced with a very small hole that allows only light from the area immediately adjacent to the planet or asteroid to get through to the detector so that background illumination is minimised. Only the light from the object being measured thereby impinges on the photoelectric cell or photodiode.

An important point is that the aperture in the diaphragm is sufficiently large to allow all of the light from the tiny star-like image of an asteroid to pass through but not so large that stray light from sky glow is admitted. The majority of apertures admit a minimum of 5 seconds of arc of sky area. Generally, the fainter the object being measured the smaller should be the aperture.

(2) *The Fabry lens.* Immediately behind the aperture diaphragm is the Fabry lens which focuses light transmitted by the objective onto the photocell or

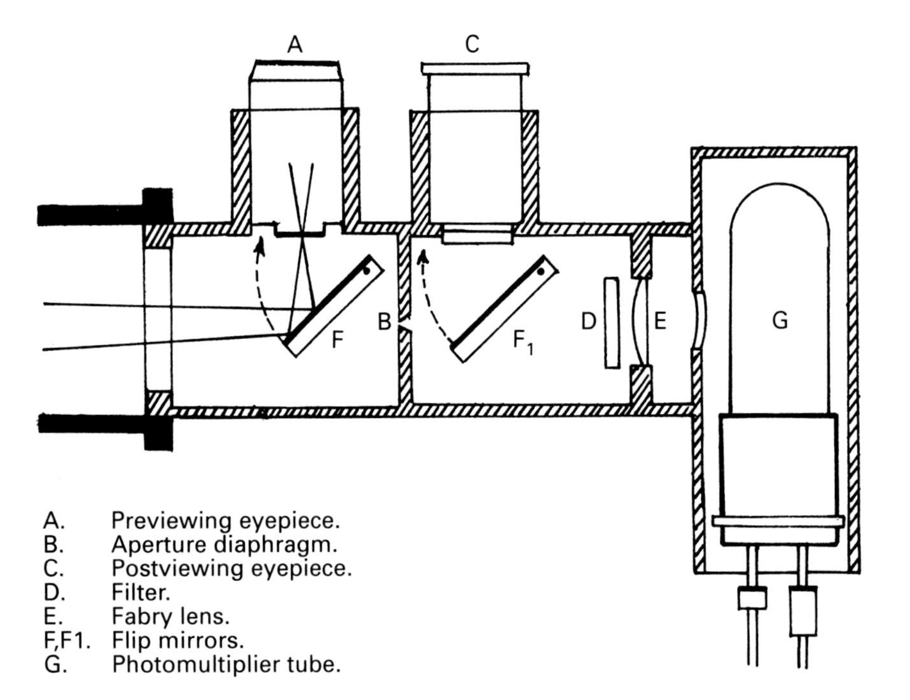

Fig. 16.1 The essentials of a photoelectric photometer: A previewing eyepiece; B aperture diaphragm; C postviewing eyepiece; D filter; E Fabry lens; F, F_1 flip mirrors; G photomultiplier tube.

photodiode. Different parts of the light-sensitive surface of some photocells and photodiodes to not respond equally to light impinging on them to give equal electrical currents; the function of the Fabry lens is to smooth and average out the light from an asteroid or star over the detector's entire surface. It also reduces scintillation that could cause the galvanometer reading to be unsteady.

(3) *Filters.* The filters used in star photometry are the UBV filters. The letters refer to the peak transmission wavelengths; U is ultraviolet, B is blue and V is visual or yellow. Two more filters may also be employed, the R (red) and the I (infrared). The whole set of filters is therefore referred to as the UBVRI system. They are all of the narrow pass band type which permit only a narrow part of the spectrum to be transmitted.

A photometer usually, but not invariably, has at least one filter placed behind the Fabry lens. If there is a single filter it will be the so-called 'V' or yellow filter which most nearly approximates to the sensitivity of the human eye. Stellar photometers have in addition a wheel of filters allowing rapid interchange of several different filters such as the UBVRI series.

(4) *Photocells or photodiodes.* The energy of the light from the object being measured is converted into electrical energy by a photocell or photodiode. The current so generated is very small and has to be amplified. A photocell that is equally sensitive to most light wavelengths is the RCA IP21. The colour sensitivity is affected by temperature; the more this is lowered the more sensitive is the photocell to the shorter wavelengths of light.

Less efficient (so far) but more recently introduced are silicon PIN photodiodes. These are not temperature-sensitive but are not as responsive to shorter wavelengths as the photocells. They are of solid state construction and so heat accumulation is minimal in normal photometric work. They look like replacing tube-type detectors. When employed at about normal room temperature (ca. 25° C) a carefully chosen photodiode will equal in response and sensitivity an inexpensive photocathode in wavelengths from 300 to 1100 nanometres.

(5) *Viewing eyepieces and 'flip' mirrors.* In order for the exact circular area to be registered by the detector to be first accurately centred in the telescope there are two viewing eyepieces beneath each of which is a hinged 'flip' mirror which intercepts and reflects the light path through the eyepiece. The previewing eyepiece is first used with the flip mirror reflecting light from the telescope objective up through it. This gives a wide field of view and so assists in locating the object of interest which is then approximately centred. The mirror is 'flipped' out of the way and the light beam now hits the diaphragm. The second eyepiece now views the much smaller field delineated by the tiny hole in the diaphragm. The reticle with one etched circle in the focal plane of the eyepiece represents the field of view of the detector and assists accurate centring of the object in the small circular field. When the objective is centred the second mirror is then flipped out of the light path allowing the light beam to activate the detector.

(6) *The signal amplifier.* The minute current (less than 0.5 microamperes) generated by the photocathode needs to be amplified to a usable current. The amplifier of a photoelectric photometer can work with AC or DC, the DC unit being most popular for amateur work.

The amplifier unit incorporates a selector for controlling gain and also offset controls for cutting down the so-called 'dark current' from the system's electronic components and the intrusive effects of sky glow.

A selector for the brightness of the asteroid or planet being measured ought to be part of any photoelectric photometer design so as to adjust the gain for the amplifier unit. A bright planet giving the maximum galvanometer needle deflection would require low gain whereas a faint asteroid near the telescope's light grasp threshold would need very high gain.

At the beginning of every measurement the galvanometer dial must be set to zero so that the meter can be zeroed and provision must be made for this.

In amateur-owned photometers there is often incorporated an offset control which allows the 'dark current' in the detector to be nullified. The offset can also be used to cancel the night sky glow near to the object being measured.

(7) *The strip chart recorder.* A graphic permanent record of the object's light curve, if variable, is obtained if the control unit includes some means of directing the current to an external readout such as a strip chart recorder. These commonly employ ink and a small stylus marks the activity registered by the detector on moving paper strips.

Although costly, the devices can be relied upon. Some can be equipped with gear trains which can be interchanged so that speeds up to 1 inch per minute are possible. Time intervals are marked on the paper if this is pressure-sensitive, by the tapping of a pointed stylus, such as once per second, depending on the gear train speed. The paper is marked by the frictional heat.

Telescopes for photoelectric photometry

No one telescope design is uniquely suited for photometry. However, some are better suited than others. The chromatic aberrations of refractors make them less than ideal for photometric work. The violet and red wavelengths come to a focus at different distances from the objective with other colours in between. Although the yellow wavelengths may be exactly focused on the detector the blue-violet and red rays will be out of focus so that the detector will not respond equally well to all of the wavelengths of light. Apochromatic refractors have even better colour correction than the ordinary achromatic refractor but the chromatic aberration is not completely eliminated even in these.

The best type of telescope to use in photoelectric photometric work is a reflector, Newtonian or Cassegrainian, because these are completely free of chromatic aberration. The popular Schmidt–Cassegrain and Maksutov–Cassegrain types of catadioptric telescopes with glass correction plates have a certain degree of chromatic aberration.

As well as being achromatic the best kind of telescope for photoelectronic photometry should also have a long focal length. The resulting increased image scale means that a smaller area of field needs to be centred in the region of the asteroid or planet being measured. There is therefore much less scatter from the background illumination such as sky glow or faint stars not visually perceptible, so that the consistency and reliability of the photometer readings will be much better. A Cassegrain telescope having a focal ratio from F/12 to F/20 is much more suitable than a Newtonian of focal ratio F/5 to F/8.

Another important matter is the type of telescope mounting. The sheer weight of a heavy detector mounted on the telescope eyepiece focusing mount can cause considerable balancing problems with a German type equatorial mount. This is most severe if a Newtonian telescope is used because the detector will be attached at the side of the tube near its upper end. The fork type of mounting is best with slow motion controls on the two axes because the balance of a detector on the bottom of the telescope can be offset easily.

Almost more important than the actual optics is that the telescope should have a massive and rigid mount otherwise the lining up of the asteroid, or other object, and comparison star images so that their light enters the pinhole in the photometer diaphragm accurately, cannot be maintained steadily when making a measurement.

Although a Cassegrainian telescope of large aperture with a focal ratio of about F/16 to F/20 is ideal for photometric work other types can be used as long as you are aware of the difficulty involved in attaching the photometer and the necessary corrections to be applied to chromatic aberration if present in the telescope.

Photoelectric photometric procedure

With the photoelectric photometric equipment commercially available the careful amateur can make brightness measurements to an accuracy of 0.01 magnitude.

Before an observing session your telescope should be uncapped and all the electronic components should be switched on at least an hour before commencing so that the current amplifier and everything else can attain thermal stability.

Photoelectric photometry of the minor planets

Admittedly, photoelectric photometry of asteroids has been and is being done by professional astronomers with more sensitive and accurate equipment than is usually found in amateur hands. However, because of the large eccentricities of some asteroid orbits it often happens that a faint asteroid usually inaccessible to amateur equipment may swing close enough to the Earth for it to come within the reach of amateur equipment. At such times valuable observations of magnitude and brightness variability may be made.

In photoelectric photometry of asteroids there are quite stringent requirements to be met in selecting a comparison star. Firstly, it should ideally be no more than 1° away from the asteroid so as to minimise any difference in conditions of the sky and atmosphere between the two bodies. If too far apart, the one nearer the horizon would be too dim owing to atmospheric light absorption. The sky background can also vary. Secondly – and obviously! – the comparison star should not be variable. However, this criterion can be quite difficult to meet. Thirdly, since most red stars are variable it is best not to select a red star as the comparison. Fourthly, the comparison star and asteroid should be similar in magnitude. The last two requirements are difficult to fulfill in practice because it is not very easy to find such a star. It is therefore best to regard these as guidelines only.

When observing the asteroid, V filtration should be used all the time. This closely matches visual magnitude as perceived by the eye. The procedure used for making estimates of asteroid brightness is the same as for visual estimation but it is still necessary to do so every 5–10 minutes. However, in the photoelectric method you should make two brightness determinations each time. The reason is that the photoelectric method is so sensitive that several factors can influence the light intensity recorded by the photoelectric photometer such as scintillation and atmospheric turbulence. To reduce these as much as possible select a comparison star as near to the asteroid as you can and if possible one whose magnitude is known to 0.01 magnitude accuracy and which is constant to within 0.01 magnitude. These values are listed in star catalogues. As soon as you get a measurement from the comparison star, shift quickly to the asteroid and measure it. This procedure can be more quickly performed if you have an assistant. You first centre the asteroid or comparison star in the viewer of the photometer while the assistant notes the readings of the photometer's galvanometer needle deflections or digital readout. This works well if the comparison star is quite close to the asteroid, say within 2° but if the distance is greater a correction must be applied for atmospheric extinction.

In the photoelectric photometer the light from the asteroid passes through the tiny hole in the opaque diaphragm before it impinges on the light detector. The hole is very small so as to exclude as much sky light as possible. It cannot be overstressed that the positioning of the asteroid or comparison star in the tiny hole in the opaque diaphragm of the photometer and being able to maintain the line-up accurately makes absolutely essential a massive and rigid mounting for the telescope.

In making a measurement the brightness reading of the sky background near the asteroid is first measured and this is subtracted from the brightness reading of the asteroid which necessarily includes some of the background sky glow. Therefore the brightness due to the asteroid alone is:

brightness of asteroid and sky – brightness of sky
= brightness of asteroid alone.

The brightness due to the comparison star alone is determined in the same way. The differential magnitude (*D*) of the asteroid, i.e., the difference between the magnitudes of asteroid and comparison star, is given by the equation:

$$D = 2.5 \log \frac{L_s}{L_a}$$

– where L_s is the photometer reading of comparison star and L_a is the photometer reading of asteroid, – which can be quickly worked out on a pocket calculator. Depending on whether the asteroid is brighter or fainter than the comparison star, *D* will be negative or positive respectively. This method is called differential photometry because it measures the difference in magnitude between the asteroid and the comparison star.

Of the asteroids which are known to be variable the only ones that are within the range of amateur-owned photometers are 433 Eros, 216 Kleopatra, 44 Nysa and 15 Eunomia. 433 Eros is difficult except when it is near to a favourable opposition. This is not to say that you should limit yourself to these. Owing to their variable axis tilts it is always possible that many of the brighter asteroids

Table 16.1. *Asteroids that vary in brightness.*

Name and number	Axial rotation period (hours)	Amplitude	Magnitude (maximum)
1 Ceres	9.08	0.04	7.0
2 Pallas	7.81	0.12–0.15	6.3
3 Juno	7.21	0.15	6.9
4 Vesta	5.34	0.10–0.14	5.5
5 Astraea	16.81	0.21–0.27	
6 Hebe	7.27	0.06–0.19	7.1
7 Iris	7.14	0.04–0.29	6.7
8 Flora	13.6	0.1–0.04	7.8
9 Metis	5.06	0.06–0.26	8.1
10 Hygeia	18.0	0.09–0.21	
11 Parthenope	10.67	0.07–0.12	
12 Victoria	8.65	0.20–0.33	
13 Egeria	7.05	0.12	
14 Irene	9.35	0.04–0.10	
15 Eunomia	6.08	0.04–0.10	7.4
16 Psyche	4.30	0.11–0.32	
17 Thetis	12.28	0.12–0.36	
18 Melpomene	11.57	0.15–0.35	
19 Fortuna	7.46	0.25	
43 Ariadne	11.47	0.50	10.4[a]
44 Nysa	6.42	0.20–0.40	9.9[a]
216 Kleopatra	5.60	0.40–1.60	10.0[a]
321 Florentina	2.87	0.30	14.4[a]
433 Eros	5.27	0.00–1.50	10.0[a]
1580 Betulia	6.13	0.50	12.0[a]

[a]Magnitudes at average oppositions. They may be brighter at favourable oppositions.

exhibit slight brightness changes that have not yet been noticed. A list of asteroids which exhibit brightness variability is given in table 16.1.

A comparison star is not really needed if you measure the light changes of an asteroid photometrically. However, it is good procedure to compare the magnitude of the asteroid with a star of known magnitude at the start of every observation period so as to establish a standard. Atmospheric light extinction is even more important to correct for when studying an asteroid for slight light changes. You should carry out such observations within two hours on either side of the meridian transit of an asteroid, i.e., 30° east or west of the meridian. Atmospheric light extinction will thereby be minimised. Some correction must still be applied, though, or the asteroid will undergo an apparent brightening and dimming as it approaches and passes the meridian. Table 16.2 shows the corrections to be applied to the apparent magnitude of an asteroid in relation to its angular elevation above the horizon.

The extinction correction needn't be applied if you use a comparison star in

Table 16.2. *Atmospheric extinction.*

When estimating the magnitude of an asteroid, atmospheric light absorption must be allowed for. The table below relates the loss of light expressed in magnitudes for various altitudes above the horizon. Light absorption can be considered negligible at altitudes of 45° and higher.

Angular height above horizon (degrees)	Light loss in magnitudes
1	3.0
2	2.5
4	2.0
10	1.0
13	0.8
15	0.7
17	0.6
21	0.4
26	0.3
32	0.2
43	0.1

the differential photometric technique. Atmospheric extinction will equally affect both the asteroid and comparison star if they are sufficiently close in the sky so that the differential magnitude will remain the same.

Colorimetric photoelectric photometry

This is used to determine reflectivity in various wavelengths of light which provides indications of the constitution of the asteroid. Some asteroids have a noticeably red colour which suggests that they may be ferruginous (iron-containing) and have Mars-like surfaces.

Filters used for this are of the narrow pass band type making up a U,B,V,R and I sequence. For success in this method you must previously know the accurate colour indices of the comparison star. If you don't you can determine only arbitrary colour indices for the asteroid relative to your instrumentation because the filters all vary in their sensitivity in different light wavelengths.

The colour index expresses the colour of the asteroid which is a relative value for the asteroid. Standard indices could be V–R, R–I and U–V. A comparison star is not needed for this work.

Photoelectric photometry of the planets and their satellites

Jupiter and Saturn

Since these two planets exhibit such large discs in the telescope, the brightness of localised areas can be measured by placing the tiny diaphragm hole in the photometer over selected areas of the planets' discs. The bright zones of Jupiter are especially suitable as they are usually clearly defined and broad.

Whole disc photometry of Jupiter and Saturn is worth doing. Significant variations have been detected in the brightnesses of Jupiter and Saturn (globe plus rings together) which have been found to be dimmer than predicted (Westfall, J. E. personal communication).

Saturn's rings

At times, there appear areas of brightness concentration on Saturn's rings. This is sometimes observed when the 'bicoloured aspect' of the rings is exhibited (chapter 10) and also during the edgewise presentation of the rings. The bicoloured aspect may be due to a combination of 'clumping' of ring material in parts of the rings, an effect due to the gravitational pull of Saturn's satellites and to atmospheric refraction. The influence of the latter is supported by the fact that the bicoloured effect is most pronounced when Saturn is close to the horizon and viewed with a red filter. Both ansae appear of the same brightness when viewed with a blue filter. The degree of atmospheric refraction may be determined by photoelectric photometry.

Of considerable importance are photometric determinations of the magnitude of the 'knots' seen at times during the edgewise ring presentation and also of the edgewise rings themselves; often, one ansa will appear as a very thin line and the other may be invisible.

Saturn's satellite Iapetus

Saturn's satellites offer an interesting field of photometric study. The brightness of the satellite Iapetus is known to be variable and has a substantial amplitude. Colorimetric work in connection with the brightness changes is scanty and this provides an area worth exploring by amateur observers. The brightness variations and colour no doubt correlate well with the albedo features of Iapetus revealed by the Voyager photographs.

Photoelectric photometry of Iapetus would therefore be an interesting study to correlate with the albedo features revealed by Voyager. When Iapetus is faintest at around magnitude 11 it is beyond the range of most amateur telescopes for effective photometry but 12- to 14-inch telescopes should collect sufficient light to give a needle deflection if the V mode is used.

Uranus

This is an obvious choice for photometry in view of the well-marked long- and short-term brightness variations exhibited by this planet. Monitoring can be done in white light or using the precision U,B,V,R,I series of filters. Details of the brightness variability characteristics of Uranus are described in chapter 11.

Neptune

During the 11-year solar cycle Neptune exhibits long term albedo and brightness variations with an amplitude of 4%. These are negatively correlated with the sun's output of ultraviolet light. Reference should be made to the paper by Lockwood and Thompson in *Science* **234**, 1543 (1986) entitled 'Long term brightness variations of Neptune and the solar cycle modulation of its albedo'.

Significant magnitude variations were found in the four planets Jupiter, Saturn, Uranus and Neptune during 1989/91 by ALPO members F. Mellilo, R. Schmude and J. E. Westfall.

Pluto and Charon

Although difficult and challenging to do, amateurs of the future with medium to large aperture telescopes (8–16 inches) should be able to monitor the light fluctuations which occur when Pluto and its satellite Charon undergo mutual transits and occultations.

This occurs when the orbit of Charon is presented edgewise to the Earth (fig. 13.2). It happens twice during one complete orbital revolution of Pluto. The transit/occultation cycles last for 5 or 6 years; the most recent series commenced in 1985, the first time that these events could be observed since Pluto's discovery. Above, I mentioned amateurs of the *future* advisedly; in view of the very long orbital revolution period of Pluto I doubt whether anyone reading these words at the time of writing (1992), however young, will be in a fit state to observe the next series of Pluto-Charon eclipses. These events won't occur until the year 2110.

Further reading

Books

'Photometry and colorimetry of planets and satellites'. Harris, D. L. Chapter 8 in: Kuiper, G. P. and Middleshurst, B. M. *Planets and Satellites.* University of Chicago Press, Chicago (1961).

Photoelectric Astronomy for Amateurs. Wood, F. B., MacMillan, New York (1963).

Astronomical Photometry. Henden, A. and Kaitchuck, R., Van Nostrand Reinhold, New York (1982).

Solar System Photometry Handbook. Genet, R. (Ed), Willmann-Bell Inc., Richmond, Virginia (1983).

Photometry of planets and satellites. Lockwood, G. W., Chapter 2 in: Genet, R. M. (Ed), *Solar System Photometry Handbook.* Willmann-Bell Inc., Richmond, Virginia (1983).

Papers and articles

Getting started in photoelectric photometry. Melillo, F. J., *Astronomy* **14(6)**: 65–8 (1986).

Backyard photoelectric photometry. Genet, R. M., *Astronomy* **11(2)**: 51–4 (1983).

Photometry of variable stars and asteroids. Melillo, F. J., *Astronomy* **14(7)**: 58–62 (1986).

Multicolor photoelectric photometry of Uranus. Appleby, J. F. and Irvine, W. M., *Astronomical Journal* **76**, 616 (1971).

NAME INDEX

Aben-Bagel 265
Adams 116
Adams, John Couch 340, 342
Adel, A. 327
Aerts, Leo 279
Ainslee, M.A. 214, 275
Airy, Sir George B. 108, 148, 281, 340
Aitkin, R. G. 273, 324
Albategnius 264
Albers, S.C. 343
Alcock, G. 72
Anderer, Joseph 88
Andrenko 287
Antoniadi, Eugene M. 78, 90, 152–4, 157, 160, 221, 268, 272, 273, 257, 277, 278, 284, 285, 286, 287, 326
Arago 111, 147
Armellini 328
Arrhenius, S. 140
Ashbrook, Joseph 105, 327
Avigliano, D. P. 277
Azout, A. 210, 269

Baade, W. 187, 188
Ball, William 269
Bardet 152
Barnard, E. E. 105, 151, 221, 273, 283, 324, 345
Bartlett 278
Bartrum, C.O. 275
Baum, Richard M. 91, 100, 110, 114, 124
Becker, W. 317, 327
Beebe, R. 220
Beer, W. 75, 116, 125, 146, 149, 156, 177
Belopolsky 285
Belton, M.J.S. 334, 346
Bertrand 104
Bessel 281, 337
Bianchini 111
Birmingham 76
Blunck 150
Bode, J.E. 10
Bond, G. 270, 271, 273
Bond, W.C. 270, 271, 272, 273, 349
Bos, Van den 185
Boscovich 104
Both, E.E. 299
Botham, J.H. 288
Bottema 117
Bouvard, A. 337, 340, 342
Bowyer, W. 274
Boyer 116
Bradley 269
Brahe, Tycho 11
Brenner, L. 273
Brown, Ernst W. 187, 358
Browning 96, 156
Bryant, R. 333
Buffham, W. 323
Burnham 72
Burton 151
Butler, F.C. 329
Butterton, M.S. 286

Camichel, H. 80, 82, 90, 221, 257, 327, 328
Cammell, B.E. 284
Campani 207, 268, 269
Campbell, L. 316, 325
Campbell, W.W. 273
Capen, C. 346, 349
Carpenter, J. 271
Cassini, G.D. 104, 111, 116, 145, 207, 209, 210, 269
Challis, Professor J. 270, 340, 344, 345
Chapman, Clark 82
Christie, W.H.M. 346
Christy, J.W. 360
Coolidge, S. 270
Copernicus, Nicolaus 29, 88, 265
Corliss, W.R. 99
Crabtree, William 105
Cragg, Thomas A. 277, 346
Cruickshank, D.P. 288, 346, 358

Daniels, S. 280
D'Arrest 341, 342
Dauvillier, A. 140

Davies, M.E. 345, 346
Dawes, W.R. 49, 147, 149, 156, 270, 271, 344, 345
De La Rue 150
Delporte, E. 185
de Mottoni, G. 150
de Vacouleurs, Gerard 150, 153
De Vico 270
di Vico, F. 344
Denning, William F. 78, 89, 209, 282, 283, 285
Derham 109
Divini, Eustachio 267, 268
Dolfuss, A. 80, 82, 90, 288
Douglas 152
Drake, S. 343
Dunham 116
Dunn, Samuel 104
Dutton, C.L.O'B. 275
Dyce, Rolf B. 87

Ebisawa, S. 157
Eddington, Sir Arthur 274
Edwards, E.A. 357
Elger, Thomas Gwynn 96, 271, 272, 283
Encke, J.F. 270, 345
Espin, T.E. 53
Eudoxus 75

Fauth, P. 273
Flammarion, C. 273, 284, 285, 344
Flamsteed 321
Focas 277
Fontana, Francesco 104, 111, 144, 267
Fouche 344
Foulkes 113
Fourier 153
Fournier, G. 274, 275, 325
Fournier, V. 275
Fowler, A. 323
Freeman, Rev. A. 284
Fritsch 96

Galileo 104, 204, 266, 267, 343
Galle, J. 271, 341, 342
Gambato, G. 280
Gassendi, Pierre 93, 105, 267
Gelinas, M.A. 126
Gentili, M. 221, 257
George III, King 321
Gill, T.B. 357
Gordon, R.W. 50
Gottingniez, Gilles-Francois 207
Graff, K. 285
Graham, D.L. 128, 329
Gray, D. 329
Green, N.A. 156, 283
Griffiths 113
Grimaldi, Francesco 267
Gruithuisen 96, 104, 112, 113
Guinan, E. 348

Haas, W. 214, 280, 288
Hall, Asaph 149, 152, 272, 282
Hall, M. 346
Halley, Edmund 107, 108
Hammel, Heidi 349
Hams, D.L. 317
Hansen, Peter A. 355
Harding, K.L. 96, 110, 183, 343
Hargreaves 214, 215
Harris, D.L. 328, 344
Hay, W.T. 286
Hayes, S.H. 334, 346
Heath, A.W. 279
Heath, M.B.B. 277, 326
Helin, Eleanor 189
Hell, Maximilian 104
Hencke, K.J. 183
Henderson, A. 324
Henry, Paul 324
Henry, Prosper 324
Herbig, G.H. 277, 279
Herschel, Sir John 269, 270, 342, 344
Herschel, Sir William 72, 75, 104, 109, 111, 116, 146, 281, 314, 317, 320, 322
Hevelius, Johannes 75, 267
Hind, J.R. 344, 345
Hipparchus 27
Hirayama, K. 187
Hodgson, R.G. 328
Hodierna, Fabri 267
Holden, E.S. 221, 272, 283
Hollis, A.J. 329
Hooke, Robert 207, 209, 210, 269
Horrebow 104
Horrocks, Jeremiah 105
Houzeau 105
Hubbard 343
Huggins, W. 96, 323
Hughes, D. 343
Humason 360
Hunt, G. 332
Hutchings, J.B. 124
Huyghens, Christian 144, 267, 268
Hyde, Raymond 212

Jacobs, W.S. 270
Jarry-Desloges, R. 286
Jeffers, H.M. 277
Johnson, H.M. 215

Kaiser, F. 147, 150
Kater, H. 270
Keeler, J.E. 270, 271, 323
Kepler, Johannes 11, 30, 105, 183
Key 113
Kirch 109
Kirkwood, Daniel 186, 271
Kitchiner, 281
Knight, J. 275
Knobel, E.B. 148, 212
Kohler 96
Kowal, C.T. 188, 343
Kozyrev, N. 82, 111
Kuehn, D. 220
Kuiper, G.P. 117, 140, 277, 317, 328, 346, 360

Lalande 343
Lampland, C.O. 357
Larson, S.M. 348

Lassell, W. 150, 270, 271, 282, 317, 322, 323, 341, 344
Lees, J.A. 125
Le Monnier 321
Lenham, A.P. 328
Le Verrier, Urbain J.J. 341, 342
Liais 152
Linfoot, E.H. 277
Littman 341
Lockwood, G.W. 348, 397
Lockyer, Sir J. Norman 147, 150, 156, 323
Lohvinenko, T.W. 126
Lomonosov, M.V. 108
Lowell, Percival 76, 113, 152, 273, 285, 325, 355, 356, 357, 358
Lyman, C.C. 108
Lyot, B. 80, 81, 90, 221, 257, 286

Mädler, von J.H. 75, 108, 116, 125, 146, 149, 156, 177, 323
Maggini, M. 275, 278
Main, Rev. 282, 345
Maraldi 134, 145, 154
Marius, Simon 204
Markwick, E.E. 325
Marsden, B. 360, 361
Marth, R.G. 282
Maunder, E.W. 151, 273
Maury, M.F. 344
Maxwell, Clerk 271
Mayer, Tobias 104, 321
McCue, J. 125
McEwen, H. 78, 113, 125
McClellan, C. 346
McIntyre, D.G. 275
McLaughlin, D.B. 140
McNeill 324
McNichol, J. 125
Meeus, 316, 342
Mellilo, F.J. 352, 398
Menzel, D.H. 326, 346
Meyer, Andreas 109
Minchin, G.M. 389
Mitani, T. 157
Mitchel 141
Molesworth, P.B. 212
Moll 96
Montaigne 104
Montalvo, A. 291
Montbaron 104
Moore, J.H. 326, 346
Moore, P.A. 124, 332, 307, 344, 357
Morov, V. 82
Morrison, D. 358
Muirden, J. 89
Muller 324

Nallino, C.A. 265
Nasmyth, James 89, 344, 345
Nelson 109
Newcomb, S. 108, 323
Newton, Sir Isaac 33
Noble, Capt. W. 323, 334

Olbers 183
Oldenberg, Henry 209
Orlow, S.W. 10
Osawa, T. 157
Ovenden, M.W. 277

Palisa 184
Palmer, J.H. 121
Peek, Bertrand M. 211, 212, 214, 215, 287
Perrotin 151
Peters, C.H.F. 184
Pettengill, G.H. 87
Phillips, T.E.R. 154, 214, 215, 275, 285, 287
Piazza 183
Pickering, E.C. 324
Pickering, William H. 152, 221, 273, 325, 356
Pierce, Benjamin 344
Pilcher, C. 358
Plummer 117
Prince 76
Proctor, Richard A. 147, 281
Ptolemy, Claudius 29, 206, 264, 265

Quenisset, M. 113

Rackham, T. 124
Ranck, O.C. 346
Redman, R.O. 277
Reid, C.D. 124
Reid, H. 275
Reid, W. 275
Reinmuth, K. 184, 185
Reitsema, H.J. 329, 348
Riccioli, Giovanni 109, 267
Richardson, Robert S. 152
Rittenhouse, David 108
Roberts, C. 272, 273, 284
Roberval 267
Robinson, J.H. 329, 289
Roedkaier 104
Roth, G.D. 126
Roth, J.G.A. 150
Rudaux 278
Ruggieri 278

Safarik, A. 323
Saheki, T. 157
Sandner, W. 126
Schabe 270
Schaeberle, J.M. 273, 324, 344
Schaer, E. 274
Schaefer, B.E. 346
Schaefer, M.W. 346
Scheuten, A. 104
Schiaparelli, Giovanni 76, 90, 148–152, 156, 323, 324
Schmude, R. 398
Schorr 105
Schroter, Johann Hieronymous 75, 92, 96, 104, 108, 109, 112, 122, 126, 146, 183, 281
Schwabe 210
Secchi, A. 323
Secchi, P. 147, 150, 156, 270
See, T.J.J. 345, 349, 356
Shaw, J.W. 348
Sheehan, W. 53

Short, James 104
Showalter, Mark 297
Sidgwick, J.B. 306
Sinton 117
Slipher, E.C. 156, 327
Slipher, V.M. 325
Smith, B.A. 329, 346, 348
Smith, C.J. 278, 288
Smyth, Admiral 104
Sola, Comas 285
Steavenson, W.H. 275, 277, 287, 325
Stone 108
Stratton, F. J. M. 105
Strom, Robert G. 75
Strong 117
Stroobant, P. 104, 273
Struve, H. 321
Struve, O. 322

Tallone, G. 280
Taylor, A. 323
Taylor, G.E. 345
Terby, M. 151, 272
Terrile, B.J. 348
Thollon 151
Thompson, D.T. 348, 397
Thomson, H. 285
Tikhoff, G.A. 326
Titius, J.B. 10
Tobal, Christopher 329
Todd, D.P. 356
Tombaugh, Clyde 357, 361
Trafton, L. 334
Trouvelot, E.L. 152, 271, 273, 278, 279, 286
Tupman 108
Tuttle, C.W. 270

Urata, T. 200

Viens, J.F. 126
Vogel, H.K. 76, 323
Von Ende 104
Von Ertborn 110

Wagman, N.E. 346
Walker, A. 361
Walker, Sears Cook 343, 344
Ward, I. 334
Wargentin 104
Waterfield, R.L. 287, 325, 326
Watson, F.G. 181, 185
Webb, T.W. 53, 110, 113, 116, 156, 323
Wegner, Gary 80
Westfall, John E. 125, 257, 352, 397, 398
Wilber, S. 291
Wildt 327
Williams, A.S. 151, 282, 284, 285, 286
Winnecke 109
Wirtz, C. 325
Wisclicenus 148
With 112
Witt, G. 185
Wolf, Max 184
Wood, R.W. 286
Wren, Sir Christopher 267
Wright, W. H. 286, 287
Wurzelbauer 96

Young, C. A. 324

Zach, Baron Franz von 183
Zander 117
Zucchi, Niccolo 207
Zupus, Giovanni 75

SUBJECT INDEX

aberration,
chromatic 32
spherical 32
Acampsis 160
accretion 8
achromatic objective 32
Adonis 185
Adrastaea 205, 225
Aethiopis 156
Aethra 185
Airy disc 50
Albert Group (asteroids) 185
almucantars 25
alpha ring, Uranus 332
altazimuth instrument 25
altitude 25
Amalthea 205, 225
Amazonis desert 158
Amor 185
Ananke 205
antidiffraction screen – see apodising screen
antipoint 50
Antoniadi's scale of seeing 70
aphelion 11
apodising screen 51–3
1862 Apollo 185, 186
apparitions of planet 15
appulse 197
areography 146, 149
A ring, Saturn 294, 295, 271–2, 273, 275, 277, 278, 280
Argyre 159
Ariel 317, 322, 324, 325, 331, 334
Aries, first point of 26
ascending node 7
ashen light, Venus 109, 128
asteroids – see minor planets
astigmatism 64
of eye 46–7
astronomical unit 4
atmosphere and seeing 69–72
atmospheric phenomena, Mars 142–4, 156–7, 174
azimuth 25

balancing of Newtonian telescope 61–2
Barlow lens 47–8
beta ring, Uranus 332
bicoloured aspect, Saturn's rings 278, 279, 280, 310–11, 397
black drop in transit of Mercury 97
blue white and white clouds and hazes, Mars 175–6
Bode's Law 10, 183
Botham's spot, Saturn 288–9
bright markings, Venus 124
brightness of planets 17
B ring, Saturn 294, 295, 271, 272, 273, 274, 275, 277, 278, 279

Callisto 204, 221, 223, 225
camcorder 379, 381
cameras, planetary 370–4
Canali 149
canals, Martian – see Canali
cardinal points 20
Carme 205
Casius 155
Cassini Division, Saturn's rings 269, 270, 271, 272, 273, 275, 277, 278, 279, 280, 293, 295, 308
CCD camera
use of 382–6
suppliers 386
celestial equator 21
celestial latitude and longitude 25–6
celestial poles 21
'Celestial Police' 183
celestial sphere 20–30
positions on 21
central meridian,
Jupiter 228
Mars 177
Ceres 183, 191
charge-coupled devices (CCD) 370, 379, 381, 382, 383, 384, 385
Charon 355, 360, 361, 362
2060 Chiron 188, 361
circles
hour 26

circles *cont.*
setting 63–4
vertical 25
circulating current, Jupiter 213–14
circumpolar stars 21
cloud belts
variations in, Jupiter 216–21
and zones, Saturn 261
clouds, blue white and yellow, Mars 143
orographic 143
collimation, of Newtonian telescope 58–61
comets 8
conjunction
inferior 13
superior planet 14
contacts in transit of Mercury 96
contrast of planetary detail 50
Copernican system 29
Coprates 157, 158
Cordelia 319, 332
Crepe ring (C ring)
Saturn 271, 272, 273, 274, 277, 278, 279, 280, 294, 295
band 279
C ring, Saturn – see Crepe ring
C-shaped markings, Venus 116, 118
C-type minor planets 182
cusp caps, Venus 124
cusp extensions, Venus 126
cylindrical projection
horizontal 364–5
parallel 366

dark areas, Mars 152–6
dark South Tropical Streak 1941–42, Jupiter 214–15
Dark Spot Two (DS2), Neptune 350
Dawes criterion of telescopic resolution 50
day, sidereal 27
daylight observation, Venus 121–2
declination 26
deferents 29
Deimos 131, 132, 149
delta ring, Uranus 332
descending node 7
diffraction disc 50
Dione 269, 296, 297
disc features, Jupiter, classification and description 232–5
discovery of
Neptune 337–43
Pluto 357–8
Uranus 320–1
drawing, planetary, video-assisted (VAD) 386–7
D ring, Saturn 278, 293
dusky markings, Venus 122–4
dust ring, Jupiter 206, 225
dust storms, Martian 156

eclipses 22
ecliptic 7
edgewise presentation, Saturn's rings 309–10
Elara 205
elements, of planetary orbits 12
elongations, greatest western and eastern 13
Enceladus 270, 296
Encke's Division (Keeler Gap) 270, 271, 272, 273, 275, 277, 278, 279, 280, 292, 295
Eosphoros 160
epicycles 29
epsilon ring, Uranus 332
equator, celestial 21
equatorial belts, Uranus 324
equinoxes
March and September 22
precession of 26
E ring, Saturn 293, 296
433 Eros 185, 193
eta ring, Uranus 332
15 Eunomia 191
Europa 221, 223, 225
evening and morning star 14
exit pupil 45
eye, peculiarities of 53
eyepiece (ocular)
Huyghenian 42
Kellner 43
Monocentric 44
Plossl 43
Ramsden 42
Tolles 44
Zeiss orthoscopic 4

filar micrometer 48–9
filters 51
finderscope (viewfinder), adjusting 63
first point of Aries 26
Flora 183
F-number – see focal ratio
focal ratio (F-number) 33
F ring, Saturn 292, 293
fundamental plane 198

Galilean satellites, Jupiter 204, 221, 223, 225, 227, 248–58
eclipses and occultations 252–7
surface features 221, 223, 225, 257–8
transits 250–2
Gallinaria Silva 160
gamma ring, Uranus 332
Ganymede 204, 221, 223, 225
gaps, Kirkwood's 186–7, 271
gas giants 4
gemination, Martian canals 151
Great Red Spot, Jupiter 209–12, 222, 223
Great White Spot of 1990, Saturn – see Wilber's Spot
Greenwich Meridian 26
grey clouds, Mars 157
G ring, Saturn 292, 293, 295–6

Halley's comet 8
Hebe 183
Hecuba family, minor planets 187
heliocentric theory 29–30
Hellas 154, 156
Hermes 185
944 Hidalgo 181, 187, 361
Himalia 205, 248

Hirayama families, minor planets 187
history of observation
  Jupiter 206–25
  Mars 144–62
  Mercury 75–82
  minor planets 183–9
  Neptune 344–51
  Saturn 264–97
  Pluto 358–61
  Uranus 321–32
  Venus 102–20
horizon 25
hour circles 26
Hygeia 183
Hygroscopic salt theory 140
Hyperion 270, 292, 297

Iapetus 269, 295, 297, 397
  photoelectric photometry of 397
Icarus 188
inferior conjunction 13
inferior planets, phases of 13
Io 204, 221, 223, 225
Iris (minor planet) 183, 191
Isidis Planitia 157, 159

Juno 181, 183
Jupiter 202–59
  central meridian 209, 228
  Circulating Current 213–14
  cloud belts 203
    variations in 216–21
  colour changes and intensity estimates of disc features 246
  dark South Tropical Streak, 1941–42 214–15
  disc drawings, strip and sectional sketches 239–41
  disc features, classification and description 232–5
    colour changes 246
    intensity estimates 246
    latitude determination of 235–9
    longitude determination of 228–32; Filar micrometer in 231–2
  dust ring 206, 225
  filter observations 248
  Galilean satellites 204, 227
    surface features 221, 223, 225, 257–8
  general 202–6
  general observing notes 247–8
  Great Red Spot 209–12, 222, 223
    history of observation 206–221; latitude determination of disc features 235–9
    longitude determination of disc features 228–32; Filar micrometer in 231–2
    observation of 243–5
    observing 226–48
    oscillating spots 215
    radio studies 258–9
    Red Spot hollow 210, 211
    rotation period determination of disc features 241–3
  photometry, photoelectric 396–7
  rotation period determination 241–3
  satellite phenomena 248–58
  satellites 204–6
    surface markings 221, 223, 225
  sidereal period 203
  'smoke stack' 213
  South Equatorial Belt:
    disappearance, 1989 218–21
    eruptions 215
  South Tropical Disturbance 212–13
  space craft exploration 222–5
  synodic period 203
  Systems I and II rotation periods 229
  visibility 225–6
  zones 203

Keeler Gap – see Encke's Division.
Kepler's Laws 11
Kirkwood's Gaps
  in minor planet zone 186–7
  in Saturn's rings 271
216 Kleopatra 394
knots, bright, in Saturn's ring 273–4

latitude, celestial – see declination.
Leda 205
lens, Barlow 47–8
Libya 154
light gathering power 50
line of nodes 7
local effects of seeing 71
longitude
  celestial – see right ascension
  determination, Martian features 177–8
    Jovian disc features 228–32
Lowell's Band, Mars 141
Lysithea 205

magnification 46–7
  empty 46
  range 46
magnitude studies, Uranus 333–4
maps
  and planispheres, constructing 364–9
    general 364
    projections: cylindrical, horizontal 364–5; cylindrical parallel 366; Mercator 367–9; orthographic horizontal 364–5; polar 369
  Mars 147–50
  Venus 118
Mare Erythraeum 148
Mare Sirenum 149
Mariner 9 mission 131
Mariner 2–10 – see space craft.
Mars 131–80
  atmospheric phenomena 142–4, 156–7, 174–7
  blue white and white clouds and hazes 175
  blue, white and yellow clouds 143
  canals 149–52
    gemination 151
  central meridian 177
  dark areas 152–6
  deserts 139
  dust storms 156
  gemination of canals 151
  general 131–3

Mars *cont.*
grey clouds 157
history of observation 144–62
longitude determination, surface features 177–8
Lowell's Band 141
maria 139
and seasonal changes 174
maps
Antoniadi 150
Beer and Mädler 147
de Vacouleurs 150
Ebisawa 150
Proctor 148
Schiaparelli 149
melt line 173–4
menisci 156
Mercator projection 367–9
Mountains of Mitchel – see Novus Mons
Nepenthes Thoth 155
Novus Mons (Mountains of Mitchel) 160, 171
observing 163–70
filters in 165–6
orbital characteristics 132–3
orographic clouds 143, 157, 160
polar caps 140–2, 145–6, 171–4
predicting oppositions 134–5
retrograde motion 135–6
seasons 137–9
space probe exploration 157–62
surface features 139–42, 144–9
longitude determination of 177–8
Tharsis bulge 157
twilight arc 156
violet clearing 174–5
violet layer 143
visibility 162–3
wave of darkening 174
W-shaped clouds 157, 176–7
yellow clouds 156, 175–6
Martian canals 149–52
dust storms 156
Mercury 73–98
axial rotation 87–8
brightness, colour and phases 88
general 73–82
history of observation 75–82
maps
Antoniadi 80
Chapman 82
Jarry-Desloges 77
Lowell 76
Rudaux 78
Schiaparelli 76
Spangenberg 80
Wegner 80
observing 88–93
surface features 90–2
transits 93–8
visibility 82
Schröter effect 75, 92–3
Meridian 21
central
Jupiter 228
Mars 177
Greenwich 26
Metis 205, 225
metal oxide semiconductors (MOS) 380
meteorites 8
micrometer, filar 48–9
in observing Jupiter 231–2
Mimas 270, 295, 296
minor planets 6, 11, 181–201
2101 Adonis 185
182 Aethra 185
Albert Group 183
1221 Amor 185
1862 Apollo 185, 186
Astraea 183
2062 Aten 186
Aten Group 186
Atlantis 149
Ceres 183, 191
Chiron 189
C-type 182
discovery and history of observation 183–9
and rediscovery 200
Eops family 187
433 Eros 191, 193
15 Eunomia 191
Flora family 183
general 181–3
Hebe 183
Hecuba family 187
Hermes 185, 186
Hidalgo 181
Hirayama families 187
Hygeia 183
Icarus 188
Iris 183, 191
Juno 181, 183
Kirkwood's gaps 186–7
216 Kleopatra 394
Koronis family 187
Maria family 187
Metis 183
Minerva, Hestia, Hilda Group 187
2090 Mizuho 200
M-type 182
44 Nysa 394
observing 191–200
occultations, stellar, by minor planets 196–200
2 Pallas 182, 183, 191
1989 PB 189
stellar occultations by 189
photometry, photoelectric 193–6
Sirona 191
stellar occultations by 196–200
S-type 182
Tereidina 191
Trojans 187
Vesta 182, 191
12 Victoria 182
visibility 189–91
Miranda 317, 328, 330, 331, 334
mirror cleaning, Newtonian telescope 66
Mons Olympica – see Olympus Mons
morning and evening stars 14, 15, 28

Mountains of Mitchel – see Novus Mons
M-type minor planets 182

1989 N1 satellite, Neptune 350
1989 N1R ring, Neptune 350
1989 N2 satellite, Neptune 350
1989 N2R ring, Neptune 350
nadir 20
Nepenthes Thoth 155
Neptune 337–54
  discovery 337–43
  general 337
  Great Dark Spot 345, 346, 349
  history of observation 344–51
  magnitude studies 352–3
  photometry, photoelectric 397
  prediscovery sightings 343–4
  ring 344–5
    plateau 350
  satellite, Nereid 346
  space craft exploration 348–51
    Dark Spot Two (DS2) 350
    globe features 349–50
    observing 352–3
    plateau ring 350
    rings 350; 1989 N1R 350; 1989 N2R 350
    S1 spot 349
    S2 spot 350
    satellites 350–351; Nereid 350; 1989 N1 350; 1989 N2 350; shepherd 350; Triton 350, 351, 352
    shepherd satellites 350
    south polar features 350
    stellar occultation studies 353
    Triton 350, 351, 352
    visibility 352
Nereid 346, 350
Nix Olympica – see Olympus Mons
Noctis Lacus 160
Nodes
  ascending 7
  descending 7
  line of 7
Novus Mons (Mountains of Mitchel) 160, 171
Nuba Lacus 155
nutation 26
44 Nysa 394

Oberon 317, 322, 324, 325, 328, 331, 334
objective, achromatic 32
observational history, planets – see history of observation
observing
  Jupiter 226–59
  Mars 163–78
  Mercury 88–98
  minor planets 191–200
  Neptune 352–3
  Pluto 362
  Saturn 298–311
  Uranus 332–5
  Venus 120–9
ocular – see eyepiece
Olympus Mons (Mons Olympica, Nix Olympica) 158, 159, 171
Ophelia (1986 U8) 319, 332
oppositions, of superior planet 14
Opus Astronomicum 265
orographic clouds 143, 157, 160
orthographic projection, horizontal 364–5
oscillating spots, Jupiter 215

2-Pallas 182, 183, 191
Pandorae Fretum 154
parallax
  effect in occultations 197
  solar 108
Pasiphae 205
1989 PB 189
perihelion 11
personal equation 198–9
perturbations 12
Phase anomaly
  Mercury 75, 92–3
  Venus 116, 125–6
Phobos 131, 132, 149
Phoebe 273
photography and videography, planetary 370–88
  camcorders 379, 381
  cameras 370–4
    CCD: use of 382–6; suppliers 386
  charge-coupled devices (CCD) 379, 380, 381, 382, 383, 384, 385
  exposure times 377–9
  film
    characteristics 375
    choice of 374
    processing, black and white 375
  general 370
  Jupiter 376
  Mars 375
  metal oxide semiconductors (MOS) 380
  prime focus 371
  projection 371–2
  Saturn 377
  video-assisted drawing (VAD), planetary 386–7
photometer, photoelectric 390–2
photometry
  minor planets 193–6
  photoelectric 389–98
    colorimetric 396
    general 389–90
    Iapetus 397
    Jupiter and Saturn 396–7
    minor planets 393–6
    Neptune 397
    Pluto and Charon 398
    Saturn globe 397
    telescopes for 392–3
    Uranus 397
phases
  inferior planets 13
  planets 14
  Mercury 73, 88–9
  Venus 125
Pioneer 11 – see space craft
Pioneer Venus 1 – see space craft

Pioneer Venus Orbiter – see space craft
pixels 380
Planet O 356
Planet P 356
Planet X 356, 357
planetary orbits, elements of 12
planetary photography and videography 379–88
Planetes Asteres 27
planetoids – see minor planets
planet, superior 14
planets
  brightness 17
  phases 14
  sidereal periods 15–17
  synodic periods 15–17
plateau ring, Neptune 350
Pluto 355–63
  Charon 355, 360, 361, 362
  discovery 357–8
  general 355
  history of observation 358–61
  observing 362
  search for 355–7
  visibility 361–2
polar caps, Mars 140–2
polar projection 369
Polaris (Pole Star) 20
poles, celestial 21
Pole Star (Polaris) 20
positional studies, Uranus 334
precession of equinoxes 26
predicting oppositions, Mars 134–5
prediscovery sightings
  Uranus 321
  Neptune 343–4
projections, map 364–9
  cylindrical parallel 366
  Mercator 367–9
  orthographic horizontal 364–5
  polar 369
psi-shaped markings, Venus 116, 118
Ptolemaic system 28–9
Puck 330

quadratures, of superior planet 14
quintuple cloud belt, Saturn 283, 284

radial streaks, on Saturn's rings 278
Rayleigh resolution limit 50
2100 Ra-Shalom 186
Red Spot Hollow, Jupiter 210, 211
residuals 355
resolution
  Rayleigh limit 50
  telescopic 49–50
retrograde motion 28
  of Mars 135–6
Rhea 269, 297
right ascension 26
Rima Australis 160
ring arcs, Uranus 332
ring divisions, Saturn 308–9
ring(s)
  Neptune 350
  Saturn 261–3
  Uranus 319, 322
Rudolphine Tables 105

S1 spot, Neptune 349
S2 spot, Neptune 350
S13, S14, S15 satellites of Saturn 295
Satellite 11, Saturn 292
satellite studies, Uranus 334–5
satellites
  Saturn 263–4, 296–7, 311
  Uranus 317–19
Saturn 260–313
  anomalous globe shadow on rings 275
  A ring 271, 272, 273, 275, 277, 278, 280
  bicoloured aspect of rings 278, 279, 280
  B ring 271, 272, 273, 274, 275, 277, 278, 279
  Cassini Division 269, 270, 271, 272, 273, 275, 277, 278, 279, 280
  cloud belts and zones 261
  colour on globe 305
  crepe ring (C ring) 271, 272, 273, 274, 277, 278, 279, 280
  crepe band 279
  D ring 278
  Encke's Division (Keeler Gap) 270, 271, 272, 273, 275, 277, 278, 279, 280
  equatorial white spots and ovals 304–5
  general 260–4
  globe 280–92
    abnormal shape of 280–2
    Botham's white spot 288–9
    cloud belts and spots 282–92
    Great White Spot of 1990 (Wilber's spot) 291
    Hay's white spot 286–7
    history of observation 264–97
    Keeler Gap – see Encke's Division
    Kirkwood's Gaps in rings 271
    knots, bright, in rings 271–4
    latitude determinations 305
    observing 298–311
    quintuple belt 283, 284
    ring divisions 308–9; Cassini's 308; Encke's 292–5
    satellites 311
  ring shadow on globe 305
  rings 261–3, 293–6
    A 271, 272, 273, 275, 277, 278, 280, 294, 295, 306, 307
    B 271, 272, 273, 274, 275, 277, 278, 279, 294, 295, 306, 307
    bicoloured aspect 310, 311, 228, 279, 279, 280
    C (crepe) 272, 273, 274, 277, 278, 279, 280, 294, 295, 306, 307
    Cassini's Division 270, 271, 272, 273, 275, 277, 278, 279, 280, 293, 295, 306
    D 278, 293
    E 293, 295, 296
    edgewise presentation 309–10
    Encke's Division 270, 271, 272, 273, 275, 277, 278, 279, 280, 292, 295
    F 292, 293, 295
    G 292, 293, 295–6

radial spokes 295
'sheep dogs' 295
satellites 263–4, 296–7, 311
Dione 296, 297, 311
Enceladus 296
Hyperion 292, 297
Iapetus 295, 297, 311, 397
Mimas 295, 296
Rhea 297, 311
S10, S11, S12 297
S13, S14 295, 297
S15 295
Tethys 292, 296, 311
Titan 297, 311
space craft exploration 292–8
Systems I and II rotation zones 304
Systems I, II and III rotation periods 261
Terby White Spot(s) 272, 279, 280, 307
visibility 298
white spots
Botham's 288–9
Great, of 1990 (Wilber's) 291
Terby's 272, 279, 280
W.T. Hay's 286–7
Schröter effect
Mercury 75, 92–3
Venus 116, 125
seasons, Martian 137–9
seeing
and atmosphere 69–72
Antoniadi's scale 70
conditions, assessing 70
effect of telescope aperture 70
local effects on 71
setting circles 63–4
shepherd satellites
Uranus 332
Neptune 350
sidereal day 27
sidereal period 156–7
sidereal time 27
sightings, prediscovery
Uranus 321
Neptune 343–4
Sinope 205
Sinus Gomer 155
Sinus Sabaeus 154
Sirona 191
'Smoke Stack', Jupiter 213
Solar System, scale model of 4–19
solar parallax 108
Solis Lacus 154, 155
South Equatorial Belt eruptions, Jupiter 215
South Polar Feature, Neptune 350
space craft
Mariner 2 117
Mariner 4 157
Mariner 5 117
Mariner 6 157
Mariner 7 157
Mariner 8 158
Mariner 9 158
Mariner 10 82, 118
Pioneer 10 222
Pioneer 11 222, 292
Pioneer Venus 1 118
Pioneer Venus Orbiter 128
Veneras 1-16 117
Viking 1 158
Voyager 1 223, 292, 296, 308
Voyager 2 223, 292, 297, 319, 330, 331, 333, 334, 337, 345, 348, 350
spokes, radial, Saturn's ring 295
South Tropical Disturbance, Jupiter 212–13
sphere, celestial 20–30
stars
evening and morning 14, 15, 28
circumpolar 21
stellar occultations by minor planets 196–200
stellar occultation studies
Uranus 334
Neptune 353
S-type minor planets 182
superior conjunction 13
superior planet 14
surface features
Mars 139–42
Mercury 76–82, 90–2
synodic period 15–17
Syntaxis 264
Syrtis Major 144, 148, 154–8

Tables, Rudolphine 105
Taylor Column 212
telescopes
accessories 31–68
aperture, effect on seeing 70
balancing of 61–2
choice of 40–1
collimating (Newtonian) 58–61
housing and care of 66
protection from dust and atmospheric pollution 65
mountings
altazimuth 53
Dobsonian 55
equatorial 55–8
tests for
correction of astigmatism 64–5
correction of spherical aberration 64
resolving power 65
types
Cassegrainian 34
catadioptric 38–40
Gregorian 34
Maksutov–Cassegrain 39
Maksutov–Gregorian 39
Newtonian 33–4
reflecting 33–4
refracting 31–3
Schmidt–Cassegrain 38
Terby Spots 272, 279, 280, 307
Tereidina 191
terminator irregularities, Venus 126
terrestrial planets 4
Tethys 269, 292, 296
Tharsis bulge, Mars 157, 159
Thebe 205, 225
Themis 273
time, sidereal 27

Titan 267, 297
Titania 317, 322, 324, 325, 328, 331, 334
Tores 274
transits, solar
  inferior planet 14
  Mercury 93–8
  Venus 105–8, 129
trans-Neptunian planet, search for 355–7
Triton 344, 346, 350–1, 352
  equatorial zone 350
  south polar cap 350
Trivium Charontis 155
Trojans 187
twilight arc 156

UBVRI filter series 391, 397
Umbriel 317, 322, 324, 325
Uranus 314–36
  alpha ring 332
  beta ring 332
  delta ring 332
  disc diameter determinations 323, 327–8
  discovery 320–1
  epsilon ring 332
  equatorial belts 324
  eta ring 332
  gamma ring 332
  general 314–19
  globe features, Voyager 2 330–1
  history of observation 321–33
  observing 332
    disc studies 332–3
    magnitude studies 333–4
  polar flattening determinations 323, 324
  positional studies 335
  prediscovery sightings 321
  ring arcs 332
  ring(s) 319, 322, 328–9, 331–2
    alpha 332
    beta 332
    delta 332
    epsilon 332
    eta 332
    gamma 332
    1985 U1R 332
  satellite studies 334–5
  satellites 317–19, 325
    Ariel 322, 324, 325
    Oberon 322, 325
    Titania 322, 324, 325
    Umbriel 322, 324
  shepherd satellites 332
  space craft exploration 330–2
  stellar occultation studies 334–5
  visibility 332
1986 U1R, Uranus ring 331, 332
Umbriel 331, 334

Vallis Marineris 157, 158
Veneras 1–16 – see space craft
Venus 99–130
  ashen light 109, 128
  'black drop' in transits 107
  bright markings 124
  cusp caps 124
  cusp extensions 126
  C-, Y- and psi-shaped markings 116, 118
  daylight observation 121–2
  dusky markings 122–4
  general 99–129
  history of observation 102–20
  map 118
  observing 120–9
  phase anomaly 116, 125–6
  phases 125
  satellite 104–5
  Schröter Effect 116, 125
  space probe exploration 117–20
  spoke-like markings 113–15
  terminator irregularities 126
  transits 105–8, 129
  ultraviolet markings 119, 122
vertical circles 25
Vesta 182, 191
12 Victoria 182
video-assisted drawing (VAD), planetary 386–7
videography and photography, planetary 370–88
viewfinder (finderscope), adjusting 63
Viking space craft – see space craft.
violet clearing, Mars 174–5
violet layer, Mars 143
visibility
  Jupiter 225–6
  Mars 162–3
  Mercury 82–6
  minor planets 189–91
  Neptune 352
  Pluto 361–2
  Saturn 298
  Uranus 332
Voyager 1 and 2 – see space craft

wave of darkening, Mars 174
white spots on Saturn
  Great White Spot of 1990 (Wilber's) 291
  Terby's, on rings 272, 279, 280
  W.T. Hay's 286–7
  Wilbur's 291
W-shaped clouds, Mars 157, 176–7

yellow clouds, Mars 156, 175–6
Y-shaped markings, Venus 116, 118

zenith 20
Zodiac 24